Human Prostate Cancer Cells: Detection, Growth and Treatment

Human Prostate Cancer Cells: Detection, Growth and Treatment

Editor: Peter Hayes

MURPHY & MOORE
www.murphy-moorepublishing.com

www.murphy-moorepublishing.com

MURPHY & MOORE

Cataloging-in-Publication Data

Human prostate cancer cells : detection, growth and treatment / edited by Peter Hayes.
p. cm.
Includes bibliographical references and index.
ISBN 978-1-63987-775-1
1. Prostate--Cancer. 2. Prostate--Cancer--Research. I. Hayes, Peter.
RC280.P7 C87 2023
616.994 63--dc23

Murphy & Moore Publishing
1 Rockefeller Plaza,
New York City,
NY 10020, USA

ISBN 978-1-63987-775-1

Contents

Preface

This book was inspired by the evolution of our times; to answer the curiosity of inquisitive minds. Many developments have occurred across the globe in the recent past which has transformed the progress in the field.

Prostate cancer occurs in the walnut shaped prostate gland, which generates seminal fluid that transports and nourishes the sperm cells in males. It can be diagnosed through various methods such as digital rectal exam, magnetic resonance imaging, prostate-specific antigen test, ultrasound imaging and prostate biopsy. Majority of people with early-stage prostate cancer can benefit from treatment options like irradiation or radical prostatectomy. Androgen deprivation therapy (ADT) is the typical systemic therapy for individuals with advanced prostate cancer. Although ADT provides a short term remission, most patients eventually develop castration-resistant prostate cancer (CRPC). Effective therapy becomes limited after prostate cancer cells lose their sensitivity to ADT. Several novel alternatives for the treatment of CRPC include taxane, CYP17 inhibitor and AR antagonist. This book contains some path-breaking studies on human prostate cancer cells. It aims to shed light on some of the unexplored aspects of their detection, growth and treatment. This book is a vital tool for all researching and studying this topic.

This book was developed from a mere concept to drafts to chapters and finally compiled together as a complete text to benefit the readers across all nations. To ensure the quality of the content we instilled two significant steps in our procedure. The first was to appoint an editorial team that would verify the data and statistics provided in the book and also select the most appropriate and valuable contributions from the plentiful contributions we received from authors worldwide. The next step was to appoint an expert of the topic as the Editor-in-Chief, who would head the project and finally make the necessary amendments and modifications to make the text reader-friendly. I was then commissioned to examine all the material to present the topics in the most comprehensible and productive format.

I would like to take this opportunity to thank all the contributing authors who were supportive enough to contribute their time and knowledge to this project. I also wish to convey my regards to my family who have been extremely supportive during the entire project.

Editor

1

Androgen Deprivation Induces Transcriptional Reprogramming in Prostate Cancer Cells to Develop Stem Cell-Like Characteristics

Shiv Verma [1,2], Eswar Shankar [1,2], F. Naz Cemre Kalayci [3], Amrita Mukunda [4], Malek Alassfar [1], Vaibhav Singh [5], E. Ricky Chan [6], Gregory T. MacLennan [7] and Sanjay Gupta [1,2,8,9,10,*]

1. Department of Urology, School of Medicine, Case Western Reserve University, Cleveland, OH 44106, USA; sxv304@case.edu (S.V.); exs334@case.edu (E.S.); mxa778@case.edu (M.A.)
2. The Urology Institute, University Hospitals Cleveland Medical Center, Cleveland, OH 44106, USA
3. The Hacettepe University Medical Center, Hacettepe University, Sihhiye, Ankara 06100, Turkey; kalaycicemre@gmail.com
4. Krieger School of Arts and Sciences, Johns Hopkins University, Baltimore, MD 21218, USA; amukund1@jhu.edu
5. Department of Inflammation and Immunity, Cleveland Clinic Foundation, Cleveland, OH 44195, USA; singhv2@ccf.org
6. Department of Population and Quantitative Health Sciences, Case Western Reserve University, Cleveland, OH 44106, USA; erc6@case.edu
7. Department of Pathology, Case Western Reserve University, Cleveland, OH 44106, USA; gtm2@case.edu
8. Department of Nutrition, Case Western Reserve University, Cleveland, OH 44106, USA
9. Division of General Medical Sciences, Case Comprehensive Cancer Center, Cleveland, OH 44106, USA
10. Department of Urology, Louis Stokes Cleveland Veterans Affairs Medical Center, Cleveland, OH 44106, USA

* Correspondence: sanjay.gupta@case.edu

Abstract: Enzalutamide, an antiandrogen, is approved for therapy of castration resistant prostate cancer. Clinical applications have shown that approximately 30% of patients acquire resistance after a short period of treatment. However, the molecular mechanisms underlying this resistance is not completely understood. To identify transcriptomic signatures associated with acquisition of drug resistance we profiled gene expression of paired enzalutamide sensitive and resistant human prostate cancer LNCaP (lymph node carcinoma of the prostate) and C4-2B cells. Overlapping genes differentially regulated in the enzalutamide resistant cells were ranked by Ingenuity Pathway Analysis and their functional validation was performed using ingenuity knowledge database followed by confirmation to correlate transcript with protein expression. Analysis revealed that genes associated with cancer stem cells, such as *POU5F1* (OCT4), *SOX2*, *NANOG*, *BMI1*, *BMP2*, *CD44*, *SOX9*, and *ALDH1* were markedly upregulated in enzalutamide resistant cells. Amongst the pathways enriched in the enzalutamide-resistant cells were those associated with RUNX2, hedgehog, integrin signaling, and molecules associated with elastic fibers. Further examination of a patient cohort undergoing ADT and its comparison with no-ADT group demonstrated high expression of POU5F1 (OCT4), ALDH1, and SOX2 in ADT specimens, suggesting that they may be clinically relevant therapeutic targets. Altogether, our approach exhibits the potential of integrative transcriptomic analyses to identify critical genes and pathways of antiandrogen resistance as a promising approach for designing novel therapeutic strategies to circumvent drug resistance.

Keywords: antiandrogens; castrate resistant prostate cancer; cancer stem cells; enzalutamide resistance; transcriptional reprogramming

1. Introduction

Androgen deprivation therapy is the standard treatment for advance-stage prostate cancer [1,2]. Recent success of second generation drugs targeting androgen receptor axis including enzalutamide and abiraterone acetate, which blocks intratumoral production of androgen, have been approved by the Food and Drug Administration for the treatment of castration-sensitive and castration-resistant prostate cancer patients [3–6]. These agents have significant impact on treatment patterns; however, the majority of patients develop resistance to these drugs, despite their initial effectiveness, which in turn reflects our limited understanding of the mechanisms [7,8]. The failure of androgen deprivation therapy is largely associated with tumor heterogeneity that results in the differentiation of distinct cell subpopulations across and within disease sites.

Cancer stem cells (CSCs) are subpopulations of cancer cells that share similar characteristics as normal stem or progenitor cells within the tumor, retain the capacity of self-renewal and multi-lineage differentiation to drive tumor growth and heterogeneity [9–11]. The embryonic origin of prostate might explain the presence of stem cells within the organ [12]. Emerging evidence support a critical role of prostate CSC-like cells in stimulating castrate resistant evolution to enzalutamide and abiraterone acetate treatment [13,14]. We have earlier demonstrated that both naïve and therapy-resistant prostate cancer cell population have distinct subpopulations of CD133+ progenitor cells exhibiting significant tumorigenic ability and drug resistance [15]. Studies have further demonstrated that the distinct microenvironment and metabolic reprograming of cancer cells regulate CSCs differentiation and acquisition of drug resistance [16,17]. At the molecular level, CSCs express genes, such as *ALDH1, POU5F1* (OCT4), *CD44, NANOG, SOX2,* and others that play an essential role in maintaining stem cell pluripotency required to reprogram the differentiated cells. These genes subsequently promote lineage plasticity of cancer cells to adapt and develop resistance to cancer therapies [18,19]. Therefore, in-depth understanding and identification of cancer stem cell differentiation genes is critical for establishing novel tumor diagnostic and therapeutic strategy. We focused our attention to understand the underlying gene network promoting enzalutamide resistance that might be helpful in better designing of new therapeutic strategies.

In the present study, we induced transcriptional reprogramming in androgen-responsive human prostate cancer LNCaP and C4-2B cells through long-term enzalutamide exposure to develop resistance. We hypothesize that androgen deprivation therapy alters the phenotype of cancer cells and differentiate to cancer stem cell-like characteristics. Here we performed Next-Gen sequencing using both cell lines and compared them with their parental enzalutamide-sensitive counterparts, which were simultaneously exposed to vehicle for the same time period. After transcriptomic analysis, we performed integrative exploration using Ingenuity Pathway Analysis and ingenuity knowledge database to identify differentially regulated gene networks and analyzed these large-scale linkages to identify subnetworks associated with acquired enzalutamide resistance. A subset of genes residing within significant network were identified and further validated employing qRT-PCR, Western blotting and immunohistochemistry using resistant cells and clinical specimens from patients who underwent androgen deprivation therapy. Taken together, our approach identified nodes of sub-networks that may be putative therapeutic targets demonstrating translational significance.

2. Results

In this study, human prostate cancer LNCaP and C4-2B cells were used to generate enzalutamide resistant cells. These cell lines were cultured in medium containing 20 μM enzalutamide for six months followed by maintenance in 5 μM enzalutamide. Under the conditions, both LNCaP and C4-2B cells underwent morphological changes such as loss of cell-to-cell tight contact with scattered growth and temporary arm-like projections (Figure 1A1-2, B1-2). The growth curves for both cell lines demonstrate a marked difference in their sensitivity to enzalutamide treatment (Supplementary Materials Figure S1). To evaluate whether these phenotype changes in these cells were linked to the induction of stem-like characteristics and plasticity, we performed staining of live cells with fluorescence-tagged alkaline phosphatase. Alkaline phosphatase has emerged as a benchmark for identifying pluripotent stem cells [20,21]. Studies demonstrate that alkaline phosphatase is highly expressed in stem cells [22,23]. Live stain of LNCaP and C4-2B enzalutamide resistant cells demonstrate increased number and intensity of alkaline phosphatase positive cells, compared to parental cell lines. Higher intensity of alkaline phosphatase staining was noted in C4-2B enzalutamide resistant cells, compared to LNCaP enzalutamide resistant cells (Figure 1A3-4, B3-4).

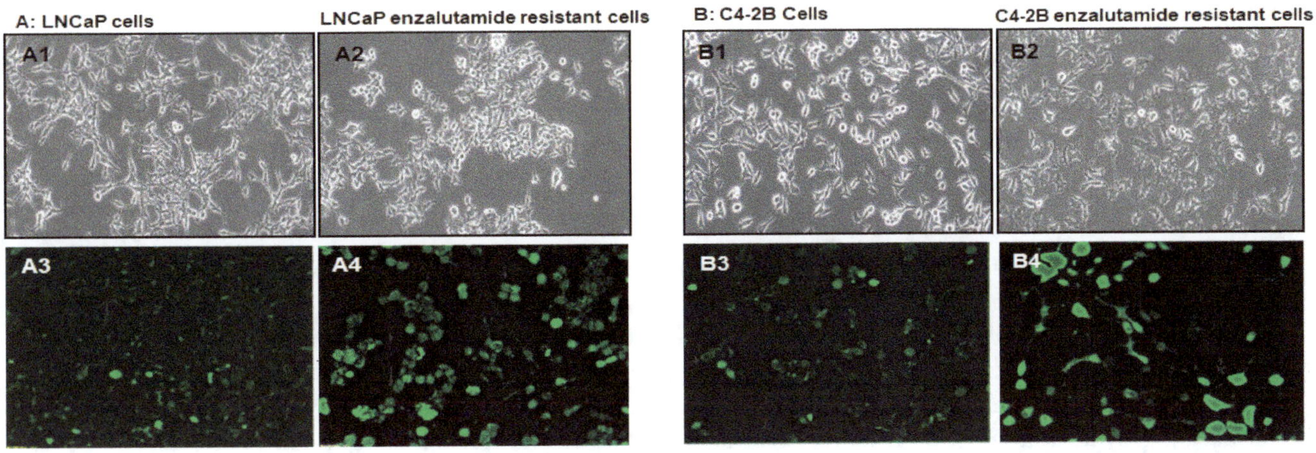

Figure 1. Self-renewal marker expression in pluripotent stem cells after enzalutamide exposure. The image shows differential staining of cancer stem-like cells in (**A**) LNCaP cells and (**B**) C4-2B cells with and without enzalutamide treatment. Morphological changes in parental cells (**A1**,**B1**) and enzalutamide treated cells (**A2**,**B2**). Staining with alkaline phosphatase, a stem cell marker in parental cells (**A3**,**B3**), and enzalutamide treated cells (**A4**,**B4**), respectively. Magnification 10×.

Next, we compared gene expression levels between enzalutamide resistant and parental cells (Figure 2). RNA-Seq analysis exhibited 35,504 expressed genes, 9409 differentially expressed genes (DEGs) were identified in LNCaP enzalutamide resistant cells, compared to parental counterpart (NCBI-GEO accession# GSE150807). In C4-2B enzalutamide resistant cells, 33,027 expressed genes and 7757 DEGs were identified, compared to parental cells (NCBI-GEO accession# GSE151083). Data visualization in the form of volcano plot display the relationship between magnitude of gene expression change (log2 fold-change; X-axis) and statistical significance of this change (−1og10 adjusted q value; Y-axis) in LNCaP and C4-2B enzalutamide resistant cells, compared with parental cell lines. Individual dots in volcano plot and scattered plot represent individual RNA gene transcripts in both cell lines (Figure 2A,B).

We next performed Ingenuity Pathway Analysis (IPA) and knowledge database on DEGs to investigate their biological relevance and pathway association. IPA analysis exhibited 4713 and 2664 differentially expressed genes (DEGs) in LNCaP and C4-2B enzalutamide resistant cells; whereas 4351 common DEGs in both cell lines (Figure 3A). The difference in the number of genes represented in the Venn-diagram (Figure 3) compared to Next Generation Sequencing (NGS) data (Figure 2) was due to the fact that some genes had infinite fold change values and were excluded. Moreover, the result of enrichment analysis analyzed through ingenuity knowledge database revealed *POU5F1* (*OCT4*), *SOX2*, and *NANOG* with the highest −log (B-H *p*-value), followed by human embryonic stem cell pluripotency, RUNX2 regulatory genes, integrin signaling, transcriptional regulation of pluripotent stem cells, molecules associated with elastic fibers, hedgehog "on" state, transcriptional regulation by RUNX2, signaling by hedgehog and, OCT4 in embryonic stem cell signaling, respectively (Figure 3B).

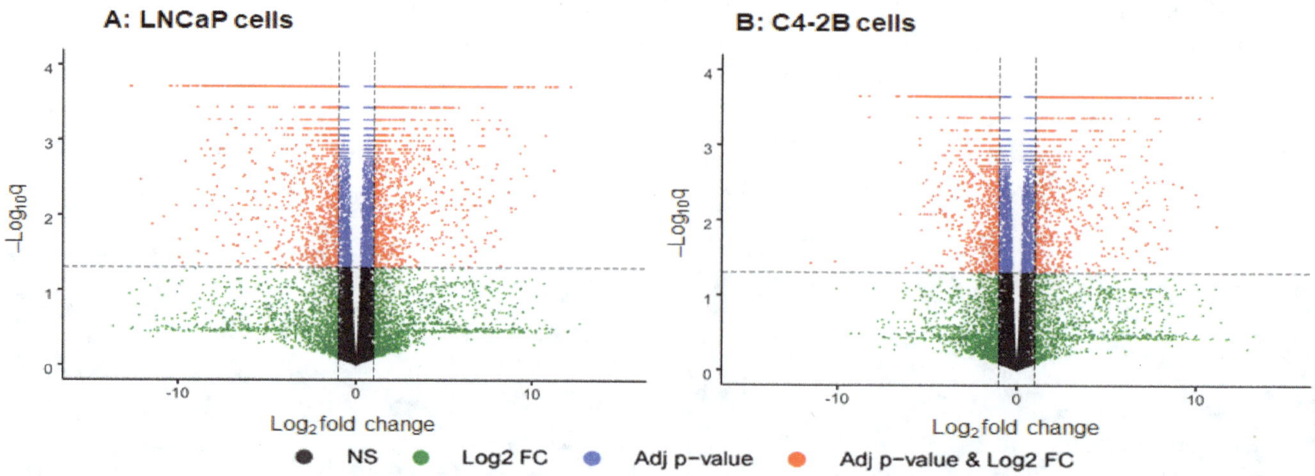

Figure 2. Volcano plots for differential gene expression analysis using R package-Enhanced volcano plot, (**A**) LNCaP cells, and (**B**) C4-2B cells with and without enzalutamide treatment. Each dot in the plot represents respective gene transcripts, red dots indicate the significantly up and downregulated genes with Log2 fold-change > 1 and adjusted p-value (FDR) < 0.05. Blue dot indicate genes that are significantly different with adjusted p-value (FDR) < 0.05 and does not meet fold change cut-off, while the grey and green color dots represent genes that did not pass the significant cut-off.

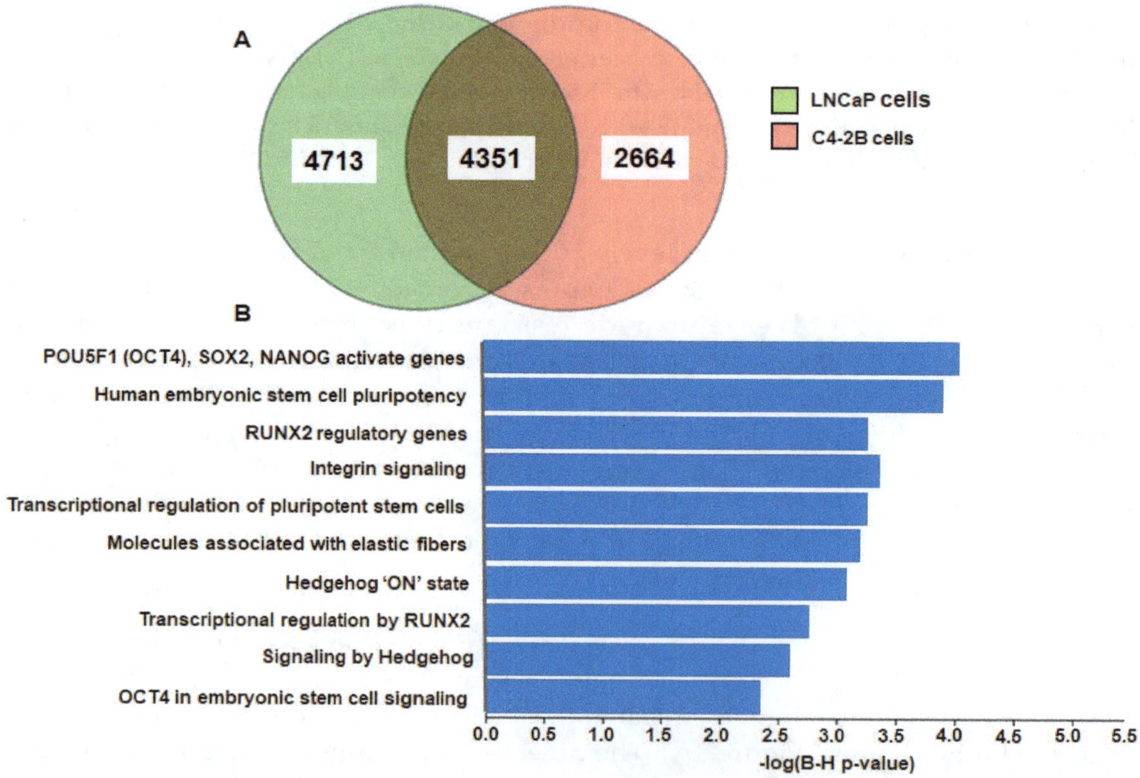

Figure 3. (**A**) Venn-diagram of differentially expressed genes (DEGs). The figure shows significantly elevated genes (p-value < 0.01) in LNCaP enzalutamide resistant cells (green) and C4-2B enzalutamide resistant cells (pink) compared to the parental cells. (**B**) Gene enrichment analysis. The DEGs were overlaid with ingenuity knowledge database of humans and to evaluate the definite overrepresented pathway(s), or to remove the chances of any randomness in data with reference to p-value, another statistical parameters of threshold value of 0.05 and Benjamin–Hochberg (B–H) was applied and represented in the form of bar graph with scale of gene and −log (B–H p value).

In the next set of experiments, we performed gene network analysis among DEGs of signaling pathway regulating stem cell pluripotency in both LNCaP and C4-2B enzalutamide resistant cells (Figure 4). The red color showed upregulation and blue color exhibited downregulation of differentially expressed genes in enzalutamide resistant cell lines. For instance, the expression of *ACVR1*, *ACVR1C*, *ACVR2B*, *AKT3*, *AXIN1*, *BMI1*, *CTNNB1*, *DLX5*, *ESRRB*, *FGFR1*, *FGFR2*, *FGFR3*, *FZD2*, *FZD10*, *HAND1*, *HESX1*, *ID2*, *ID4*, *INHBA*, *ISL1*, *JAK1*, *JAK3*, *JARID2*, *MAPK14*, *MEIS1*, *OTX1*, *PAX6*, *PCGF6*, *RAF1*, *RIF1*, *SMAD3*, *SMARCAD1*, *TCF7*, *WNT5A*, *WNT5B*, *WNT6*, and *WNT9A* were upregulated; whereas the expression of *AKT1*, *BMPR1B*, *DVL1*, *FZD1*, *FZD4*, *FZD5*, *FZD9*, *ID1*, *ID3*, *IGF1*, *IGF1R*, *INHBB*, *KLF4*, *KRAS*, *LIFR*, *MAPK11*, *MAPK12*, *MYC*, *PCGF2*, *PIK3CB*, *PIK3CD*, *PIK3R2*, *SMAD2*, *STAT3*, *TBX3*, *WNT10B*, *WNT7B*, *WNT8B*, and *ZFHX3* were downregulated in LNCaP enzalutamide resistant cells (Figure 4A). Similarly, the expression of *ACVR2B*, *BMPR1A*, *FGFR1*, *FGFR3*, *FZD1*, *FZD8*, *ID1*, *ID2*, *ID3*, *ID4*, *INHBB*, *JAK3*, *KAT6A*, *MAPK11*, *MAPK12*, *PAX6*, *PCGF1*, *PCGF6*, *SMAD3*, and *SMAD9* were upregulated; whereas *ACVR1*, *BMPR1B*, *FGFR2*, *FZD2*, *FZD4*, *FZD9*, *IGF1*, *IGF1R*, *INHBE*, *LIFR*, *MAP2K1*, *PCGF5*, *PIK3CB*, *PIK3CD*, *PIK3R2*, *PIK3R3*, *SMAD1*, *STAT3*, *WNT10A*, *WNT7B*, and *ZFHX3* are downregulated in C4-2B enzalutamide resistant cells (Figure 4B).

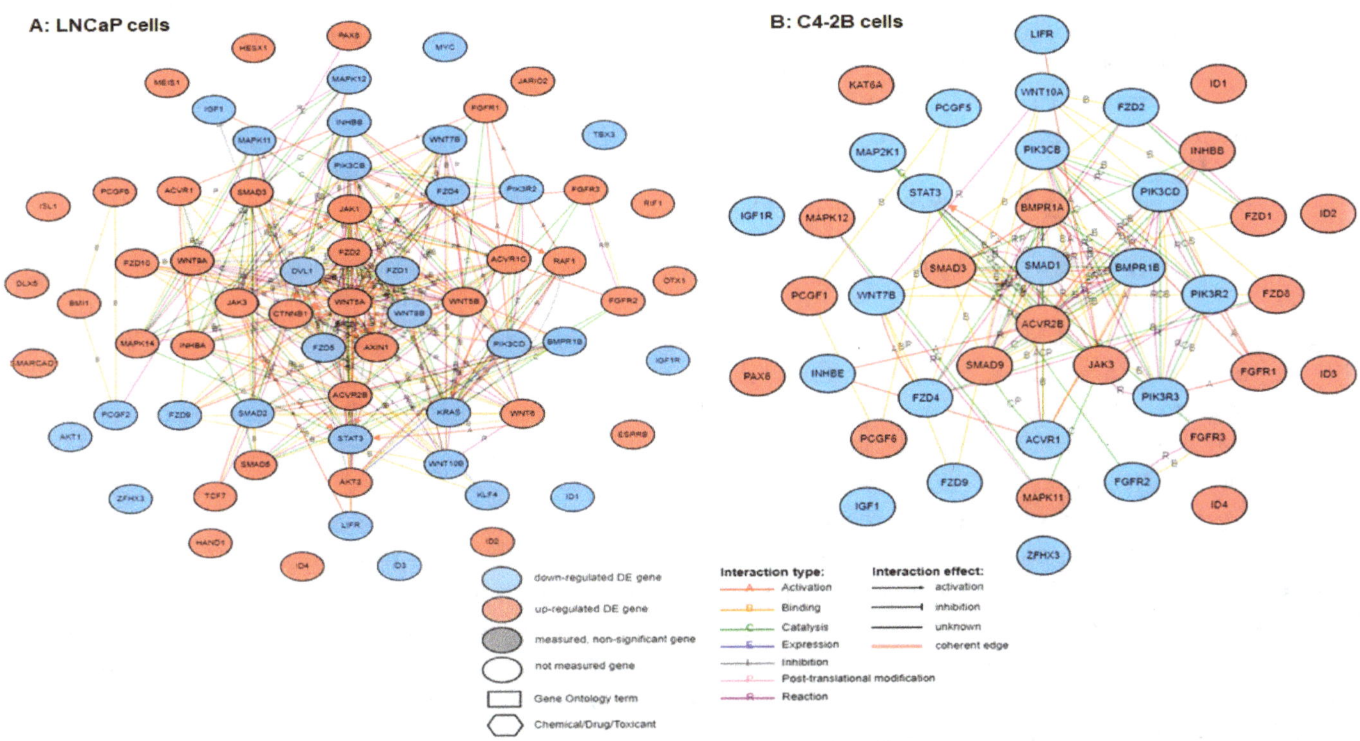

Figure 4. Gene network interaction of cancer stem cell signaling pathway in (**A**) LNCaP cells and (**B**) C4-2B cells with and without enzalutamide treatment. Red color indicates upregulated genes, blue indicates downregulated genes.

In order to validate our findings obtained from the RNA-Seq data, we selected a subset of 10 stem cell regulatory genes *viz.* *ALDH1*, *BMI1*, *BMP2*, *CD44*, *POU3F2*, *POU5F1*, *POU6F1*, *SOX2*, *SOX8*, and *SOX9* for their differential expression in LNCaP and C4-2B enzalutamide resistant cells in comparison with parental cells. In LNCaP enzalutamide resistant cells, *ALDH1*, *BMI1 BMP2*, *POU6F1*, *SOX2, and SOX9* were expressed in higher levels, compared with LNCaP cells at p value < 0.05. The fold change expression of *SOX9* (76.1), *ALDH1* (50), *POU6F1* (41.1), *BMP2* (26.6), *SOX2* (18.8), *BMI1* (10.5), *POU5F1* (9.3), *SOX8* (8.3), *POU3F2* (3.6), and *CD44* (3.1) was noted in the LNCaP enzalutamide resistant cells compared to LNCaP parental cells (Figure 5A). Similar trend of high gene expression was observed in C4-2B enzalutamide cells compared to C4-2B parental cells. The fold change expression of *ALDH1* (54), *POU3F2* (12.2), *BMP2* (11.5), *BMI1* (10), *SOX9* (8.7), *SOX2* (8.2),

POU5F1 (6.1), *POU6F1* (5.4), *CD44* (5), and *SOX8* (3.3) was noted in the enzalutamide resistant cells compared to C4-2B parental cells (Figure 5B).

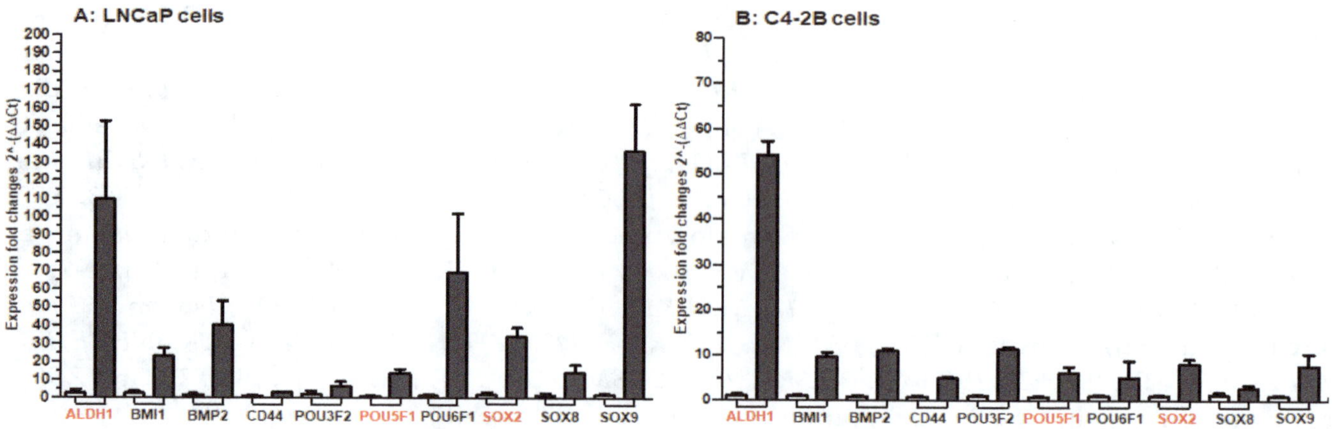

Figure 5. Real-time PCR analysis of stem cell marker genes in (**A**) LNCaP cells and (**B**) C4-2B cells with and without enzalutamide treatment. The abundance of transcripts ALDH1, BMI1, BMP2, CD44, POU3F2, POU5F1, POU6F1, SOX2, SOX8, and SOX9 were quantified in both resistant cells and their parental counterparts. GAPDH and actin were used as reference genes and set as baseline value to which all transcript levels were normalized. Y-axis in the bar represents expression fold change (2^-(ΔΔCt) and X-axis shows the name of the transcript. Error bars refer to mean ± standard deviation for three technical and three biological replicates.

Based on the gene validation analysis, we identified a subset of differentially expressed genes *viz.* *ALDH1*, *POU5F1* (OCT4), and *SOX2* in both enzalutamide resistant cells. Next, we aimed to identify target genes based on Next-Gen sequencing curated databases. All nodes were sorted by degree, interaction types (activation, binding etc.) and interaction effect (activation, inhibition). Data showed the target genes, which were upregulated by *POU5F1*, included DNA methyltransferase 1 (*DNMT1*), Baculoviral IAP Repeat Containing 5 (*BIRC5*), FRAT regulator of WNT signaling pathway 2 (*FRAT2*), GATA Binding Protein 6 (*GATA6*), Forkhead Box D3 (*FOXD3*) and Zinc Finger E-Box Binding Homeobox 1 (*ZEB1*); whereas Myogenic Differentiation 1 (*MYOD1*), and Snail Family Transcriptional Repressor (*SNAI1*) were downregulated after binding with the homeodomain octamer motif (5′-ATTTGCAT-3′) of POU5F1 (Figure 6A). Similarly, *SOX2* upregulated target genes include hedgehog acyltransferase (*HHAT*), *FOXD3*, VRK Serine/Threonine Kinase 1 (*VRK1*), Fibroblast Growth Factor Receptor 1 (*FGFR1*), and *DNMT1* respectively, whereas *ALDH1A*, Chromodomain Helicase DNA Binding Protein 1(*CDH1*), Cyclin D1(*CCND1*), and Reversion inducing cysteine rich protein with kazal motifs (*RECK*) were downregulated after binding with SOX2 motif. (Figure 6B). Lastly, *ALDH1* upregulates 2 target genes that include NIMA Related Kinase 2 (*NEK2*) and *SOX9* (Figure 6C).

Next, we performed immunohistochemistry for ALDH1, POU5F1, and SOX2 in clinical specimens of the prostate from patients who had undergone androgen deprivation therapy and the tissues were obtained post-ADT treatment from the tumors. The grade-matched tumor samples, which did not received any adjuvant treatment, were included in the study, and served as controls. The ALDH1 expression was specifically detected in luminal cancer cells and exhibited broad variations, ranging from negative to focal, diffuse, or positive—albeit patchy and milder in samples with no-ADT, compared to the ADT specimens. Typically, ALDH1 expression was higher and more frequently expressed in tumor cells. Immunohistochemistry with POU5F1 antibody on specimens with no-ADT exhibited weak positive staining or no staining in luminal cancer cells, whereas positive cytoplasmic staining was noted in basal cells. In sharp contrast, specimens with ADT exhibited moderate POU5F1 staining patterns with strong nuclear presence in both the luminal and basal cells. IHC performed for SOX2 in no-ADT specimens exhibited a mixed basal and luminal epithelial cell staining in cancer with higher percentage of SOX2 positive cells, specifically in basal cells. A marked increase with uniform strong

SOX2 expression was noted in specimens with ADT (Figure 7A). The total number of ALDH1, POU5F1, and SOX2 positive cells were quantified in both no-ADT and ADT specimens in 10–12 fields from each specimen, and compared using unpaired Student's t-test (Figure 7B).

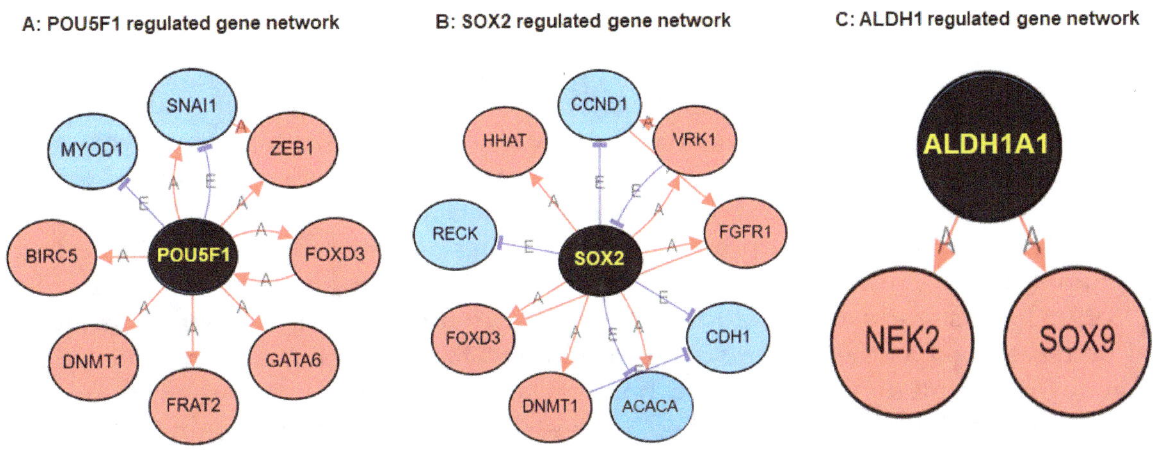

Figure 6. Differentially regulated genes (DEGs) as downstream target genes identified in LNCaP and C4-2B cells with and without enzalutamide treatment. (**A**) DEGs genes downstream of POU5F1 (OCT4), (**B**) DEGs genes downstream of SOX2, and (**C**) DEGs genes downstream of ALDH1, respectively. Red color indicates upregulated genes; blue indicates downregulated genes. Arrow head in red showed activation while blue head showed inhibition.

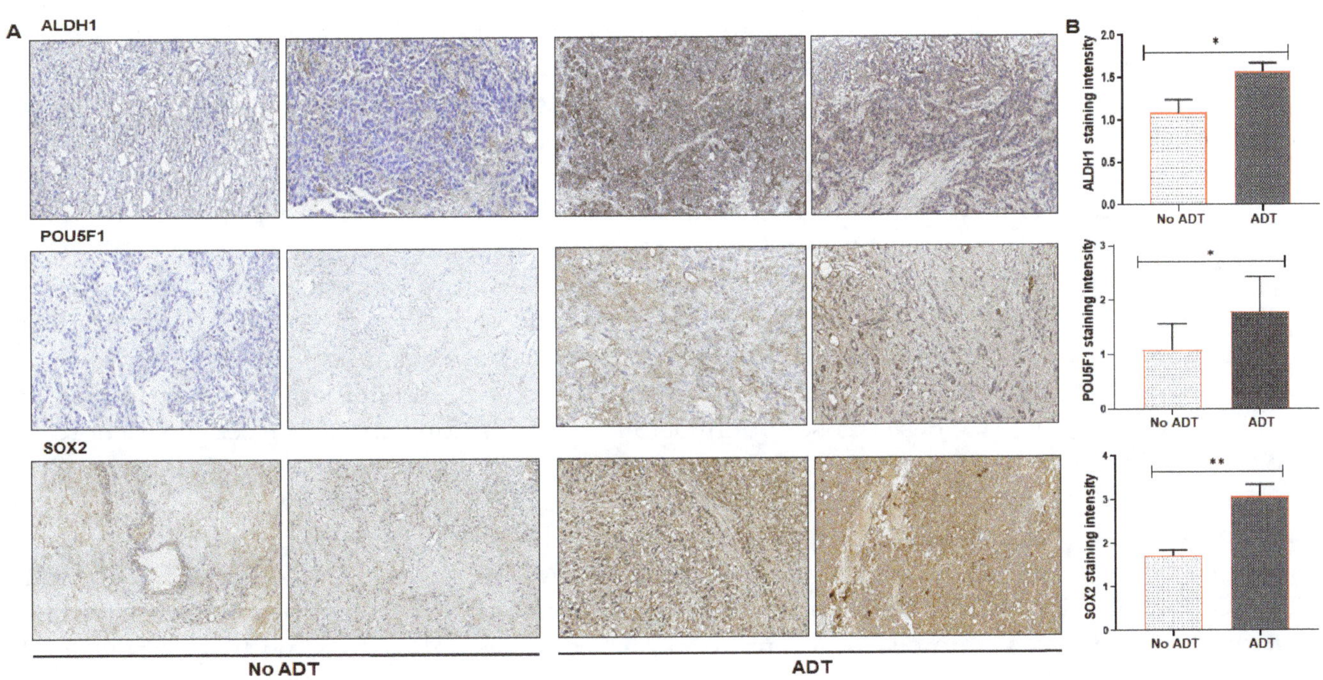

Figure 7. (**A**) Immunohistochemistry in paraffin-embedded tissue sections from patients who had undergone androgen deprivation therapy ($n = 23$) and specimens of grade-matched cancer tissues ($n = 14$) for protein expression of POU5F1 (OCT4), SOX2, and ALDH1 using anti-ALDH1, anti-POU5F1, and anti-SOX2 antibodies. (**B**) Statistical significance of immunohistochemistry was in lieu of the form of bar graph, error bars represent standard deviation * $p < 0.05$ and ** $p < 0.001$, compared to ADT versus no-ADT. Magnification 10×.

To further confirm whether antiandrogen treatment resulted in higher expression of stem cell markers, we performed Western blotting for AR, AR-v7, ALDH1, POU5F1, and SOX2 in LNCaP and C4-2B enzalutamide resistant cells and their parental counterparts. The AR protein expression were significantly decreased in both enzalutamide resistant cells whereas AR-v7 expression was increased in C4-2B enzalutamide resistant cells. A marked increase in the protein expression of ALDH1, POU5F1, and SOX2 were noted in both enzalutamide resistant cells, compared to their parental counterparts (Supplementary Materials Figure S2).

3. Discussion

In phase III clinical trials, the second-generation AR antagonist enzalutamide has been demonstrated to improve survival for patients with metastatic CRPC [24–26]. However, patients who initially respond to enzalutamide ultimately develop chemo-resistance and undergo disease progression [3,27]. Because of this, there is an urgent need to identify mechanisms of drug resistance. Approximately 30% of advanced-stage prostate cancer patients' exhibit cellular plasticity and acquisition of altered phenotypes often associated with the loss of AR signaling and/or alteration in AR splice variants [28,29]. In this study, we used the IPA to identify mechanisms of enzalutamide resistance that could be developed as therapeutic drug target(s). The IPA analysis identified 4351 common subset of genes differentially expressed in both resistant cell lines. Our data demonstrate that the subset of overlapping genes is enriched with stem cell marker genes, and these show high levels of expression in both LNCaP and C4-2B enzalutamide-resistant cell lines. In particular, the LNCaP enzalutamide-resistant cell line exhibited stem cell marker genes associated with embryonic stem cells, mesenchymal stem cells, hematopoietic stem cells, and more neural stem cells (Supplementary Materials Figure S3A). The C4-2B enzalutamide-resistant cell lines showed higher percentage of embryonic stem cells, hematopoietic stem cells, mammary stem cells and mesenchymal stem cells (Supplementary Materials Figure S3B). These subsets of cell populations within the tumor change the microenvironment and may account for resistance against antiandrogens.

Next, we focused on pathway analysis using IPA knowledge database as this approach can identify genes implicated in specific signaling networks. Our study showed enriched signaling pathways that include hedgehog in 'off' and 'on' state, transcriptional regulation of RUNX2, molecules associated with elastic fibers, integrin signaling and transcriptional regulation of pluripotent stem cells. We observed enrichment of human embryonic stem cell pluripotency genes along with transcriptional regulation of pluripotent stem cells that might lead to activation of *POU5F1* (OCT4), *SOX2, CD44, NANOG,* and other genes associated with self-renewal and drug resistance of cancer cells. Among these signaling pathways, regulating stem cell pluripotency were involved in self-renewal, having potential to differentiate various cell types of the 3 germinal layer. These signaling pathways converge to activate the core transcription factors, such as OCT4, SOX2, NANOG, and others, which are involved in self-renewal of cells [30–32]. These transcription factors and their downstream target genes coordinate to promote self-renewal, pluripotency, and drug resistance (Supplementary Materials Figure S4).

Gene network analysis of signaling pathways that regulate pluripotency of stem cells showed upstream and downstream regulatory gene interactions [33,34]. The analysis showed molecules associated with WNT signaling pathway such as WNT5A, WNT6, WNT9A, and others, which include AKT3, FGFR1, FGFR2, FGFR3, BMI1 (upregulated in nodes of LNCaP enzalutamide resistant cells) that promote embryonic stem cell differentiation and self-renewal. Along with the above EP300, SMAD3, SMAD4, and others are additional subsets of gene regulatory network, which regulate the TGF-β signaling pathway. In C4-2B enzalutamide resistant cells, SMAD3, SMAD9, FGFR1, FGFR3, BMPR1A, and others were upregulated, acting in concert with the downstream molecules. Furthermore, the TGF-β signaling pathway activates BMP that binds with BMPR activating SMAD-like protein, along with SOX2 and OCT4 for their involvement in the differentiation, self-renewal, and drug resistance (Supplementary Materials Figure S5).

Several transcription factors have been shown to be involved in the initiation and maintenance of stemness in prostate cancer [35,36]. Examination of castration-resistant and hormone-naïve prostate cancer datasets demonstrate similar functionally validated genes expressed in post-androgen deprivation therapy specimens [32,37,38]. Several studies have shown that ALDH1 is expressed to different degrees in most human tumors [39,40]. In mantle cell lymphoma, a small population of CSCs expressing ALDH showed resistance to a wide range of chemotherapeutic agents [41]. Moreover, silencing of ALDH1A1 using nano-liposomal siRNA sensitized both taxane-and platinum resistant cell lines to chemotherapy in ovarian cancer [42]. High expression of ALDH1 at the transcript or protein level in tumors has been implicated in drug resistance and disease relapse [43,44]. A positive association between ALDH1 expression and poor prognosis was observed in core biopsies and tissue microarrays of prostate cancer patients [45]. Our study also demonstrates similar results with increase in ALDH1 levels in drug resistant cells and prostate tumors with ADT. Our analysis further identified the mitotic kinase NEK2 and stem cell transcription factor SOX9 as downstream targets of ALDH1, which might be involved in the development of multi-drug resistance in prostate cancer cells. Additional studies are required to validate these findings.

Elevated expression of POU5F1 (OCT4) both at the transcript and protein levels confers self-renewal of CSCs, and may lead towards enzalutamide resistance [46]. Likewise, high expression of OCT4 was observed in breast CSCs, and correlates with therapeutic resistance [47]. OCT4 regulates stem cell pluripotency and differentiation through regulation of DNMT1 [48,49]. OCT4 and NANOG upregulate DNMT1 through direct binding to its promoter to prevent the cells from reverting to an undifferentiated state [50]. Other OCT4 target genes including FRAT2, FOXD3, GATA6, and ZEB1 have been shown to play a significant role in pluripotency and self-renewal of CSCs [51,52]. Furthermore, increased expression of OCT4 target gene NAIP/BIRC5 has been associated with therapeutic resistance in various human cancers [53,54]. We found upregulation of these genes in enzalutamide resistant cell lines, which could function as determinants of multidrug resistance and tumor aggressiveness. These molecules may serve as potential new molecular targets to overcome enzalutamide resistance in prostate cancer.

Studies have shown that SOX2 overexpression leads to increase cell proliferation and promotes invasion, migration, and metastasis in various human cancers [55–57]. In prostate cancer, SOX2 has been shown to promote resistance to antiandrogen therapy by initiating lineage plasticity [58]. Studies reference SOX2 as an androgen repressed gene where AR binds the enhancer element within the SOX2 promoter to regulate its expression [28]. Aberrant SOX2 expression is noted, particularly in castration-resistant prostate cancer cells where ligand activation of AR promotes a decrease in SOX2 expression as a result of direct binding of the AR to the SOX2 cis-enhancer region [28,59]. In fact, SOX2 appears to promote castration resistance via mechanisms that do not involve the re-expression of embryonic stem cell SOX2-target genes but provides survival advantage through the loss of tumor suppressors including p53 and Rb genes [60,61]. In our study, we identified *DNMT1* and *FOXD3* as common subsets of genes upregulated by *POU5F1* (OCT4) and *SOX2*. Further studies, however, are required to confirm these findings.

4. Materials and Methods

4.1. Chemicals and Reagents

Enzalutamide (Cat# A10562) was purchased from Adooq Bioscience, Irvine, CA, USA, and all other reagents purchased were of GLP grade. Antibodies including Anti-AR (Cat# 5153), anti-AR-v7 (Cat# 19672), anti-POU5F1 (OCT4) (Cat# 2750S) were purchased from Cell Signaling Technologies, Beverly, MA, USA. The anti-ALDH1 (Cat# SC-166362), anti-SOX2 (Cat# SC-365823), anti-α-GAPDH (Cat# SC-47724), goat anti-mouse IgG-HRP (Cat# SC-2005), bovine anti-goat IgG-HRP (Cat# SC-2350), and goat anti-rabbit IgG-HRP (Cat# SC-2004) antibodies were purchased from Santa Cruz Biotechnology, Dallas, TX, USA.

4.2. Cell Culture

Human prostate cancer LNCaP and C4-2B cells were grown in RPMI 1640 (Cat# SH30027.01, GE Healthcare, Marlborough, MA, USA) supplemented with 10% fetal bovine serum, 50 U/mL penicillin and 50 μg/mL streptomycin in 100-mm tissue culture plates at 37 °C in a humidified atmosphere (5% CO_2). These cells were used for generating enzalutamide resistant clones by exposing the cells to 20 μM enzalutamide for a minimum of six months and maintaining in media containing 5 μM enzalutamide. The absence of mycoplasma contamination was tested using PCR-based assay (Cat# MP0025; Sigma-Aldrich, St. Louis, MO, USA). The parental cells were maintained in the drug vehicle for the same time period and served as corresponding controls.

4.3. *Alkaline Phosphatase Staining*

LNCaP and C4-2B enzalutamide resistant and parental cells were stained for alkaline phosphatase activity as a measure of pluripotency using alkaline phosphatase live stain reagent (Cat# 14353, Thermo Fisher Scientific, Grand Island, NY, USA). The dye is a non-toxic, cell-permeable fluorescent substrate for alkaline phosphatase that diffuses out over the course of two hours. The cells were incubated with the substrate for 20–30 min and washed twice with DMEM/F-12 media to remove excess reagent. Following the final wash, fresh media was added and images were captured within 30–60 min after staining. Visualization of fluorescent-labeled cells were observed under fluorescent microscopy using a standard FITC filter.

4.4. *Library Preparation and Next Generation Sequencing (NGS)*

Total RNA was extracted from both LNCaP and C4-2B enzalutamide resistant and sensitive cells continuously exposed to enzalutamide (5 μM), using RNA RNeasy kit (Qiagen, Maryland, MD, USA). The total RNA integrity (RIN) was assessed using an RNA 6000 nanochip (Agilent Technologies, Santa Clara, CA, USA) on a Bioanalyzer 2100 (Agilent Technologies). Libraries were prepared using the Illumina TruSeq Stranded Total RNA Sample Preparation kit according to the manufacturer's protocol. The 50 bp single-end sequencing was performed on pooled libraries using an Illumina HiSeq 2500 platform. Library preps and sequencing were completed by the Case Western Reserve University Genomics Core Facility.

4.5. *NGS Data Analysis and Visualization*

Sequencing reads generated from the Illumina platform were assessed for quality using FastQC. Illumina HiSeq 2500 reads were trimmed and clipped for quality control in TrimGalore v0.4.3 a wrapper script for cutAdapt and FastQC. Alignment of the data was performed using STAR Aligner v2.5.3 using the human reference genome GRCh38 and the GENCODE transcript annotation v25. Differential expression was determined using Cufflinks v2.2.1. Differentially expressed genes were identified using a multiple testing corrected q-value < 0.05. Mitochondrial chromosome and the non-chromosomal sequences were excluded from the analysis. The Next-Gen sequencing data of LNCaP and C4-2B enzalutamide resistant cells were submitted to NCBI-GEO having accession number GSE150807 and GSE151083, respectively.

4.6. *Pathway and Gene Set Enrichment Analysis*

Pathway analysis was performed using Ingenuity Pathway Analysis v 5.0 (IPA, Qiagen), the differentially expressed genes (DEGs) were imported into the IPA software and were subjected to functional annotations and regulatory network analysis. The DEGs were overlaid with ingenuity knowledge database of humans, and, to evaluate the definite overrepresented pathway(s), or to remove the chances of any randomness in data with reference to p-value, another statistical parameter of threshold value of 0.05 and Benjamin–Hochberg (B–H) was applied and represented in the form of bar graph, with scale of gene and −log (B–H p value).

4.7. Quantitative Real-Time PCR

Total RNA was isolated from enzalutamide resistant and respective parental cell lines, and RNA quality was analyzed using NanoDrop ND-1000 Spectrophotometer (NanoDrop, Wilmington, DC, USA). 1 µg total RNA was used for cDNA synthesis (Applied Biosystems™, Foster City, CA, USA) using High-Capacity cDNA Reverse Transcription Kit (Thermo Fisher, Waltham, MA). To quantify and amplify the gene oligonucleotides designed by Integrated DNA Technologies (Coralville, IA, USA) were used. The list of the genes probed are mentioned in Table 1, and GAPDH (NM_008084), and Actin (NM_007393) were used as internal control in the reaction. All reactions were performed in triplicate (three biological and three technical replicates) along with no template controls (NTC). The reaction for qRT was setup accordingly; 2.5 µl of SYBR green (Radiant™ SYBR Green low-ROX qPCR, Alkali Scientific, Fort Lauderdale, FL, USA) of 5× sample were added for a total 10 µL volume with thermal cycler program used started at 50 °C for 2 min then proceeded with 95 °C for 10 min for initial denaturing, followed by 40 cycles of 95 °C for 15 s, 60 °C for 40 s, and 72 °C for 35 s to collect cycle threshold (Ct) values, along with dissociation curve cycle. The 2-ΔΔCt method was used to calculate relative expression of each gene as previously described [62].

Table 1. List of primers.

Gene	Forward Primer (5′ to 3′)	Reverse Primer (5′ to 3′)
ALDH1	GTCAAACCAGCAGAGCAAAC	GGCCCATAACCAGGAACAATA
BMI1	ATCAGTCACCAGAGAGATGGA	GGGCTAGGCAAACAAGAAGA
BMP2	CAGCTGTAAGAGACACCCTTTG	GCATTCTCCGTGGCAGTAAA
CD44	GCAGGTATGGGTTCATAGAAGG	GGTGTTGGATGTGAGGATGT
POU3F2	CTGGAGAGCCATTTCCTCAAA	AAACCAAACTCTCACCACCTC
POU5F1	GGAGGAAGCTGACAACAATGA	CTCTCACTCGGTTCTCGATACT
POU6F1	CTCCACAGCACCACTCAATA	GGTTACAGTGAGGCGAGATT
SOX2	CGTTCATCGACGAGGCTAAG	CTTCTTCATGAGCGTCTTGGT
SOX8	GTGTCGCAGGTGCTCAA	TTCATGGGCCGCTTCAC
SOX9	TCTGGAGACTTCTGAACGAGAG	CGCGGCTGGTACTTGTAATC

4.8. Western Blotting

The protein content in the cell lysates from LNCaP and C4-2B enzalutamide resistant cells and parental counterparts was determined using the DC Bio-Rad protein assay kit. The 40 µg of protein was resolved using 4–20% polyacrylamide gels (Novex, Carlsbad, CA, USA) and transferred to a nitrocellulose membrane. The blot was blocked in blocking buffer (5% nonfat dry milk/1% Tween 20; in 20 mM TBS, pH 7.6) for 2 h at room temperature, incubated with appropriate primary antibody for 2 h at room temperature or overnight at 4 °C, followed by incubation with the appropriate IgG secondary antibody conjugated to horseradish peroxidase (Amersham-Pharmacia, Piscataway, NJ, USA), and detected by ECL-chemiluminescence (Alkali Scientific Inc., Fort Lauderdale, FL, USA) and autoradiography using XAR-5 film (Eastman Kodak, Rochester, NY, USA).

4.9. Clinical ADT and Non-ADT Prostate Tissue Specimens

Specimens of post-surgical human prostate tissue from patients who had undergone androgen deprivation therapy were procured from the Tissue Procurement Facility at the University Hospitals Cleveland Medical Center. Consent was not prerequisite for these discarded tissues per hospital policies and the Institutional Review Board. Approval for this study was confirmed by the Institutional Review Board at the University Hospitals (STUDY20200324). Additional grade matched tumor tissue specimens were obtained from patients having undergone surgical procedures for prostatic disease without having received any adjuvant therapy.

4.10. Immunohistochemistry

Samples from patients undergone androgen deprivation therapy (n = 23) and specimens of grade-matched cancer tissues (n = 14) were sliced in 4 μm sections, de-paraffinized, rehydrated, and immersed in a target retrieval solution and blocked for endogenous peroxidase activity. Sections were permeabilized in TNB-BB (100 mM Tris, pH 7.5, 150 mM NaCl, 0.5% blocking agent, 0.3% Triton-X and 0.2% saponin) and incubated with anti-ALDH1, anti-POU5F1, and anti-SOX2 antibodies overnight at 4 °C, followed by incubation with biotinylated secondary antibody. The immunoreactive complexes were detected using diaminobenzidine substrate-chromogen on an inverted Olympus BX51 microscope. Images were acquired with Olympus MicroSuite^{TM} Five Software (Soft Imaging System, Lakewood, CO, USA) and the staining intensity was graded semi-quantitatively. Each specimen was assigned a score on a scale from 0 to 3 designated as 0 (no staining), 1 = weak staining, 2 = moderate staining, 3 = strong staining, reported as 0–3 staining score.

4.11. Statistical Analysis

Unpaired two-tailed student's t-test was used to compare gene expression LNCaP and C4-2B enzalutamide resistant cells and parental counterparts and results were presented as mean ± SD, p value < 0.05 was considered statistically significant.

5. Conclusions

Our study demonstrates high expression of ALDH1, POU5F1 (OCT4), and SOX2 during androgen deprivation therapy maintain pluripotency and self-renewal through regulation of several target genes and gene encoding components of key signaling pathways. Further studies targeting CSCs and their key signaling pathways is a promising approach for designing novel therapeutic strategies to circumvent chemo-resistance in castration resistant prostate cancer.

Supplementary Materials
Figure S1: (A) Sensitivity to enzalutamide treatment in LNCaP and C4-2B parental and enzalutamide resistant cells. (B) Methylene blue assay in LNCaP and C4-2B parental and enzalutamide resistant cells, Figure S2: Representative image of protein expression of AR, AR-v7, POU5F1 (OCT4), SOX2 and ALDH1 in LNCaP and C4-2B cells with and without enzalutamide treatment, Figure S3: *In-silico* stemness signatures for (A) LNCaP and (B) C4-2B enzalutamide resistant cell lines represented in radar chart, Figure S4: Signaling pathway regulating pluripotency of stem-cell leading to pluripotency maintenance, Figure S5: Human embryonic stem cell pluripotency signaling pathway.

Author Contributions: Conceptualization, S.V., E.S., and S.G.; methodology, S.V., E.S., F.N.C.K., A.M., M.A., V.S., and G.T.M.; software, S.V., V.S., and E.R.C.; validation, S.V, E.S., and E.R.C.; formal analysis, E.R.C., and V.S.; investigation, S.V., and E.S.; resources, S.G.; data curation, S.V.; writing—original draft preparation, S.V.; writing—review and editing, S.V., E.S., G.T.M., and S.G.; visualization, S.V. and S.G.; supervision, S.G.; project administration, S.G.; funding acquisition, S.G. All authors have read and agreed to the published version of the manuscript.

Acknowledgments: This research was supported by the Genomics Core Facility of the Case Western Reserve University School of Medicine's Genetics and Genome Sciences Department.

Abbreviations

ADT	Androgen deprivation therapy

ALDH	Aldehyde dehydrogenase
BIRC5	Baculoviral IAP Repeat Containing 5
BMI1	Polycomb complex protein
BMP2	Bone Morphogenetic Protein 2
CCND1	Cyclin D1
CDH1	Chromodomain Helicase DNA Binding Protein 1
CSCs	Cancer stem cells
DEGs	Differentially expressed genes
DNMT1	DNA methyltransferase 1
FGFR	Fibroblast Growth Factor Receptor
FOXD3	Forkhead Box D3
FRAT2	FRAT regulator of WNT signaling pathway 2
GATA6	GATA Binding Protein 6
HHAT	Hedgehog acyltransferase
LNCaP	Lymph node carcinoma of the prostate
MYOD1	Myogenic Differentiation 1
NEK2	NIMA Related Kinase 2
OCT4	Octamer-binding transcription factor 4
RECK	Reversion inducing cysteine rich protein with kazal motifs
RUNX2	Runt-related transcription factor 2
SNAI	Snail Family Transcriptional Repressor
VRK1	VRK Serine/Threonine Kinase 1
ZEB1	Zinc Finger E-Box Binding Homeobox 1

References

1. Magnan, S.; Zarychanski, R.; Pilote, L.; Bernier, L.; Shemilt, M.; Vigneault, E.; Fradet, V.; Turgeon, A.F. Intermittent vs Continuous Androgen Deprivation Therapy for Prostate Cancer: A Systematic Review and Meta-analysis. *JAMA Oncol.* **2015**, *1*, 1261–1269. [CrossRef] [PubMed]
2. Klotz, L.; Toren, P. Androgen deprivation therapy in advanced prostate cancer: Is intermittent therapy the new standard of care? *Curr. Oncol.* **2012**, *19*, S13–S21. [CrossRef] [PubMed]
3. Attard, G.; Borre, M.; Gurney, H.; Loriot, Y.; Andresen-Daniil, C.; Kalleda, R.; Pham, T.; Taplin, M.E. Abiraterone Alone or in Combination with Enzalutamide in Metastatic Castration-Resistant Prostate Cancer With Rising Prostate-Specific Antigen During Enzalutamide Treatment. *J. Clin. Oncol.* **2018**, *36*, 2639–2646. [CrossRef] [PubMed]
4. Chopra, A.; Georgieva, M.; Lopes, G.; Yeo, C.M.; Haaland, B. Abiraterone or Enzalutamide in Advanced Castration-Resistant Prostate Cancer: An Indirect Comparison. *Prostate* **2017**, *77*, 639–646. [CrossRef]
5. Becker, D.J.; Iyengar, A.D.; Punekar, S.R.; Ng, J.; Zaman, A.; Loeb, S.; Becker, K.D.; Makarov, D. Treatment of Metastatic Castration-resistant Prostate Cancer With Abiraterone and Enzalutamide Despite PSA Progression. *Anticancer Res.* **2019**, *39*, 2467–2473. [CrossRef]
6. Matsubara, N.; Yamada, Y.; Tabata, K.I.; Satoh, T.; Kamiya, N.; Suzuki, H.; Kawahara, T.; Uemura, H.; Yano, A.; Kawakami, S.; et al. Abiraterone Followed by Enzalutamide Versus Enzalutamide Followed by Abiraterone in Chemotherapy-naive Patients With Metastatic Castration-resistant Prostate Cancer. *Clin. Genitourin. Cancer* **2018**, *16*, 142–148. [CrossRef]
7. Patil, T.; Bernard, B. Complications of Androgen Deprivation Therapy in Men With Prostate Cancer. *Oncology* **2018**, *32*, 470–474.
8. Pal, S.K.; Patel, J.; He, M.; Foulk, B.; Kraft, K.; Smirnov, D.A.; Twardowski, P.; Kortylewski, M.; Bhargava, V.; Jones, J.O. Identification of mechanisms of resistance to treatment with abiraterone acetate or enzalutamide in patients with castration-resistant prostate cancer (CRPC). *Cancer* **2018**, *124*, 1216–1224. [CrossRef]
9. Ayob, A.Z.; Ramasamy, T.S. Cancer stem cells as key drivers of tumour progression. *J. Biomed. Sci.* **2018**, *25*, 20. [CrossRef]

10. Woodward, W.A.; Hill, R.P. Cancer Stem Cells. *Recent Results Cancer Res.* **2016**, *198*, 25–44. [CrossRef]
11. Friedmann-Morvinski, D.; Verma, I.M. Dedifferentiation and reprogramming: Origins of cancer stem cells. *EMBO Rep.* **2014**, *15*, 244–253. [CrossRef] [PubMed]
12. Kasper, S. Exploring the origins of the normal prostate and prostate cancer stem cell. *Stem Cell Rev.* **2008**, *4*, 193–201. [CrossRef]
13. Ojo, D.; Lin, X.; Wong, N.; Gu, Y.; Tang, D. Prostate Cancer Stem-like Cells Contribute to the Development of Castration-Resistant Prostate Cancer. *Cancers* **2015**, *7*, 2290–2308. [CrossRef] [PubMed]
14. Li, Q.; Deng, Q.; Chao, H.P.; Liu, X.; Lu, Y.; Lin, K.; Liu, B.; Tang, G.W.; Zhang, D.; Tracz, A.; et al. Linking prostate cancer cell AR heterogeneity to distinct castration and enzalutamide responses. *Nat. Commun.* **2018**, *9*, 3600. [CrossRef]
15. Kanwal, R.; Shukla, S.; Walker, E.; Gupta, S. Acquisition of tumorigenic potential and therapeutic resistance in CD133+ subpopulation of prostate cancer cells exhibiting stem-cell like characteristics. *Cancer Lett.* **2018**, *430*, 25–33. [CrossRef] [PubMed]
16. Shen, Y.A.; Pan, S.C.; Chu, I.; Lai, R.Y.; Wei, Y.H. Targeting cancer stem cells from a metabolic perspective. *Exp. Biol. Med.* **2020**, *245*, 465–476. [CrossRef] [PubMed]
17. Abad, E.; García-Mayea, Y.; Mir, C.; Sebastian, D.; Zorzano, A.; Potesil, D.; Zdrahal, Z.; Lyakhovich, A.; Lleonart, M.E. Common Metabolic Pathways Implicated in Resistance to Chemotherapy Point to a Key Mitochondrial Role in Breast Cancer. *Mol. Cell. Proteom.* **2019**, *18*, 231–244. [CrossRef]
18. van Schaijik, B.; Davis, P.F.; Wickremesekera, A.C.; Tan, S.T.; Itinteang, T. Subcellular localisation of the stem cell markers OCT4, SOX2, NANOG, KLF4 and c-MYC in cancer: A review. *J. Clin. Pathol.* **2018**, *71*, 88–91. [CrossRef]
19. Sánchez, B.G.; Bort, A.; Vara-Ciruelos, D.; Díaz-Laviada, I. Androgen Deprivation Induces Reprogramming of Prostate Cancer Cells to Stem-Like Cells. *Cells* **2020**, *9*, 1441. [CrossRef]
20. Singh, U.; Quintanilla, R.H.; Grecian, S.; Gee, K.R.; Rao, M.S.; Lakshmipathy, U. Novel live alkaline phosphatase substrate for identification of pluripotent stem cells. *Stem Cell Rev. Rep.* **2012**, *8*, 1021–1029. [CrossRef]
21. Lu, H.E.; Tsai, M.S.; Yang, Y.C.; Yuan, C.C.; Wang, T.H.; Lin, X.Z.; Tseng, C.P.; Hwang, S.M. Selection of alkaline phosphatase-positive induced pluripotent stem cells from human amniotic fluid-derived cells by feeder-free system. *Exp. Cell Res.* **2011**, *317*, 1895–1903. [CrossRef] [PubMed]
22. Vêncio, E.F.; Nelson, A.M.; Cavanaugh, C.; Ware, C.B.; Milller, D.G.; Garcia, J.C.; Vêncio, R.Z.; Loprieno, M.A.; Liu, A.Y. Reprogramming of prostate cancer-associated stromal cells to embryonic stem-like. *Prostate* **2012**, *72*, 1453–1463. [CrossRef] [PubMed]
23. Zhang, Y.; Chen, B.; Xu, P.; Liu, C.; Huang, P. Reprogramming Prostate Cancer Cells into Induced Pluripotent Stem Cells: A Promising Model of Prostate Cancer Stem Cell Research. *Cell. Reprogram.* **2020**, *22*, 262–268. [CrossRef] [PubMed]
24. Armstrong, A.J.; Szmulewitz, R.Z.; Petrylak, D.P.; Holzbeierlein, J.; Villers, A.; Azad, A.; Alcaraz, A.; Alekseev, B.; Iguchi, T.; Shore, N.D.; et al. ARCHES: A Randomized, Phase III Study of Androgen Deprivation Therapy With Enzalutamide or Placebo in Men With Metastatic Hormone-Sensitive Prostate Cancer. *J. Clin. Oncol.* **2019**, *37*, 2974–2986. [CrossRef]
25. Davis, I.D.; Martin, A.J.; Stockler, M.R.; Begbie, S.; Chi, K.N.; Chowdhury, S.; Coskinas, X.; Frydenberg, M.; Hague, W.E.; Horvath, L.G.; et al. Enzalutamide with Standard First-Line Therapy in Metastatic Prostate Cancer. *N. Engl. J. Med.* **2019**, *381*, 121–131. [CrossRef]
26. Fizazi, K.; Scher, H.I.; Miller, K.; Basch, E.; Sternberg, C.N.; Cella, D.; Forer, D.; Hirmand, M.; de Bono, J.S. Effect of enzalutamide on time to first skeletal-related event, pain, and quality of life in men with castration-resistant prostate cancer: Results from the randomised, phase 3 AFFIRM trial. *Lancet Oncol.* **2014**, *15*, 1147–1156. [CrossRef]
27. Annala, M.; Vandekerkhove, G.; Khalaf, D.; Taavitsainen, S.; Beja, K.; Warner, E.W.; Sunderland, K.; Kollmannsberger, C.; Eigl, B.J.; Finch, D.; et al. Circulating Tumor DNA Genomics Correlate with Resistance to Abiraterone and Enzalutamide in Prostate Cancer. *Cancer Discov.* **2018**, *8*, 444–457. [CrossRef]
28. Kregel, S.; Kiriluk, K.J.; Rosen, A.M.; Cai, Y.; Reyes, E.E.; Otto, K.B.; Tom, W.; Paner, G.P.; Szmulewitz, R.Z.; Vander Griend, D.J. Sox2 is an androgen receptor-repressed gene that promotes castration-resistant prostate cancer. *PLoS ONE* **2013**, *8*, e53701. [CrossRef]

29. Lunardi, A.; Ala, U.; Epping, M.T.; Salmena, L.; Clohessy, J.G.; Webster, K.A.; Wang, G.; Mazzucchelli, R.; Bianconi, M.; Stack, E.C.; et al. A co-clinical approach identifies mechanisms and potential therapies for androgen deprivation resistance in prostate cancer. *Nat. Genet.* **2013**, *45*, 747–755. [CrossRef]
30. Gu, G.; Yuan, J.; Wills, M.; Kasper, S. Prostate cancer cells with stem cell characteristics reconstitute the original human tumor in vivo. *Cancer Res.* **2007**, *67*, 4807–4815. [CrossRef]
31. Mathieu, J.; Zhang, Z.; Zhou, W.; Wang, A.J.; Heddleston, J.M.; Pinna, C.M.; Hubaud, A.; Stadler, B.; Choi, M.; Bar, M.; et al. HIF induces human embryonic stem cell markers in cancer cells. *Cancer Res.* **2011**, *71*, 4640–4652. [CrossRef] [PubMed]
32. Federer-Gsponer, J.R.; Müller, D.C.; Zellweger, T.; Eggimann, M.; Marston, K.; Ruiz, C.; Seifert, H.H.; Rentsch, C.A.; Bubendorf, L.; Le Magnen, C. Patterns of stemness-associated markers in the development of castration-resistant prostate cancer. *Prostate* **2020**, *80*, 1108–1117. [CrossRef] [PubMed]
33. Li, H.; Sun, Y.; Zhan, M. Exploring pathways from gene co-expression to network dynamics. *Methods Mol. Biol.* **2009**, *541*, 249–267. [CrossRef] [PubMed]
34. Strunz, S.; Wolkenhauer, O.; de la Fuente, A. Network-Assisted Disease Classification and Biomarker Discovery. *Methods Mol. Biol.* **2016**, *1386*, 353–374. [CrossRef] [PubMed]
35. Chang, Y.T.; Lin, T.P.; Campbell, M.; Pan, C.C.; Lee, S.H.; Lee, H.C.; Yang, M.H.; Kung, H.J.; Chang, P.C. REST is a crucial regulator for acquiring EMT-like and stemness phenotypes in hormone-refractory prostate cancer. *Sci. Rep.* **2017**, *7*, 42795. [CrossRef]
36. Zhang, L.; Jiao, M.; Li, L.; Wu, D.; Wu, K.; Li, X.; Zhu, G.; Dang, Q.; Wang, X.; Hsieh, J.T.; et al. Tumorspheres derived from prostate cancer cells possess chemoresistant and cancer stem cell properties. *J. Cancer Res. Clin. Oncol.* **2012**, *138*, 675–686. [CrossRef]
37. Toropainen, S.; Niskanen, E.A.; Malinen, M.; Sutinen, P.; Kaikkonen, M.U.; Palvimo, J.J. Global analysis of transcription in castration-resistant prostate cancer cells uncovers active enhancers and direct androgen receptor targets. *Sci. Rep.* **2016**, *6*, 33510. [CrossRef]
38. Rajan, P.; Sudbery, I.M.; Villasevil, M.E.; Mui, E.; Fleming, J.; Davis, M.; Ahmad, I.; Edwards, J.; Sansom, O.J.; Sims, D.; et al. Next-generation sequencing of advanced prostate cancer treated with androgen-deprivation therapy. *Eur. Urol.* **2014**, *66*, 32–39. [CrossRef]
39. Ginestier, C.; Hur, M.H.; Charafe-Jauffret, E.; Monville, F.; Dutcher, J.; Brown, M.; Jacquemier, J.; Viens, P.; Kleer, C.G.; Liu, S.; et al. ALDH1 is a marker of normal and malignant human mammary stem cells and a predictor of poor clinical outcome. *Cell Stem Cell* **2007**, *1*, 555–567. [CrossRef]
40. Tomita, H.; Tanaka, K.; Tanaka, T.; Hara, A. Aldehyde dehydrogenase 1A1 in stem cells and cancer. *Oncotarget* **2016**, *7*, 11018–11032. [CrossRef]
41. Brennan, S.K.; Meade, B.; Wang, Q.; Merchant, A.A.; Kowalski, J.; Matsui, W. Mantle cell lymphoma activation enhances bortezomib sensitivity. *Blood* **2010**, *116*, 4185–4191. [CrossRef] [PubMed]
42. Landen, C.N., Jr.; Goodman, B.; Katre, A.A.; Steg, A.D.; Nick, A.M.; Stone, R.L.; Miller, L.D.; Mejia, P.V.; Jennings, N.B.; Gershenson, D.M.; et al. Targeting aldehyde dehydrogenase cancer stem cells in ovarian cancer. *Mol. Cancer Ther.* **2010**, *9*, 3186–3199. [CrossRef] [PubMed]
43. Kulsum, S.; Sudheendra, H.V.; Pandian, R.; Ravindra, D.R.; Siddappa, G.; Nisheena, R.; Chevour, P.; Ramachandran, B.; Sagar, M.; Jayaprakash, A.; et al. Cancer stem cell mediated acquired chemoresistance in head and neck cancer can be abrogated by aldehyde dehydrogenase 1 A1 inhibition. *Mol. Carcinog.* **2017**, *56*, 694–711. [CrossRef] [PubMed]
44. Ayub, T.H.; Keyver-Paik, M.D.; Debald, M.; Rostamzadeh, B.; Thiesler, T.; Schröder, L.; Barchet, W.; Abramian, A.; Kaiser, C.; Kristiansen, G.; et al. Accumulation of ALDH1-positive cells after neoadjuvant chemotherapy predicts treatment resistance and prognosticates poor outcome in ovarian cancer. *Oncotarget* **2015**, *6*, 16437–16448. [CrossRef]
45. Nishida, S.; Hirohashi, Y.; Torigoe, T.; Kitamura, H.; Takahashi, A.; Masumori, N.; Tsukamoto, T.; Sato, N. Gene expression profiles of prostate cancer stem cells isolated by aldehyde dehydrogenase activity assay. *J. Urol.* **2012**, *188*, 294–299. [CrossRef]
46. Vaddi, P.K.; Stamnes, M.A.; Cao, H.; Chen, S. Elimination of SOX2/OCT4-Associated Prostate Cancer Stem Cells Blocks Tumor Development and Enhances Therapeutic Response. *Cancers* **2019**, *11*, 1331. [CrossRef]
47. Rodriguez, D.; Ramkairsingh, M.; Lin, X.; Kapoor, A.; Major, P.; Tang, D. The Central Contributions of Breast Cancer Stem Cells in Developing Resistance to Endocrine Therapy in Estrogen Receptor (ER)-Positive Breast Cancer. *Cancers* **2019**, *11*, 1028. [CrossRef]

48. Chen, Q.; Qiu, C.; Huang, Y.; Jiang, L.; Huang, Q.; Guo, L.; Liu, T. Human amniotic epithelial cell feeder layers maintain iPS cell pluripotency by inhibiting endogenous DNA methyltransferase 1. *Exp. Ther. Med.* **2013**, *6*, 1145–1154. [CrossRef]
49. Schmidt, C.S.; Bultmann, S.; Meilinger, D.; Zacher, B.; Tresch, A.; Maier, K.C.; Peter, C.; Martin, D.E.; Leonhardt, H.; Spada, F. Global DNA hypomethylation prevents consolidation of differentiation programs and allows reversion to the embryonic stem cell state. *PLoS ONE* **2012**, *7*, e52629. [CrossRef]
50. Tsai, C.C.; Su, P.F.; Huang, Y.F.; Yew, T.L.; Hung, S.C. Oct4 and Nanog directly regulate Dnmt1 to maintain self-renewal and undifferentiated state in mesenchymal stem cells. *Mol. Cell* **2012**, *47*, 169–182. [CrossRef]
51. Yang, J.; Gao, C.; Chai, L.; Ma, Y. A novel SALL4/OCT4 transcriptional feedback network for pluripotency of embryonic stem cells. *PLoS ONE* **2010**, *5*, e10766. [CrossRef] [PubMed]
52. Zuo, J.; Guo, Y.; Peng, X.; Tang, Y.; Zhang, X.; He, P.; Li, S.; Wa, Q.; Li, J.; Huang, S.; et al. Inhibitory action of pristimerin on hypoxia-mediated metastasis involves stem cell characteristics and EMT in PC-3 prostate cancer cells. *Oncol. Rep.* **2015**, *33*, 1388–1394. [CrossRef] [PubMed]
53. Cao, L.; Li, C.; Shen, S.; Yan, Y.; Ji, W.; Wang, J.; Qian, H.; Jiang, X.; Li, Z.; Wu, M.; et al. OCT4 increases BIRC5 and CCND1 expression and promotes cancer progression in hepatocellular carcinoma. *BMC Cancer* **2013**, *13*, 82. [CrossRef] [PubMed]
54. Du, L.; Sun, W.; Zhang, H.; Chen, D. BDE-209 inhibits pluripotent genes expression and induces apoptosis in human embryonic stem cells. *J. Appl. Toxicol.* **2016**, *36*, 659–668. [CrossRef]
55. Wuebben, E.L.; Rizzino, A. The dark side of SOX2: Cancer—A comprehensive overview. *Oncotarget* **2017**, *8*, 44917–44943. [CrossRef]
56. Chaudhary, S.; Islam, Z.; Mishra, V.; Rawat, S.; Ashraf, G.M.; Kolatkar, P.R. Sox2: A Regulatory Factor in Tumorigenesis and Metastasis. *Curr. Protein Pept. Sci.* **2019**, *20*, 495–504. [CrossRef]
57. Schaefer, T.; Lengerke, C. SOX2 protein biochemistry in stemness, reprogramming, and cancer: The PI3K/AKT/SOX2 axis and beyond. *Oncogene* **2020**, *39*, 278–292. [CrossRef]
58. Tuerff, D.; Sissung, T.; Figg, W.D. Cellular identity crisis: Antiandrogen resistance by lineage plasticity. *Cancer Biol. Ther.* **2017**, *18*, 841–842. [CrossRef]
59. Handle, F.; Erb, H.H.; Luef, B.; Hoefer, J.; Dietrich, D.; Parson, W.; Kristiansen, G.; Santer, F.R.; Culig, Z. SOCS3 Modulates the Response to Enzalutamide and Is Regulated by Androgen Receptor Signaling and CpG Methylation in Prostate Cancer Cells. *Mol. Cancer Res.* **2016**, *14*, 574–585. [CrossRef]
60. Ku, S.Y.; Rosario, S.; Wang, Y.; Mu, P.; Seshadri, M.; Goodrich, Z.W.; Goodrich, M.M.; Labbé, D.P.; Gomez, E.C.; Wang, J.; et al. Rb1 and Trp53 cooperate to suppress prostate cancer lineage plasticity, metastasis, and antiandrogen resistance. *Science* **2017**, *355*, 78–83. [CrossRef]
61. Mu, P.; Zhang, Z.; Benelli, M.; Karthaus, W.R.; Hoover, E.; Chen, C.C.; Wongvipat, J.; Ku, S.Y.; Gao, D.; Cao, Z.; et al. SOX2 promotes lineage plasticity and antiandrogen resistance in TP53- and RB1-deficient prostate cancer. *Science* **2017**, *355*, 84–88. [CrossRef] [PubMed]
62. Verma, S.; Shukla, S.; Pandey, M.; MacLennan, G.T.; Gupta, S. Differentially Expressed Genes and Molecular Pathways in an Autochthonous Mouse Prostate Cancer Model. *Front Genet.* **2019**, *10*, 235. [CrossRef] [PubMed]

2

Crosstalk between Prostate Cancer Cells and Tumor-Associated Fibroblasts Enhances the Malignancy by Inhibiting the Tumor Suppressor PLZF

Kum Hee Noh [1,2,†], Ae Jin Jeong [1,2], Haeri Lee [1,2], Song-Hee Lee [1,2], Eunhee Yi [1], Pahn-Shick Chang [3,4,5], Cheol Kwak [6,7] and Sang-Kyu Ye [1,2,8,9,*]

1 Department of Pharmacology and Biomedical Sciences, Seoul National University College of Medicine, Seoul 03080, Korea; nkhys1209@snu.ac.kr (K.H.N.); lovej89@snu.ac.kr (A.J.J.); hrlee519@snu.ac.kr (H.L.); 24happy92@snu.ac.kr (S.-H.L.); Eunhee.Yi@jax.org (E.Y.)
2 Biomedical Science Project (BK21PLUS), Seoul National University College of Medicine, Seoul 03080, Korea
3 Department of Agricultural Biotechnology, Seoul National University, Seoul 08826, Korea; pschang@snu.ac.kr
4 Center for Food and Bioconvergence, Seoul National University, Seoul 08826, Korea
5 Research Institute of Agriculture and Life Sciences, Seoul National University, Seoul 08826, Korea
6 Department of Urology, Seoul National University Hospital, Seoul 03080, Korea; mdrafael@snu.ac.kr
7 Department of Urology, Seoul National University Hospital, Seoul National University College of Medicine, Seoul 03080, Korea
8 Ischemic/Hypoxic Disease Institute, Seoul National University College of Medicine, Seoul 03080, Korea
9 Neuro-Immune Information Storage Network Research Center, Seoul National University College of Medicine, Seoul 03080, Korea
* Correspondence: sangkyu@snu.ac.kr
† The author is first author.

Abstract: Although prostate cancer is clinically manageable during the early stages of progression, metastatic progression severely compromises the prognosis and leads to mortality. Constitutive activation of STAT3 has been connected to prostate cancer malignancy, and abolishing the STAT3 activity may diminish tumor growth and metastasis. However, its suppressor genes and pathways have not been well established. In this study, we show that promyelocytic leukemia zinc finger (PLZF) has a putative tumor-suppressor function in prostate cancer by inhibiting phosphorylation of STAT3. Compared with a benign prostate, high-grade prostate cancer patient tissue was negatively correlated with PLZF expression. PLZF depletion accelerated proliferation and survival, migration, and invasion in human prostate cancer cells. Mechanistically, we demonstrated a novel role of PLZF as the transcriptional regulator of the tyrosine phosphatase SHP-1 that inhibits the oncogenic JAKs–STAT3 pathway. These results suggest that the collapse of PLZF expression by the CCL3 derived from fibroblasts accelerates the cell migration and invasion properties of prostate cancer cells. Our results suggest that increasing PLZF could be an attractive strategy for suppressing prostate cancer metastasis as well as for tumor growth.

Keywords: promyelocytic leukemia zinc finger; STAT3; prostate cancer; tumor-associated fibroblast; tumor-suppressor gene

---

1. Introduction

Prostate cancer is the most frequently diagnosed malignancies and the sixth leading cause of cancer-related death in men worldwide [1]. In general, patients with organ-confined prostate cancer are treated successfully. However, uncontrolled metastasis indicates disease relapse, which is virtually incurable and results in considerable disease mortality [2–4]. In addition, although much research has been done to determine the cause of prostate cancer metastasis, our understanding is still incomplete. For this reason, a better understanding of prostate cancer metastasis would lead to new approaches and targets for preventing or treating metastasis [5].

Tumors can be considered as complex organs composed of tumor cells and various non-malignant stromal cells that form the tumor microenvironment (TME). These stromal cells include endothelial cells, pericytes, immune inflammatory cells, and fibroblasts, all of which are genetically stable and typically not malignant [6–8]. However, they are affected by interactions with tumor cells via soluble factors and modify tumor cells to favor tumor development, tumor growth, and invasion [8–10]. Among them, fibroblasts are crucial regulators of prostate tumor metastatic progression [11,12]. Several researches have investigated that the normal fibroblasts have a role in maintaining epithelial homeostasis by inhibiting the proliferation and carcinogenic potential of the adjacent epithelium; but, following neoplastic transformation of the epithelium, tumor-associated fibroblasts promote tumor growth and remodels the extracellular matrix (ECM). That is, normal fibroblasts can be trained by carcinoma cells to become tumor-associated fibroblasts [13–15].

The mechanism by which local prostate cancer progresses into a lethal disease is unclear. One predictable mechanism is the loss or alteration of tumor-suppressor genes. Promyelocytic leukemia zinc finger (PLZF), also known as Zinc finger and BTB domain-containing 16 (ZBTB16), binds to specific DNA sequences with its carboxy-terminal zinc finger domain and protein–protein interaction [16,17]. PLZF is involved in diverse cellular signaling, particularly in stem cells self-renewal or differentiation, hematopoiesis, and immune regulation [18–22]. Lately, PLZF was reported to potentially play a tumor-suppressor role, which reduces cell growth and survival in numerous solid tumors, including melanoma, malignant mesothelioma, and non-small cell lung cancer cells [23–26]. Besides, recent studies also suggested that PLZF is implicated in prostate cancer as a tumor-suppressor protein. In prostate cancer, PLZF expression is reduced or lost in high grade tumor and castration-resistant prostate cancer (CRPC) [17,27]. PLZF inhibits prostate cancer cell growth through its inhibitory effects on AR, AKT, mTOR, and MAPK signaling [28,29]. Research of PLZF is actively underway as it has a high potential as a target for prostate cancer, but the relationship between prostate tumor microenvironment and PLZF is not yet well known. Therefore, we aim to clarify the role of PLZF in the association between prostate cancer and the tumor microenvironment. Restoring PLZF expression or reactivation also could be a novel strategy for prostate cancer therapy.

Surprisingly, we discovered that tumor-associated fibroblasts inhibited PLZF expression. In this study, we hypothesized that PLZF is suppressed by tumor-associated fibroblasts, thereby promoting metastatic prostate cancer progression. As a result, we found that tumor-induced impetus stimulated fibroblasts to produce CCL3, which promoted prostate cancer growth and metastasis through the reduction of PLZF/tyrosine phosphatase SHP-1. By studying the PLZF/SHP-1/STAT3 signaling pathway involved in CCL3-induced malignancy, we may find a potential therapeutic target for prostate cancer.

2. Results

2.1. PLZF Inhibits the Activation of STAT3, Which Has an Inverse Correlation with PLZF

We first evaluated the protein levels of PLZF and Tyr705 phosphorylation STAT3 (pY-STAT3) in 40 prostate cancer patients and 10 benign patient tissues, which were categorized according to Gleason scores (GS). Immunohistochemistry (IHC) assays showed that the levels of pY-STAT3 increased with the progression of prostate cancer, as opposed to the decrease in PLZF levels (Figure 1A, Figure S1A). Pearson's correlation analyses revealed a significant negative correlation between PLZF and pY-STAT3 levels (Figure 1B). The Prostate Specific Antigen (PSA) from prostate cancer patients also has been shown to increase with Gleason scores (Figure S1B, Table S1). The pY-STAT3 protein levels are correlated, but the PLZF protein levels are inversely correlated with PSA (Figure 1C). Moreover, prostate cancer patients from The Cancer Genome Atlas (TCGA) database were divided into PLZF- and pY-STAT3-low/high groups. Kaplan–Meier analysis indicated that patients with PLZF-low and pY-STAT3-high groups had poorer recurrence-free survival than those with PLZF-high and pY-STAT3-low expression (Figure 1D). In addition, the higher-PLZF group also showed higher survival rates than the lower groups in the CRPC patient group with the poorest prediction for prostate cancer (Figure S1C). We compared the PLZF mRNA expression between a normal, primary tumor, and metastatic tumor using the TCGA database, and found that PLZF expression was more decreased in metastatic tumors (Figure 1E). We examined PLZF, pY-STAT3, and STAT3 protein/mRNA levels by Western blotting/RT-PCR in prostate cancer cell lines (Figure 1F, Figure S1D). Interestingly, PLZF was not affected by exogenous constitutive activation (CA) of STAT3, but it was discovered that pY-STAT3 was decreased by PLZF. Based on the gene-silencing efficiency (Figure S1E), we chose the #5 siRNA directed against Human PLZF (SI03090346). It was also confirmed that knockdown of PLZF increased the phosphorylation of STAT3 (Figure 1G). Reducing pY-STAT3, a well-known oncogene in prostate cancer, suggests that PLZF is likely involved in progression of prostate cancer. Therefore, we hypothesized that PLZF acts as the tumor-suppressor gene to improve the poor prognosis in prostate cancer patients.

2.2. PLZF Induces the Cell Cycle Arrest and Apoptosis Effects by Suppression of STAT3 Signaling

To investigate the tumor-suppressing role of PLZF in prostate cancer, we overexpressed PLZF in DU145 cells. Overexpression of PLZF resulted in significantly reduced proliferating cell nuclear antigen (PCNA) protein expression, and inhibited cell growth. In contrast, knockdown of endogenous PLZF increased cell viability in LNCaP cells (Figure 2A,B, Figure S2A). Moreover, to explain that PLZF functions as a tumor suppressor, cell cycle distribution was detected. Cell cycle examined by flow cytometry analysis revealed that 10.825% more cells increased in the sub-G1 phase proportion and 14.75% more cells accumulated in the G0/G1 phase compartment with PLZF-overexpressed DU145 cells ($n = 3$) (Figure 2C). To confirm the molecular mechanism of PLZF, the expression levels of the cell cycle arrest regulators, including c-MYC, cyclin D1, cyclin D3, CDK4, p21, and p27, were tested. As a result, the G0/G1 phase arrest is confirmed by PLZF (Figure 2D,E). In addition to promoting cell cycle arrest, PLZF triggered prostate cancer cell apoptosis, with effectively increasing the apoptosis proportion in the Annexin V-FITC/PI staining assay. Compared to the controls, an increase in the percentage of early and late apoptotic cells was observed in PLZF-overexpressed DU145 (early, from 7.43% to 14.83%; late, from 3.47% to 10.69%; Figure 2F). The mRNA/protein levels of the BCL-2 family apoptotic markers were inhibited in PLZF-overexpressed cells, but they were increased in PLZF-knockdown cells (Figure 2G,H). As a result, these findings indicated that the increase in PLZF expression in prostate cancer cells induced cell cycle arrest and apoptosis.

2.3. PLZF Ablation in Prostate Cancer Promotes Cell Migration and Invasion via Activation of STAT3

We next evaluated the migration and invasion ability by wound healing assays, Transwell cell migration assays, and Matrigel invasion assays. Quantitative analysis of the wound healing assay revealed that PLZF-transfected cells delayed the closure of the wound gap as well as delayed the STAT3-knockdown in DU145 cells. (Figure 3A, Figure S2B). Consistently, a similar effect of PLZF was observed in the Transwell cell migration assays. As with the siSTAT3-transfected cells, the number of migrated cells with ectopically expressed PLZF showed an apparent reduction compared to the control cells. In contrast, knockdown of endogenous PLZF in LNCaP cells by siRNA increased the cell migration ability. The Matrigel invasion assays demonstrated that the cell invasion ability is significantly attenuated by PLZF-overexpressed DU145 cells but is increased by PLZF-knockdown LNCaP cells (Figure 3B). The mRNA/protein levels of the epithelial marker (E-cadherin) were increased in PLZF-overexpressed cells but inhibited in PLZF-knockdown cells. In mesenchymal markers (N-cadherin, Vimentin and Fibronectin), the opposite was observed (Figure 3C,D). Furthermore, consistent with an epithelial–mesenchymal transition (EMT) phenotype, DU145 cells exhibited diminished ECM degradation as evidenced by decreased levels of MMP2 and MMP9 expression

(Figure S2C). The data presented above suggested PLZF could suppress prostate cancer migratory and invasive abilities by preventing the EMT process.

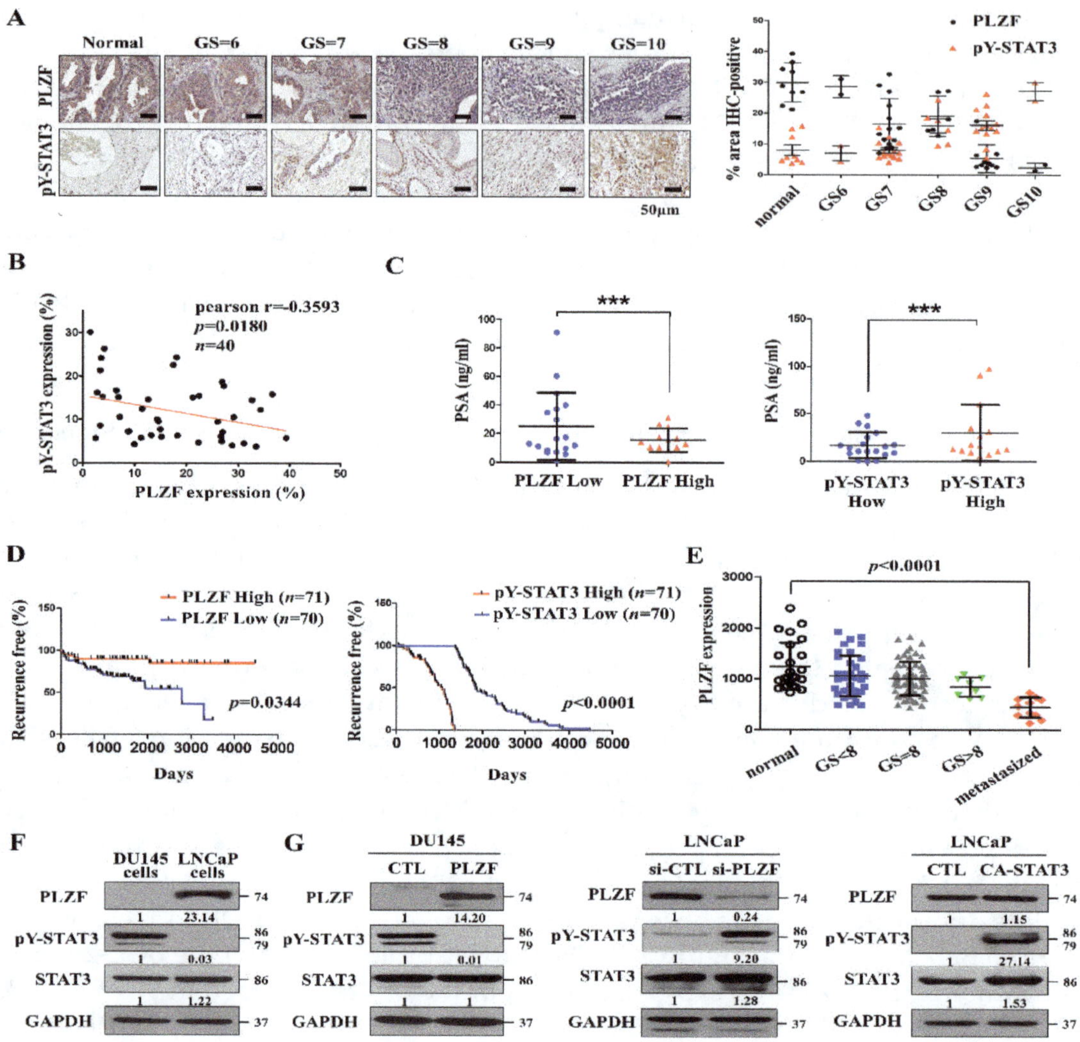

Figure 1. Promyelocytic leukemia zinc finger (PLZF) inhibits the activation of STAT3, which has an inverse correlation with PLZF. (**A**) IHC of PLZF, pY-STAT3 in tissue arrays of surrounding normal prostate tissues, and human prostate cancer specimens correlates with Gleason scores (GS). Quantification of PLZF-positive expression according to the GS in normal tissues adjacent to the benign tumors ($n = 9$) and malignant tumors ($n = 40$) (right). (**B**) Associations between expression of PLZF and pY-STAT3. Scatter plots showing the linear correlation determined by Pearson correlation coefficient calculation of those genes that were statistically significant. Pearson correlation coefficient r and p-values are given in each scatter plot. (**C**) PSA levels were measured according to the PLZF and pY-STAT3 protein levels, $n = 40$. (**D**) Kaplan–Meier recurrence-free survival analysis of prostate cancer patients according to PLZF (* $p = 0.0344$) and pY-STAT3 (* $p < 0.0001$) expression. (**E**) Quantification of PLZF mRNA expression according to the GS and metastasis in prostate cancer patients' samples, *** $p < 0.0001$. (**F**) PLZF, pY-STAT3, STAT3, and GAPDH protein expression by Western blotting in the prostate cancer cell lines DU145 and LNCaP. GAPDH was used as a loading control. (**G**) Western blotting was performed in PLZF, CA-STAT3 plasmid, and siRNA-transfected cells. The uncropped blots and molecular weight markers of Figure 1 are shown in Figure S5

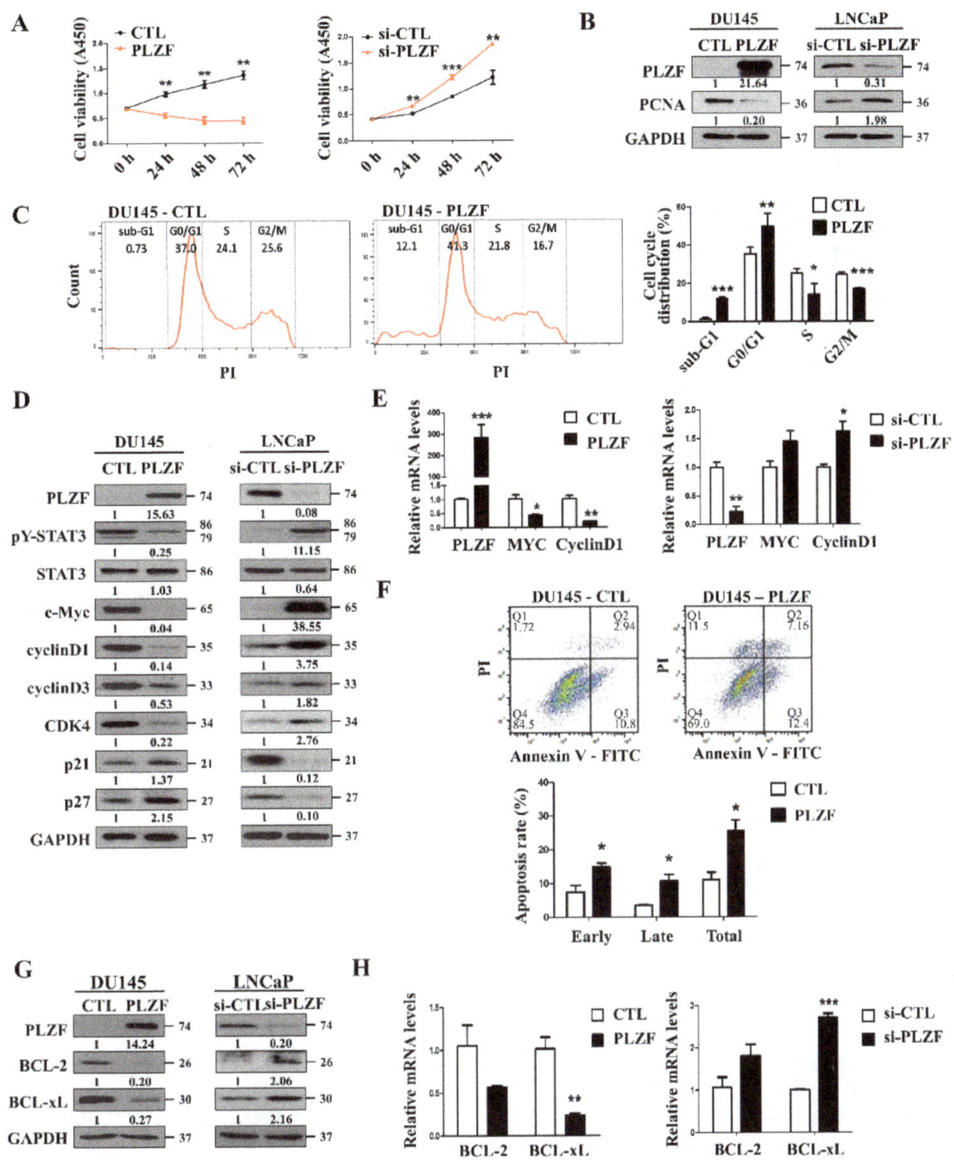

Figure 2. PLZF induces cell cycle arrest and apoptosis effects by suppression of STAT3 signaling. (**A**) CCK assay was performed by transfecting DU145 and LNCaP cells with plasmid and siRNA, followed by culture for 1–3 days. (**B**) Western blotting was performed in PLZF plasmid and siRNA-transfected cells. (**C**) Effect of cell cycle distribution of MOCK- and PLZF-transfected DU145 cells was detected by flow cytometry analysis. Representative histograms of cell cycle alteration. Summarized results from three independent experiments were quantified as mean ± SD (right). (**D**) Protein expression levels of indicated cell cycle regulators were detected by Western blotting. (**E**) mRNA expression levels of PLZF, MYC, and CyclinD1 were examined by qRT-PCR in DU145 and LNCaP cells transfected with PLZF plasmid and siRNA. (**F**) Apoptosis assay of PLZF plasmid and siRNA transfected DU145 and LNCaP cells was detected by Annexin V-FITC/PI staining. Representative histograms of cell cycle alteration. Summarized results from three independent experiments were quantified as mean ± SD (right). (**G**) Protein expression levels of indicated apoptosis regulators were detected by Western blotting. (**H**) mRNA expression levels of BCL2 and BCLxL were examined by qRT-PCR in DU145 and LNCaP cells transfected with PLZF plasmid and siRNA. In (A) and (B), data are presented as the mean ± SD; * $p < 0.05$, ** $p < 0.01$, *** $p < 0.001$. The uncropped blots and molecular weight markers of Figure 2 are shown in Figure S6

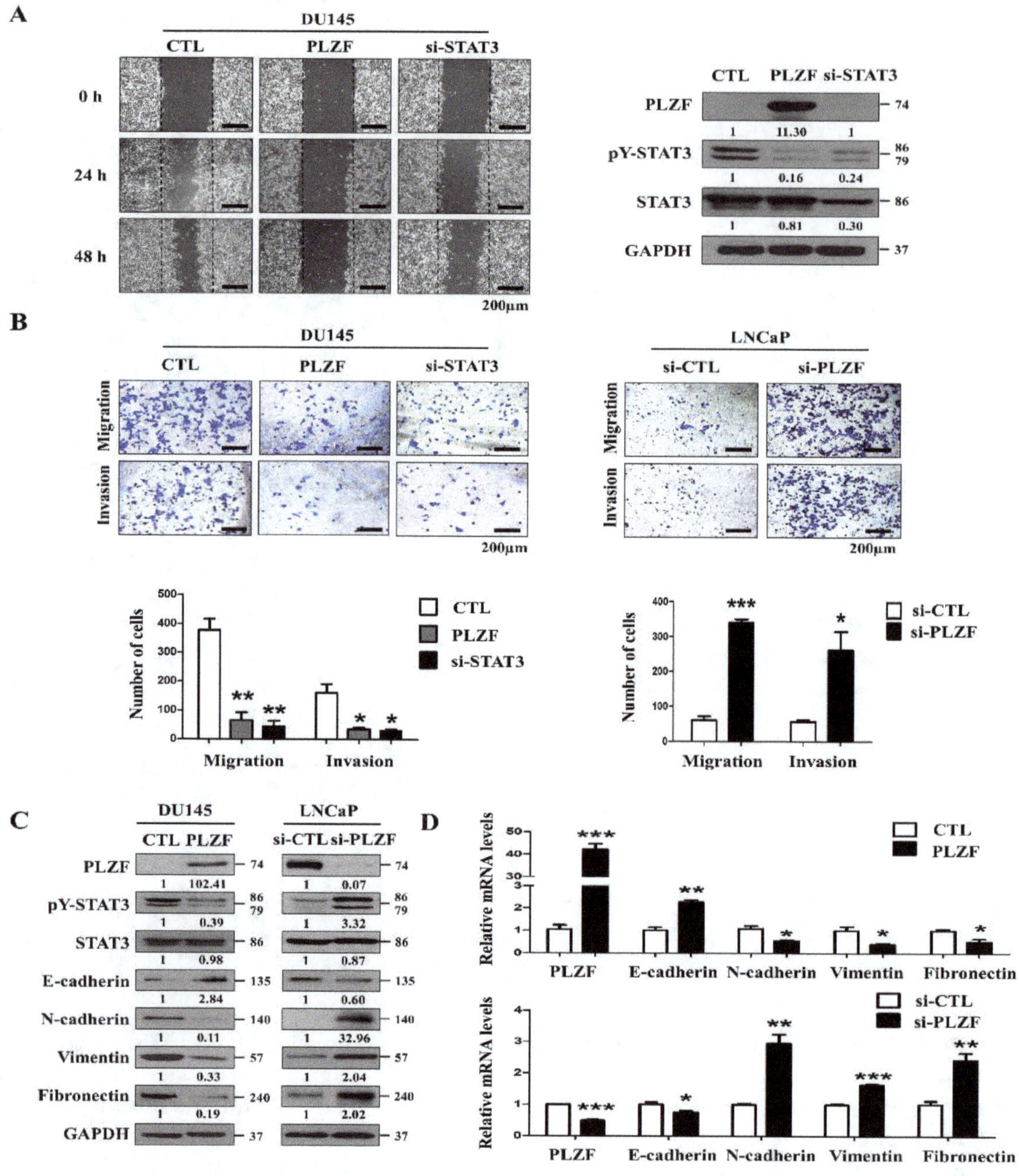

Figure 3. PLZF ablation in prostate cancer promotes cell migration and invasion via activation of STAT3. (**A**) Wound healing assay (left) and Western blotting (right) was conducted in PLZF plasmid and siSTAT3-transfected DU145 cells. (**B**) Transwell migration and Matrigel invasion assays were conducted in MOCK-/PLZF-/siSTAT3- and siNT-/siPLZF-transfected cells. The relative cell numbers are shown (bottom). (**C**) Protein expression of the epithelial/mesenchymal related markers were compared by Western blotting in MOCK-/PLZF-transfected cells and siNT-/siPLZF-transfected cells. (**D**) mRNA expression levels of epithelial/mesenchymal related markers were examined by qRT-PCR in DU145 and LNCaP cells transfected with PLZF plasmid and siRNA. In (A) and (B) representative images from three independent experiments were quantified as the mean ± SD; * $p < 0.05$, ** $p < 0.01$, *** $p < 0.001$. The uncropped blots and molecular weight markers of Figure 3 are shown in Figure S7

2.4. *Tyrosine Phosphatase SHP-1 Is a Direct Target of PLZF and Inhibits the Tyrosine Phosphorylation of JAKs–STAT3 Signaling*

To interrogate the molecular mechanism of PLZF, which has been demonstrated as a tumor-suppressor gene in human prostate cancer cells, and whether it is associated with the oncogenes JAK1, JAK2, and JAK3 as well as the TYK2(JAKs)–STAT signaling molecules, we ectopically transfected PLZF in DU145 cells. As a result, PLZF inhibits the phosphorylation of the JAKs–STAT3 signaling pathway molecules but not in STAT1 and STAT5. Especially, tyrosine phosphorylation of the JAKs–STAT3 proteins was inhibited in PLZF-overexpressed DU145 cells. In contrast, serine phosphorylation and the total JAKs–STAT proteins remained unchanged (Figure 4A). The mRNA levels of STAT3 were also not significantly changed in PLZF-overexpressed cells (Figure S2D). Besides, pY-STAT3 and total STAT3 proteins did not bind with PLZF (Figure S2E). If only tyrosine, not serine, was reduced by PLZF during phosphorylation of STAT3, the status of the phosphorylation would be questioned when phosphorylated STAT3 was translocated into the nucleus. We performed cytosol–nuclear fractionation with PLZF-overexpressed DU145 cells. The nuclear and cytoplasmic proteins were dissociated, and the Western blot showed that exogenous overexpression of PLZF significantly decreased only tyrosine phosphorylation of STAT3, not serine phosphorylation of STAT3 in both the cytosol and nucleus compared to the control (Figure S2F). Since PLZF suppresses tyrosine phosphorylation only, it is assumed that PLZF can dephosphorylate JAKs–STAT3 through tyrosine phosphatase. To identify the tyrosine phosphatases related to STAT3, the mRNA/protein levels of protein tyrosine phosphatase SHP-1 were increased in PLZF-overexpressed cells, but they were inhibited in PLZF-knockdown cells (Figure 4B,C). Based on the gene-silencing efficiency (Figure S2G), we chose the #10 siRNA directed against Human SHP-1 (SI04436831). PLZF overexpression significantly induced SHP-1 expression, whereas tyrosine-phosphorylation of JAKs–STAT3 was abolished by ectopic-PLZF despite SHP-1 ablation (Figure 4D).

To determine whether the SHP-1 promoter region contributes to transcriptional regulation of the PLZF gene, we constructed a pGL3-luciferase reporter plasmid containing the SHP-1 promoter fragment. In prostate cancer cell lines, SHP-1 promoter activity was increased by PLZF overexpression, while the activity was diminished by PLZF knockdown (Figure 4E). The IHC quantification of SHP-1 protein expression demonstrated that the protein levels are lower in high GS (Figure 4F). Pearson's correlation analyses revealed a prominently positive correlation between SHP-1 and PLZF levels (Figure 4G). The SHP-1 protein levels of prostate cancer patient tissues are inversely correlated in PSA (Figure 4H). Moreover, prostate cancer patients from the TCGA database were divided into the SHP-1 high (n = 71)/low (n = 70) expression groups. Kaplan–Meier analysis indicated that patients of low SHP-1 levels had poorer recurrence-free survival than those with high SHP-1 levels (Figure 4I). These results indicate that PLZF suppressed the tyrosine phosphorylation of JAKs–STAT3 by promoting SHP-1 transcription, acting as a tumor-suppressor gene.

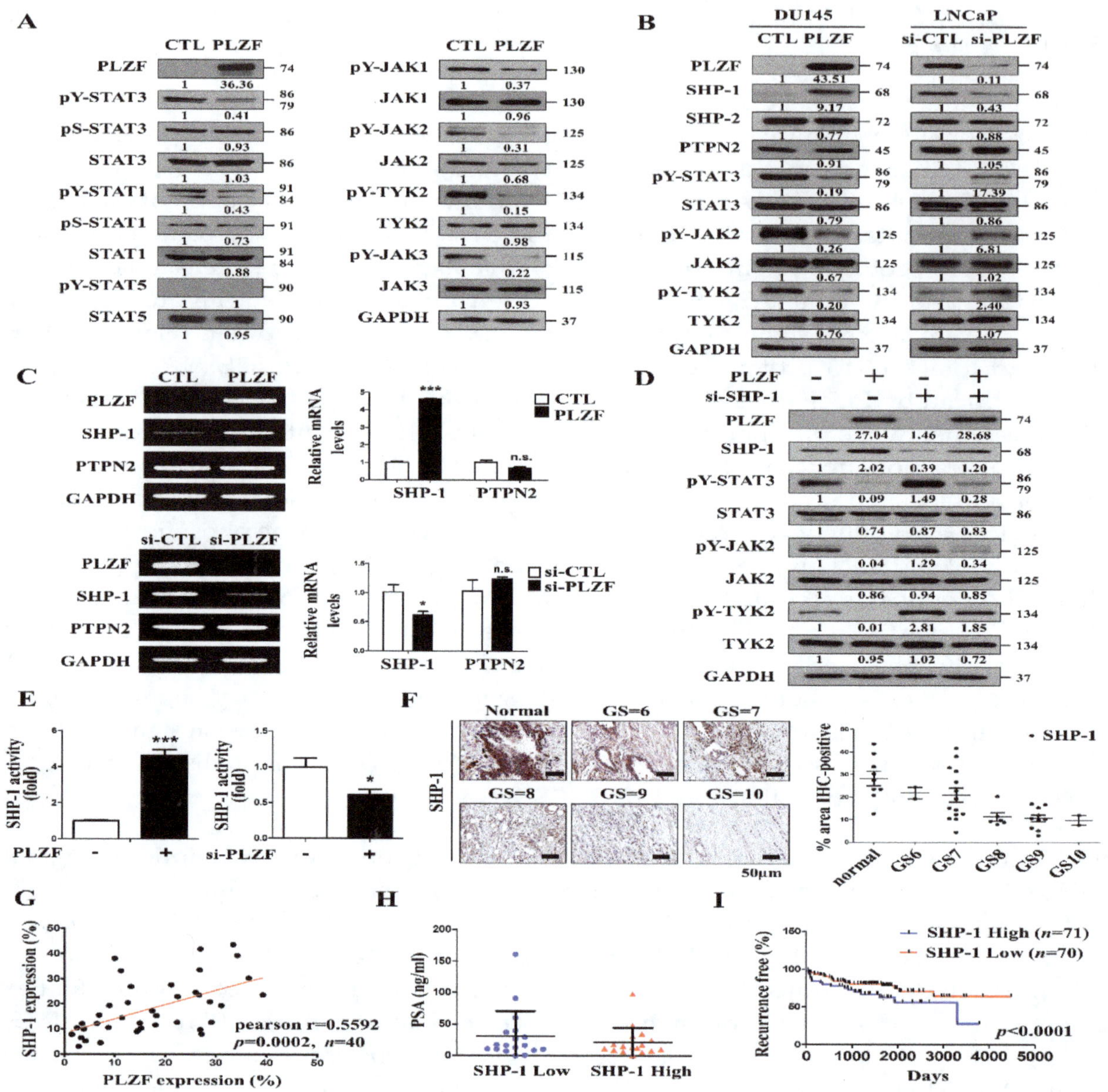

Figure 4. Tyrosine phosphatase SHP-1 is a direct target of PLZF and inhibits the tyrosine phosphorylation of JAKs–STAT3 signaling. (**A**) Western blotting was performed in PLZF-transfected DU145 cells. (**B**) Protein expression levels of the tyrosine phosphatase-related markers STAT3, JAK2, and TYK2 were detected by Western blotting. (**C**) mRNA expression levels of tyrosine phosphatases were examined by RT-PCR and qRT-PCR in DU145 and LNCaP cells. (**D**) Western blotting of SHP-1, STAT3, JAK2, and TYK2 in cells that were co-transfected with PLZF and si-SHP-1. (**E**) SHP-1 reporter assay in DU145 and LNCaP cells transfected with PLZF plasmid and siRNA. (**F**) IHC of SHP-1 in tissue arrays of surrounding normal prostate tissues and human prostate cancer specimens correlates with GS. Quantification of SHP-1-positive expression according to the GS scores in benign (n = 9) and malignant tumors (n = 40) (right). (**G**) Associations between SHP-1 and PLZF. Scatter plots showing the linear correlation determined by Pearson correlation coefficient calculation of those genes that were statistically significant. Pearson correlation coefficient r and p-values are given in each scatter plot. (**H**) PSA levels were measured according to the SHP-1 protein levels, n = 40. (**I**) Kaplan–Meier recurrence-free survival analysis of prostate cancer patients according to SHP-1 (* $p < 0.0001$) expression. In C, E, and F data are presented as the mean ± SD; * $p < 0.05$, *** $p < 0.001$. The uncropped blots and molecular weight markers of Figure 4 are shown in Figure S8

2.5. PLZF Is Reduced in Prostate Cancer Due to the Prostate Tumor-Associated Fibroblasts

Fibroblasts are one of the most abundant cell types in the prostate stroma. Fibroblasts synthesize many components of ECM (collagens I, II, and V, and fibronectin) and are responsible for the remodeling of ECM by secreting matrix metalloproteinases (MMPs). Fibroblasts also secrete growth factors and cytokines to regulate differentiation and homeostasis of adjacent epithelium. In some physiological (inflammation) or pathological (cancer) conditions, fibroblasts have an activated phenotype due to stimulation by growth factors or cytokines [30]. In prostate cancer progressions, interactions with various tumor microenvironment cells, including cancer cells and fibroblasts, are known to play an important role [8,11,12]. In this study, we hypothesized that prostate tumor-associated fibroblasts suppress PLZF expression and enhance malignant behaviors. To examine the cells affecting cancer cells among various stromal cells in the tumor microenvironment, LNCaP cells were co-cultured with lymphocytes (Jurkat), NK cells (NK92), monocytes (THP-1), and fibroblasts. Interestingly, only the LNCaP sample co-cultured with the fibroblast suppressed PLZF levels (Figure S3A). To determine the influence of fibroblasts on prostate cancer, we adopted a co-culturing system using Transwell chambers (Figure 5A). Surprisingly, we found that both protein and RNA expression of PLZF were more reduced by direct co-culturing of fibroblast and cancer cells than fibroblast CM (FCM). The direct co-culturing systems enable the sharing of medium and soluble factors throughout the entire culture well. These results indicated that co-culturing with fibroblast was expected to promote the malignancy of prostate cancer cells (Figure 5B,C). In addition, both protein and RNA expression of PLZF were confirmed with the same results in PLZF-overexpressed DU145 cells as in LNCaP cells (Figure S3B,C). These results indicated that co-culturing with fibroblast was expected to promote the malignancy of prostate cancer cells. To understand the functional role of FCM and direct co-culturing in prostate cancer, we examined cell viability, migration, and invasion assays. LNCaP cells viability was further increased by the co-cultured CM (=Co-CM) than FCM through the CCK assay (Figure 5D). Next, Transwell cell migration and Matrigel invasion assays were examined, and Co-CM significantly promoted cell migration and invasion in LNCaP cells rather than FCM (Figure 5E, Figure S3D). The mRNA/protein levels of the mesenchymal markers were more increased in the LNCaP cells treated with Co-CM, whereas E-cadherin was more decreased than FCM (Figure 5F,G). To confirm that normal fibroblasts increased the malignancies of cancer cells, non-normal cells, the normal prostatic epithelial cell line (PNT2), and LNCaP cells were co-cultured with fibroblasts. In PLZF-overexpressed PNT2 cells, the PLZF levels were not significantly affected by co-culturing with fibroblasts; however, PLZF levels were substantially inhibited by co-culturing with fibroblasts in LNCaP cells (Figure 5H). Taken together, these findings further support the functional significance of fibroblasts in the tumor microenvironment, in that they remarkably promote the malignant behavior of prostate cancer cells.

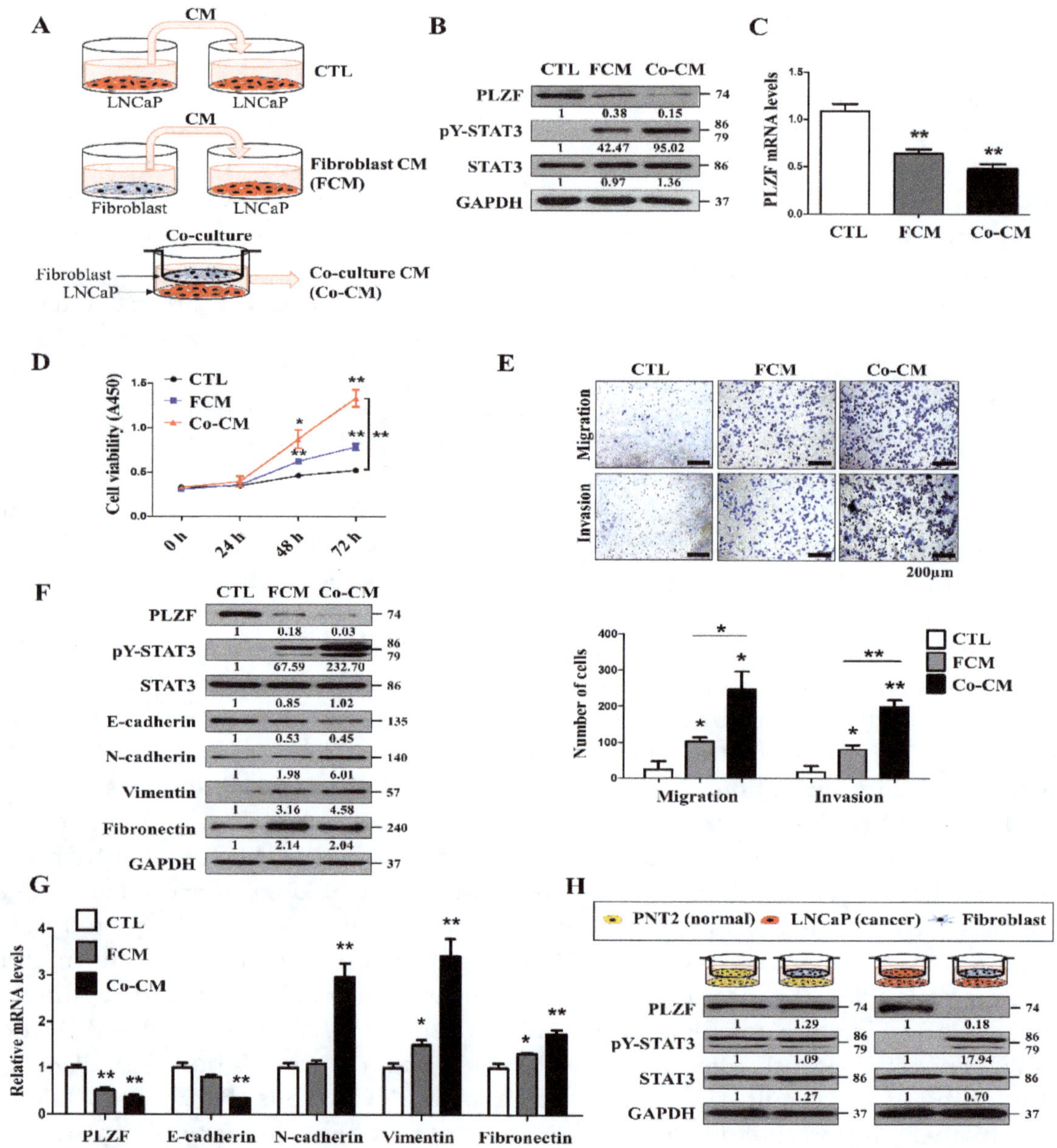

Figure 5. PLZF is reduced in prostate cancer due to the prostate tumor-associated fibroblasts. (**A**) Schematic of the experimental set-up for conditioned media (CMs) treatment and the Transwell co-culture system. Conditioned media (CM) were collected from fibroblast (FCM) that had been incubated for 48 h. LNCaP cells were cultured in a mixture of conditioned media and fresh media for 24 h (=FCM). LNCaP cells were co-cultured with fibroblast using a cell culture chamber for 24 h and the cells in the lower chamber were lysed. (**B**) LNCaP cells with FCM or co-cultured with fibroblast for 24 h. LNCaP cells were lysed for Western blotting. (**C**) mRNA expression levels of PLZF were examined by qRT-PCR in LNCaP cells with FCM or co-culture. (**D**) Cell growth rates between FCM-treated cells and co-culture CM (Co-CM) were compared using a CCK assay. (**E**) Transwell migration and Matrigel invasion assays were conducted in LNCaP cells with FCM or Co-CM. The relative cell numbers are shown (bottom). Representative images from three independent experiments were quantified as the mean ± SD. (**F**) Protein expression levels of EMT markers in LNCaP cells with FCM or co-culture. (**G**) mRNA expression levels of epithelial–mesenchymal-transition (EMT) markers were examined by qRT-PCR in LNCaP cells with FCM or co-culture. (**H**) Prostate normal cells (PNT2) and prostate cancer cells (LNCaP) were co-cultured with fibroblast for 24 h. PLZF, pY-STAT3, STAT3, and GAPDH protein expression by Western blotting in LNCaP cells. * $p < 0.05$, ** $p < 0.01$. The uncropped blots and molecular weight markers of Figure 5 are shown in Figure S9

2.6. Loss of CCL3 Inhibits the Fibroblast-Induced Prostate Cancer Cells Migration and Invasion

To confirm whether the fibroblast-derived factors in CM is a polypeptide, Co-CM was treated or heated with mixture of CM and fresh medium. Consequently, the fibroblast-inducing effect was practically abolished by heat inactivation (Figure S3E). To explore the soluble mediators released by the fibroblast and fibroblast co-cultured with prostate cancer cells, we analyzed by using protein arrays the levels of multiple cytokines, chemokines, and growth factors (Figure 6A, Figure S4A). These candidates that were more enriched in co-cultured CM (Co-CM) than prostate cancer cell CM (CTL CM) and fibroblast CM (FCM) were identified: C–C motif chemokine ligand 3/4 (CCL3/4), serpinE1, and urokinase-type plasminogen activator receptor (uPAR). To determine the cytokine responsible for PLZF suppression, each of the four candidates was knocked down in fibroblast and cancer cells using siRNAs. Knockdown of CCL3 or uPAR, but not CCL4 or serpinE1, attenuated the PLZF protein and mRNA expression suppression by FCM and Co-CM (Figure S4B,C). Based on the gene-silencing results, in order to evaluate the increase in the identified cytokines, fibroblasts were co-cultured with the normal prostate cells (PNT2) or the prostate cancer cells (LNCaP, DU145). In contrast to no effect on CCL4 and SerpinE1 mRNA expression, fibroblasts visibly induced CCL3 and uPAR mRNA production by co-culturing with prostate cancer cells (Figure 6B). To confirm the PLZF-suppressing effects of the two cytokines (CCL3, uPAR), neutralizing antibodies were administered to FCM and Co-CM. Only anti-CCL3 alone abolished the PLZF-inhibitory effect (Figure 6C). We next performed the ELISAs; CCL3 is secreted at higher levels in Co-CM than CTL and fibroblasts alone (Figure 6D). In this study, PLZF acts as a transcription factor of the SHP-1 gene (Figure 4). Therefore, to investigate whether CCL3 inhibits PLZF-mediated transcription of the SHP-1 gene, prostate cancer cells were treated with a recombinant CCL3. As a result, SHP-1-driven transcription was more inhibited with Co-CM than FCM by CCL3 (Figure 6E, Figure S4D). These results suggest that if fibroblasts present near cancer cells, prostate tumor-associated fibroblasts are expected to produce CCL3 more abundantly than normal fibroblast cells. Moreover, by neutralizing CCL3, cell migration and invasion abilities were further abolished by Co-CM, more so than FCM (Figure 6F). As a result of the confirmed mRNA/protein levels reinforcing the EMT assays, the alteration in EMT markers expression was abolished by the antibody of CCL3 (Figure S4E,F). Collectively, these results indicate that CCL3 secreted by fibroblasts co-cultured with prostate cancer cells inhibits tumor suppressor PLZF expression in prostate cancer.

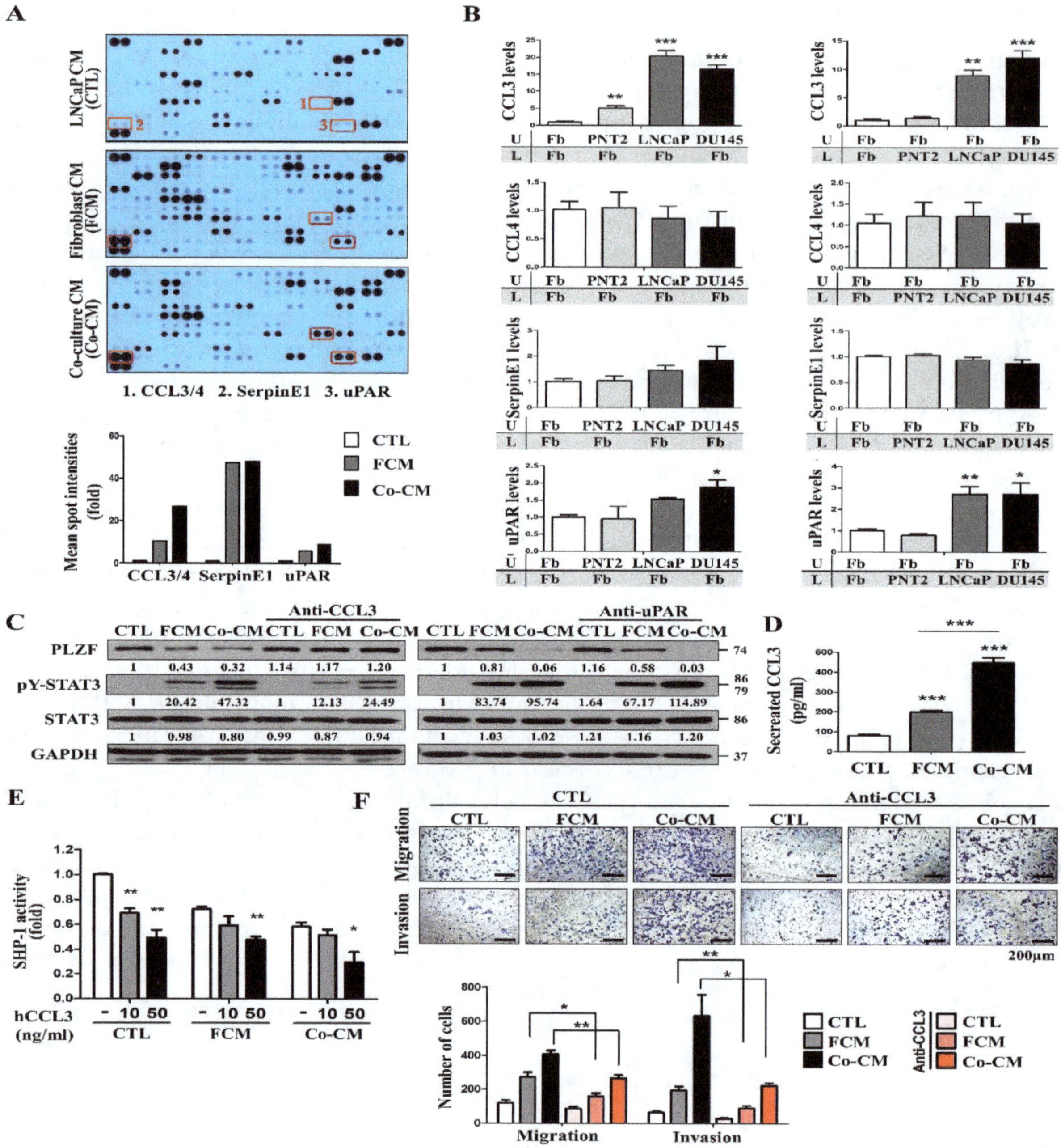

Figure 6. Loss of CCL3 inhibits the fibroblast-induced prostate cancer cells migration and invasion. (**A**) Each CM was applied to the cytokine arrays (top). The cytokines in boxes are more enriched in co-cultured CM than in fibroblast CM. Mean intensities of cytokines are plotted (bottom). (**B**) PNT2, LNCaP, and DU145 cells were co-cultured with fibroblast in a cell culture chamber for 24 h. Cells in a lower chamber were lysed for qRT-PCR. (**C**) LNCaP cells were treated with the indicated conditioned media with the addition of an anti-CCL3 antibody or anti-uPAR antibody and then subjected to Western blotting. (**D**) Concentrations of CCL3 in the indicated conditioned media were analyzed using an ELISA kit. Summarized results from three independent experiments were quantified as the mean ± SD. (**E**) Cells, which had been transfected with the luciferase plasmid, were pre-treated with recombinant protein CCL3 for 4 h, incubated under FCM or Co-CM for 24 h. (**F**) LNCaP cells, which had been pre-treated with anti-CCL3 antibody for 4 h, were incubated in the indicated CMs for 24 h. The relative cell numbers are shown (right). In B, D, E, and F, representative images from three independent experiments were quantified as the mean ± SD; * $p < 0.05$, ** $p < 0.01$, *** $p < 0.001$. The uncropped blots and molecular weight markers of Figure 6 are shown in Figure S10

3. Discussion

In this study, we discovered a new role for fibroblast in the prostate-tumor microenvironment to inhibit the tumor suppressor PLZF. While numerous reports have described the coevolution of stroma and prostate cancer cells in genotypic and phenotypic characters, some studies have elaborated on the contribution of tumor-associated fibroblasts to overall changes in gene and protein expression in prostate cancer microenvironments [31,32]. We identified for the first time a mechanism by which fibroblast-derived CCL3 suppresses PLZF expression, which promotes the synthesis of SHP-1 at the transcriptional levels, and then activates the JAKs–STAT3 pathway. PLZF was also significantly reduced in prostate cancer patient tissues, and low PLZF expression was found to correlate with GS and recurrence-free survival. Accordingly, down-regulation of PLZF activities can promote metastasis progression through improved growth and invasion of prostate cancer.

Cancer cells make use of the tremendous plastic nature of stromal cells, such as fibroblasts and macrophages, and produce multiple signals that generate a tumor-promoting microenvironment [33–35]. Previous studies have indicated that co-culture or tumor cell CM could activate stromal fibroblasts, and tissue around fibroblasts were suggested as precursors for tumor-associated fibroblasts, activated by tumor cells [34–36]. In addition, fibroblasts are constantly exposed to different stimuli in the tumor microenvironment, facilitating unique features, such as excessive and specific secretion and the ECM remodeling phenotype [37,38]. Activated fibroblasts, i.e., cancer-associated fibroblasts (CAF), are involved in tumor invasiveness and metastasis by expressing a series of proteins that are not expressed by normal fibroblasts, such as α-smooth muscle actin (α-SMA) [39,40]. Moreover, contact between cancer cells and normal fibroblasts can facilitate the CAF phenotype in breast cancer through Notch signaling [41]. Various inflammatory modulators can promote CAF activation by interleukin-1 (IL-1), which acts through NF-κB and IL-6, which mainly acts on signal transducer and activator of transcription (STAT) transcription factors [15]. Crosstalk and positive feedback involved in Janus kinase (JAK)–STAT signaling transduction, the contractile cytoskeleton, and alterations in histone acetylation further promote CAF activation [42,43]. It has been reported that tumor-associated fibroblasts promote tumor growth by directly stimulating tumor cell proliferation, play an important role in wound healing, and mediate other aspects of complex processes such as extracellular matrix remodeling [10,15,44]. Moreover, tumor-associated fibroblasts contribute to tumor progression by providing cancer cells with growth factors and pro-inflammatory tumor-promoting microenvironments [15,45,46].

CCL3, which belongs to the C–C chemokine family, is produced by fibroblast, macrophage, T cell, monocytes, and epithelial cells [47,48]. Previously, several evidence models suggested that fibroblast-derived CCL3 could activate tumor progression. Deficiency of the CCL3/CCR5 system resulted in a remarkably decreased tumor formation and lung metastasis. CCL3/CCR5 accumulation of tumor-associated fibroblasts also have been reported to be important in colitis-related carcinogenesis and oral tumor formation [46,49–51]. However, the fundamental mechanisms of CCL3 in the tumor microenvironment and the signaling pathways are not well known. We demonstrated that the expression of the marker PLZF was significantly increased in the absence of CCL3.

We have identified how various oncogenes related to prostate tumor survival are altered by PLZF. It was surprisingly found that only tyrosine phosphorylation of JAKs–STAT3 was reduced, so we focused on tyrosine phosphatase SHP-1 related to STAT3. The cytoplasmic protein tyrosine phosphatase (PTP), characterized by containing two Src homology 2 (SH2) N-terminus, a single catalytic domain, and a C-terminal PTP domain, is referred to as SHP-1. The expression of SHP-1 in prostate cancer is inversely correlated with tumor stage and malignancy, as well as with biochemical recurrence after prostatectomy [51–54]. In addition, SHP-1 is known to have a tumor-suppressor role in various cancers due to the SHP-1-negative regulation of JAKs–STAT activation via growth factors and cytokines [54–57]. We found that SHP-1 transcription was promoted by PLZF, which reduced tyrosine phosphorylation of JAK2, TYK2, and STAT3. Collectively, these results suggest that PLZF can function with SHP-1 to reduce prostate cancer progression.

As a result, our current findings suggest a crucial mechanism involving positive feedback between fibroblast and cancer cells, and this mechanism is essential for the progression and metastasis of prostate cancer. To our knowledge, we have identified for the first that the CCL3/PLZF/SHP-1/pY-STAT3 signaling cascades may be a target for combating prostate cancer cells. In conclusion, considering its interactions with several key signaling pathways in the prostate tumor microenvironment, understanding PLZF may help develop new biomarkers and therapeutic strategies for prostate cancer.

4. Materials and Methods

4.1. Cell Culture and Culture Conditions

The human prostate cancer (LNCaP) cell line was purchased from the Korean Cell Line Bank (Seoul, Korea). Another human prostate cancer (DU145) and normal human dermal fibroblast cell lines were kindly provided by the German Collection of Microorganisms (DSMZ, Braunschweig, Germany) and Modern Cell & Tissue Technologies (Seoul, Korea). The normal prostate epithelial PNT2, Jurkat-T, NK92 and THP-1 were purchased from American Type Culture Collection (Manassas, VA). DU145 cells was maintained in Dulbecco's Modified Eagles medium (Capricorn scientific, Ebsdorfergrund, Germany). LNCaP, PNT2, Jurkat-T, NK-92, THP-1 and normal human dermal fibroblast cells were maintained in RPMI-1640 medium (Life Technologies, Carlsbad, CA, USA). All the cells were maintained in the medium supplemented with 10% heat-inactivated fetal bovine serum (FBS, Life Technologies, Carlsbad, CA, USA) and 1% penicillin/streptomycin solution (Life Technologies, Carlsbad, CA, USA) at 37 °C with 5% CO_2 in a humidified incubator.

4.2. Immunohistochemistry Staining and Scoring

Briefly, the arrays were rehydrated and autoclaved at 121 °C for 10 min in a citrate buffer (Fisher Scientific, Rockford, IL, USA) to retrieve the antigens. They were incubated with 3% H_2O_2 for 10 min and with 10% bovine serum for 1 h to block non-specific signals. Sections were incubated overnight at 4 °C with antibodies against PLZF (1:200; abCam, Cambridge, MA, USA), pY-STAT3 (1:50; cell signaling, Danvers, MA, USA), and SHP-1 (1:200; abCam). Immune complexes were visualized using the Vectastatin ABC kit (Vector Laboratories, Burlingame, CA, USA) and the DAB detection kit (Dako, Carpinteria, CA, USA). Finally, all sections were counterstained with hematoxylin. Positive staining cells were counted on 5 randomly chosen visual fields at 400× magnification by means of ImageJ (National Institutes of Health, Bethesda, MD, USA).

4.3. Human Prostate Cancer Patients Tissues

Human prostate cancer tissue arrays were purchase from SuperBioChips Lab (Seoul, Korea). The tissue array contained 40 tumor specimens, as well as 10 normal tissues adjacent to the cancer from prostate cancer patients, whose clinical data, including age, sex, TNM stage, and Gleason score, were informed by the supplier. The tumor tissues were fixed with formalin, paraffin-embedded, and sectioned by a microtome to a 4-μm thickness and put on Superfrost plus slides.

4.4. Cell Migration and Invasion Assays

Transwell assay was used to evaluate the migration and invasion abilities of the prostate cancer cells. Twenty-four-well Transwell chambers with an 8 μm pore size polycarbonate membrane (Corning Inc., Corning, NY, USA) were used in these assays. For the Transwell assays, the PLZF expression vector and PLZF siRNA were transfected into DU145 or LNCaP cells and resuspended at a concentration of 5×10^4 cells in the top well with 200 μL serum-free RPMI 1640 medium. The lower chamber was filled with a 700 μL mixture of fresh and CMs (1:1) with 10% fetal bovine serum or neutralizing antibody CCL3 (R&D System, Minneapolis, MN, USA) were added as a chemoattractant. For cell migration assay, DU145 or LNCaP cells in a serum-free medium were seeded into the top chamber and incubated at 37 °C for 24 h. For cell invasion assay, the membrane was coated with Matrigel (Corning Inc.,

Corning, NY, USA) and incubated at 37 °C for 36 h. Cells on the top surface of the interface membrane were removed using a cotton swab. Migrated and invaded cells on the lower surface of the membrane were fixed with 10% formaldehyde, stained with a hematoxylin and eosin kit (Sysmex Corporation, Kobe, Japan), and counted under an optical microscope (100×, Nikon, Tokyo, Japan) from four random fields using ImageJ software (National Institutes of Health, Bethesda, MD, USA).

4.5. Luciferase Gene Report Assay

For the PLZF target reporter assay, DU145 and LNCaP cells were transiently co-transfected with 1 μg of the pGL3-SHP-1 promoter-luciferase reporter construct, pcDNA3.1-PLZF or si-PLZF, and pCMV-β-galactosidase. After 48h, cells were lysed and luciferase activity was determined using Luciferase reporter assay system (Promega, Madison, WI, USA). β-galactosidase activity in each lysate was assayed to normalize transfection efficiency. Three independent experiments were performed with triplicate samples.

4.6. Conditioned Media Preparation

Conditioned media (CM) for treating LNCaP cells was prepared by incubating normal human dermal fibroblast, Jurkat, NK92 and THP1 cells in 10mL dishes for 48 h with RPMI-1640 medium (Life Technologies, Carlsbad, CA, USA) containing only 1% penicillin/streptomycin solution (Life Technologies, Carlsbad, CA, USA). Supernatant was collected and centrifuged at 13,000rpm for 3 min to remove the cell debris. Cells were counted and CMs were normalized by adding media to a final volume that would represent 1×10^6 cells per mL. LNCaP cells were cultured in a mixture of indicated CMs and fresh media for 24 h.

4.7. Conditioned Human Proteome Profiler Cytokine Arrays

Human cytokine array kits were used according to the manufacturer's protocol (R&D System, Minneapolis, MN, USA). Although membranes were blocked for 1 h, culture supernatants from human dermal fibroblast cells were added to the membranes and incubated with cytokine arrays membranes overnight at 4 °C. Membranes were washed and incubated with detection biotinylated antibody cocktail (1:50) at room temperature for 1 h and incubated with horseradish-peroxidase-conjugated streptavidin (1:2000) at room temperature for 30 min, developed using a chemi-luminescence-type solution, and exposed to X-ray film. Spot intensities were analyzed using Adobe Photoshop.

4.8. ELISA Assays

Human CCL3 levels in supernatants of LNCaP cells, normal human dermal fibroblasts, and co-cultured LNCaP/fibroblasts were analyzed by commercial kits purchased from R&D Systems (Minneapolis, MN, USA), according to the manufacturer's instructions. Positive controls were supplied in the kit.

4.9. Microscopy, Image Capture, and Analysis

Samples were visualized and analyzed using a Leica Aperio AT Turbo microscope and the Leopard program included with the microscope (BX53; Olympus, Tokyo, Japan), using the Leica Application Suite software. Brightness and contrast were adjusted equally in all images presented.

4.10. Statistical Analysis

All experiments were performed with Microsoft Excel 2016 or GraphPad Prism 5 analytical tools (GraphPad software, Inc.), and results were presented as the means ± SD from least three independent samples. Student's unpaired two-tailed t-test for independent analysis was applied to evaluate the differences. Pearson's correlation was used to evaluate the correlation between PLZF, pY-STAT3, and SHP-1 protein expression. Survival rate analyses were performed by drawing curves and calculating log-rank p test using the Kaplan–Meier method. A p-value less than 0.05 was considered statistically significant. * $p < 0.05$, ** $p < 0.01$, *** $p < 0.001$.

5. Conclusions

In conclusion, our results strongly suggest that PLZF acts as a prostate-tumor suppressor, blocking cell viability, migration, and invasion. Furthermore, prostate cancer cooperates with tumor-associated fibroblasts to collapse PLZF that occurred by CCL3 derived from the tumor-associated fibroblasts. Mechanistic studies revealed that loss of PLZF by CCL3 activates the oncogenic JAK/STAT3 pathways through reduction of SHP-1 (Figure 7). Therefore, PLZF may act as a potential therapeutic target for prostate cancer.

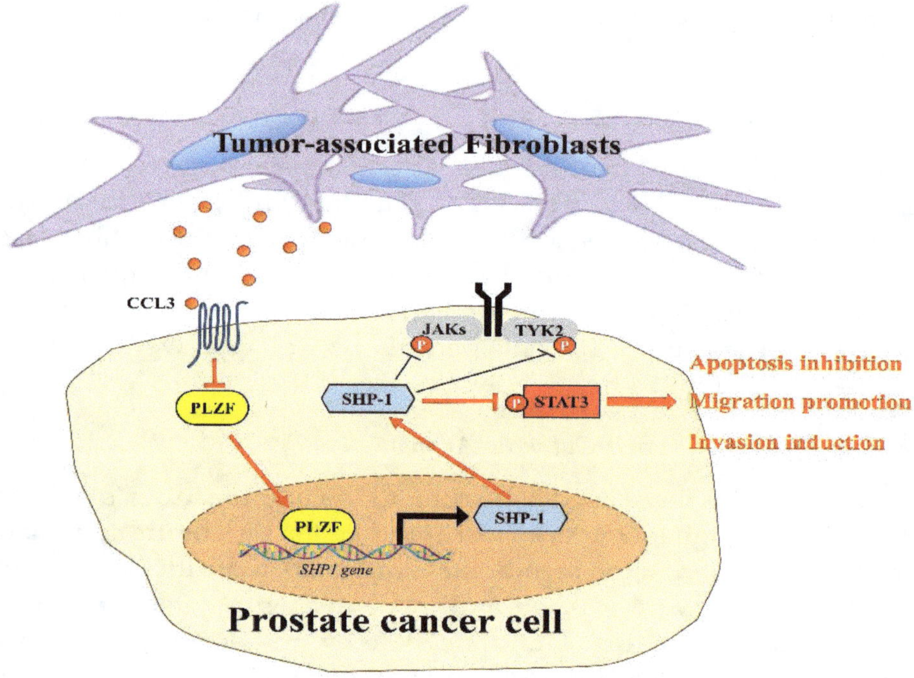

Figure 7. A schematic illustration of the tumor suppressor PLZF actions in prostate cancer progression.

Supplementary Materials
Figure S1: PLZF was downregulated in prostate tumors and negatively correlated with pY-STAT3, Figure S2: Exogenous transfection with PLZF reduces cell proliferation, migration ability, and MMP9/11, and which ECM component degradation enzymes, Figure S3: Prostate cancer cells are affected by fibroblasts to produce soluble factors and reduce PLZF expression, Figure S4: Identification of tumor-associated fibroblast-derived cytokines, Figure S5: Original blots for western blot analyses shown in Figure 1, Figure S6: Original blots for western blot analyses shown in Figure 2, Figure S7: Original blots for western blot analyses shown in Figure 3, Figure S8: Original blots for western blot analyses shown in Figure 4, Figure S9: Original blots for western blot analyses shown in Figure 5, Figure S10: Original blots for western blot analyses shown in Figure 6, Figure S11: Original blots for western blot analyses shown in Figure S1 and S2, Figure S12: Original blots for western blot analyses shown in Figure S3, Figure S13: Original blots for western blot analyses shown in Figure S4, Table S1: PSA levels information of the prostate cancer patients.

Author Contributions: Conceptualization, supervision, writing—review and editing, K.H.N. and S.-K.Y.; investigation, methodology, K.H.N. and E.Y.; validation, visualization, software, data curation, project administration, manuscript writing—original draft, K.H.N.; formal analysis, K.H.N., A.J.J., H.L. and S.-H.L.; resources, P.-S.C. and C.K. All authors have read and agreed to the published version of the manuscript.

Acknowledgments: This work was supported by grants from the National Research Foundation of Korea (NRF) Grant funded by the Korea government (MSIT) (NRF-2018R1A5A2025964), and the Seoul National University Hospital (SNUH) Research Fund (0320190210).

References

1. Siegel, R.L.; Miller, K.D.; Jemal, A. Cancer statistics, 2020. *CA Cancer J. Clin.* **2020**, *70*, 7–30. [CrossRef] [PubMed]
2. Gundem, G.; Van Loo, P.; Kremeyer, B.; Alexandrov, L.B.; Tubio, J.M.; Papaemmanuil, E.; Brewer, D.S.; Kallio, H.M.L.; Högnäs, G.; Annala, M.; et al. The evolutionary history of lethal metastatic prostate cancer. *Nature* **2015**, *520*, 353–357. [CrossRef] [PubMed]
3. Harryman, W.L.; Hinton, J.P.; Rubenstein, C.P.; Singh, P.; Nagle, R.B.; Parker, S.J.; Knudsen, B.S.; Cress, A.E. The cohesive metastasis phenotype in human prostate cancer. *Biochim. Biophys. Acta* **2016**, *1866*, 221–231. [CrossRef] [PubMed]
4. Cheng, Y.; Gao, X.H.; Li, X.J.; Cao, Q.H.; Zhao, D.D.; Zhou, J.R.; Wu, H.X.; Wang, Y.; You, L.J.; Yang, H.B.; et al. Depression promotes prostate cancer invasion and metastasis via a sympathetic-cAMP-FAK signaling pathway. *Oncogene* **2018**, *37*, 2953–2966. [CrossRef]
5. Birnbaum, M.D.; Zhao, N.; Moorthy, B.T.; Patel, D.M.; Kryvenko, O.N.; Heidman, L.; Kumar, A.; Morgan, W.M.; Ban, Y.; Reis, I.M.; et al. Reduced Arginyltransferase 1 is a driver and a potential prognostic indicator of prostate cancer metastasis. *Oncogene* **2019**, *38*, 838–851. [CrossRef]
6. Hanahan, D.; Weinberg, R.A. Hallmarks of cancer: The next generation. *Cell* **2011**, *144*, 646–674. [CrossRef]
7. Hanahan, D.; Coussens, L.M. Accessories to the crime: Functions of cells recruited to the tumor microenvironment. *Cancer Cell* **2012**, *21*, 309–322. [CrossRef]
8. Cheteh, E.H.; Augsten, M.; Rundqvist, H.; Bianchi, J.; Sarne, V.; Egevad, L.; Bykov, V.J.; Östman, A.; Wiman, K.G. Human cancer-associated fibroblasts enhance glutathione levels and antagonize drug-induced prostate cancer cell death. *Cell Death Dis.* **2017**, *8*, e2848. [CrossRef]
9. Qiu, W.; Hu, M.; Sridhar, A.; Opeskin, K.; Fox, S.; Shipitsin, M.; Trivett, M.; Thompson, E.R.; Ramakrishna, M.; Gorringe, K.L.; et al. No evidence of clonal somatic genetic alterations in cancer-associated fibroblasts from human breast and ovarian carcinomas. *Nat. Genet.* **2008**, *40*, 650–655. [CrossRef]
10. Allinen, M.; Beroukhim, R.; Cai, L.; Brennan, C.; Lahti-Domenici, J.; Huang, H.; Porter, D.; Hu, M.; Chin, L.; Richardson, A.; et al. Molecular characterization of the tumor microenvironment in breast cancer. *Cancer Cell* **2004**, *6*, 17–32. [CrossRef]
11. Gascard, P.; Tlsty, T.D. Carcinoma-associated fibroblasts: Orchestrating the composition of malignancy. *Genes Dev.* **2016**, *30*, 1002–1019. [CrossRef] [PubMed]
12. Bu, L.; Baba, H.; Yoshida, N.; Miyake, K.; Yasuda, T.; Uchihara, T.; Tan, P.; Ishimoto, T. Biological heterogeneity and versatility of cancer-associated fibroblasts in the tumor microenvironment. *Oncogene* **2019**, *38*, 4887–4901. [CrossRef] [PubMed]
13. Trimboli, A.J.; Cantemir-Stone, C.Z.; Li, F.; Wallace, J.A.; Merchant, A.; Creasap, N.; Thompson, J.C.; Caserta, E.; Wang, H.; Chong, J.L.; et al. Pten in stromal fibroblasts suppresses mammary epithelial tumours. *Nature* **2019**, *461*, 1084–1091. [CrossRef] [PubMed]
14. Bhowmick, N.A.; Neilson, E.G.; Moses, H.L. Stromal fibroblasts in cancer initiation and progression. *Nature* **2004**, *432*, 332–337. [CrossRef]
15. Erez, N.; Truitt, M.; Olson, P.; Arron, S.T.; Hanahan, D. Cancer-associated fibroblasts are activated in incipient neoplasia to orchestrate tumor-promoting inflammation in an NF-kappaB-dependent manner. *Cancer Cell* **2010**, *17*, 135–147. [CrossRef]
16. Suliman, B.A.; Xu, D.; Williams, B.R. The promyelocytic leukemia zinc finger protein: Two decades of molecular oncology. *Front. Oncol.* **2012**, *17*, 2–74. [CrossRef]
17. Xiao, G.Q.; Unger, P.; Yang, Q.; Kinoshita, Y.; Singh, K.; McMahon, L.; Nastiuk, K.; Sha, K.; Krolewski, J.; Burstein, D. Loss of PLZF expression in prostate cancer by immunohistochemistry correlates with tumor aggressiveness and metastasis. *PLoS ONE* **2015**, *10*, e0121318. [CrossRef]
18. Liu, T.M.; Lee, E.H.; Lim, B.; Shyh-Chang, N. Concise review: Balancing stem cell self-renewal and differentiation with PLZF. *Stem Cells* **2016**, *34*, 277–287. [CrossRef]
19. Shen, H.; Zhan, M.; Zhang, Y.; Huang, S.; Xu, S.; Huang, X.; He, M.; Yao, Y.; Man, M.; Wang, J. PLZF inhibits proliferation and metastasis of gallbladder cancer by regulating IFIT2. *Cell Death Dis.* **2018**, *9*, 71. [CrossRef]
20. Kovalovsky, D.; Uche, O.U.; Eladad, S.; Hobbs, R.M.; Yi, W.; Alonzo, E.; Chua, K.; Eidson, M.; Kim, H.J.; Im, J.S.; et al. The BTB-zinc finger transcriptional regulator PLZF controls the development of invariant natural killer T cell effector functions. *Nat. Immunol.* **2008**, *9*, 1055–1064. [CrossRef]

21. Savage, A.K.; Constantinides, M.G.; Han, J.; Picard, D.; Martin, E.; Li, B.; Lantz, O.; Bendelac, A. The transcription factor PLZF directs the effector program of the NKT cell lineage. *Immunity* **2008**, *29*, 391–403. [CrossRef] [PubMed]
22. Eidson, M.; Wahlstrom, J.; Beaulieu, A.M.; Zaidi, B.; Carsons, S.E.; Crow, P.K.; Yuan, J.; Wolchok, J.D.; Horsthemke, B.; Wieczorek, D.; et al. Altered development of NKT cells, gd T cells, CD8 T cells and NK cells in a PLZF deficient patient. *PLoS ONE* **2011**, *6*, e24441. [CrossRef] [PubMed]
23. McConnell, M.J.; Chevallier, N.; Berkofsky-Fessler, W.; Giltnane, J.M.; Malani, R.B.; Staudt, L.M.; Licht, J.D. Growth suppression by acute promyelocytic leukemiaassociated protein PLZF is mediated by repression of c-myc expression. *Mol. Cell. Biol.* **2003**, *23*, 9375–9388. [CrossRef] [PubMed]
24. Felicetti, F.; Bottero, L.; Felli, N.; Mattia, G.; Labbaye, C.; Alvino, E.; Peschle, C.; Colombo, M.P.; Carè, A. Role of PLZF in melanoma progression. *Oncogene* **2004**, *23*, 4567–4576. [CrossRef] [PubMed]
25. Cheung, M.; Pei, J.; Pei, Y.; Jhanwar, S.C.; Pass, H.I.; Testa, J.R. The promyelocytic leukemia zinc-finger gene, PLZF, is frequently downregulated in malignant mesothelioma cells and contributes to cell survival. *Oncogene* **2010**, *29*, 1633–1640. [CrossRef] [PubMed]
26. Wang, X.; Wang, L.; Guo, S.; Bao, Y.; Ma, Y.; Yan, F.; Xu, K.; Xu, Z.; Jin, L.; Lu, D.; et al. Hypermethylation reduces expression of tumor-suppressor PLZF and regulates proliferation and apoptosis in non-small-cell lung cancers. *FASEB J.* **2013**, *27*, 4194–4203. [CrossRef] [PubMed]
27. Hsieh, C.L.; Botta, G.; Gao, S.; Li, T.; Van Allen, E.M.; Treacy, D.J.; Cai, C.; He, H.H.; Sweeney, C.J.; Brown, M.; et al. PLZF, a tumor suppressor genetically lost in metastatic castration-resistant prostate cancer, is a mediator of resistance to androgen deprivation therapy. *Cancer Res.* **2015**, *75*, 1944–1948. [CrossRef]
28. Cao, J.; Zhu, S.; Zhou, W.; Li, J.; Liu, C.; Xuan, H.; Yan, J.; Zheng, L.; Zhou, L.; Yu, J.; et al. PLZF mediates the PTEN/AKT/FOXO3a signaling in suppression of prostate tumorigenesis. *PLoS ONE* **2013**, *8*, e77922. [CrossRef]
29. Stopsack, K.H.; Gerke, T.; Tyekucheva, S.; Mazzu, Y.Z.; Lee, G.M.; Chakraborty, G.; Abida, W.; Mucci, L.A.; Kantoff, P.W. Low Expression of the Androgen-Induced Tumor Suppressor Gene PLZF and Lethal Prostate Cancer. *Cancer Epidemiol. Biomark. Prev.* **2019**, *28*, 707–714. [CrossRef]
30. Kalluri, R.; Zeisberg, M. Fibroblasts in cancer. *Nat. Rev. Cancer* **2006**, *6*, 392–401. [CrossRef]
31. Kansy, B.A.; Dissmann, P.A.; Hemeda, H.; Bruderek, K.; Westerkamp, A.M.; Jagalski, V.; Schuler, P.; Kansy, K.; Lang, S.; Dumitru, C.A.; et al. The bidirectional tumor–mesenchymal stromal cell interaction promotes the progression of head and neck cancer. *Stem Cell Res. Ther.* **2014**, *5*, 95. [CrossRef] [PubMed]
32. Hughes, R.M.; Simons, B.W.; Khan, H.; Miller, R.; Kugler, V.; Torquato, S.; Theodros, D.; Haffner, M.C.; Lotan, T.; Huang, J.; et al. Asporin Restricts Mesenchymal Stromal Cell Differentiation, Alters the Tumor Microenvironment, and Drives Metastatic Progression. *Cancer Res.* **2019**, *79*, 3636–3650. [CrossRef] [PubMed]
33. Wang, W.; Yang, X.; Dai, J.; Lu, Y.; Zhang, J.; Keller, E.T. Prostate cancer promotes a vicious cycle of bone metastasis progression through inducing osteocytes to secrete GDF15 that stimulates prostate cancer growth and invasion. *Oncogene* **2019**, *38*, 4540–4559. [CrossRef] [PubMed]
34. Sharon, Y.; Raz, Y.; Cohen, N.; Ben-Shmuel, A.; Schwartz, H.; Geiger, T.; Erez, N. Tumor-derived osteopontin reprograms normal mammary fibroblasts to promote inflammation and tumor growth in breast cancer. *Cancer Res.* **2015**, *75*, 963–973. [CrossRef]
35. Sato, N.; Maehara, N.; Goggins, M. Gene expression profiling of tumor-stromal interactions between pancreatic cancer cells and stromal fibroblasts. *Cancer Res.* **2004**, *64*, 6950–6956. [CrossRef]
36. Augsten, M. Cancer-associated fibroblasts as another polarized cell type of the tumor microenvironment. *Front. Oncol.* **2014**, *4*, 62. [CrossRef]
37. Alkasalias, T.; Moyano-Galceran, L.; Arsenian-Henriksson, M.; Lehti, K. Fibroblasts in the Tumor Microenvironment: Shield or Spear? *Int. J. Mol. Sci.* **2018**, *19*, 1532. [CrossRef]
38. Kalluri, R. The biology and function of fibroblasts in cancer. *Nat. Rev. Cancer* **2016**, *16*, 582–598. [CrossRef]
39. Iacopino, F.; Angelucci, C.; Sica, G. Interactions between normal human fibroblasts and human prostate cancer cells in a co-culture system. *Anticancer Res.* **2012**, *32*, 1579–1588.
40. De Wever, O.; Demetter, P.; Mareel, M.; Bracke, M. Stromal myofibroblasts are drivers of invasive cancer growth. *Int. J. Cancer* **2008**, *123*, 2229–2238. [CrossRef]
41. Strell, C.; Paulsson, J.; Jin, S.B.; Tobin, N.P.; Mezheyeuski, A.; Roswall, P.; Mutgan, C.; Mitsios, N.; Johansson, H.; Wickberg, S.M.; et al. Impact of Epithelial-Stromal Interactions on Peritumoral Fibroblasts in Ductal Carcinoma in Situ. *J. Natl. Cancer Inst.* **2019**, *111*, 983–995. [CrossRef] [PubMed]

42. Sahai, E.; Astsaturov, I.; Cukierman, E.; DeNardo, D.G.; Egeblad, M.; Evans, R.M.; Fearon, D.; Greten, F.R.; Hingorani, S.R.; Hunter, T.; et al. A framework for advancing our understanding of cancer-associated fibroblasts. *Nat. Rev. Cancer* **2020**, *20*, 174–186. [CrossRef]
43. Albrengues, J.; Bertero, T.; Grasset, E.; Bonan, S.; Maiel, M.; Bourget, I.; Philippe, C.; Herraiz Serrano, C.; Benamar, S.; Croce, O.; et al. Epigenetic switch drives the conversion of fibroblasts into proinvasive cancer-associated fibroblasts. *Nat. Commun.* **2015**, *6*, 10204. [CrossRef] [PubMed]
44. Kisseleva, T.; Brenner, D.A. Mechanisms of fibrogenesis. *Exp. Biol. Med. Maywood* **2008**, *233*, 9–22. [CrossRef] [PubMed]
45. Neufert, C.; Becker, C.; Tureci, O.; Waldner, M.J.; Backert, I.; Floh, K.; Atreya, I.; Leppkes, M.; Jefremow, A.; Vieth, M.; et al. Tumor fibroblast-derived epiregulin promotes growth of colitis-associated neoplasms through ERK. *J. Clin. Investig.* **2013**, *123*, 1428–1443. [CrossRef] [PubMed]
46. Tanabe, Y.; Sasaki, S.; Mukaida, N.; Baba, T. Blockade of the chemokine receptor, CCR5, reduces the growth of orthotopically injected colon cancer cells via limiting cancer-associated fibroblast accumulation. *Oncotarget* **2016**, *7*, 48335–48345. [CrossRef]
47. Menten, P.; Wuyts, A.; Van Damme, J. Macrophage inflammatory protein-1. *Cytokine Growth Factor Rev.* **2002**, *13*, 455–481. [CrossRef]
48. da Silva, J.M.; Moreira Dos Santos, T.P.; Sobral, L.M.; Queiroz-Junior, C.M.; Rachid, M.A.; Proudfoot, A.E.I.; Garlet, G.P.; Batista, A.C.; Teixeira, M.M.; Leopoldino, A.M.; et al. Relevance of CCL3/CCR5 axis in oral carcinogenesis. *Oncotarget* **2017**, *8*, 51024–51036. [CrossRef]
49. Wu, Y.; Li, Y.Y.; Matsushima, K.; Baba, T.; Mukaida, N. CCL3-CCR5 axis regulates intratumoral accumulation of leukocytes and fibroblasts and promotes angiogenesis in murine lung metastasis process. *J. Immunol.* **2008**, *181*, 6384–6393. [CrossRef]
50. Sasaki, S.; Baba, T.; Shinagawa, K.; Matsushima, K.; Mukaida, N. Crucial involvement of the CCL3/CCR5 axis-mediated fibroblast accumulation in colitis-associated carcinogenesis in mice. *Int. J. Cancer* **2014**, *135*, 1207–1306. [CrossRef]
51. Chong, Z.Z.; Maiese, K. The Src homology 2 domain tyrosine phosphatases SHP-1 and SHP-2: Diversified control of cell growth, inflammation, and injury. *Histol. Histopathol.* **2007**, *22*, 1251–1267. [PubMed]
52. Nunes-Xavier, C.E.; Mingo, J.; López, J.I.; Pulido, R. The role of protein tyrosine phosphatase in prostate cancer biology. *Biochim. Biophys. Acta Mol. Cell Res.* **2019**, *1866*, 102–113. [CrossRef] [PubMed]
53. Cariaga-Martinez, A.E.; Lorenzati, M.A.; Riera, M.A.; Cubilla, M.A.; De La Rossa, A.; Giorgio, E.M.; Tiscornia, M.M.; Gimenez, E.M.; Rojas, M.E.; Chaneton, B.J.; et al. Tumoral prostate shows different expression pattern of somatostatin receptor 2 (SSTR2) and phosphotyrosine phosphatase SHP-1 (PTPN6) according to tumor progression. *Adv. Urol.* **2009**, *2009*, 8. [CrossRef] [PubMed]
54. Zapata, P.D.; Ropero, R.M.; Valencia, A.M.; Buscail, L.; López, J.I.; Martín-Orozco, R.M.; Prieto, J.C.; Angulo, J.; Susini, C.; López-Ruiz, P.; et al. Autocrine regulation of human prostate carcinoma cell proliferation by somatostatin through the modulation of the SH2 domain containing protein tyrosine phosphatase (SHP)-1. *J. Clin. Endocrinol. Metab.* **2002**, *87*, 915–926. [CrossRef]
55. Tassidis, H.; Culig, Z.; Wingren, A.G.; Härkönen, P. Role of the protein tyrosine phosphatase SHP-1 in Interleukin-6 regulation of prostate cancer cells. *Prostate* **2010**, *70*, 1491–1500. [CrossRef]
56. Tassidis, H.; Brokken, L.J.; Jirstrom, K.; Ehrnstrom, R.; Ponten, F.; Ulmert, D.; Bjartell, A.; Härkönen, P.; Wingren, A.G. Immunohistochemical detection of tyrosine phosphatase SHP-1 predicts outcome after radical prostatectomy for localized prostate cancer. *Int. J. Cancer* **2010**, *126*, 2296–2307. [CrossRef]
57. Huang, T.T.; Su, J.C.; Liu, C.Y.; Shiau, C.W.; Chen, K.F. Alteration of SHP-1/p-STAT3 signaling: A potential target for anticancer therapy. *Int. J. Mol. Sci.* **2017**, *18*, 1234. [CrossRef]

3

Polyphenolic Compounds from Lespedeza Bicolor Root Bark Inhibit Progression of Human Prostate Cancer Cells via Induction of Apoptosis and Cell Cycle Arrest

Sergey A. Dyshlovoy [1,2,3,*], Darya Tarbeeva [4], Sergey Fedoreyev [4], Tobias Busenbender [1], Moritz Kaune [1], Marina Veselova [4], Anatoliy Kalinovskiy [4], Jessica Hauschild [1], Valeria Grigorchuk [5], Natalya Kim [4], Carsten Bokemeyer [1], Markus Graefen [3], Petr Gorovoy [4] and Gunhild von Amsberg [1,3]

1 Department of Oncology, Hematology and Bone Marrow Transplantation with Section Pneumology, Hubertus Wald-Tumorzentrum, University Medical Center Hamburg-Eppendorf, 20251 Hamburg, Germany; tobias.busenbender@gmx.de (T.B.); moritz.kaune@stud.uke.uni-hamburg.de (M.K.); j.hauschild@uke.de (J.H.); c.bokemeyer@uke.de (C.B.); g.von-amsberg@uke.de (G.v.A.)

2 Laboratory of Biologicaly Active Compounds, School of Natural Sciences, Far Eastern Federal University, Vladivostok 690091, Russia

3 Martini-Klinik, Prostate Cancer Center, University Hospital Hamburg-Eppendorf, 20251 Hamburg, Germany; graefen@martini-klinik.de

4 G.B. Elyakov Pacific Institute of Bioorganic Chemistry, Far-East Branch, Russian Academy of Sciences, Vladivostok 690022, Russia; tarbeeva1988@mail.ru (D.T.); fedoreev-s@mail.ru (S.F.); veselmv@mail.ru (M.V.); kaaniv@piboc.dvo.ru (A.K.); natalya_kim@mail.ru (N.K.); petrgorovoy@gmail.com (P.G.)

5 Federal Scientific Center of the East Asia Terrestrial Biodiversity (Institute of Biology and Soil Science), Far Eastern Branch, Russian Academy of Sciences, Vladivostok 690022, Russia; kera1313@mail.ru

* Correspondence: dyshlovoy@gmail.com or s.dyshlovoy@uke.de

Abstract: From a root bark of *Lespedeza bicolor* Turch we isolated two new (7 and 8) and six previously known compounds (1–6) belonging to the group of prenylated polyphenols. Their structures were elucidated using mass spectrometry, nuclear magnetic resonance and circular dichroism spectroscopy. These natural compounds selectively inhibited human drug-resistant prostate cancer in vitro. Prenylated pterocarpans 1–3 prevented the cell cycle progression of human cancer cells in S-phase. This was accompanied by a reduced expression of mRNA corresponding to several human cyclin-dependent kinases (CDKs). In contrast, compounds 4–8 induced a G1-phase cell cycle arrest without any pronounced effect on CDKs mRNA expression. Interestingly, a non-substituted hydroxy group at C-8 of ring D of the pterocarpan skeleton of compounds 1–3 seems to be important for the CDKs inhibitory activity.

Keywords: Polyphenolic compounds; *Lespedeza bicolor*; pterocarpans; prostate cancer; apoptosis; cell cycle

---

1. Introduction

Lespedeza bicolor Turcz. is a shrub plant belonging to the Leguminosae family. It is widely distributed in East Asia, including the Primorye region of the Russian Far East [1]. In traditional folk medicine, this plant is used for the treatment of nephritis, azotemia, inflammation, hyperpigmentation, energy depletion, diabetes and diuresis [2–6]. In addition, *L. bicolor* extracts possess antioxidant, anti-tyrosinase, anti-inflammatory, estrogenic, antimicrobial and antifungal activities [6–8].

Recently, a group from Korea has shown that *L. bicolor* extracts exert a potent memory-enhancing effect when treating cognitive dysfunction induced by amyloid β peptide (25–35) in mice models [9]. Additionally, this extract has been described as a promising therapeutic tool to prevent diabetic nephropathy in methylglyoxal (MGO)-induced models both in vitro and in vivo [10]. Remarkably, the *L. bicolor* extract reduced hyperglycemia-induced hepatic damage, hepatic oxidative stress, and inflammation, as well as liver fibrosis [11]. Finally, this extract inhibited the growth of lung carcinoma LU-1 and prostate cancer LNCaP cells [12]. However, only a little information is available on the compounds responsible for the biological activity of the extract and even less is known about mechanisms of action of these substances.

We have previously isolated several polyphenolic compounds from *L. bicolor* stem bark, which were able to inhibit the growth of human cancer HTB-19, Kyse-30, and HEPG-2 cells [13]. First insights into the mechanism of action were reported for pterocarpans, coumestans, and arylbenzofurans recently isolated from *L. bicolor.* These natural compounds were found to promote cell death via induction of a G1 cell cycle arrest, reduction of Bcl-2 levels, and induction of PARP cleavage in Jurkat blood cancer cells [14]. However, to date no significant reports on the mechanisms of action of the purified *L. bicolor* metabolites are available.

In the current study, we further investigated the metabolites of *L. bicolor* root bark collected in the Primorye Region (Russian Federation). Therefore, we isolated several new as well as previously known prenylated polyphenolic compounds, investigated their cytotoxic properties and the mechanism of action in human drug-resistant prostate cancer cells.

2. Materials and Methods

2.1. General Experimental Procedures

Optical rotations were measured on a PerkinElmer 343 polarimeter (Perkin Elmer, Waltham, MA, USA). The UV spectra were obtained using a UV-1601 PC spectrophotometer (Shimadzu, Kyoto, Japan). The CD spectra were obtained on a Chirascan-plus Quick Start CD Spectrometer (Applied Photophysics Limited, Leatherhead, UK) (acetonitrile, 20 °C). The IR spectrum was recorded on a Vector 22 fourier-transform infrared spectrometer spectrophotometer (Bruker, Rheinstetten, Germany). The ^{1}H, ^{13}C, and two-dimensional (2D) NMR spectra were recorded in $CDCl_3$ at 30 °C using NMR Bruker AVANCE III DRX-700 and DRX-500 instruments (Bruker, Karlsruhe, Germany). The chemical shift values (δ) and the coupling constants (*J*) are given in parts per million and Hz, respectively.

2.2. Plant Material

L. bicolor was collected in Khasansky District (Andreevka village) of the Primorye Region (The Russian Federation) from a grassy dry meadow in August 2016 by academician P.G. Grovoy. Voucher specimen No. 103608 is preserved in the herbarium of the Laboratory of Chemotaxonomy (G.B. Elyakov Pacific Institute of Bioorganic Chemistry, FEB RAS).

2.3. Analytical and Preparative HPLC

The analytical HPLC was carried out using an Agilent Technologies 1100 series HPLC system (Agilent Technologies, Waldbronn, Germany) equipped with a VWD detector (λ = 280 nm). The extracts were analyzed using a Supelco Analytical HS-C18 (Supelco Analytical, Bellefonte, PA, USA) column (3 μm, 4.6, 75 mm) thermostated at 30 °C. The mobile phase consisted of 1% aqueous acetic acid (A) and acetonitrile containing 1% acetic acid (B). For the analysis, the following gradient steps were programmed: 0–2 min—5% B, 2–4 min—5–20% B, 5–17 min—20–50% B, 18–23 min—50–90% B, 24–25 min—90–100% B, 16–27 min—100% B, 28–33 min—100–5% B. The flow rate was 0.8 mL/min. The data were analyzed with the ChemStation software (v. 09, Agilent Technologies, Waldbronn, Germany).

The preparative HPLC was carried out using a Shimadzu HPLC system equipped with an LC-20AT pump and SPD-20A detector (λ = 280 nm) (Shimadzu, Kyoto, Japan). The polyphenolic compounds were purified using a Supelco Analytical HS-C18 (Supelco Analytical, Bellefonte, PA, USA) column (3 μm, 4.6, 75 mm). The mobile phase consisted of 1% aqueous acetic acid (A) and acetonitrile containing 1% acetic acid (B). For the analysis, the following gradient steps were programmed: 0–2 min—5–10% B, 2–4 min—10–20% B, 4–7 min—20–30% B, 7–10 min—30–40% B, 10–12 min—40–50% B, 12–17 min—50–100% B, 17–25 min—100–5% B. The flow rate was 1 mL/min.

The data were analyzed with the LabSolutions software (v. 1.0, Shimadzu, Kyoto, Japan).

2.4. *HR-ESI-MS*

HR-ESI-MS experiments were carried out using a Shimadzu hybrid ion trap–time of flight mass spectrometer (Shimadzu, Kyoto, Japan). The operating settings of the instrument were as follows: electrospray ionization (ESI) source potential, −3.8 and 4.5 kV for negative and positive polarity ionization, respectively, drying gas (N_2) pressure—200 kPa, nebulizer gas (N_2) flow—1.5 L/min, temperature for the curved desolvation line (CDL) and heat block—200 °C, detector voltage—1.5 kV and the range of detection—100–900 m/z. The mass accuracy was below 4 ppm. The data were acquired and processed using Shimadzu LCMS Solution software (v.3.60.361, Shimadzu, Kyoto, Japan).

2.5. *Extraction and Isolation*

Air-dried root bark of *L. bicolor* (150 g) was extracted twice under reflux with a $CHCl_3$–EtOH mixture at a volume ratio of 3:1 for 3 h (60 °C). The obtained extract of *L. bicolor* root bark (1.7 g) was chromatographed on a polyamide column (80 g, 50–160 μm, Sigma-Aldrich, St. Louis, MI, USA). The column was eluted with a hexane–$CHCl_3$ solution system with gradually increasing $CHCl_3$ amounts (hexane/$CHCl_3$, 1:0, 100:1, 50:1, 40:1) to give fractions 1–13 and then with a $CHCl_3$–EtOH solution system with gradually increasing EtOH amounts ($CHCl_3$/EtOH, 1:0, 100:1, 50:1, 40:1, *v*/*v*) to give fractions 14–20. The fractions containing polyphenolic compounds according to TLC HPLC data were subjected to further purification. TLC plates were treated with a 3% solution of $FeCl_3$ in ethanol to reveal the presence of polyphenolic compounds. Fraction 17 (287 mg) eluted with $CHCl_3$–EtOH (50:1) was chromatographed twice over a silica gel column (40–63 μm, Sigma-Aldrich, St. Louis, MI, USA). The column was eluted with a benzene–ethylacetate solution system with gradually increasing ethylacetate amounts (benzene/EtOAc, 1:0, 200:1, 100:1, 50:1, 40:1, *v*/*v*) to obtain compound 1 (10.7 mg). Fraction 18 (176 mg), eluted with $CHCl_3$–EtOH (40:1), was also chromatographed on a silica gel column using the same solution system to obtain compounds 1 (5.5 mg), 2 (9.2 mg). Fraction 20 (281 mg), eluted with $CHCl_3$ – EtOH (20:1), was also chromatographed on a silica gel column using the same solution system to obtain compounds 6 (15.0 mg) and 8 (2.5 mg).

Fraction 14 (82 mg), eluted with $CHCl_3$, was subsequently chromatographed over a silica gel column (40–63 μm) eluted twice with a benzene–ethylacetate solution system with gradually increasing ethylacetate amounts (benzene/EtOAc, 1:0, 200:1, 100:1, 50:1, 40:1, *v*/*v*) resulting in compound 3 (6.6 mg). Fraction 10 (67 mg), eluted with $CHCl_3$, was subsequently chromatographed over a silica gel (40–63 μm) column twice using the same solution system to afford compounds 4 (6.1 mg), 5 (8.5 mg), and 7 (3.5 mg).

2.5.1. (6aR,11aR)-8-O-Methyl-6a,11a-Dihydrolespedezol A_2 (7)

White, amorphous powder; $[\alpha]_D^{24}$ – 201°; (1 mg/mL, MeOH); UV (MeOH), λ_{max} 206, 288, 333 nm; CD (1.89 × 10^{-4} M, CH_3CN) λ_{max} ($\Delta\varepsilon$) 192 (+80899), 206 (–12011), 216 (–109803), 234 (–59308), 292 (+19668); ^1H and ^{13}C NMR data, see Table 1; HR-ESI-MS *m/z* 421.2007 $[M-H]^-$ (calcd for $[C_{26}H_{29}O_5]^-$ 421.2020,), *m/z* 423.2157 $[M+H]^+$ (calcd for $[C_{26}H_{31}O_5]^+$ 423.2166).

Table 1. ^{1}H (700 MHz), ^{13}C (175 MHz), and HMBC NMR data for compound 7 (δ in ppm, J in Hz).

N	δ_H (J in Hz)	δ_C	HMBC	ROESY
1	7.40 d (8.4), 1H	132.4	C-3, 4, 4a	2
2	6.55dd (2.6, 8.4), 1H	109.5	C-3, 4, 11b	1
3		156.7		
4	6.40 dd (2.6), 1H	103.5	C-2, 3, 4a, 11b	
4a		156.6		
6	4.22 dd (5.0, 11,0), 1H	66.6	C-4a, 6a, 6b, 11a	6, 6a, 7
	3.66 t (11.0), 1H	66.6	C-4a, 6a, 6b, 11a	6
6a	3.48 m, 1H	40.7	C-6, 6b, 10a	6, 11a
6b		115.6		
7	6.66 s, 1H	105.1	C-6, 6a, 6b, 8, 9, 10, 10a	6, OCH_3-8
8		141.0		
9		144.5		
10		111.8		
10a		152.4		
11a	5.40 d (7.0), 1H	77.4	C-1, 4a, 6, 6a, 6b, 11b	6a
11b		113.3		
1′	3.32 m, 2H	23.0	C-8, 9, 10, 2′, 3′	2′, 9′
2′	5.30 t (7.2), 1H	121.5	C-10, 1′, 4′, 9′	1′, 4′
3′		135.6		
4′	1,96 m, 2H	39.8	C-2′, 3′, 5′, 6′, 9′	2′, 6′
5′	2.04 m (8.3, 7.4, 7.2), 2H	26.7	C-3′, 4′, 6′, 7′	6′
6′	5.07 t (6.8), 1H	124.4	C-5′, 8′, 10′	4′, 5′
7′		131.1		
8′	1.64 s, 3H	25.6	C-6′, 7′, 10′	
9′	1.76 s, 3H	16.1	C-2′, 3′, 4′	1′
10′	1.56 s, 3H	17.6	C-6′, 7′, 8	
OH-3	4.78 bs, 1H			
OCH_3-8	3.85, 3H	56.9	C-8	7
OH-9	5.69 s, 1H		C-8, 9, 10, 10[a]	

2.5.2. Lespebicolin A (**8**)

Yellow, amorphous powder; $[\alpha]_D^{24}$ – 95°; (1 mg/mL, MeOH); UV (MeOH), λ_{max} 204, 269, 311 nm; CD (2.46×10^{-4} M, CH_3CN) λ_{max} ($\Delta\varepsilon$) 198 (+40706), 211 (–37771), 235 (–33085), 257(+1494), 271 (–7833), 294 (–7058), 329 (–10749), 392 (+3822); ^{1}H and ^{13}C NMR data, see Table 2; HR-ESI-MS *m/z* 811.3488 $[M-H]^-$ (calcd for $[C_{50}H_{51}O_{10}]^-$ 811.3488,), *m/z* 813.3634 $[M+H]^+$ (calcd for $[C_{50}H_{53}O_{10}]^+$ 813.3633).

Table 2. ^{1}H (700 MHz), ^{13}C (175 MHz) and HMBC NMR data for compound **8** (δ in ppm, J in Hz).

N	δ_H (J in Hz)	δ_C	HMBC
		Upper	
2		153.7	
3		119.5	
4		194.6	
5	6.82 s, 1H	103.4	C-6, 7, 8, 9, 10
6		141.5	
7		147.3	
8		111.0	
9		141.9	
10		116.9	
1′		110.7	
2′		156.6	
3′	6.50 d (2.3), 1H	105.0	C-1′, 2′, 4′, 5′
4′		158.6	
5′	6.41 dd (2.3, 8.5), 1H	108.6	C-1′, 3′
6′	7.26 d (8.5), 1H	132.2	C-2, 2′, 4′
1″	3.69 d (7.3), 2H	23.4	C-7, 8, 9, 2″, 3″
2″	5.42 t (7.3), 1H	120.4	C-4″, 9″
3″		139.3	
4″	2.11 m, 2H	39.7	C-2″, 3″, 5″, 6″, 9″
5″	2.13 m, 2H	26.4	C-4″, 6″, 7″
6″	5.07 m, 1H	123.8	
7″		132.1	
8″	1.68 s, 3H	25.7	C-6″, 7″, 10″
9″	1.86 s, 3H	16.3	C-2″, 3″, 4″
10″	1.60 s, 3H	17.7	C-6″, 7″, 8″
		Lower	
1	7.38 d (8.5), 1H	132.4	C-3, 4a,11a
2	6.56 dd (2.4, 8.5), 1H	110.1	C-4, 11b
3		157.2	
4	6.38 d (2.4), 1H	103.6	C-2, 3, 4a, 11b
4a		156.4	
6	4.01 m, 1H	66.3	
	3.56 m, 1H	66.3	
6a	3.46 m, 1H	39.6	
6b		118.4	
7	7.34 s, 1H	127.1	C-4 (upper), 6a, 9, 10a
8		113.4	
9		164.8	

Table 2. *Cont.*

10		112.3	
10a		164.8	
11a	5.58 d (7.3), 1H	79.2	C-1, 4a, 6, 11b
11b		112.5	
1‴	3.32 d (7.2), 2H	22.2	C-9, 10, 10a, 2‴, 3‴
2‴	5.26 t (7.2), 1H	121.0	C-10, 1‴, 4‴, 9‴
3‴		136.0	
4‴	1.98 m, 2H	39.8	C-2‴, 3‴, 5‴, 6‴, 9‴
5‴	2.04 m, 2H	29.7	C-3‴, 4‴, 6‴, 7‴
6‴	5.07 m, 1H	124.4	C-8‴, 10‴
7‴		131.3	
8‴	1.64 s, 3H	25.6	C-6‴, 7‴, 10‴
9‴	1.77 s, 3H	16.2	C-2‴, 3‴, 4‴
10‴	1.56 s, 3H	17.6	C-6‴, 7‴, 8‴
OH-9	12.83, s, 1H		

2.6. Cell Lines and Conditions

The human prostate cancer cell lines PC-3 [docetaxel-resistant, androgen-independent, AR-FL(-), AR-V7(-)], 22Rv1 [docetaxel-sensitive, androgen-independent, AR-FL(+), AR-V7(+)], and the human non-cancer fibroblast cell line MRC-9 were used. PC-3 and 22Rv1 were purchased from ATCC (Manassas, VA, USA). MRC-9 cells were kindly donated by Prof. Dr. med. Sonja Loges (University Medical Center Hamburg-Eppendorf, Hamburg, Germany). Cells were cultured according to the manufacturers' protocol in 10% FBS/RPMI (PC-3 and 22Rv1) or 10% FBS/DMEM (MRC-9) medium.

2.7. MTT Assay

In vitro drug sensitivity MTT assay was performed as previously described [15]. In total, 6000 cells/well were seeded in 96-well plates, incubated overnight, and treated with the investigated drugs for 48 h.

2.8. Flow Cytometry Analysis

The effect of the compounds on apoptosis induction and cell cycle progression was analyzed by flow cytometry technique using PI staining as reported before [15]. Cells (0.2×10^6 cells/well) were seeded in six-well plates, incubated overnight, and treated with the investigated drugs for 48 h. The cells were further harvested, fixed with 70% EtOH/H_2O, stained with propidium iodide (PI), and analyzed using a FACS Calibur machine (BD Bioscience, San Jose, CA, USA). The results were quantified using BD Bioscience Cell Quest Pro software (v.5.2.1., BD Bioscience).

2.9. Quantitative Real-Time PCR (qPCR)

The CDKs gene expression was measured using qPCR technique. PC-3 cells were seeded in Petri dishes (1×10^6 cells per ø 6 cm dish in 5 mL) in standard culture media, incubated overnight and treated with the investigated drugs in the fresh culture media for 24 h. Cells were harvested using the cell scraper, washed with PBS and homogenized using QIAshredder (QIAGEN, Hilden, Germany). The total RNA was isolated using PureLink® RNA Mini Kit (Invitrogen, Carlsbad, CA, USA) and the on-column DNA digestion using PureLink™ DNase (Invitrogen). Correspondent RNA (2 μg in 30 μL) and was transcribed into cDNA using Maxima First Strand cDNA Synthesis Kit for RT-qPCR (Thermo Scientific, Vilnius, Lithuania). The following qPCR was performed using 2X KAPA SYBR FAST qPCR

Master Mix Optimized for Roche LightCycler 480 (KAPA biosystems, Worburn, MA, USA) according to the manufacturer's protocol. In total, 20 ng of the template cDNA and 2 pmol of primers were used for each reaction. The PCR conditions were 30 s at 95 °C, followed by 40 cycles of 15 s at 95 °C, 5 s at melting temperature (Tm), and 26 s at 72 °C (fluorescence measurement). Melting curve analysis was performed under the following conditions directly after the PCR run: 10 s at 95 °C, 60 s at 65 °C and 1 s at 97 °C. Relative gene expression was calculated using the $2^{-\Delta\Delta CT}$ method. The expression of CDKs was normalized to GAPDH gene expression. The analysis was performed using primers, purchased from Eurofins MWG-Biotech AG (Ebersberg, Germany). Primers sequences and melting temperatures (Tm) is presented in Table 3.

Table 3. The sequences and melting temperatures (Tm) of the primers used for qPCR.

Primer	Sequence	Tm [°C]
CDK1_for	ACAGGTCAAGTGGTAGCCATGA	60
CDK1_rev	ACCTGGAATCCTGCATAAGCA	
CDK2_for	TTCTCATCGGGTCCTCCACC	61
CDK2_rev	TCGGTACCACAGGGTCACCA	
CDK4_for	CTGTGCCACATCCCGAACTG	61
CDK4_rev	GCCTCTTAGAAACTGGCGCA	
CDK5_for	CCACAACATCCCTGGTGAACGT	62
CDK5_rev	CCTCTTCTGCTGAGATACGCTG	
CDK6_for	CCGAAGTCTTGCTCCAGTCC	61
CDK6_rev	GGGAGTCCAATCACGTCCAA	
CDK7_for	TCACATCTTCAGTGCAGCAGG	59
CDK7_rev	TGGCAGCTGACATCCAGGT	
CDK8_for	AGCGGGTCGAGGACCTGTTT	62
CDK8_rev	CATGCCGACATAGAGATCCCAG	
CDK3_for	TCGCTGCTCAAGGAACTGAAGC	62
CDK3_rev	GTCCTGGCTGAGGAACTCAAAC	
CDK9_for	CCATTACAGCCTTGCGGGAGAT	62
CDK9_rev	CAGCAAGGTCATGCTCGCAGAA	

3. Results and Discussion

3.1. Isolation and Structure Elucidation of Compounds **1–8**

We used a column chromatography on polyamide sorbent to isolate the fractions of the bioactive prenylated polyphenolic compounds. The fractions were tested for the presence of polyphenolics using TLC plates treated with $FeCl_3$ and an HPLC–PDA–MS technique. The fractions evaluated positively in a $FeCl_3$ test were subsequently separated using silica gel column. Individual compounds were further purified using preparative HPLC. Compounds 1–6 were identified by comparison of their HPLC-PDA-MS and NMR spectra with previously published data [13]. Apart from prenylated polyphenolic compounds 1–6, we isolated and identified two new prenylated pterocarpan derivatives 7 and 8 (Figure 1).

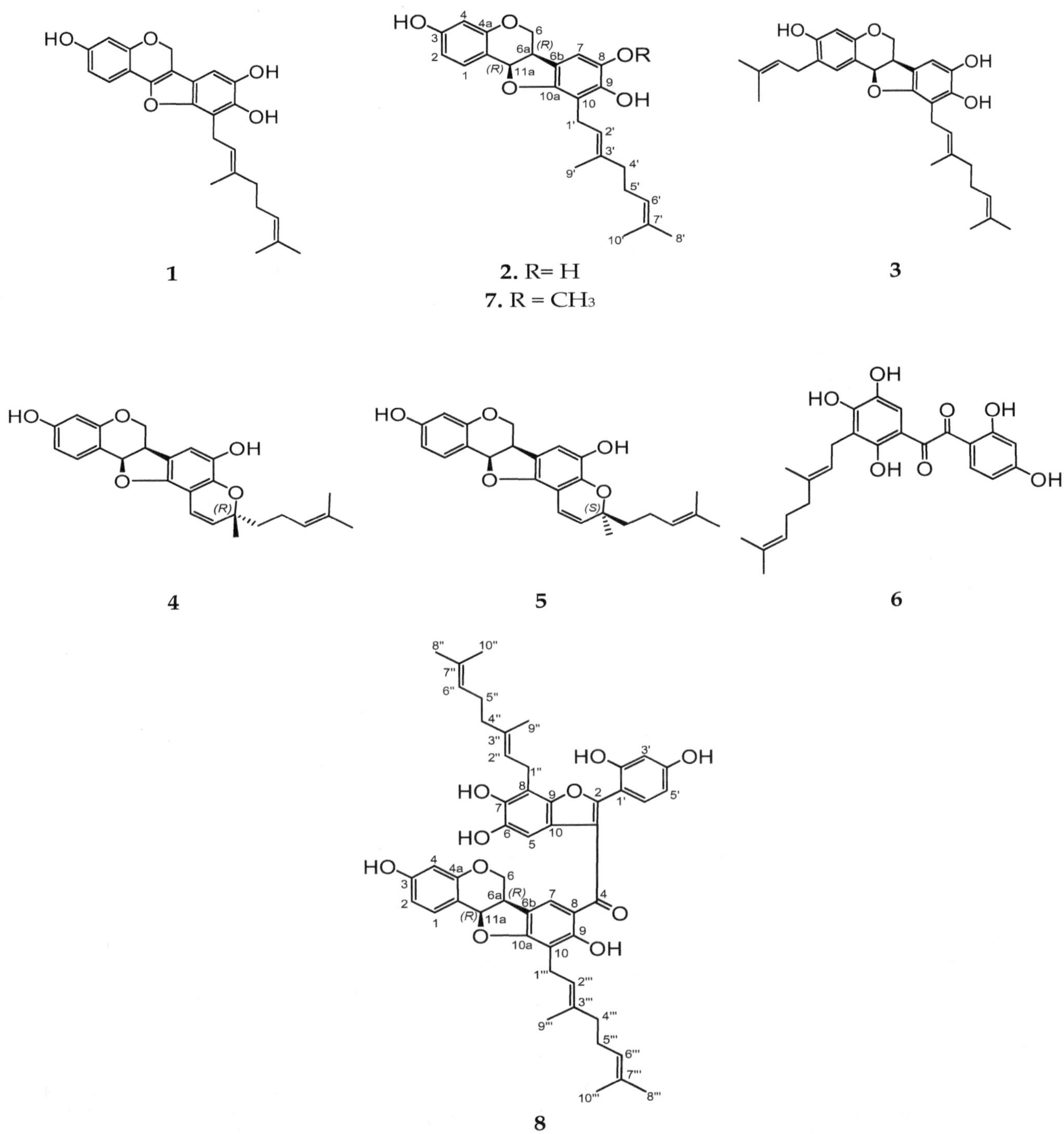

Figure 1. The structures of polyphenolic compounds **1–8** isolated from *L. bicolor* root bark and stem bark.

The structures of the isolated compounds were elucidated using mass spectrometry, NMR and CD spectroscopy (for the detailed experimental spectral data of the new compounds please see Supplementary data). Optically active prenylated polyphenolic compound **7** was isolated from *L. bicolor* root bark as a white amorphous powder. Its molecular formula was determined as $C_{26}H_{30}O_5$ based on HRESI-MS data. We observed the $[M-H]^-$ ion at *m/z* 421.2007 (calc. for $[C_{26}H_{29}O_5]^-$ 421.2020) in the negative ion mode, whereas the spectrum in the positive ion mode revealed the presence of the $[M+H]^+$ ion at *m/z* 423.2157 (calc. 423.2166 for $[C_{26}H_{31}O_5]^-$). The ^{13}C NMR spectrum contained 26 carbon atoms: 15 carbon atoms of the pterocarpan skeleton, 10 carbon atoms of the geranyl side chain and one carbon atom of the methoxy group (Table 1). ^1H NMR spectrum also revealed characteristic signals of protons of the pterocarpan skeleton at δ_H 3.48 (1H, m), 3.66 (1H, t, *J* = 11.0), 4.22 (1H, dd, *J* = 5.0, 11.0) and 5.40 (1H, d, *J* = 7.0) assigned to H-6a, two H-6, and H-11a protons, respectively. The signals at δ_H 7.40 (1H, d, *J* = 8.4), 6.55 (1H, dd, *J* = 2.6, 8.4), and 6.40 (1H, d, *J* = 2.6) belonged to H-1, H-2, and H-4 protons of ring A, respectively, and formed a spin system. The singlet signal at δ_H 6.66 (1H, s) was assigned to the aromatic proton at C-7 of ring D. The location of the geranyl side chain at C-10 of the pterocarpan skeleton was determined based on the correlation between the proton signal of 2H-1′ at δ_H 3.32 and carbon signals of C-9, C-10 иC-10a at δ_C 144.5, 111.8, and 152.4, respectively, in the HMBC spectrum of **7**. The position of the methoxy group was determined as C-8 based on the observed correlation fbetween the signal of its protons at δ_H 3.85 and the signal of C-8 at δ_C 141.0 in the HMBC spectrum of **7** (Table 1). The signals in the ^{1}H and ^{13}C spectra were completely assigned using HSQC, ROESY, and HMBC data (Table 1). Negative optical rotation value ($[\alpha]_D^{24}$ – 201°) [16], as well as characteristic Cotton effects in the CD spectrum of **7** (positive bands at λ 192 nm and 292 nm, and negative bands at 216 and 234 nm), confirmed the 6a*R*,11a*R* configuration of the asymmetric centers in **7** [2,17]. Thus, the structure of **7** was determined as (6a*R*,11a*R*)-8-*O*-methyl-6a,11a-dihydrolespedezol A_2.

Compound **8** was obtained as a yellow amorphous powder. Its HR-ESI-MS spectrum contained the $[M\text{-}H]^-$ ion at *m/z* 811.3469 in the negative ion mode (calc. 811.3488 for $[C_{50}H_{51}O_{10}]^-$) and the $[M+H]^+$ ion at *m/z* 813.3634 (calc. 813.3633 for $[C_{50}H_{53}O_{10}]^+$) in the positive ion mode. Thus, the molecular formula of **8** was determined as $C_{50}H_{52}O_{10}$. Both positive and negative MS/MS spectra revealed the intensive signals of the ions $[M+H\text{-}H_2O]^+$ at *m/z* 795.3521 (calc. 795.3528) and 793.3360 $[M\text{-}H\text{-}H_2O]^-$ (calc. 793.3382) due to water molecule loss (Figures S11 and S13, Supplementary data). The presence of the pterocarpan and 2-arylbenzofuran fragments, linked with the carbonyl group, was confirmed using the tandem mass spectrometry technique. The negative MS/MS product ions at *m/z* 419.1476 with the composition $[C_{25}H_{23}O_6]^-$ (calc. 419.1500) and at *m/z* 393.1663 with the composition $[C_{24}H_{25}O_5]^-$ (calc. 393.1707) resulted from the carbonyl bridge degradation and indicated a neutral loss of a pterocarpan fragment (Figures S11 and S12, Supplementary data). The signals at *m/z* 419.1870 (calc. 419.1853) and *m/z* 377.1711 (calc. 377.1747) corresponded to the ions $[C_{26}H_{27}O_5]^+$ and $[C_{24}H_{25}O_4]^-$, respectively, and indicated the loss of a neutral pterocarpan fragment. The ^{13}C NMR spectrum revealed the presence of 50 carbon atoms: 15 atoms constituted the pterocarpan skeleton, 14 carbon atoms formed the 2-arylbezofuran fragment, 20 atoms belonged to the two geranyl side chains, and one carbon atom belonged to the carbonyl group. The presence of the pterocarpan fragment in **8** was confirmed by the proton signals at δ_H 3.46 (1H, m), 3.56 (1H, m), 4.01 (1H, m), and 5.58 (1H, d, *J* = 7.3) assigned to H-6a, two H-6, and H-11a protons, respectively. The chemical shift values of these signals were very similar to the signals of H-6a, two H-6, and H-11a protons in the ^{1}H spectrum of compounds **1–5** and **7**. The geranyl side chain was attached to C-10 of the pterocarpan moiety, which was confirmed by the presence of the HMBC-correlation between the signal of 2H-1‴ at δ_H 3.32 and the carbon signals of C-9, C-10 иC-10a at δ_C 164.8, 112.3, and 164.8, respectively.

In the ^{1}H spectrum of **8**, we have observed a singlet signal at δ_H 6.82 assigned to the aromatic proton H-5 and the signals of the ABX proton system in the 2-arylbezofuran fragment at δ_H 6.50 (1H, d, *J* = 2.3), 6.41 (1H, dd, *J* = 2.3, 8.5), and 7.26 (1H, d, *J* = 8.5) of H-3′, H-5′, and H-6′ protons, respectively. The second geranyl side chain was located at C-8 of the 2-arylbezofuran fragment, which was demonstrated by the presence of the HMBC-correlation between the signals of 2H-1″ at δ_H 3.69 (2H, d, *J* = 7.3) and the signals of C-7, C-8 and C-9 at δ_c 147.3, 111.0, and 141.9, respectively. The chemical shift values of the other atoms of 2-arylbezofuran fragment in ^{1}H and ^{13}C NMR spectra were similar to those for lespecyrtin H_2 [18].

We observed a correlation between the proton signal of H-7 at δ_H 7.34 (1H, s) of the pterocarpan fragments and the carbon signal of C-4 of the carbonyl group at δ_C 194.6 in the HMBC spectrum of **8**. Thus, we concluded that the carbonyl group was attached to C-8 of a pterocarpan fragment. The signal of the hydroxy group proton at C-9 of the pterocarpan fragment had a δ_H of 12.83, which also confirmed that the carbonyl group was located at C-8 and formed a hydrogen bond with this hydroxy group. The signals in ^{1}H and ^{13}C NMR spectra of **8** were assigned using HSQC, HMBC, and ROESY experiments. Thus, compound **8**, named lespebicolin A, was assumed to be a dimeric flavonoid consisting of the 2-arylbenzofuran and the pterocarpan fragments linked via carbonyl group.

The absolute configurations of the asymmetric centers at C-6a and C-11a in the pterocarpan fragment of **8** were determined based on the negative optical rotation value ($[\alpha]_D^{24}$ − 95°) and CD spectral data similar to those of lespecyrtins H_1–H_4 [18].

Thus, in addition to the six polyphenolic compounds 1–6 previously isolated from *L. bicolor* stem bark [13], we isolated two new pterocarpans 7 and 8 from *L. bicolor* root bark. A comparison of the HPLC profiles of *L. bicolor* stem bark and root bark showed that compounds **7** and **8** are also present in *L. bicolor* stem bark but in smaller amounts than in root bark (Figure 2).

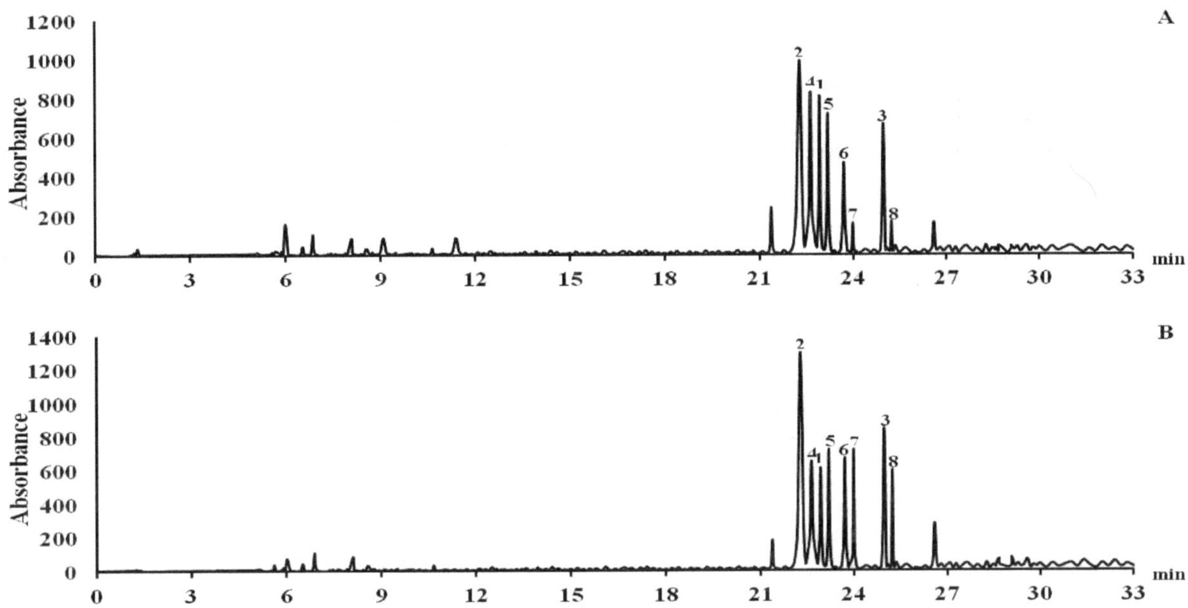

Figure 2. HPLC profiles of *L. bicolor* stem bark (**A**) and root bark (**B**) ethanolic extracts.

Recently, Korean colleagues isolated five pterocarpans, two new coumestans, and two new arylbenzofurans with prenyl and geranyl substituents in their structures [14]. These isolated compounds contain a methoxy group at C-1 and exhibited antiproliferative effects on human leukemia cells. Of note, we could not find these compounds in the stem bark and root bark of *L. bicolor* harvested in the South of the Primorskiy region of the Russian Far East. The polyphenolic compounds isolated by us from the stem bark and root bark of *L. bicolor* also contained the pterocarpan skeleton and geranyl side chains in their structures. However, they did not have a methoxy group at C-1. Moreover, for the first time, we have isolated a new dimeric flavonoid lespebicolin A (**8**) harboring two geranyl side chains. Previously, the related dimeric flavonoids had been found in *L. homoloba*, *L. floribunda* and *L. cyrtobotria* [18,19]. Thus, the chemical compositions of polyphenolic compounds of *L. bicolor* growing in the Primorskiy region and South Korea differed significantly. This may be because *L. bicolor* samples collected in the Primorye Region (the South of the Russian Far East) and South Korea belong to different variants of this species [20].

3.2. *Investigation of Anticancer In Vitro Activity of the Compounds* **1–8**

Next, we examined the cytotoxic activity of the isolated compounds in prostate cancer cell lines as well as in non-cancer cells. To get a first impression on the impact of these novel compounds for the treatment of advanced prostate cancer, human drug-resistant cell lines PC-3 and 22Rv1 were used. Both cell lines are known to be castration-resistant (androgen-independent). In addition, the cell lines are known to exhibit resistance to novel second-generation androgen receptor (AR) targeting drugs, e.g., abiraterone and enzalutamide, due to the loss or alternative splicing of the AR, respectively [21,22].

Moreover, PC-3 cells have been reported to be rather resistant to docetaxel, and therefore, are known as one of the most aggressive prostate cancer cell lines used as in vitro and in vivo models. Notably, all eight isolated compounds exhibited a cytotoxic activity in both prostate cancer cell lines in the micromolar range, while non-malignant MRC-9 cells were less affected under the treatment (Table 4). The selectivity index evaluation (SI; MRC-9 vs. PC-3 cells) has revealed compounds 1, 2, 3 as well as 6 and 8 to be the most promising and selective in human drug-resistant prostate cancer cells (SI = 2.5 ~ 8) (Table 4). Remarkably, a well-established cytotoxic chemotherapeutic drug cisplatin applied for the treatment of different cancer types and used in the current research as a positive control was less selective (SI = 0.72) (Table 4). For further investigations on the mechanisms of action, we have chosen the aggressive and drug-resistant prostate cancer PC-3 cell line.

Table 4. Inhibition concentrations 50% (IC_{50}s) of the investigated compounds in different cell lines. The cells were treated for 48 h. Cisplatin was used as a positive control. Selectivity index (SI) was calculated as [IC_{50} in MRC-9 cells] / [IC_{50} in PC-3 cells].

		IC_{50}, µM								
	Compound	1	2	3	4	5	6	7	8	Cisplatin
Cell line	PC-3	6.73	5.24	2.1	16.37	15.23	19.26	14.86	9.24	21.0
	22Rv1	20.14	13.85	6.38	29.53	35.81	46.9	30.51	17.51	4.53
	MRC-9	48.51	37.57	16.82	27.18	24.61	>50	24.4	23.95	15.12
	Selectivity index (MRC-9 vs. PC-3)	7.21	7.17	7.99	1.66	1.62	>2.6	1.64	2.59	0.72

To investigate the mechanism of action contributing to the cytotoxicity in PC-3 cells, we investigated the effects on DNA fragmentation, a well-established marker of cellular apoptosis. Therefore, we evaluated the effects of the isolated compounds at four different concentrations of 0.5 µM, 1 µM, 5 µM, and 10 µM by flow cytometry (Figure 3). All the substances apart from structurally different lespebicolin A (**8**) revealed a significant apoptosis induction at the tested concentrations (Figure 3). Therefore, prenylated dimeric flavonoid **8** may either reveal a non-apoptotic character of cell death or a predominantly antiproliferative rather than cytotoxic effect. Note, compound **3** exhibited the most pronounced apoptosis-inducing effect.

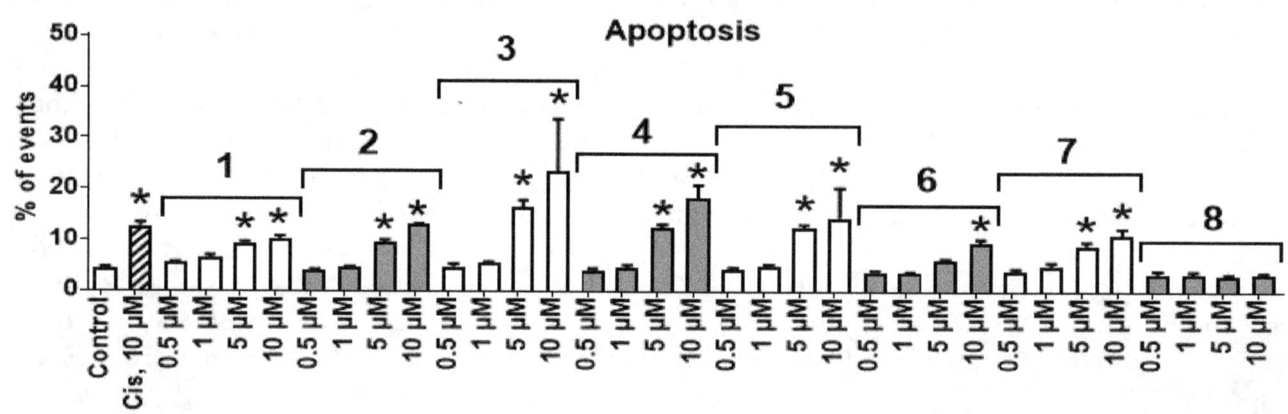

Figure 3. Effect of natural compounds on DNA fragmentation of PC-3 cells. Cells were treated with the investigated compounds for 24 h. The effects were analyzed by flow cytometry. Cells detected as sub-G1 population were considered to be apoptotic. Cisplatin (Cis) was used as a positive control. * $p \leq 0.05$ (Student's t-test).

Next, we examined the effects of the isolated compounds on the cell cycle progression of PC-3 cells. For compounds **1–3**, a significant accumulation of the cells in S-phase as well as slight accumulation in G2/M-phase was observed (Figure 4). Thus, these compounds mainly induced a S-phase arrest, whereas the compounds **4–8** led to a pronounced G1-phase cell cycle arrest (Figure 4).

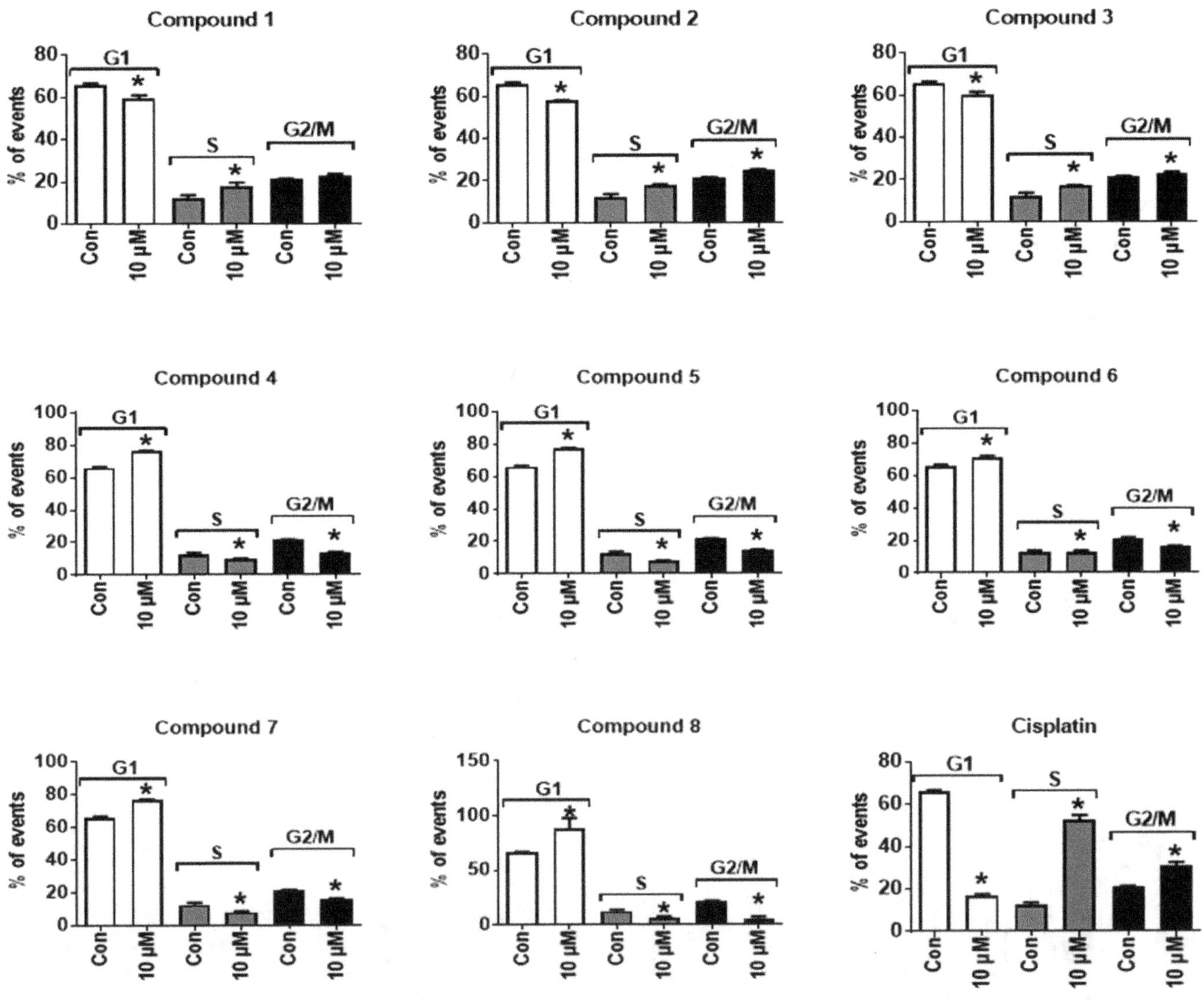

Figure 4. Treatment effects on the progression of the cell cycle of PC-3 cells. Cells were treated with the investigated compounds for 24 h. The effects were analyzed by flow cytometry. Cisplatin was used as a positive control. * $p \leq 0.05$ (Student's t-test).

Due to the significant alterations of the cell cycle, we further investigated the effects of the isolated compounds on the cyclin-dependent kinases (CDKs). These proteins are known to control proliferation as well as some apoptotic processes. We have examined the expression of mRNAs corresponding to the nine most studied CDKs known to be involved in the cell cycle progression of cancer cells, namely, CDK1–CDK9. Indeed, we detected the inhibitory effects of compounds **1**, **2**, and **3** on the expression of mRNA corresponding to several essential CDKs (Figure 5). The most pronounced effects were observed for CDK1, CDK2, CDK4, and CDK5. Remarkably, these effects correlated well with the cytotoxicity of the isolated compounds. Thus, the strongest inhibitory effect on the corresponding mRNA expression was observed with the most cytotoxic compound **3** (IC_{50} = 2.1 µM in PC-3 cells, Table 3).

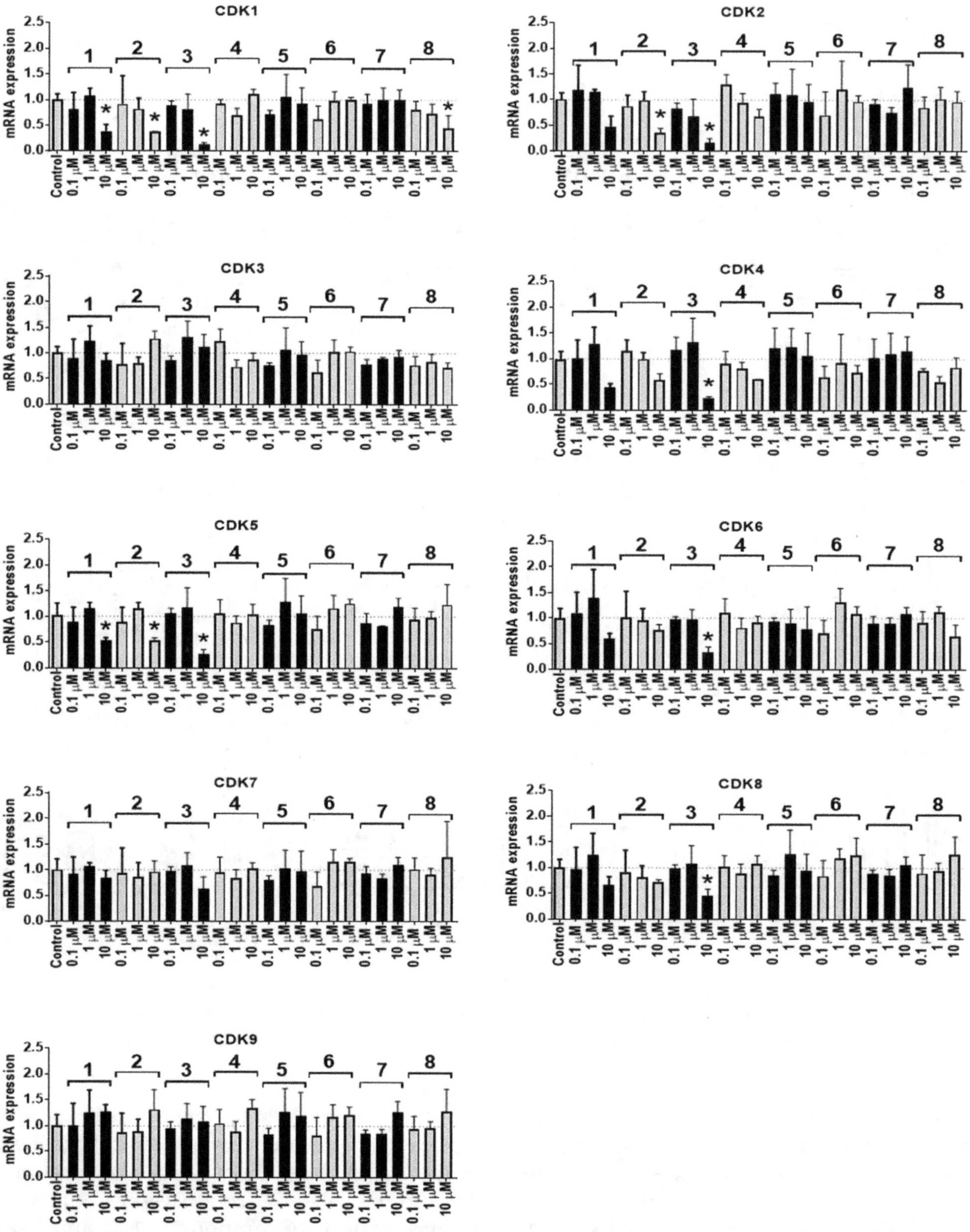

Figure 5. Expression of mRNA correspondent to CDK1-9 in PC-3 cells treated with the isolated compounds **1–8** for 24 h. The target gene expression was normalized to GAPDH gene expression. * $p \leq 0.05$ (Student's t-test).

CDKs are known to play an important role in the development and progression of several cancer types, including human prostate cancer. These kinases are involved in the cell cycle control and are often overexpressed or mutated in cancer cells [23]. Therefore, CDKs are an attractive target in anticancer therapy and are currently actively explored [23]. In particular, CDK1 and CDK6 are known to phosphorylate AR and activate its transcriptional activity at different phosphorylation sites finally leading to a poorer clinical prognosis [24–26]. Expression and activity of CDK2 are associated with prostate cancer relapse in patients and with cancer cell invasion [27]. In addition, the particular importance and clinical relevance of CDK5 for prostate cancer growth have been highlighted in different publications. Thus, it was reported to phosphorylate and stabilize AR leading to the promotion of its transcriptional activity [28,29]. Moreover, CDK5 stimulates the growth of AR-negative prostate cancer cells via Akt activation [30]. Of note, the CDK4/6 inhibitor is currently undergoing clinical trials in metastatic castration-resistant prostate cancer. They can promote cell senescence as well as disruption of cancer cells in vivo by the induction of cytotoxic T cell-mediated cell death. Moreover, inhibition of CDK4/6 may cause S-phase cell cycle arrest, which has been observed for compounds 1–3 (Figure 4) [31,32].

Of note, the other compounds 4–8 did not exhibit pronounced effects on CDKs expression (Figure 5) but still inhibited cell cycle progression, however, in the G1 phase (Figure 4). This effect may result from the other processes related either to apoptosis induction or inhibition of different CDKs at the post-transcriptional (inhibition of CDKs protein expression) or even post-translational levels (inhibition of CDKs activity). Indeed, a free hydroxyl group at C-8 (e.g., pterocarpans 1–3) seems to be important for the inhibition of CDKs mRNA expression by these natural compounds. Hence, compound 2, containing 8-OH was fairly active in this experiment, however, very structurally similar new compound **7** containing a methylated hydroxyl group (8-OCH_3) did not exhibit this activity; moreover, it was less cytotoxic and had a different effect on cell cycle progression (i.e., G1-phase arrest instead of S-phase arrest). In addition, new compound 8 has a pterocarpan fragment related to compounds 1, 2 and 3. However, this compound is substituted at C-8 and therefore similar to **7** did not exhibit any pronounced effect on CDKs mRNA expression and could not induce an S-phase cell cycle arrest.

4. Conclusions

In conclusion, we isolated two new (**7** and **8**) and six recently reported (**1–6**) polyphenolic compounds from the root bark of *L. bicolor*. Structures were established using mass spectrometry and NMR spectroscopy. The natural compounds were active and selective in human drug-resistant prostate cancer cells in vitro. Prenylated pterocarpans **1–3** effectively inhibited cell cycle progression of human cancer cells in S-phase. This was associated with the inhibition of mRNA expression corresponding to several human CDKs. Compounds **4–8** induced G1-phase cell cycle arrest without any pronounced effect on CDK mRNA expression. The non-substituted hydroxy group at C-8 in ring D of the pterocarpan skeleton seems to be important for the CDKs inhibitory activity.

Author Contributions: Conceptualization, S.A.D., D.T., and S.F.; methodology, S.A.D., D.T., S.F., J.H., and G.v.A.; validation, S.A.D., D.T., S.F., T.B., M.K., M.V., A.K., J.H., V.G., and N.K.; formal analysis, all authors; investigation, S.A.D., D.T., S.F., T.B., M.K., M.V., A.K., J.H., V.G., and N.K., V.G.; resources, P.G., C.B., M.G., and G.v.A.; data curation, S.A.D., D.V., S.F., T.B., M.K., M.V., A.K., J.H., V.G., N.K., and V.G.; writing—original draft preparation, S.A.D., D.T., S.F., and G.v.A.; writing—review and editing, all authors; visualization, S.A.D., D.T., and S.F.; supervision, S.F. and G.v.A.; project administration, S.F., and G.v.A.; funding acquisition, D.T., C.B., M.G., and G.v.A. All authors have read and agreed to the published version of the manuscript.

References

1. Sun, H.; Lei, Y.N.; Deng, W.; Wang, H.M.; Teng, Y.O.; Zhao, H.Y.; Yao, Q.W.; Yu, P. First Total Synthesis and Cytotoxicity of Naturally Occurring Lespedezol E1. *Chem. Nat. Compd.* **2016** *52*, 896–898. [CrossRef]

2. Woo, H.S.; Kim, D.W.; Curtis-Long, M.J.; Lee, B.W.; Lee, J.H.; Kim, J.Y.; Kang, J.E.; Park, K.H. Potent inhibition of bacterial neuraminidase activity by pterocarpans isolated from the roots of Lespedeza bicolor. *Bioorg. Med. Chem. Lett.* **2011**, *21*, 6100–6103. [CrossRef] [PubMed]
3. Maximov, O.B.; Kulesh, N.I.; Stepanenko, L.S.; Dmitrenok, P.S. New prenylated isoflavanones and other constituents of Lespedeza bicolor. *Fitoterapia* **2004**, *75*, 96–98. [CrossRef] [PubMed]
4. Glyzin, V.I.; Ban'kovskii, A.I.; Zhurba, O.V.; Sheichenko, V.I. Flavonoids of *Lespedeza bicolor*. *Chem. Nat. Compd.* **1970**, *6*, 487–488. [CrossRef]
5. Jae-Hak, T.A.L.; Jin-Woo, T.A.J. Antioxidant activity of different parts of *Lespedeza bicolor* and isolation of antioxidant compound. *Korean J. Food Sci. Technol.* **2012**, *44*, 763–771.
6. Do, M.H.; Lee, J.H.; Wahedi, H.M.; Pak, C.; Lee, C.H.; Yeo, E.J.; Lim, Y.; Ha, S.K.; Choi, I.; Kim, S.Y. Lespedeza bicolor ameliorates endothelial dysfunction induced by methylglyoxal glucotoxicity. *Phytomedicine* **2017**, *36*, 26–36. [CrossRef] [PubMed]
7. Lee, Y.-S.; Chang, Z.; Park, S.-C.; Rim, N.-R.; Kim, N.-W. Antioxidative activity and irritation response of Lespedeza bicolor. *Toxicol. Res.* **2005**, *21*, 115–119.
8. Lee, S.J.; Hossaine, M.D.; Park, S.C. A potential anti-inflammation activity and depigmentation effect of Lespedeza bicolor extract and its fractions. *Saudi J. Biol. Sci.* **2016**, *23*, 9–14. [CrossRef]
9. Ko, Y.H.; Shim, K.Y.; Kim, S.K.; Seo, J.Y.; Lee, B.R.; Hur, K.H.; Kim, Y.J.; Kim, S.E.; Do, M.H.; Parveen, A.; et al. Lespedeza bicolor Extract Improves Amyloid Beta25–35-Induced Memory Impairments by Upregulating BDNF and Activating Akt, ERK, and CREB Signaling in Mice. *Planta Med.* **2019**, *85*, 1363–1373. [CrossRef]
10. Do, M.H.; Lee, J.H.; Cho, K.; Kang, M.C.; Subedi, L.; Parveen, A.; Kim, S.Y. Therapeutic Potential of Lespedeza bicolor to Prevent Methylglyoxal-Induced Glucotoxicity in Familiar Diabetic Nephropathy. *J. Clin. Med.* **2019**, *8*, 1138. [CrossRef]
11. Kim, Y.; Lee, H.; Kim, S.Y.; Lim, Y. Effects of Lespedeza Bicolor Extract on Regulation of AMPK Associated Hepatic Lipid Metabolism in Type 2 Diabetic Mice. *Antioxidants* **2019**, *8*, 599. [CrossRef] [PubMed]
12. Ullah, S. Methanolic extract from Lespedeza bicolor: Potential candidates for natural antioxidant and anticancer agent. *J. Tradit. Chin. Med.* **2017**, *37*, 444–451. [CrossRef]
13. Tarbeeva, D.V.; Fedoreyev, S.A.; Veselova, M.V.; Blagodatski, A.S.; Klimenko, A.M.; Kalinovskiy, A.I.; Grigorchuk, V.P.; Berdyshev, D.V.; Gorovoy, P.G. Cytotoxic polyphenolic compounds from Lespedeza bicolor stem bark. *Fitoterapia* **2019**, *135*, 64–72. [CrossRef] [PubMed]
14. Thuy, N.T.T.; Lee, J.E.; Yoo, H.M.; Cho, N. Antiproliferative Pterocarpans and Coumestans from Lespedeza bicolor. *J. Nat. Prod.* **2019**, *82*, 3025–3032. [CrossRef]
15. Dyshlovoy, S.A.; Tabakmakher, K.M.; Hauschild, J.; Shchekaleva, R.K.; Otte, K.; Guzii, A.G.; Makarieva, T.N.; Kudryashova, E.K.; Fedorov, S.N.; Shubina, L.K.; et al. Guanidine alkaloids from the marine sponge *Monanchora pulchra* show cytotoxic properties and prevent EGF-induced neoplastic transformation in vitro. *Mar. Drugs* **2016**, *14*, 133. [CrossRef]
16. Goel, A.; Kumar, A.; Raghuvanshi, A. Synthesis, stereochemistry, structural classification, and chemical reactivity of natural pterocarpans. *Chem. Rev.* **2013**, *113*, 1614–1640. [CrossRef]
17. Veitch, N.C. Isoflavonoids of the leguminosae. *Nat. Prod. Rep.* **2013**, *30*, 988–1027. [CrossRef]
18. Mori-Hongo, M.; Yamaguchi, H.; Warashina, T.; Miyase, T. Melanin synthesis inhibitors from Lespedeza cyrtobotrya. *J. Nat. Prod.* **2009**, *72*, 63–71. [CrossRef]
19. Miyase, T.; Sano, M.; Yoshino, K.; Nonaka, K. Antioxidants from Lespedeza homoloba (II). *Phytochemistry* **1999**, *52*, 311–319. [CrossRef]
20. Lee, T.B. *Illustrated Flora of Korea Seoul*; Hyangmunsa: Seoul, Korea, 1993.
21. Dyshlovoy, S.A.; Otte, K.; Tabakmakher, K.M.; Hauschild, J.; Makarieva, T.N.; Shubina, L.K.; Fedorov, S.N.; Bokemeyer, C.; Stonik, V.A.; von Amsberg, G. Synthesis and anticancer activity of the derivatives of marine compound rhizochalin in castration resistant prostate cancer. *Oncotarget* **2018**, *9*, 16962–16973. [CrossRef]
22. Miao, L.; Yang, L.; Li, R.; Rodrigues, D.N.; Crespo, M.; Hsieh, J.T.; Tilley, W.D.; de Bono, J.; Selth, L.A.; Raj, G.V. Disrupting Androgen Receptor Signaling Induces Snail-Mediated Epithelial-Mesenchymal Plasticity in Prostate Cancer. *Cancer Res.* **2017**, *77*, 3101–3112. [CrossRef] [PubMed]
23. Vijayaraghavan, S.; Moulder, S.; Keyomarsi, K.; Layman, R.M. Inhibiting CDK in Cancer Therapy: Current Evidence and Future Directions. *Target. Oncol.* **2018**, *13*, 21–38. [CrossRef] [PubMed]
24. Willder, J.M.; Heng, S.J.; McCall, P.; Adams, C.E.; Tannahill, C.; Fyffe, G.; Seywright, M.; Horgan, P.G.; Leung, H.Y.; Underwood, M.A.; et al. Androgen receptor phosphorylation at serine 515 by Cdk1 predicts biochemical relapse in prostate cancer patients. *Br. J. Cancer* **2013**, *108*, 139–148. [CrossRef] [PubMed]

25. Chen, S.; Xu, Y.; Yuan, X.; Bubley, G.J.; Balk, S.P. Androgen receptor phosphorylation and stabilization in prostate cancer by cyclin-dependent kinase 1. *Proc. Natl. Acad. Sci. USA* **2006**, *103*, 15969–15974. [CrossRef]
26. Lim, J.T.; Mansukhani, M.; Weinstein, I.B. Cyclin-dependent kinase 6 associates with the androgen receptor and enhances its transcriptional activity in prostate cancer cells. *Proc. Natl. Acad. Sci. USA* **2005**, *102*, 5156–5161. [CrossRef]
27. Yin, X.; Yu, J.; Zhou, Y.; Wang, C.; Jiao, Z.; Qian, Z.; Sun, H.; Chen, B. Identification of CDK2 as a novel target in treatment of prostate cancer. *Future Oncol.* **2018**, *14*, 709–718. [CrossRef]
28. Hsu, F.N.; Chen, M.C.; Chiang, M.C.; Lin, E.; Lee, Y.T.; Huang, P.H.; Lee, G.S.; Lin, H. Regulation of androgen receptor and prostate cancer growth by cyclin-dependent kinase 5. *J. Biol. Chem.* **2011**, *286*, 33141–33149. [CrossRef]
29. Oner, M.; Lin, E.; Chen, M.-C.; Hsu, F.-N.; Shazzad Hossain Prince, G.M.; Chiu, K.-Y.; Teng, C.-L.J.; Yang, T.-Y.; Wang, H.-Y.; Yue, C.-H.; et al. Future Aspects of CDK5 in Prostate Cancer: From Pathogenesis to Therapeutic Implications. *Int. J. Mol. Sci.* **2019**, *20*, 3881. [CrossRef]
30. Lindqvist, J.; Imanishi, S.Y.; Torvaldson, E.; Malinen, M.; Remes, M.; Orn, F.; Palvimo, J.J.; Eriksson, J.E. Cyclin-dependent kinase 5 acts as a critical determinant of AKT-dependent proliferation and regulates differential gene expression by the androgen receptor in prostate cancer cells. *Mol. Biol. Cell* **2015**, *26*, 1971–1984. [CrossRef]
31. Schettini, F.; De Santo, I.; Rea, C.G.; De Placido, P.; Formisano, L.; Giuliano, M.; Arpino, G.; De Laurentiis, M.; Puglisi, F.; De Placido, S.; et al. CDK 4/6 Inhibitors as Single Agent in Advanced Solid Tumors. *Front. Oncol.* **2018**, *8*, 608. [CrossRef]
32. Batra, A.; Winquist, E. Emerging cell cycle inhibitors for treating metastatic castration-resistant prostate cancer. *Expert Opin. Emerg. Drugs* **2018**, *23*, 271–282. [CrossRef] [PubMed]

4

Gas Plasma-Treated Prostate Cancer Cells Augment Myeloid Cell Activity and Cytotoxicity

Sander Bekeschus [1,*], Verena Ressel [1,2], Eric Freund [1,3], Nadine Gelbrich [2], Alexander Mustea [4] and Matthias B. Stope [4]

1. ZIK plasmatis, Leibniz Institute for Plasma Science and Technology (INP), 17489 Greifswald, Germany; verena-ressel@outlook.de (V.R.); Eric.freund@inp-greifswald.de (E.F.)
2. Department of Urology, University Medicine Greifswald, 17475 Greifswald, Germany; nadine.gelbrich@uni-greifswald.de
3. Department of General, Visceral and Thoracic Surgery, University Medicine Greifswald, 17475 Greifswald, Germany
4. Department of Gynecology and Gynecological Oncology, University Hospital Bonn, 53127 Bonn, Germany; alexander.mustea@ukbonn.de (A.M.); matthias.stope@ukbonn.de (M.B.S.)

* Correspondence: sander.bekeschus@inp-greifswald.de

Abstract: Despite recent improvements in cancer treatment, with many of them being related to foster antitumor immunity, tumor-related deaths continue to be high. Novel avenues are needed to complement existing therapeutic strategies in oncology. Medical gas plasma technology recently gained attention due to its antitumor activity. Gas plasmas act via the local deposition of a plethora of reactive oxygen species (ROS) that promote the oxidative cancer cell death. The immunological consequences of plasma-mediated tumor cell death are only poorly understood, however. To this end, we exposed two prostate cancer cell lines (LNCaP, PC3) to gas plasma in vitro, and investigated the immunomodulatory effects of the supernatants in as well as of direct co-culturing with two human myeloid cell lines (THP-1, HL-60). After identifying the cytotoxic action of the kINPen plasma jet, the supernatants of plasma-treated prostate cancer cells modulated myeloid cell-related mitochondrial ROS production and their metabolic activity, proliferation, surface marker expression, and cytokine release. Direct co-culture amplified differentiation-like surface marker expression in myeloid cells and promoted their antitumor-toxicity in the gas plasma over the untreated control conditions. The results suggest that gas plasma-derived ROS not only promote prostate cancer cell death but also augment myeloid cell activity and cytotoxicity.

Keywords: HL-60; immunomodulation; kINPen; LNCaP; PC3; plasma medicine; reactive oxygen species; ROS; THP-1

---

1. Introduction

Despite recent improvements in cancer treatment, with many of them being related to foster antitumor immunity, tumor-related deaths continue to be high. With over 1.5 million new cases worldwide in 2018 and a high mortality rate [1], this is also seen in prostate cancer patients. While the 5-year survival rate of loco-regional prostate cancer is nearly 100%, it drops to 31% in the case of tumor metastasis [2]. In the US, prostate-cancer related death ranks second in men behind lung cancer. Standard treatment for loco-regional prostate cancer is surgery or radiation, while metastatic prostate cancer is targeted using androgen deprivation therapy as first-line treatment, with docetaxel as an emerging drug [3]. However, the disease is still fatal in many patients, exemplifying the need for novel therapeutic avenues against prostate cancer.

Medical gas plasma technology recently gained attention due to its antitumor activity against many types of cancers [4–6]. Specifically, plasma was shown to inactivate, for instance, malignant melanoma [7], squamous cell carcinoma [5,8], lung cancer [9], colon cancer [10], pancreatic cancer [11], osteosarcoma [12], glioblastoma [13], and hepatocellular carcinoma [14]. The technology uses electrical discharge to partially ionize a neutral gas, and is designed in a way that it promotes biomedical effects in cells and tissues without thermal harm [15]. Instead, the gas plasmas act via the local deposition of a plethora of reactive oxygen species (ROS) [16] that promote cancer cell death [17]. Previous studies suggested that plasma-induced tumor toxicity also has an immunogenic component [18–20], leading to antitumor immunity in vivo [21–23]. Specifically, it was observed that plasma treatment leads to an increased expression of calreticulin, a protein that dictates the immunogenicity of cells [24]. In vivo, plasma treatment not only limits tumor growth but also is associated with an increase in tumor-infiltrating leukocytes [25,26]. Prostate cancer also has an immunological dimension [27], and especially myeloid cells are frequently high-jacked by tumor cells to render them into tumor-promoting phenotypes [28].

To this end, we here investigated not only the cytotoxic potential of medical gas plasma technology in targeting human prostate cancer cells but also the immunomodulatory consequences of the latter towards human myeloid cells. Using both supernatant and co-culture assay systems, we identified an augmentation of myeloid cell activity and cytotoxicity towards gas plasma-treated prostate cancer cells in vitro.

2. Materials and Methods

2.1. Cell Culture

The prostate cancer cell lines LNCaP (ATCC, Manassas, VA, USA; ATCC CRL-1740) and PC3 (ATCC CRL-1435) as well as the myeloid cell lines THP-1 (ATCC TIB-202) and HL-60 (ATCC CCL-240) were used in this study. The prostate cancer cell lines were retrieved from metastatic lesions, with LNCaP as an androgen-sensitive cell line derived from a lymph node metastasis and the androgen-insensitive PC3 cell line derived from bone metastasis. The cells were cultured in Roswell Park Memorial Institute (RPMI) 1640 medium supplemented with 10% fetal bovine serum, 2% glutamine, and 1% penicillin and streptomycin (all Sigma-Aldrich, Taufkirchen, Germany). Cells were sub-cultured twice or thrice per week and maintained under standard culture conditions (37 °C, 5% CO_2, and 95% humidity).

2.2. Gas Plasma Treatment of Tumor Cells

The atmospheric pressure plasma jet kINPen (neoplas tools, Greifswald, Germany) was operated at 3.0 standard liters of argon per minute and 1.1 MHz. Quality control of the plasma generation was done as previously described [29]. A thorough description of the physical principles of plasma jets was given recently [30]. For cell growth kinetics, 3×10^4 (LNCaP) or 1×10^4 (PC3) cells in 500 µL of fully supplemented cell culture medium were seeded into 24-well plates and exposed to different plasma treatment times. For the generation of tumor cell culture supernatants, 2×10^5 cells in 500 µL of fully supplemented cell culture medium were seeded into 24-well plates and exposed to plasma. In initial experiments, argon gas only (plasma: off) treatment was used as a mock control. Argon is an inert gas with no known direct biological effects.

2.3. Tumor Cell Supernatant Incubation of Myeloid Cells and Setup of Co-Culture Assays

Four hours after plasma treatment, tumor cell culture supernatants were collected, centrifuged at 1000× g for 5 min to discard residual cells and debris, and stored at −20 °C until use. For experiments, 80 µL of this supernatant was added to wells of a flat-bottom 96-well plate (Eppendorf, Hamburg, Germany). To each well, 20 µL of a cell suspension containing 1×10^4 of either THP-1 or HL-60 cells was added. The myeloid cells were incubated for up to 96 h. The Eppendorf 96-well plates have an outer rim that was filled with deionized water to prevent excessive evaporation from the outer wells, as observed with extensive culture durations. For the co-culture of prostate cancer and myeloid cells, the

96-well plates were coated with 0.01% poly-l-lysine. 2.5×10^4 (LNCaP) or 1.25×10^4 (PC3) in 100 μL of cell culture medium cells were labeled with trace violet (Thermo Fisher Scientific, Dreieich, Germany) before being added to each well. The cells were exposed to the gas plasma and incubated for another 30 min in the incubator. Subsequently, 2.5×10^4 or 1.25×10^4 THP-1 or HL-60 cells were labeled with cell trace red (Thermo Fisher Scientific, Dreieich, Germany) and added to the tumor cells together with sytox green (Thermo Fisher Scientific, Dreieich, Germany) for the identification of terminally dead cells.

2.4. *Cell Counting, Metabolic Activity, and Cell Viability*

For the counting of cells in growth kinetic experiments, a CASY cell counter and analyzer model TT (Roche Applied Science, Mannheim, Germany) with a 150 μm capillary was used. For analyzing the metabolic activity of cells, resazurin (100 μM; Alfa Aesar, Haverhill, MA, USA) was added. The non-fluorescent resazurin is converted intracellularly only in metabolically active cells via an NADPH-dependent reduction to the fluorescent product resorufin. After 2 h of incubation, resorufin fluorescence was quantified using a microplate reader (F200; Tecan, Männedorf, Switzerland) at λ_{ex}535 nm and λ_{em}590 nm. To determine the absolute number of viable myeloid cells, flow cytometry (CytoFLEX S; Beckman-Coulter, Brea, CA, USA) was used. Live-dead discrimination was done by the addition of 4′,6-diamidino-2-phenylindole (DAPI, 1 μM; Sigma-Aldrich, Taufkirchen, Germany), and only live cells were gated for subsequent analysis.

2.5. *Analysis of Reactive Oxygen Species (ROS), Oxidation, and Mitochondria*

Hydrogen peroxide (H_2O_2) was quantified using the amplex ultra red (Thermo Fisher Scientific, Dreieich, Germany) assay according to the manufacturer's instructions as described in detail before [29]. To analyze the generation or deposition of ROS to the intracellular compartment, the myeloid cells were stained with either chloromethyl 2′,7′-dichlorodihydrofluorescein diacetate (CM-H_2DCF-DA, 1 μM; Thermo Fisher Scientific, Dreieich, Germany) converted intracellularly to the fluorescent dichlorodihydrofluorescein (DCF) or 3′-(p-aminophenyl) fluorescein (APF, 1 μM; (Thermo Fisher Scientific, Dreieich, Germany) before the addition of tumor cell culture supernatants to 2×10^4 labeled myeloid cells per well. Subsequently, the mean fluorescent intensity (MFI) of the dyes in the cells was analyzed by flow cytometry. To analyze the effect of the supernatants on the mitochondrial membrane potential, the myeloid cells were labeled with mitotracker orange (MTO, 1 μM; Thermo Fisher Scientific, Dreieich, Germany) before the addition of the tumor cell culture supernatants, and the MFI of MTO was analyzed using flow cytometry. In all assays, DAPI was used to gate on the live-cell population only for analysis of oxidation and mitochondria.

2.6. *Cell Surface Marker Analysis*

To analyze the expression of several myeloid cell surface markers simultaneously, multicolor flow cytometry was used. At 96 h, the myeloid cells were harvested into 96-well v-bottom plates using accutase. The cells were washed and stained against a selection of the monoclonal antibodies used in this study for targeting several cell surface receptors (Table 1). DAPI was added to quantify the MFI of each marker only in the live-cell population. After 15 min of incubation at room temperature in the dark, the cells were washed and acquired using a CytoFLEX S cytometer (Beckman-Coulter, Brea, CA, USA). The same procedure was performed for both the monoculture and co-culture assay systems. Discrimination of myeloid cells from tumor cells was done based on CD55 vs. HLA-ABC expression (THP-1 cells had higher baseline intensities for both parameters compared to prostate cancer cells) for THP-1 cells, and based on forward-scatter (tumor cells had higher signal intensities) vs. CD32 (HL-60 cells had higher signal intensities) for HL-60 cells.

2.7. *Quantification of Chemokines and Cytokines*

The quantification of chemokines and cytokines was performed in supernatants derived from prostate cancer cells (4 h after plasma treatment), myeloid cells incubated with and without supernatants of prostate cancer cells (at 96 h), and co-cultures of myeloid cells with cancer cells (at 96 h). To this end, a 12-target, bead-based sandwich immunoassay (LegendPLEX; BioLegend, London, UK) was utilized. Quantification was done against an internal standard for chemokine (C-C motif) ligand 17 (CCL17), chemokine (C-X-C motif) ligand 1 (CXCL1), CXCL10, interferon-gamma (IFNγ), interleukin 1-beta (IL1β), IL6, IL8, IL10, IL12p70, tumor growth factor-beta (TGFβ), tumor necrosis factor-alpha (TNFα), and vascular endothelial growth factor (VEGF).

Table 1. Antibodies and clones used in this study.

Target	Clone	Vendor
CD11b	ICRF44	BD
CD11c	118/A5	eBioscience
CD14	TÜK4	Miltenyi Biotec
CD16	CLB-gran11.5	BD
CD32	6C4	eBioscience
CD45RA	HI100	BD
CD55	IA10	BD
CD69	FN50	BD
CD71	M-A712	BD
CD163	GHI/61	BD
CD271	C40-1457	BD
HLA-ABC	G46-2.6	BD
P2Y2	E-3	SantaCruz

2.8. *Live-Cell High Content Imaging*

Prostate cancer cells were labeled with cell trace violet (1 µM; Thermo Fisher Scientific, Dreieich, Germany) and cultured together with cell trace red-labeled myeloid cells (1 µM; Thermo Fisher Scientific, Dreieich, Germany) as described above for co-culture. Sytox green (0.1 µM; Thermo Fisher Scientific, Dreieich, Germany) was added for the detection of terminally dead cells. Live-cell imaging was performed using an Operetta CLS (PerkinElmer, Hamburg, Germany) high content imaging device equipped with a CO_2 and temperature control chamber. Protection from evaporation was ensured since the rim of the 96-well plate was filled with deionized water. Imaging was performed at 24 h, 48 h, 72 h, and 96 h post plasma exposure using a 20× (NA 0.4) air objective (Zeiss, Jena, Germany) for 25 fields of view per well, time point, and channel. Channels were brightfield, digital phase contrast, cell trace violet (λ_{ex}405 nm, λ_{em}465 nm), sytox green (λ_{ex}475 nm, λ_{em}525 nm), and cell trace red (λ_{ex}630 nm, λ_{em}680 nm). A total of 25,000 images were acquired and analyzed in this study.

2.9. *Statistical and Software Analysis*

Statistical analysis was performed using prism 8.4 (Graphpad Software, San Diego, CA, USA). Either *t*-test, ratio-paired *t*-test, or one-way analysis of variances with *Dunnett* posthoc testing was used. The level of significance is indicated as follows: $p < 0.05$ (*), $p < 0.01$ (**), and $p < 0.001$ (***). At least three independent experiments with several technical replicates each were done for each assay. Flow cytometric analysis and fluorescence compensation were done using Kaluza analysis 2.1.1 (Beckman-Coulter, Brea, CA, USA). About 2000 individual measurements were analyzed in the present study. Quantitative image analysis was performed using Harmony 4.9 (PerkinElmer, Hamburg, Germany). The cells were segmented based on their digital phase-contrast signal, and the type of cell was determined by using a threshold on relative signal intensities from cell trace violet or cell trace red for prostate cancer and myeloid cells, respectively.

3. Results

3.1. *The Plasma Jet Treatment Decelerated Prostate Cancer Cell Growth*

Medical gas plasma treatment using the kINPen atmospheric pressure plasma jet was performed against prostate cancer cells in vitro. Two experimental schemes were used to identify the potential effects of that treatment in prostate cancer cells on myeloid cells (Figure 1a). In the first setup, prostate cancer cells were exposed to the plasma, and the cell culture supernatants were collected after four hours and added to myeloid cells. At 1 h, immediate effects such as ROS production and mitochondrial activity were determined. At 24 h, the metabolic activity and cell proliferation of the myeloid cells were investigated. At 96 h, the cell surface marker expression profile and cytokine release were assessed. In the direct co-culture approach, the prostate cancer cells were exposed to the plasma, and myeloid cells were added 1 h later. The co-cultures were imaged at 24 h, 48 h, 72 h, and 96 h, with an additional investigation of the cell surface marker profile of myeloid cells and the cytokine release in the supernatants of the co-cultures at 96 h. An image of the plasma treatment procedure is shown in Figure 1b. Based on previous studies, it was hypothesized that the plasma treatment decelerates tumor cell growth. Investigating the impact on the metabolic activity of the cancer cells 4 h after plasma treatment, the treatment led to a significant reduction in both prostate cancer cell lines investigated (Figure 1c). The plasma labeling denotes cells that were treated with the gas plasma directly. The argon control indicates conditions where the inert argon gas was not ignited into plasma but instead was merely blown onto the cell suspension as a mock treatment to exclude any biological effects of the noble gas alone. Similar to other plasma jets, the kINPen generates ROS in the plasma gas phase that subsequently diffuse into the treated liquid [31]. From there, the species further diffuse to cells [16] with some of the species possibly accumulating in the cytosol, as has been suggested for H_2O_2 passing through aquaporin channels in the membrane [32]. As the purpose was to investigate the immunomodulatory potential of supernatants from the untreated and plasma-treated prostate cancer cells, we also investigate residual ROS in these suspensions. Only a minor presence of ROS was observed in the plasma-treated LNCaP cultures 1 h after exposure, while residual ROS were absent for plasma-treated PC3 cells at 1 h (Figure 1d). At 4 h after plasma treatment, supernatants of the prostate cancer cells were collected and used for subsequent experiments with myeloid cells. Altogether, plasma treatment decelerated prostate cancer cell growth, and the majority of plasma-derived ROS have reacted with the cancer cells.

3.2. *Prostate Cancer Cell Supernatants Affected the Intracellular Oxidative Milieu and Metabolic Activity of Myeloid Cells*

The myeloid cell lines THP-1 and HL-60 were incubated with regular cell culture medium, supernatants from untreated prostate cancer cells, or supernatants from plasma-treated prostate cancer cells. Many cellular processes are subject to redox control, and we investigated the intracellular oxidative milieu of the myeloid cells using the redox-sensitive dyes DCF and APF that accumulate in the cytosol of cells upon staining. DCF is a pan-ROS sensor [33], while a certain degree of specificity towards three types of ROS (hypochlorous acid, peroxynitrite, and hydroxyl radical) is attributed to APF [34]. In THP-1 cells, the plasma-treated supernatants did not change DCF (Figure 2a) but APF intensities, indicative of more intracellular ROS (Figure 2b). The difference was significant when compared against both the use of untreated cancer cell supernatants and medium control only. Mitochondria can be a source of intracellular ROS following response to a stimulus. To this end, the mitochondrial membrane potential was investigated (Figure 2c) and a significant decrease was observed in response to incubation with all types of prostate cancer cell supernatants when compared against the medium control (Figure 2d). In HL-60 cells, and similarly to THP-1, DCF did not show significant changes in fluorescence for any of the conditions investigated (Figure 2e). In contrast to THP-1 cells, this was also the case for APF in HL-60 cells (Figure 2f). For mitochondria, however, a decrease in the membrane potential was observed (Figure 2g), which was significant following incubation with all types of prostate cancer cell supernatants when compared against the medium control (Figure 2h). These data suggest that the tumor cell supernatants are sensed and recognized by the myeloid cells, as evident in the mitochondria response and partially also in the ROS production. Mitochondrial activity is closely linked to metabolic activity. To this end, the effect of the cancer cell supernatants on the metabolic activity and proliferation of myeloid cells was investigated 24 h after incubation. For THP-1 cells, a significant increase in metabolic activity was observed (Figure 2i), while cell numbers differed only to a minor extent from that of the cell culture medium control (Figure 2j). The metabolic activity per cell was significantly increased for supernatants from PC3 cells (Figure 2k). For HL-60 cells, a decrease in metabolic activity was found, which was significant for supernatants from LNCaP cells (Figure 2l). The absolute cell numbers indicative of changes in proliferation did not vary to a great extent (Figure 2m), which in sum led to a significantly decreased metabolic activity per cell for supernatants from LNCaP on HL-60 cells (Figure 2n). In summary, it can be concluded for the metabolic activity assays that the supernatants of the prostate cancer cells had an impact on the myeloid cell activity, especially in THP-1 cells. The plasma treatment, however, did not provoke effects much different from those observed with supernatants from untreated prostate cancer cells.

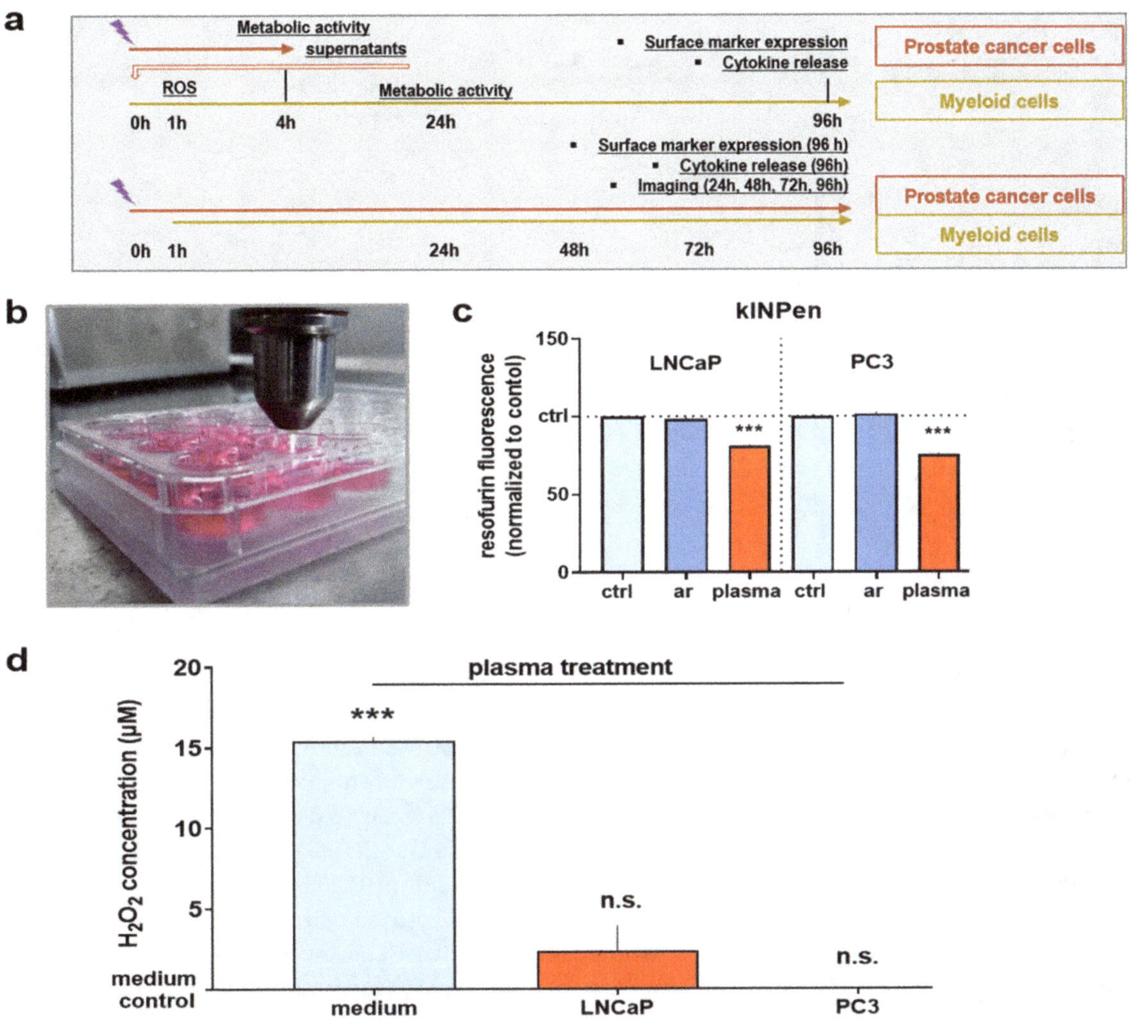

Figure 1. The kINPen gas plasma treatment decelerated the growth of prostate cancer cells. (**a**) outline of the experimental procedures and time lines performed in this study; (**b**) image of the kINPen atmospheric pressure argon plasma jet treating cells in a well of a 24-well plate; (**c**) metabolic activity of LNCaP and PC3 cells 4 h after kINPen plasma treatment; (**d**) residual reactive oxygen species (ROS) (H_2O_2) 1 h after kINPen plasma treatment of LNCaP and PC3 cells as well as fully supplemented cell culture medium compared to untreated medium control. Data are presented as mean ± SEM of three independent experiments. Statistical analysis was performed using one-way analysis of variances with posthoc testing after *Dunnett*. The level of significance is indicated as follows: $p < 0.001$ (***). ctrl = control; ar = inert argon gas mock treatment (plasma off).

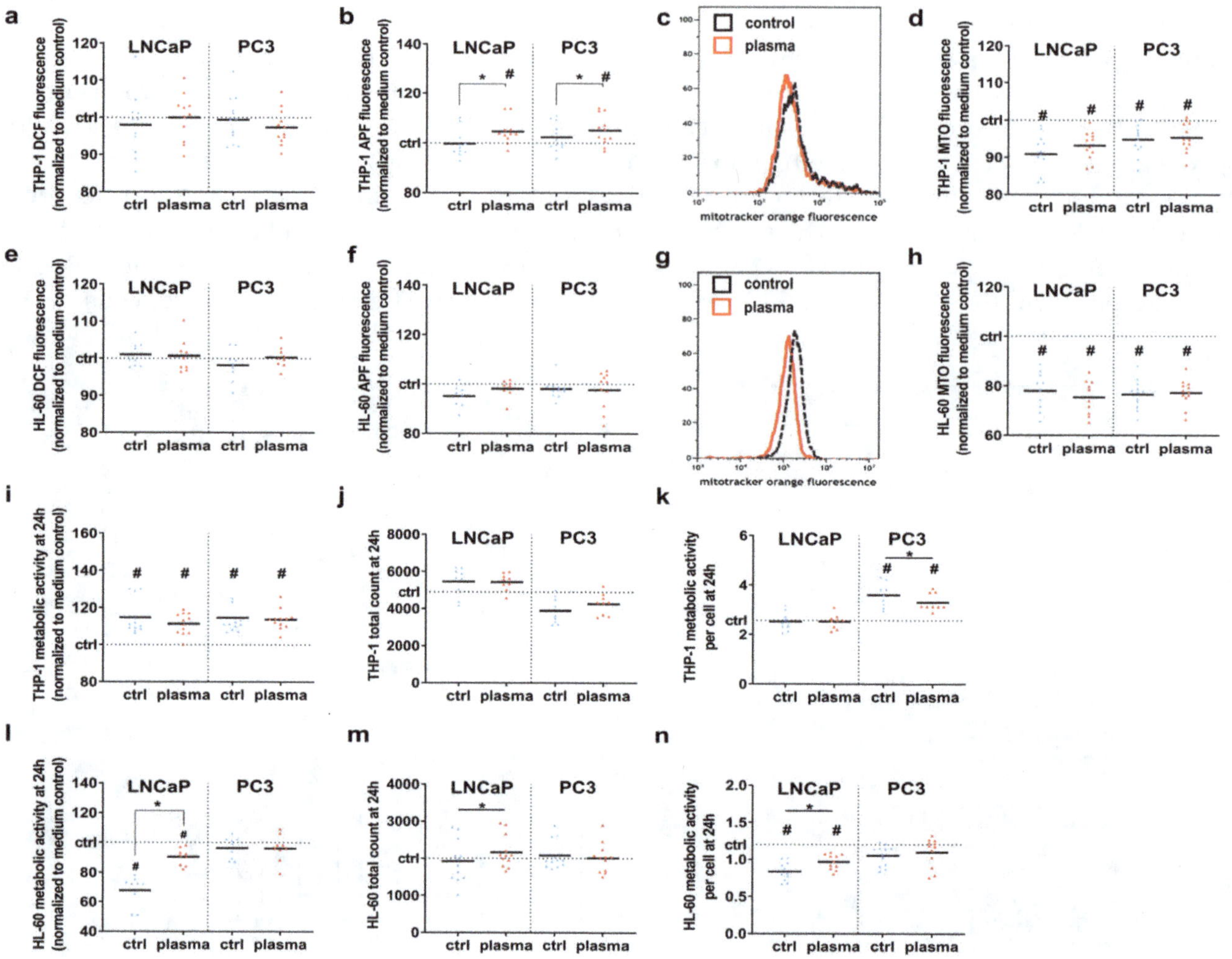

Figure 2. Oxidation and metabolic activity of myeloid cells following exposure to supernatants of kINPen plasma-treated prostate cancer cells. (**a**,**b**) normalized mean fluorescent intensity of THP-1 cells stained with (**a**) DCF and (**b**) APF, two dyes indicating intracellular redox changes, and exposed to control cell culture medium (dashed line), untreated supernatants of LNCaP or PC3 cells (ctrl), or supernatants of plasma-treated LNCaP or PC3 cells (plasma); (**c**,**d**) overlay histogram (**c**) and quantification (**d**) of THP-1 cells stained with mitotracker orange (MTO) and exposed to control cell culture medium (dashed line), untreated supernatants of LNCaP or PC3 cells (ctrl), or supernatants of plasma-treated LNCaP or PC3 cells (plasma); (**e**,**f**) normalized mean fluorescent intensity of HL-60 cells stained with (**e**) DCF and (**f**) APF, and exposed to control cell culture medium (dashed line), untreated supernatants of LNCaP or PC3 cells (ctrl), or supernatants of plasma-treated LNCaP or PC3 cells (plasma); (**g**,**h**) overlay histogram (**g**) and quantification (**h**) of HL-60 cells stained with mitotracker orange (MTO) and exposed to control cell culture medium (dashed line), untreated supernatants of LNCaP or PC3 cells (ctrl), or supernatants of plasma-treated LNCaP or PC3 cells (plasma); (**i**–**k**) metabolic activity (**i**) and absolute cell counts (**j**) as well as ratio of metabolic activity over cell counts (**k**) of THP-1 cells exposed to control cell culture medium (dashed line), untreated supernatants of LNCaP or PC3 cells (ctrl), or supernatants of plasma-treated LNCaP or PC3 cells (plasma) at 24 h; (**l**–**n**) metabolic activity (**l**) and absolute cell counts (**m**) as well as ratio of metabolic activity over cell counts (**n**) of HL-60 cells exposed to control cell culture medium (dashed line), untreated supernatants of LNCaP or PC3 cells (ctrl), or supernatants of plasma-treated LNCaP or PC3 cells (plasma) at 24 h. Data are presented as mean of three to five independent experiments with several technical replicates each. Statistical analysis was performed using ratio-paired t-test. ctrl = control. The level of significance is indicated as follows: $p < 0.05$ (*) or (#).

3.3. Plasma-Treated Prostate Cancer Cell Supernatants Promoted Immunomodulation in Myeloid Cells

The activity and differentiation of immune cells can be judged based on their expression of surface molecules and the release of pro and anti-inflammatory mediators such as chemokines and cytokines. Tumor cells secrete immunomodulatory mediators into the liquid environment. It was hypothesized that plasma treatment changes the composition and concentration of these mediators, which would lead to a different myeloid cell phenotype. To this end, we investigated both the surface receptor expression profile and chemokine and cytokine release of myeloid cells cultured with supernatants of untreated or plasma-treated prostate cancer cells at 96 h. Representative overlay flow cytometry histograms illustrate the distribution of a selection of surface receptor staining intensities (Figure 3a). For THP-1 cells, there was a significant increase observed with supernatants of plasma-treated over untreated prostate cancer cells for the cell surface markers CD11b, CD11c, and CD45RA (Figure 3c), pointing to increased activation of the cells. Other significant changes observed were a decrease for CD55, an increase in CD69, an increase in CD163, a decrease in CD271, and an increase in both cell size and granularity, while the expression of HLA-ABC was unchanged. Most changes were observed for supernatants of LNCaP (8 out of 10 markers), while significant changes with PC3 cell supernatants were less prominent (2 out of 10 markers). For HL-60 cells, incubation with supernatants of plasma-treated prostate cancer cells increased cell size and granularity but not the expression of CD11b, CD11c, CD14, CD32, and CD 71 (Figure 3c). For P2Y2, a significant increase was observed for plasma-treated LNCaP cells and their supernatants. In general, the markers investigated are indicators of myeloid cell activation and differentiation, and the effect of the supernatants was more pronounced in THP-1 compared to HL-60 cells.

To elucidate the immunomodulatory potential of the supernatants further, 12 different chemokines and cytokines were quantified in the prostate cancer cell as well as myeloid cell supernatants (Figure 4). This included CCL17 (Figure 4a), CXCL1 (Figure 4b), CXCL10 (Figure 4c), IFNγ (Figure 4d), IL1β (Figure 4e), IL6 (Figure 4f), IL8 (Figure 4g), IL10 (Figure 4h), IL12p70 (Figure 4i), TGFβ (Figure 4j), TNFα (Figure 4k), and VEGF (Figure 4l). For LNCaP, only IL6, IL1β, and VEGF were detectable, while for PC3 cells, CXCL1, CXCL10, and IL8 were also found. Plasma treatment significantly decreased the levels of all targets released from PC3 cells, while for LNCaP, a decrease in IL8 and VEGF, and an increase in IL6 was observed. VEGF is a tumor-promoting factor [35], and its decrease with plasma exposure points to a potentially beneficial effect. For THP-1 and HL-60 cells alone without tumor cell supernatants, only IL6, IL8, TGFβ, and VEGF were found. For the supernatants of plasma-treated vs. control prostate cancer cells alone (not added to myeloid cells), a significant decrease was found for IL6 and VEGF. The same was found for HL-60 cells, with other significant decreases observed for CXCL1 and IL8. Notably, there was no significant difference for supernatants of LNCaP cells incubated on myeloid cells. This is in line with the surface marker expression data for HL-60 but not THP-1 cells, with the latter showing a marked modulation that, however, was not reflected in the chemokine and cytokine spectrum.

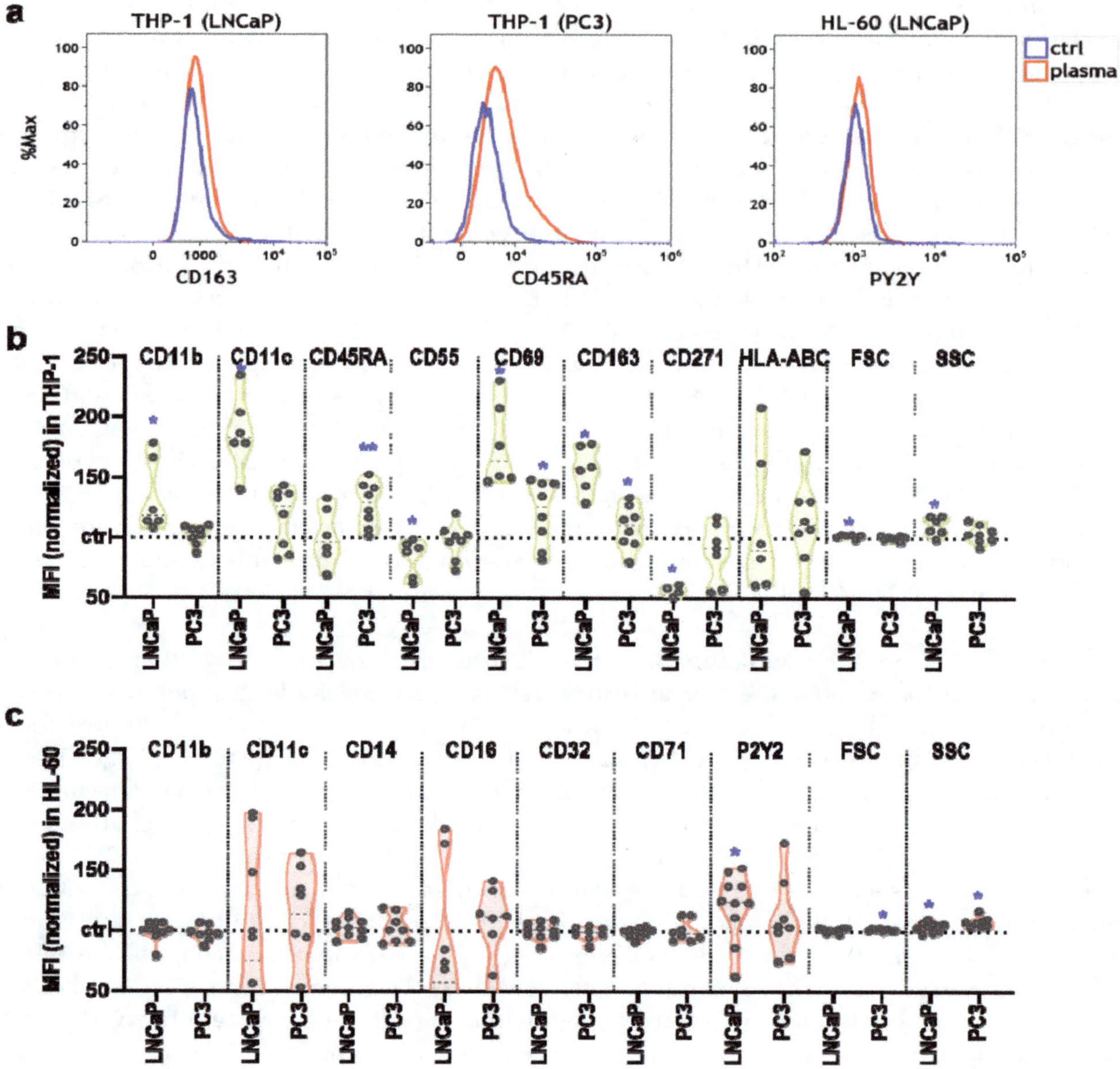

Figure 3. The cell surface marker profiles of myeloid cells following exposure to supernatants of kINPen plasma-treated prostate cancer cells. (**a**) representative overlay histograms of the flow cytometry analysis; (**b**) normalized mean fluorescent intensities at 96 h of CD11b, CD11c, CD45RA, CD55, CD69, CD163, CD271, HLA-ABC, forward scatter, and side scatter of THP-1 cells exposed to untreated supernatants of LNCaP or PC3 cells (ctrl, dashed line) or supernatants of plasma-treated LNCaP or PC3 cells; (**c**) normalized mean fluorescent intensities at 96 h of forward scatter, side scatter, CD11b, CD11c, CD14, CD16, CD32, CD71, and P2Y2 of HL-60 cells exposed to untreated supernatants of LNCaP or PC3 cells (ctrl, dashed line) or supernatants of plasma-treated LNCaP or PC3 cells. Data are presented as violin plots with median of three to five independent experiments with several technical replicates each. Statistical analysis was performed using ratio-paired t-test. The level of significance is indicated as follows: $p < 0.05$ (*) and $p < 0.01$ (**).

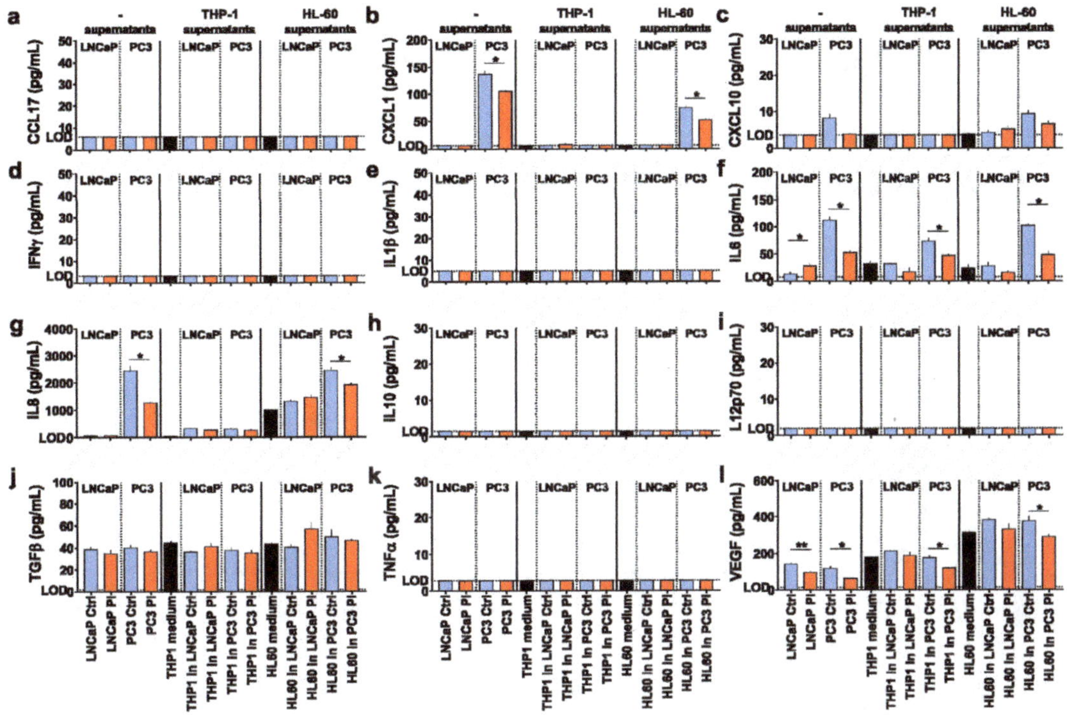

Figure 4. The cytokine profiles of myeloid cells following exposure to supernatants of kINPen plasma-treated prostate cancer cells. (**a–l**) quantification of the levels of CCL17 (**a**), CXCL1 (**b**), CXCL10 (**c**), IFNγ (**d**), IL1β (**e**), IL6 (**f**), IL8 (**g**), IL10 (**h**), IL12p70 (**i**), TGFβ (**j**), TNFα (**k**), and vascular endothelial growth factor (VEGF) (**l**) in untreated and kINPen plasma-treated prostate cancer cells (collected at 4 h) as well as THP-1 and HL-60 cell culture supernatants at 96 h with or without the addition of untreated or kINPen plasma-treated tumor cell culture supernatants. Data show the mean + S.E. of data from supernatants retrieved from at least three independent experiments. Statistical analysis was performed using ratio-paired *t*-test. The level of significance is indicated as follows: $p < 0.05$ (*) and $p < 0.01$ (**). ctrl = control; Pl = plasma; LOD = limit of detection.

3.4. Plasma-Treated Prostate Cancer Cells Promoted Myeloid Cell Differentiation and Tumor-Toxic Activity

The supernatants of untreated and plasma-treated prostate cancer cells led to a modulation of intracellular ROS, metabolic activity, cell surface marker expression, and cytokine release in myeloid cells. The next question was to understand how the direct cell–cell interaction between plasma-treated tumor and myeloid cells induced immunomodulation in myeloid cells. Representative overlay flow cytometry histograms illustrate the distribution of selected signal intensities (Figure 5a). A range of cell surface receptors was investigated in THP-1 cells, including CD11b, CD11c, CD45RA, CD55, CD69, CD163, HLA-ABC, as well as the percentage of viable cells, and cell size and granularity (Figure 5b). Except for CD11c and granularity, a significant increase was observed for all markers for one or both of the prostate cancer cell co-cultures investigated. This was even more pronounced in HL-60 cells, where all of the studied parameters were significantly regulated in one or both of the prostate cancer cell co-cultures. This included cell viability, size, and granularity as well as expression of CD11b, CD14, CD16, CD32, CD71, and P2Y2 (Figure 5c). This suggested that the plasma-induced prostate cancer cell death was driving differentiation-like responses in human myeloid as compared to those observed in co-cultures with untreated prostate cancer cells.

To investigate the immunomodulatory potential of the secretory products of plasma-treated prostate cancer cells on myeloid cells, chemokine, and cytokine quantification of the same targets as above (Figure 4) was performed (Figure 6). For THP-1 prostate cancer co-cultures, a significant regulation of CXCL1, CXCL10, IL8, and VEGF was observed with plasma treatment. For HL-60,

CXCL1, IL8, and VEGF were significantly regulated. With the exception of IL6 but not CXL10 being significantly regulated (Figure 4), the co-culture results were mostly congruent with those of the supernatant cultures except that the concentrations identified were several-fold higher.

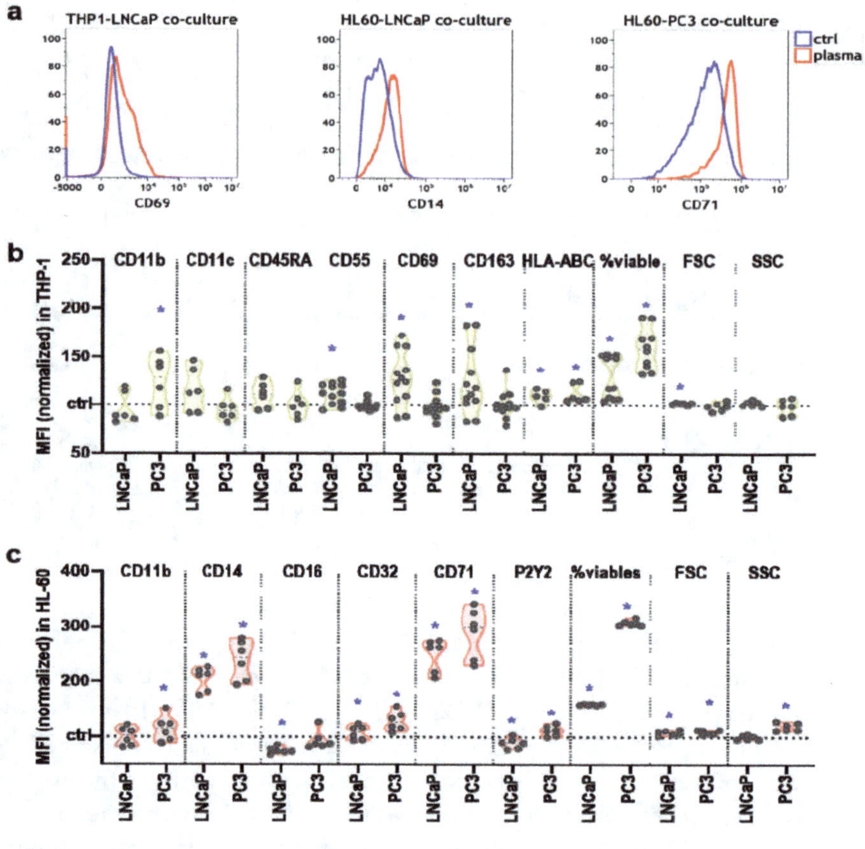

Figure 5. The cell surface marker profiles of myeloid cells co-cultured with kINPen plasma-treated prostate cancer cells. (**a**) representative overlay histograms of the flow cytometry analysis; (**b**) normalized mean fluorescent intensities at 96 h of CD11b, CD11c, CD45RA, CD55, CD69, CD163, HLA-ABC, viable cells, forward scatter, and side scatter of THP-1 cells cultured with either untreated (ctrl, dashed line) or plasma-treated LNCaP or PC3 cells; (**c**) normalized mean fluorescent intensities at 96 h of forward scatter, side scatter, CD11b, CD14, CD16, CD32, CD71, P2Y2, and viable HL-60 cells cultured with either untreated (ctrl, dashed line) or plasma-treated LNCaP or PC3 cells. Data are presented as violin plots and median of three to five independent experiments with several technical replicates each. Statistical analysis was performed using ratio-paired t-test. The level of significance is indicated as follows: $p < 0.05$ (*).

Plasma-treated prostate cancer cells co-cultured with myeloid cells spurred the expression of activation markers and modulated the cytokine release profiles. The next question was how these observed findings related to the tumor-toxic behavior of the myeloid cells. To this end, we utilized differential fluorescent staining and live-cell imaging to investigate several parameters using algorithm-driven quantitative image analysis. For THP-1 cells co-cultured with untreated or plasma-treated prostate cancer cells (Figure 7a), a significant increase in dead tumor cells was observed for LNCaP and in tendency also for PC3 cells (Figure 7b). Plasma-induced cell death is usually found within the first 24 h of incubation, but the number of dead cells in co-cultures was elevated throughout the time course up to 96 h, arguing for an auto-amplification of the tumor-toxic effects by THP-1 cells. This is supported by the enhanced phagocytosis index of THP-1 cells taking up prostate cancer cells (Figure 7c), suggesting clearance of dead prostate cancer cells. To investigate the effect of plasma treatment on the cell–cell interaction between the myeloid and prostate cancer cells, the mean distance between myeloid cells and cancer cells was calculated using algorithm-driven quantitative analysis.

A decrease was observed (Figure 7d), indicating that the myeloid cells migrated closer to the tumor cells if the latter had been plasma-treated. We also calculated the THP-1 over the prostate cancer cell ratio to model growth and survival behavior in the co-cultures over the 96 h (Figure 7e). An increase for the THP-1 to tumor cell ratio was observed for the plasma treatment conditions that was significant for the 72 h time point. The same set of experiments was performed for HL-60 cells (Figure 8a). Here, a continuous elevation of dead tumor cells was less pronounced as compared to the results obtained for THP-1 cells, but significantly elevated cell death was observed for PC3 cells (Figure 8b). Tumor cell phagocytosis was increased in tendency in all conditions, being significant only for 72 h in co-cultures with PC3 cells (Figure 8c). The HL-60 to tumor cell distance was unchanged for LNCaP but significantly decreased for PC3 cells, pointing to enhanced migratory activity of HL-60 to PC3 cells. Lastly, the HL-60 to tumor cell ratio was decreased for HL-60-LNCaP co-cultures in the plasma conditions (Figure 8d), suggesting a growth advantage of the cancer cells over the myeloid cells in this setting. For plasma-treated PC3 cells, however, a significant increase of that ratio was observed (Figure 8e). From the results and the amplitude of changes observed, it can be concluded that THP-1 cells actively responded to plasma-treated LNCaP cells and mostly also to PC3 cells in terms of favorable repression of tumor cell growth and survival. By contrast, HL-60 cells seemed not to propagate the cytotoxic plasma effects when co-cultured with plasma-treated LNCaP cells, while responses to PC3 prostate cancer cells were more pronounced.

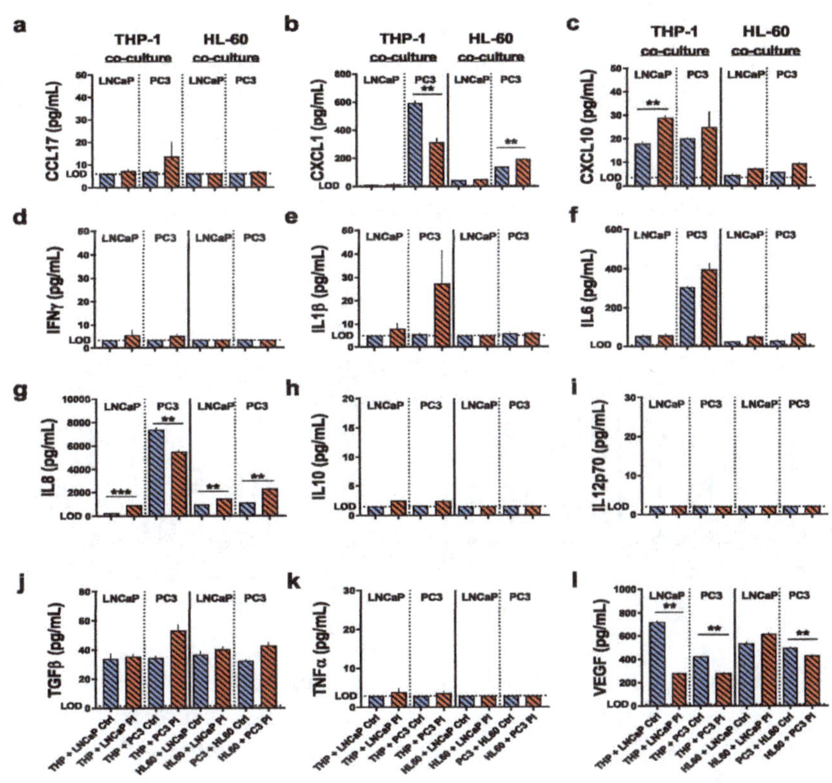

Figure 6. The cytokine profiles of myeloid cells co-cultured with kINPen plasma-treated prostate cancer cells. (**a–m**) quantification of the levels of CCL17 (**a**), CXCL1 (**b**), CXCL10 (**c**), IFNγ (**d**), IL1β (**e**), IL6 (**f**), IL8 (**g**), IL10 (**h**), IL12p70 (**i**), TGFβ (**j**), TNFα (**k**), and VEGF (**l**) retrieved after 96 h from supernatants of THP-1 and HL-60 cell culture supernatants co-cultured with untreated or kINPen plasma-treated prostate cancer cells. Data show the mean + S.E. of data from supernatants retrieved from at least three independent experiments. Statistical analysis was performed using ratio-paired *t*-test. The level of significance is indicated as follows: $p < 0.01$ (**) and $p < 0.001$ (***). ctrl = control; Pl = plasma; LOD = limit of detection.

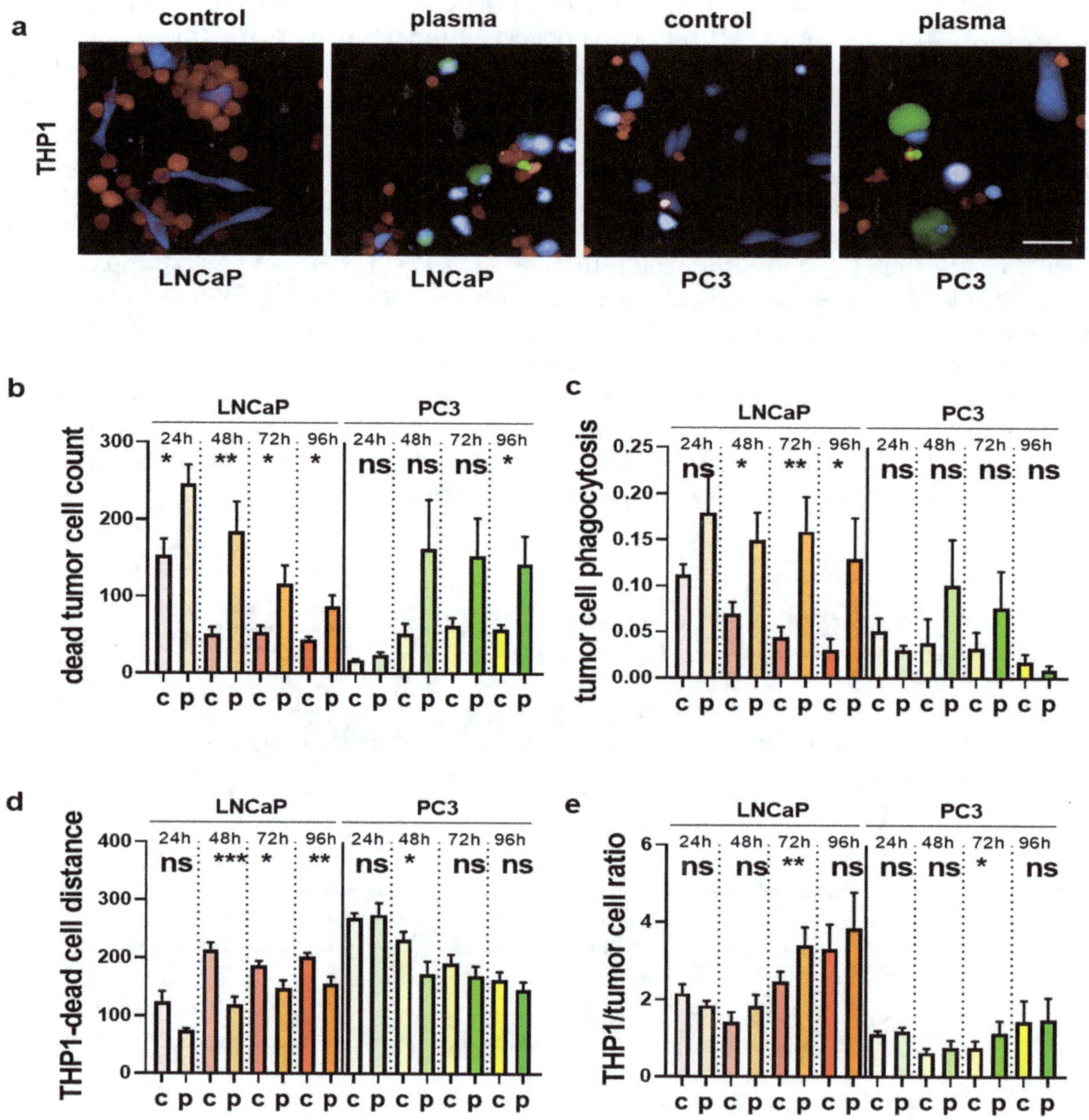

Figure 7. Quantitative image analysis of THP-1 cells co-cultured with plasma-treated prostate cancer cells. (**a**) prostate cancer cells were labeled with cell trace violet (blue) and were left untreated or exposed to the plasma, followed by addition of cell trace red (red) labeled THP-1 cells one hour later with sytox green (green) serving as dead cell marker; (**b**–**e**) algorithm-driven quantitative image analysis per image area of the total number of dead tumor cells per well (**b**), tumor cell phagocytosis (**c**), mean distance (in µm) between THP-1 cells and tumor cells (**d**), and the THP-1/tumor cell ratio (**e**). Data are presented as mean + S.E. of three independent experiments. Statistical analysis was performed using *t*-test. The level of significance is indicated as follows: $p < 0.05$ (*), $p < 0.01$ (**), and $p < 0.001$ (***). Scale bar is 50 µm. c = control; p = plasma; ns = not significant.

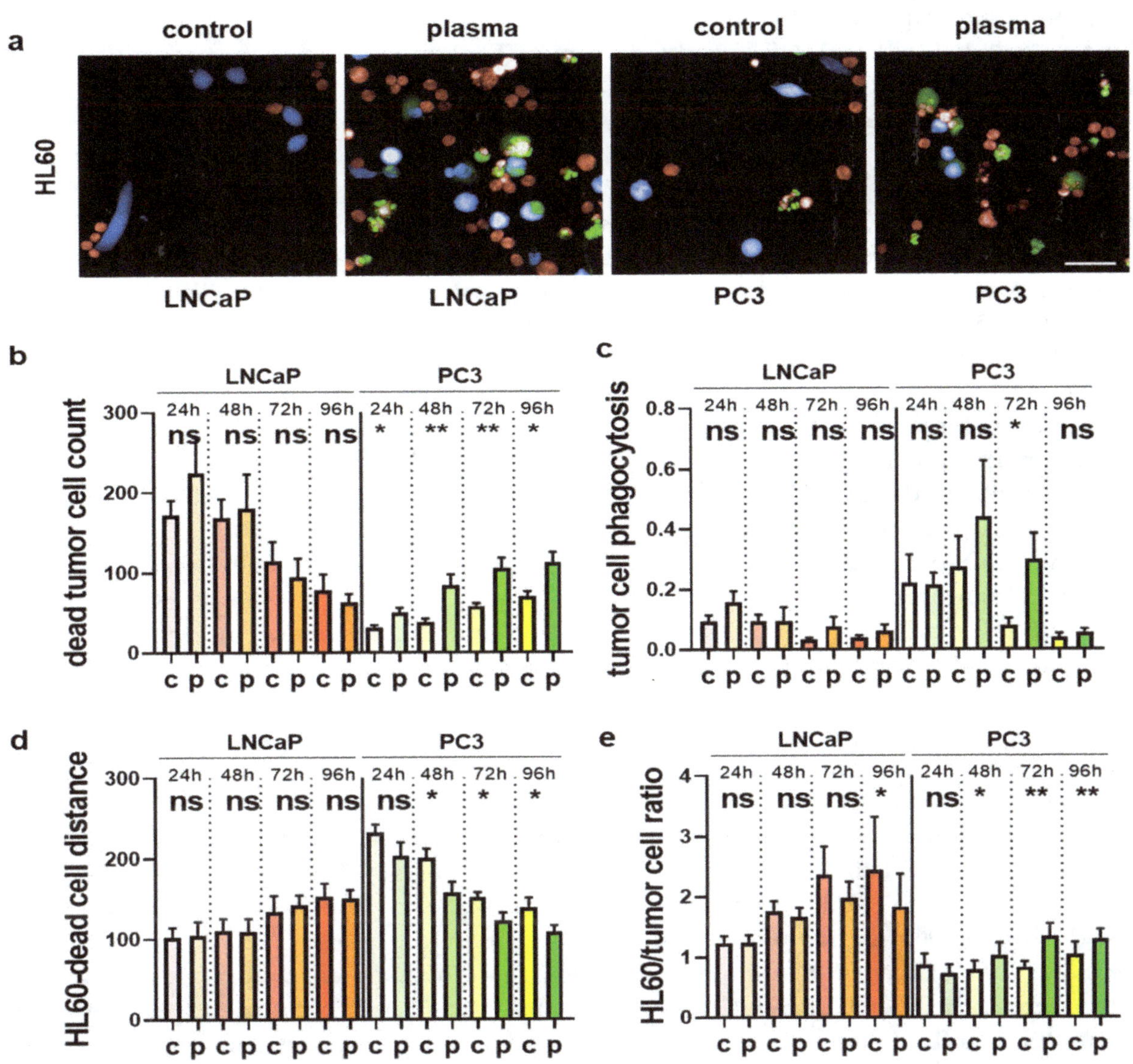

Figure 8. Quantitative image analysis of HL-60 cells co-cultured with plasma-treated prostate cancer cells. (**a**) prostate cancer cells were labeled with cell trace violet (blue) and were left untreated or exposed to the plasma, followed by addition of cell trace red (red) labeled HL-60 cells one hour later with sytox green (green) serving as dead cell marker; (**b**–**e**) algorithm-driven quantitative image analysis per image area of the total number of dead tumor cells per well (**b**), tumor cell phagocytosis (**c**), mean distance (in µm) between HL-60 cells and tumor cells (**d**), and the HL-60/tumor cell ratio (**e**). Data are presented as mean + S.E. of three independent experiments. Statistical analysis was performed using t-test. The level of significance is indicated as follows: $p < 0.05$ (*) and $p < 0.01$ (**). Scale bar is 50 µm. c = control; p = plasma; ns = not significant.

4. Discussion

The aim of this study was to investigate the immunomodulatory effects of two human plasma-treated prostate cancer cell lines towards two human myeloid cell lines in vitro. In comparison to untreated prostate cancer cells, the plasma treatment provoked significant changes in surface marker expression, cytokine release, and cytotoxic activity of the myeloid cells towards the tumor cells.

Supernatants of and co-cultures with cancer and myeloid cells modulated the surface marker expression and cytokine release upon plasma treatment. Nearly all markers investigated were significantly changed in at least one plasma-condition. Similar to CD11b [36,37], CD11c is associated with myeloid cell maturation in both cell types [38,39]. Interestingly, CD11c expression is regulated via AP-1 [40], a transcription factor we have recently described to respond to plasma-derived redox modulation in THP-1 cells [41]. Similarly, de novo CD69 expression is induced during the early activation of myeloid cells [42]. An increase in HLA-ABC expression, and cell size and granularity has been described for myeloid cell activation as well [39,43,44]. In terms of polarization, our marker panel was not designed for a detailed view on this aspect. We found CD163 to be increased in THP-1 cells, a marker commonly associated with M2 macrophages [45] but also found in inflammatory macrophages [46]. For HL-60 cells, a marked upregulation of CD14, CD32, and CD71 was determined only in co-culture experiments but not in the supernatant conditions related to plasma treatment. CD14, part of the LPS receptor, is associated with activation and maturation of HL-60 cells [47,48]. The same is true for CD32 [49,50], an Fcγ receptor. The transferrin receptor CD71, however, usually is internalized in activated HL-60 cells [51], leading to a decrease in surface receptor expression. Nevertheless, CD71 was also reported to be associated with macrophage development [52,53]. Although not all markers were regulated in each of the settings to a similar extent, a clear tendency towards activation was seen for both the supernatant and co-culture experiments. This suggests that plasma-induced cytotoxicity in prostate cancer cells leads to myeloid cell activation and differentiation, which often is also associated with changes in the secretion profile of cytokines [54].

Cytokine release reflects the inflammatory milieu generated by cells. In the supernatant experiments, a significant decrease was observed for CXCL1, IL6, IL8, and VEGF in the plasma conditions. Both IL6 and IL8 are markers of activation and differentiation in myeloid cells, at least when using mitogens as stimulating agents [55]. Along similar lines, an increase in CXCL1 is found in M1 macrophages differentiated with PMA [56]. Increased VEGF is observed in M2 polarized macrophages [57]. This suggests that activation and differentiation were lower with plasma treatment of tumor cells. However, our data also show that the levels of CXCL1, IL6, IL8, and VEGF for myeloid cells were similar to that of the tumor cells alone. Hence, it can be speculated that the contribution of the myeloid cells to the cytokine levels was not pronounced in the supernatant experiments, despite the significant changes in surface marker expression. One reason for that might be the low initial cell number used, and the ability of the cells to internalize extracellular chemokines and cytokines produced during the 96 h-period of incubation. The unchanged levels of TGFβ and the significantly decreased levels of VEGF need to be emphasized because both are potent tumor-promoting effectors in the tumor microenvironment [58,59]. This was also observed in our co-culture experiments, pointing to a potentially beneficial effect of the plasma treatment. In contrast to the supernatant experiment, a marked increase for CXCL10 was observed in the plasma conditions. CXCL10 is released upon tumor-toxic M1 macrophage polarization [60]. IL8 also was significantly increased in three of the four plasma-conditions investigated in the co-culture system, and the chemokine is associated with chemo-attraction of phagocytes and increased inflammation. Several other pro-inflammatory targets leading to potentially tumor-toxic effects, such as CCL17, IFNγ, IL1β, and IL6, were increased in tendency. It was previously reported that myeloid cell-derived IL1β abrogated proliferation of LNCaP cells [61]. Elevated IL6 levels may also be a consequence of a feedback loop between cancer and myeloid cells, as least for LNCaP and THP-1 cells [62]. Our data on only little induction of IL1β and IL6 in myeloid cells by prostate cancer cell-derived supernatants are in line with previous findings [63]. A noticeable increase in TNFα, as identified in a previous study being the principal tumor-toxic agent of macrophages [64], was not observed, however. In general, it is still a matter of debate of how macrophages exert cytotoxic effects against cancer cells. Besides the reported TNFα [63], activated M1 but not M2 macrophages were previously found to produce family 18 chitinases that pose potent tumor-toxic acitivity [65]. GM-CSF was also noted to enhance the tumor toxic activity of macrophages in vitro [66], while others reported direct antibody-dependent cytotoxicity and phagocytosis of tumor cells as mechanisms of action [67].

Nevertheless, we observed cytotoxic effects in prostate cancer cells co-cultured with myeloid cells that extended much further to the initial plasma treatment at 0 h. This was supported by our findings on increased tumor cell uptake and migration towards tumor cells in the plasma conditions. Enhanced tumor-toxicity in co-culture systems was reported before with activated vs. non-activated THP-1 cells [68]. In our study, we have several indices of macrophage activation through direct interaction with plasma-treated prostate cancer cells and their secretory products. At least in THP-1 cells, which showed greater tumor-toxic properties compared to HL-60 cells, this was underlined by an increase in intracellular ROS, a hallmark of monocyte activation [69] and macrophage differentiation [70]. Both myeloid cell lines showed a significant decrease in the mitochondrial membrane potential, which is associated with increased ROS production [71]. Whether the differences observed with DCF and APF point to differences in ROS entities and functional consequences in myeloid cells remains an open question. Notwithstanding, especially the HL-60-LNCaP co-culture showed the least cytotoxic activity on the cancer cells following plasma exposure. From the cell surface and cytokine data, it seems that this condition did not promote HL-60 differentiation and activation to such an extent as observed for PC3 cells, for instance. CD11b, CD14, CD32, CD71, and side scatter were markedly less increased in co-culture with LNCaP compared to PC3 cells, which supports this view. CD16, a marker of differentiation in HL-60 cells [72], even was significantly decreased in HL-60 cells co-cultured with plasma-treated LNCaP over untreated LNCaP cells. Moreover, this co-culture condition was the only one among the four myeloid cell–tumor cell co-cultures that showed a non-significant increase instead of a significant decrease in VEGF. VEGF can polarize myeloid cells towards a non-inflammatory phenotype [73], which might explain the lower killing activity of HL-60 cells in co-culture with the LNCaP cells.

Our study focused on myeloid cell activation and inflammatory markers in response to plasma-killed prostate cancer cells. Monocyte-to-macrophage differentiation and polarization, mainly to tumor-associated macrophages (TAM), is a clinical trait of prostate cancer associated with poor prognosis and tumor recurrence [74,75]. Mechanistically, tumors constantly attract myeloid cells from the bloodstream via several secreted factors, such as VEGF, which in turn drives pro-tumorigenic macrophage polarization [76,77]. Plasma treatment decreased VEGF levels in prostate cancer cell supernatants and co-cultures with myeloid cells. This suggests that plasma treatment might be able to modulate the tumor microenvironment in disfavor of tumor growth. This question should be targeted in future studies using more complex prostate tumor models and primary myeloid cells, both being a limitation of our study. Nevertheless, THP-1 cells are a good model for studying monocyte activation and differentiation, and compared genereally well with primary monocytes and especially M1 polarization in various previous studies [78–83]. Another question is how plasma treatment might be applied clinically in prostate cancer patients as already done in a fraction of head and neck cancer patients [84,85]. One option would be to expose to resection-margins after surgery to the gas plasma. This way, micro-metastasis that often lead to tumor recurrence might be decreased. This might also be achieved using plasma-treated liquids that were recently proposed to have anticancer properties as well [86,87]. Such liquid might also be injected directly into the tumor, using guided ultrasound microinjection. Tumor-toxic injections with plasma-treated liquids have been recently demonstrated in vivo using a model of pancreatic cancer [88]. Finally, another timely question is whether plasma treatment might be able to revert chemotherapeutic resistance, but apart from first studies on drug-transporter expression in cancer cells after plasma treatment [89,90], such studies are yet awaited to come.

5. Conclusions

Plasma-treated prostate cancer cells were superior in activating myeloid cells as compared to the tumor cell supernatants. This suggests that not only soluble factors but also direct cell–cell contact to be a factor promoting the cytotoxicity and differentiation of tumor-toxic myeloid cells. The promising nature of these findings in terms of gas plasma treatment of prostate cancer needs to be further evaluated in experimental disease models in the future.

Author Contributions: Conceptualization, S.B. and M.B.S.; Formal analysis, V.R. and E.F.; Funding acquisition, S.B. and A.M.; Investigation, V.R. and S.B.; Methodology, S.B. and E.F.; Project administration, S.B. and M.B.S.; Resources,

S.B. and A.M.; Supervision, S.B. and M.B.S.; Validation, V.R. and N.G.; Visualization, S.B.; Writing—original draft, S.B.; Writing—review and editing, V.R., E.F., N.G., A.M., and M.B.S. All authors have read and agreed to the published version of the manuscript.

Acknowledgments: The technical assistance of Felix Nießner is greatly appreciated.

References

1. Rawla, P. Epidemiology of Prostate Cancer. *World J. Oncol.* **2019**, *10*, 63–89. [CrossRef]
2. Pernar, C.H.; Ebot, E.M.; Wilson, K.M.; Mucci, L.A. The Epidemiology of Prostate Cancer. *Cold Spring Harb. Perspect. Med.* **2018**, *8*, a030361. [CrossRef] [PubMed]
3. Litwin, M.S.; Tan, H.J. The Diagnosis and Treatment of Prostate Cancer: A Review. *JAMA* **2017**, *317*, 2532–2542. [CrossRef]
4. Bekeschus, S.; Lippert, M.; Diepold, K.; Chiosis, G.; Seufferlein, T.; Azoitei, N. Physical plasma-triggered ROS induces tumor cell death upon cleavage of HSP90 chaperone. *Sci. Rep.* **2019**, *9*, 4112. [CrossRef] [PubMed]
5. Hasse, S.; Seebauer, C.; Wende, K.; Schmidt, A.; Metelmann, H.R.; von Woedtke, T.; Bekeschus, S. Cold Argon Plasma as Adjuvant Tumour Therapy on Progressive Head and Neck Cancer: A Preclinical Study. *Appl. Sci.* **2019**, *9*, 2061. [CrossRef]
6. Sagwal, S.K.; Pasqual-Melo, G.; Bodnar, Y.; Gandhirajan, R.K.; Bekeschus, S. Combination of chemotherapy and physical plasma elicits melanoma cell death via upregulation of SLC22A16. *Cell Death Dis.* **2018**, *9*, 1179. [CrossRef] [PubMed]
7. Pasqual-Melo, G.; Gandhirajan, R.K.; Stoffels, I.; Bekeschus, S. Targeting malignant melanoma with physical plasmas. *Clin. Plasma Med.* **2018**, *10*, 1–8. [CrossRef]
8. Kang, S.U.; Cho, J.H.; Chang, J.W.; Shin, Y.S.; Kim, K.I.; Park, J.K.; Yang, S.S.; Lee, J.S.; Moon, E.; Lee, K.; et al. Nonthermal plasma induces head and neck cancer cell death: The potential involvement of mitogen-activated protein kinase-dependent mitochondrial reactive oxygen species. *Cell Death Dis.* **2014**, *5*, e1056. [CrossRef] [PubMed]
9. Hou, J.; Ma, J.; Yu, K.N.; Li, W.; Cheng, C.; Bao, L.; Han, W. Non-thermal plasma treatment altered gene expression profiling in non-small-cell lung cancer A549 cells. *BMC Genom.* **2015**, *16*, 435. [CrossRef]
10. Bekeschus, S.; Lin, A.; Fridman, A.; Wende, K.; Weltmann, K.D.; Miller, V. A Comparison of Floating-Electrode DBD and kINPen Jet: Plasma Parameters to Achieve Similar Growth Reduction in Colon Cancer Cells Under Standardized Conditions. *Plasma Chem. Plasma Process.* **2018**, *38*, 1–12. [CrossRef]
11. Liedtke, K.R.; Diedrich, S.; Pati, O.; Freund, E.; Flieger, R.; Heidecke, C.D.; Partecke, L.I.; Bekeschus, S. Cold Physical Plasma Selectively Elicits Apoptosis in Murine Pancreatic Cancer Cells In Vitro and In Ovo. *Anticancer Res.* **2018**, *38*, 5655–5663. [CrossRef] [PubMed]
12. Gumbel, D.; Bekeschus, S.; Gelbrich, N.; Napp, M.; Ekkernkamp, A.; Kramer, A.; Stope, M.B. Cold Atmospheric Plasma in the Treatment of Osteosarcoma. *Int. J. Mol. Sci.* **2017**, *18*, 2004. [CrossRef] [PubMed]
13. Cheng, X.; Sherman, J.; Murphy, W.; Ratovitski, E.; Canady, J.; Keidar, M. The effect of tuning cold plasma composition on glioblastoma cell viability. *PLoS ONE* **2014**, *9*, e98652. [CrossRef] [PubMed]
14. Duan, J.; Lu, X.; He, G. The selective effect of plasma activated medium in an in vitro co-culture of liver cancer and normal cells. *J. Appl. Phys.* **2017**, *121*, 013302. [CrossRef]
15. Bekeschus, S.; Brüggemeier, J.; Hackbarth, C.; von Woedtke, T.; Partecke, L.-I.; van der Linde, J. Platelets are key in cold physical plasma-facilitated blood coagulation in mice. *Clin. Plasma Med.* **2017**, *7–8*, 58–65. [CrossRef]
16. Wende, K.; von Woedtke, T.; Weltmann, K.D.; Bekeschus, S. Chemistry and biochemistry of cold physical plasma derived reactive species in liquids. *Biol. Chem.* **2018**, *400*, 19–38. [CrossRef]
17. Privat-Maldonado, A.; Schmidt, A.; Lin, A.; Weltmann, K.D.; Wende, K.; Bogaerts, A.; Bekeschus, S. ROS from Physical Plasmas: Redox Chemistry for Biomedical Therapy. *Oxidative Med. Cell. Longev.* **2019**, *2019*, 9062098. [CrossRef]

18. Rödder, K.; Moritz, J.; Miller, V.; Weltmann, K.-D.; Metelmann, H.-R.; Gandhirajan, R.; Bekeschus, S. Activation of Murine Immune Cells upon Co-culture with Plasma-treated B16F10 Melanoma Cells. *Appl. Sci.* **2019**, *9*, 660. [CrossRef]
19. Lin, A.; Truong, B.; Patel, S.; Kaushik, N.; Choi, E.H.; Fridman, G.; Fridman, A.; Miller, V. Nanosecond-Pulsed DBD Plasma-Generated Reactive Oxygen Species Trigger Immunogenic Cell Death in A549 Lung Carcinoma Cells through Intracellular Oxidative Stress. *Int. J. Mol. Sci.* **2017**, *18*, 966. [CrossRef]
20. Bekeschus, S.; Clemen, R.; Metelmann, H.-R. Potentiating anti-tumor immunity with physical plasma. *Clin. Plasma Med.* **2018**, *12*, 17–22. [CrossRef]
21. Freund, E.; Liedtke, K.R.; van der Linde, J.; Metelmann, H.R.; Heidecke, C.D.; Partecke, L.I.; Bekeschus, S. Physical plasma-treated saline promotes an immunogenic phenotype in CT26 colon cancer cells in vitro and in vivo. *Sci. Rep.* **2019**, *9*, 634. [CrossRef] [PubMed]
22. Lin, A.; Gorbanev, Y.; De Backer, J.; Van Loenhout, J.; Van Boxem, W.; Lemiere, F.; Cos, P.; Dewilde, S.; Smits, E.; Bogaerts, A. Non-Thermal Plasma as a Unique Delivery System of Short-Lived Reactive Oxygen and Nitrogen Species for Immunogenic Cell Death in Melanoma Cells. *Adv. Sci.* **2019**, *6*, 1802062. [CrossRef] [PubMed]
23. Liedtke, K.R.; Freund, E.; Hackbarth, C.; Heidecke, C.-D.; Partecke, L.-I.; Bekeschus, S. A myeloid and lymphoid infiltrate in murine pancreatic tumors exposed to plasma-treated medium. *Clin. Plasma Med.* **2018**, *11*, 10–17. [CrossRef]
24. Obeid, M.; Tesniere, A.; Ghiringhelli, F.; Fimia, G.M.; Apetoh, L.; Perfettini, J.L.; Castedo, M.; Mignot, G.; Panaretakis, T.; Casares, N.; et al. Calreticulin exposure dictates the immunogenicity of cancer cell death. *Nat. Med.* **2007**, *13*, 54–61. [CrossRef] [PubMed]
25. Lin, A.G.; Xiang, B.; Merlino, D.J.; Baybutt, T.R.; Sahu, J.; Fridman, A.; Snook, A.E.; Miller, V. Non-thermal plasma induces immunogenic cell death in vivo in murine CT26 colorectal tumors. *Oncoimmunology* **2018**, *7*, e1484978. [CrossRef]
26. Bekeschus, S.; Clemen, R.; Nießner, F.; Sagwal, S.K.; Freund, E.; Schmidt, A. Medical Gas Plasma Jet Technology Targets Murine Melanoma in an Immunogenic Fashion. *Adv. Sci.* **2020**, 1903438. [CrossRef]
27. Bryant, G.; Wang, L.; Mulholland, D.J. Overcoming Oncogenic Mediated Tumor Immunity in Prostate Cancer. *Int. J. Mol. Sci.* **2017**, *18*, 1542. [CrossRef]
28. Sass, S.N.; Ramsey, K.D.; Egan, S.M.; Wang, J.; Cortes Gomez, E.; Gollnick, S.O. Tumor-associated myeloid cells promote tumorigenesis of non-tumorigenic human and murine prostatic epithelial cell lines. *Cancer Immunol. Immunother.* **2018**, *67*, 873–883. [CrossRef]
29. Bekeschus, S.; Schmidt, A.; Niessner, F.; Gerling, T.; Weltmann, K.D.; Wende, K. Basic Research in Plasma Medicine—A Throughput Approach from Liquids to Cells. *J. Vis. Exp.* **2017**, e56331. [CrossRef]
30. Winter, J.; Brandenburg, R.; Weltmann, K.D. Atmospheric pressure plasma jets: An overview of devices and new directions. *Plasma Sources Sci. Technol.* **2015**, *24*, 064001. [CrossRef]
31. Jablonowski, H.; von Woedtke, T. Research on plasma medicine-relevant plasma–liquid interaction: What happened in the past five years? *Clin. Plasma Med.* **2015**, *3*, 42–52. [CrossRef]
32. Bienert, G.P.; Moller, A.L.; Kristiansen, K.A.; Schulz, A.; Moller, I.M.; Schjoerring, J.K.; Jahn, T.P. Specific aquaporins facilitate the diffusion of hydrogen peroxide across membranes. *J. Biol. Chem.* **2007**, *282*, 1183–1192. [CrossRef] [PubMed]
33. Oparka, M.; Walczak, J.; Malinska, D.; van Oppen, L.; Szczepanowska, J.; Koopman, W.J.H.; Wieckowski, M.R. Quantifying ROS levels using CM-H2DCFDA and HyPer. *Methods* **2016**, *109*, 3–11. [CrossRef] [PubMed]
34. Setsukinai, K.; Urano, Y.; Kakinuma, K.; Majima, H.J.; Nagano, T. Development of novel fluorescence probes that can reliably detect reactive oxygen species and distinguish specific species. *J. Biol. Chem.* **2003**, *278*, 3170–3175. [CrossRef] [PubMed]
35. Linde, N.; Lederle, W.; Depner, S.; van Rooijen, N.; Gutschalk, C.M.; Mueller, M.M. Vascular endothelial growth factor-induced skin carcinogenesis depends on recruitment and alternative activation of macrophages. *J. Pathol.* **2012**, *227*, 17–28. [CrossRef]
36. Zeng, C.; Wang, W.; Yu, X.; Yang, L.; Chen, S.; Li, Y. Pathways related to PMA-differentiated THP1 human monocytic leukemia cells revealed by RNA-Seq. *Sci. China Life Sci.* **2015**, *58*, 1282–1287. [CrossRef]
37. Drayson, M.T.; Michell, R.H.; Durham, J.; Brown, G. Cell proliferation and CD11b expression are controlled independently during HL60 cell differentiation initiated by 1,25 alpha-dihydroxyvitamin D(3) or all-trans-retinoic acid. *Exp. Cell Res.* **2001**, *266*, 126–134. [CrossRef]

38. Barber, N.; Belov, L.; Christopherson, R.I. All-trans retinoic acid induces different immunophenotypic changes on human HL60 and NB4 myeloid leukaemias. *Leuk. Res.* **2008**, *32*, 315–322. [CrossRef]
39. Czernek, L.; Chworos, A.; Duechler, M. The Uptake of Extracellular Vesicles is Affected by the Differentiation Status of Myeloid Cells. *Scand. J. Immunol.* **2015**, *82*, 506–514. [CrossRef]
40. Lopez-Rodriguez, C.; Kluin-Nelemans, H.C.; Corbi, A.L. AP-1 regulates the basal and developmentally induced transcription of the CD11c leukocyte integrin gene. *J. Immunol.* **1996**, *156*, 3780–3787.
41. Schmidt, A.; Rodder, K.; Hasse, S.; Masur, K.; Toups, L.; Lillig, C.H.; von Woedtke, T.; Wende, K.; Bekeschus, S. Redox-regulation of activator protein 1 family members in blood cancer cell lines exposed to cold physical plasma-treated medium. *Plasma Process. Polym.* **2016**, *13*, 1179–1188. [CrossRef]
42. Vella, A.; Sartoris, S.; Bambara, L.; Ortolani, R.; Carletto, A.; Biasi, D.; Stefani, E.; Tridente, G. Cell contact-dependent PMN HLA-DR and CD69 membrane expression induced by autologous mono-lymphocytes and cell lines. *Inflammation* **2002**, *26*, 143–152. [CrossRef] [PubMed]
43. Bekeschus, S.; Schmidt, A.; Bethge, L.; Masur, K.; von Woedtke, T.; Hasse, S.; Wende, K. Redox Stimulation of Human THP-1 Monocytes in Response to Cold Physical Plasma. *Oxidative Med. Cell. Longev.* **2016**, *2016*, 5910695. [CrossRef] [PubMed]
44. Koski, G.K.; Schwartz, G.N.; Weng, D.E.; Gress, R.E.; Engels, F.H.; Tsokos, M.; Czerniecki, B.J.; Cohen, P.A. Calcium ionophore-treated myeloid cells acquire many dendritic cell characteristics independent of prior differentiation state, transformation status, or sensitivity to biologic agents. *Blood* **1999**, *94*, 1359–1371. [CrossRef]
45. Jensen, T.O.; Schmidt, H.; Moller, H.J.; Hoyer, M.; Maniecki, M.B.; Sjoegren, P.; Christensen, I.J.; Steiniche, T. Macrophage markers in serum and tumor have prognostic impact in American Joint Committee on Cancer stage I/II melanoma. *J. Clin. Oncol.* **2009**, *27*, 3330–3337. [CrossRef] [PubMed]
46. Choi, K.D.; Vodyanik, M.A.; Slukvin, I.I. Generation of mature human myelomonocytic cells through expansion and differentiation of pluripotent stem cell-derived lin-CD34+CD43+CD45+ progenitors. *J. Clin. Investig.* **2009**, *119*, 2818–2829. [CrossRef]
47. Gonelli, A.; Milani, D.; Rimondi, E.; Voltan, R.; Grill, V.; Celeghini, C. Activation of PKC-epsilon counteracts maturation and apoptosis of HL-60 myeloid leukemic cells in response to TNF family members. *Eur. J. Histochem.* **2009**, *53*, e21. [CrossRef]
48. Li, C.; Wang, Y.; Gao, L.; Zhang, J.; Shao, J.; Wang, S.; Feng, W.; Wang, X.; Li, M.; Chang, Z. Expression of toll-like receptors 2 and 4 and CD14 during differentiation of HL-60 cells induced by phorbol 12-myristate 13-acetate and 1 alpha, 25-dihydroxy-vitamin D(3). *Cell Growth Differ. Mol. Boil. J. Am. Assoc. Cancer Res.* **2002**, *13*, 27–38.
49. Brackman, D.; Lund-Johansen, F.; Aarskog, D. Expression of leukocyte differentiation antigens during the differentiation of HL-60 cells induced by 1,25-dihydroxyvitamin D3: Comparison with the maturation of normal monocytic and granulocytic bone marrow cells. *J. Leukoc. Biol.* **1995**, *58*, 547–555. [CrossRef]
50. Mollinedo, F.; Lopez-Perez, R.; Gajate, C. Differential gene expression patterns coupled to commitment and acquisition of phenotypic hallmarks during neutrophil differentiation of human leukaemia HL-60 cells. *Gene* **2008**, *419*, 16–26. [CrossRef]
51. Santos-Beneit, A.M.; Mollinedo, F. Expression of genes involved in initiation, regulation, and execution of apoptosis in human neutrophils and during neutrophil differentiation of HL-60 cells. *J. Leukoc. Biol.* **2000**, *67*, 712–724. [CrossRef] [PubMed]
52. Bauvois, B.; Laouar, A.; Rouillard, D.; Wietzerbin, J. Inhibition of Gamma-Glutamyl-Transpeptidase Activity at the Surface of Human Myeloid Cells Is Correlated with Macrophage Maturation and Transforming Growth-Factor-Beta Production. *Cell Growth Differ. Mol. Boil. J. Am. Assoc. Cancer Res.* **1995**, *6*, 1163–1170.
53. Van Leeuwen-Stok, E.A.; Jonkhoff, A.R.; Visser-Platier, A.W.; Drager, L.M.; Teule, G.J.; Huijgens, P.C.; Schuurhuis, G.J. Cell cycle dependency of 67gallium uptake and cytotoxicity in human cell lines of hematological malignancies. *Leuk Lymphoma* **1998**, *31*, 533–544. [CrossRef] [PubMed]
54. Italiani, P.; Boraschi, D. From Monocytes to M1/M2 Macrophages: Phenotypical vs. Functional Differentiation. *Front. Immunol.* **2014**, *5*, 514. [CrossRef] [PubMed]
55. Kang, J.W.; Park, Y.S.; Lee, D.H.; Kim, J.H.; Kim, M.S.; Bak, Y.; Hong, J.; Yoon, D.Y. Intracellular interaction of interleukin (IL)-32alpha with protein kinase Cepsilon (PKCepsilon) and STAT3 protein augments IL-6 production in THP-1 promonocytic cells. *J. Biol. Chem.* **2012**, *287*, 35556–35564. [CrossRef]

56. Chanput, W.; Mes, J.J.; Savelkoul, H.F.; Wichers, H.J. Characterization of polarized THP-1 macrophages and polarizing ability of LPS and food compounds. *Food Funct.* **2013**, *4*, 266–276. [CrossRef]
57. Wang, Y.; Chang, T.; Wu, T.; Xu, W.; Dou, G.; Wang, Y.; Guo, C. M2 macrophages promote vasculogenesis during retinal neovascularization by regulating bone marrow-derived cells via SDF-1/VEGF. *Cell Tissue Res.* **2020**. [CrossRef]
58. Mariathasan, S.; Turley, S.J.; Nickles, D.; Castiglioni, A.; Yuen, K.; Wang, Y.; Kadel, E.E., III; Koeppen, H.; Astarita, J.L.; Cubas, R.; et al. TGFbeta attenuates tumour response to PD-L1 blockade by contributing to exclusion of T cells. *Nature* **2018**, *554*, 544–548. [CrossRef]
59. Ichim, T.E.; Li, S.; Ma, H.; Yurova, Y.V.; Szymanski, J.S.; Patel, A.N.; Kesari, S.; Min, W.P.; Wagner, S.C. Induction of tumor inhibitory anti-angiogenic response through immunization with interferon Gamma primed placental endothelial cells: ValloVax. *J. Transl. Med.* **2015**, *13*, 90. [CrossRef]
60. Genin, M.; Clement, F.; Fattaccioli, A.; Raes, M.; Michiels, C. M1 and M2 macrophages derived from THP-1 cells differentially modulate the response of cancer cells to etoposide. *BMC Cancer* **2015**, *15*, 577. [CrossRef]
61. Culig, Z.; Hobisch, A.; Herold, M.; Hittmair, A.; Thurnher, M.; Eder, I.E.; Cronauer, M.V.; Rieser, C.; Ramoner, R.; Bartsch, G.; et al. Interleukin 1beta mediates the modulatory effects of monocytes on LNCaP human prostate cancer cells. *Br. J. Cancer* **1998**, *78*, 1004–1011. [CrossRef] [PubMed]
62. Lee, G.T.; Kwon, S.J.; Lee, J.H.; Jeon, S.S.; Jang, K.T.; Choi, H.Y.; Lee, H.M.; Kim, W.J.; Lee, D.H.; Kim, I.Y. Macrophages induce neuroendocrine differentiation of prostate cancer cells via BMP6-IL6 Loop. *Prostate* **2011**, *71*, 1525–1537. [CrossRef] [PubMed]
63. Lundholm, M.; Hagglof, C.; Wikberg, M.L.; Stattin, P.; Egevad, L.; Bergh, A.; Wikstrom, P.; Palmqvist, R.; Edin, S. Secreted Factors from Colorectal and Prostate Cancer Cells Skew the Immune Response in Opposite Directions. *Sci. Rep.* **2015**, *5*, 15651. [CrossRef] [PubMed]
64. Kaushik, N.K.; Kaushik, N.; Min, B.; Choi, K.H.; Hong, Y.J.; Miller, V.; Fridman, A.; Choi, E.H. Cytotoxic macrophage-released tumour necrosis factor-alpha (TNF-alpha) as a killing mechanism for cancer cell death after cold plasma activation. *J. Phys. D Appl. Phys.* **2016**, *49*, 084001. [CrossRef]
65. Pan, X.Q. The mechanism of the anticancer function of M1 macrophages and their use in the clinic. *Chin. J. Cancer* **2012**, *31*, 557–563. [CrossRef]
66. Grabstein, K.H.; Urdal, D.L.; Tushinski, R.J.; Mochizuki, D.Y.; Price, V.L.; Cantrell, M.A.; Gillis, S.; Conlon, P.J. Induction of macrophage tumoricidal activity by granulocyte-macrophage colony-stimulating factor. *Science* **1986**, *232*, 506–508. [CrossRef]
67. Oberling, F. Monocyte/macrophages as effector cells in cancer immunotherapy. *Transfus. Sci.* **1997**, *18*, 243–250. [CrossRef]
68. Kimball, E.S.; Kovacs, E.; Clark, M.C.; Schneider, C.R. Activation of cytokine production and adhesion molecule expression on THP-1 myelomonocytic cells by macrophage colony-stimulating factor in combination with interferon-gamma. *J. Leukoc. Biol.* **1995**, *58*, 585–594. [CrossRef]
69. Traore, K.; Trush, M.A.; George, M., Jr.; Spannhake, E.W.; Anderson, W.; Asseffa, A. Signal transduction of phorbol 12-myristate 13-acetate (PMA)-induced growth inhibition of human monocytic leukemia THP-1 cells is reactive oxygen dependent. *Leuk. Res.* **2005**, *29*, 863–879. [CrossRef]
70. Nishi, K.; Oda, T.; Takabuchi, S.; Oda, S.; Fukuda, K.; Adachi, T.; Semenza, G.L.; Shingu, K.; Hirota, K. LPS induces hypoxia-inducible factor 1 activation in macrophage-differentiated cells in a reactive oxygen species-dependent manner. *Antioxid. Redox Signal.* **2008**, *10*, 983–995. [CrossRef]
71. Calabriso, N.; Gnoni, A.; Stanca, E.; Cavallo, A.; Damiano, F.; Siculella, L.; Carluccio, M.A. Hydroxytyrosol Ameliorates Endothelial Function under Inflammatory Conditions by Preventing Mitochondrial Dysfunction. *Oxidative Med. Cell. Longev.* **2018**, *2018*, 9086947. [CrossRef] [PubMed]
72. Baran, Y.; Gur, B.; Kaya, P.; Ural, A.U.; Avcu, F.; Gunduz, U. Upregulation of multi drug resistance genes in doxorubicin resistant human acute myelogeneous leukemia cells and reversal of the resistance. *Hematology* **2007**, *12*, 511–517. [CrossRef]
73. Roskoski, R., Jr. Vascular endothelial growth factor (VEGF) signaling in tumor progression. *Crit. Rev. Oncol. Hematol.* **2007**, *62*, 179–213. [CrossRef] [PubMed]
74. Lanciotti, M.; Masieri, L.; Raspollini, M.R.; Minervini, A.; Mari, A.; Comito, G.; Giannoni, E.; Carini, M.; Chiarugi, P.; Serni, S. The role of M1 and M2 macrophages in prostate cancer in relation to extracapsular tumor extension and biochemical recurrence after radical prostatectomy. *BioMed Res. Int.* **2014**, *2014*, 486798. [CrossRef]

75. Nonomura, N.; Takayama, H.; Nakayama, M.; Nakai, Y.; Kawashima, A.; Mukai, M.; Nagahara, A.; Aozasa, K.; Tsujimura, A. Infiltration of tumour-associated macrophages in prostate biopsy specimens is predictive of disease progression after hormonal therapy for prostate cancer. *BJU Int.* **2011**, *107*, 1918–1922. [CrossRef] [PubMed]
76. Lee, H.W.; Choi, H.J.; Ha, S.J.; Lee, K.T.; Kwon, Y.G. Recruitment of monocytes/macrophages in different tumor microenvironments. *Biochim. Biophys. Acta* **2013**, *1835*, 170–179. [CrossRef] [PubMed]
77. Solis-Martinez, R.; Cancino-Marentes, M.; Hernandez-Flores, G.; Ortiz-Lazareno, P.; Mandujano-Alvarez, G.; Cruz-Galvez, C.; Sierra-Diaz, E.; Rodriguez-Padilla, C.; Jave-Suarez, L.F.; Aguilar-Lemarroy, A.; et al. Regulation of immunophenotype modulation of monocytes-macrophages from M1 into M2 by prostate cancer cell-culture supernatant via transcription factor STAT3. *Immunol. Lett.* **2018**, *196*, 140–148. [CrossRef] [PubMed]
78. Daigneault, M.; Preston, J.A.; Marriott, H.M.; Whyte, M.K.; Dockrell, D.H. The identification of markers of macrophage differentiation in PMA-stimulated THP-1 cells and monocyte-derived macrophages. *PLoS ONE* **2010**, *5*, e8668. [CrossRef] [PubMed]
79. Heil, T.L.; Volkmann, K.R.; Wataha, J.C.; Lockwood, P.E. Human peripheral blood monocytes versus THP-1 monocytes for in vitro biocompatibility testing of dental material components. *J. Oral Rehabil.* **2002**, *29*, 401–407. [CrossRef]
80. Chanput, W.; Mes, J.J.; Wichers, H.J. THP-1 cell line: An in vitro cell model for immune modulation approach. *Int. Immunopharmacol.* **2014**, *23*, 37–45. [CrossRef]
81. Stokes, R.W.; Doxsee, D. The receptor-mediated uptake, survival, replication, and drug sensitivity of Mycobacterium tuberculosis within the macrophage-like cell line THP-1: A comparison with human monocyte-derived macrophages. *Cell. Immunol.* **1999**, *197*, 1–9. [CrossRef]
82. Shiratori, H.; Feinweber, C.; Luckhardt, S.; Linke, B.; Resch, E.; Geisslinger, G.; Weigert, A.; Parnham, M.J. THP-1 and human peripheral blood mononuclear cell-derived macrophages differ in their capacity to polarize in vitro. *Mol. Immunol.* **2017**, *88*, 58–68. [CrossRef]
83. Qin, Z. The use of THP-1 cells as a model for mimicking the function and regulation of monocytes and macrophages in the vasculature. *Atherosclerosis* **2012**, *221*, 2–11. [CrossRef]
84. Metelmann, H.-R.; Seebauer, C.; Miller, V.; Fridman, A.; Bauer, G.; Graves, D.B.; Pouvesle, J.-M.; Rutkowski, R.; Schuster, M.; Bekeschus, S.; et al. Clinical experience with cold plasma in the treatment of locally advanced head and neck cancer. *Clin. Plasma Med.* **2018**, *9*, 6–13. [CrossRef]
85. Metelmann, H.R.; Seebauer, C.; Rutkowski, R.; Schuster, M.; Bekeschus, S.; Metelmann, P. Treating cancer with cold physical plasma: On the way to evidence-based medicine. *Contrib. Plasma Phys.* **2018**, *58*, 415–419. [CrossRef]
86. Liedtke, K.R.; Bekeschus, S.; Kaeding, A.; Hackbarth, C.; Kuehn, J.P.; Heidecke, C.D.; von Bernstorff, W.; von Woedtke, T.; Partecke, L.I. Non-thermal plasma-treated solution demonstrates antitumor activity against pancreatic cancer cells in vitro and in vivo. *Sci. Rep.* **2017**, *7*, 8319. [CrossRef] [PubMed]
87. Adachi, T.; Tanaka, H.; Nonomura, S.; Hara, H.; Kondo, S.; Hori, M. Plasma-activated medium induces A549 cell injury via a spiral apoptotic cascade involving the mitochondrial-nuclear network. *Free Radic. Biol. Med.* **2015**, *79*, 28–44. [CrossRef]
88. Sato, Y.; Yamada, S.; Takeda, S.; Hattori, N.; Nakamura, K.; Tanaka, H.; Mizuno, M.; Hori, M.; Kodera, Y. Effect of Plasma-Activated Lactated Ringer's Solution on Pancreatic Cancer Cells In Vitro and In Vivo. *Ann. Surg. Oncol.* **2018**, *25*, 299–307. [CrossRef]
89. Bekeschus, S.; Eisenmann, S.; Sagwal, S.K.; Bodnar, Y.; Moritz, J.; Poschkamp, B.; Stoffels, I.; Emmert, S.; Madesh, M.; Weltmann, K.D.; et al. xCT (SLC7A11) expression confers intrinsic resistance to physical plasma treatment in tumor cells. *Redox Biol.* **2020**, *30*, 101423. [CrossRef]
90. Liedtke, K.R.; Freund, E.; Hermes, M.; Oswald, S.; Heidecke, C.D.; Partecke, L.I.; Bekeschus, S. Gas Plasma-Conditioned Ringer's Lactate Enhances the Cytotoxic Activity of Cisplatin and Gemcitabine in Pancreatic Cancer In Vitro and In Ovo. *Cancers* **2020**, *12*, 123. [CrossRef]

5

Progression-Related Loss of Stromal Caveolin 1 Levels Mediates Radiation Resistance in Prostate Carcinoma via the Apoptosis Inhibitor TRIAP1

Julia Ketteler [1], Andrej Panic [1,2], Henning Reis [3], Alina Wittka [1], Patrick Maier [4], Carsten Herskind [4], Ernesto Yagüe [5], Verena Jendrossek [1] and Diana Klein [1,*]

1 Institute of Cell Biology (Cancer Research), University of Duisburg-Essen, University Hospital, Virchowstrasse 173, 45122 Essen, Germany; Julia.Ketteler@uk-essen.de (J.K.); Andrej.Panic@uk-essen.de (A.P.); Alina.Wittka@uk-essen.de (A.W.); Verena.Jendrossek@uk-essen.de (V.J.)

2 Department of Urology and Urooncology, University of Duisburg-Essen, University Hospital, Essen, Hufelandstr. 55, 45122 Essen, Germany

3 Institute of Pathology, University of Duisburg-Essen, University Hospital, Hufelandstr. 55, 45122 Essen, Germany; henning.reis@uk-essen.de

4 Department of Radiation Oncology, University Medical Center Mannheim, Medical Faculty Mannheim, Heidelberg University, Theodor-Kutzer-Ufer 1-3, 68167 Mannheim, Germany; Patrick.Maier@medma.uni-heidelberg.de (P.M.); Carsten.Herskind@medma.uni-heidelberg.de (C.H.)

5 Cancer Research Center, Division of Cancer, Imperial College London, Hammersmith Hospital Campus, London W12 0NN, UK; ernesto.yague@imperial.ac.uk

* Correspondence: Diana.Klein@uk-essen.de

Abstract: Tumour resistance to chemo- and radiotherapy, as well as molecularly targeted therapies, limits the effectiveness of current cancer treatments. We previously reported that the radiation response of human prostate tumours is critically regulated by CAV1 expression in stromal fibroblasts and that loss of stromal CAV1 expression in advanced tumour stages may contribute to tumour radiotherapy resistance. Here we investigated whether fibroblast secreted anti-apoptotic proteins could induce radiation resistance of prostate cancer cells in a CAV1-dependent manner and identified TRIAP1 (TP53 Regulated Inhibitor of Apoptosis 1) as a resistance-promoting CAV1-dependent factor. TRIAP1 expression and secretion was significantly higher in CAV1-deficient fibroblasts and secreted TRIAP1 was able to induce radiation resistance of PC3 and LNCaP prostate cancer cells *in vitro*, as well as of PC3 prostate xenografts derived from co-implantation of PC3 cells with TRIAP1-expressing fibroblasts *in vivo*. Immunohistochemical analyses of irradiated PC3 xenograft tumours, as well as of human prostate tissue specimen, confirmed that the characteristic alterations in stromal-epithelial CAV1 expression were accompanied by increased TRIAP1 levels after radiation in xenograft tumours and within advanced prostate cancer tissues, potentially mediating resistance to radiation treatment. In conclusion, we have determined the role of CAV1 alterations potentially induced by the CAV1-deficient, and more reactive, stroma in radio sensitivity of prostate carcinoma at a molecular level. We suggest that blocking TRIAP1 activity and thus avoiding drug resistance may offer a promising drug development strategy for inhibiting resistance-promoting CAV1-dependent signals.

Keywords: Caveolin-1; TP53-regulated inhibitor of apoptosis 1; tumour stroma; tumour microenvironment; fibroblast; CAF; resistance; prostate cancer; radiotherapy

---

1. Introduction

Cancer therapeutic resistance occurs through many different mechanisms, including specific genetic and epigenetic changes in the cancer cell itself and/or the respective microenvironment.

The tumour stroma is now recognized as a key player in cancer cell invasiveness, progression and therapy resistance [1–3]. Activated fibroblasts (cancer associated fibroblasts, CAF) are capable of preventing cancer cell apoptosis and induce proliferation, as well as invasion, of surrounding cancer cells via direct stroma-tumour interactions by secreting extracellular matrix components, growth factors and matrix metalloproteinases, among others [4]. Although the exact mechanisms of fibroblast activation remain elusive, the activation or repression of specific genes or proteins within stromal cells has also been correlated with clinical outcome. Within that scenario, the membrane protein Caveolin-1 (CAV1) came into focus as it is highly expressed in many tumours and high CAV1 levels in tumour cells, as well as the downregulation of stromal CAV1, were shown to correlate with cancer progression, invasion and metastasis and thus, a worse clinical outcome [4,5]. Loss of stromal CAV1 can even be used as a prognostic marker, for example, in breast and prostate cancer patients [6–9]. Data on the CAV1-dependent epithelia-stroma crosstalk indicates that stromal CAV1 possesses tumour-suppressor properties, whereas loss of stromal CAV1 fosters malignant epithelial cell resistance by evading apoptosis [5,10]. Stromal loss of CAV1 is particularly prominent in epithelial prostate cancer, where loss of CAV1 in the stroma correlates with high Gleason score, presence of metastasis and pronounced resistance to chemotherapy and radiotherapy [6,8,11,12]. However, a detailed mechanism explaining how CAV1-deficient fibroblasts foster therapy resistance of malignant prostate cancer cells remains elusive. An improved understanding of the molecular basis of resistance will inevitably lead to the clinical assessment of rational drug combinations in selected patient populations.

An important mechanism by which cancer cells acquire drug resistance is by apoptosis evasion [3] and apoptosis inhibiting proteins have been described in both the development of cancer [13] and drug resistance [14]. TP53-regulated inhibitor of apoptosis 1 (TRIAP1, also known as p53-inducible cell-survival factor, p53CSV) is a small, 76 amino acids long, evolutionary conserved protein [15]. TRIAP1 was first characterized as a p53-inducible cell survival factor [16]. A genetic screen further identified TRIAP1 as a pathway-specific regulator of the cellular response to p53 activation [17]. Mechanistically, TRIAP1 modulates the apoptotic pathways through interaction with HSP70, inhibition of the interaction of cytochrome c with the apoptotic protease activating factor 1 and activation of the downstream caspase-9, thus resulting in increased resistance by inhibiting apoptosis and permitting DNA damage repair [15,16].

In this study, we aimed at determining the role of CAV1 alterations potentially induced by stromal CAV1-deficiency for the radio sensitivity of prostate cancer on molecular level and identified the apoptosis inhibitor TRIAP1 as a CAV1-dependent fibroblastic secreted factor, fostering radio resistance of malignant prostate epithelial cells.

2. Material and Methods

2.1. Reagents and Antibodies

Antibodies against CAV1 (N-20: sc-894) and XIAP1 (H-202: sc-11426) were from Santa Cruz (Santa Cruz, CA, USA), against CCND1 (92G2: #2978) and GFP (D5.1: #2956) from Cell Signalling Technology (Denvers, MA, Germany), against PCNA (PC10: GTX20029) from GeneTex (Irvine, CA, USA), against TRIAP1 from ProteinTech Group [15351-1-AP, (WB) Rosemont, IL, USA] and LSBio [LS-C346398-100, (Histology) Seattle, WA, USA), against SURVIVIN (NB500-201) from Novus Biologicals (Centennial, CO, USA) and against β-actin (clone AC-74, A2228) from Sigma-Aldrich (St. Louis, MO, USA). The rabbit anti human ASA antibody BE#3 was previously described [18] and the goat anti ASM antibody was kindly provided by Prof. K. Sandhoff (Bonn, Germany) [19].

2.2. Cell Culture Conditions

The human prostate cancer cell lines PC3, DU145 and LNCaP, the human skin fibroblast cell line HS5 and the human prostate fibroblast cell line WPMY-1 were from ATCC (Manassas, VA, USA) and cultured in RPMI Medium (Gibco, ThermoFisher, Waltham, MA, USA) supplemented with 10% foetal bovine serum and 100 U/mL Penicillin/Streptomycin under standard cell culture conditions (37 °C, 5% CO_2, 95% humidity) and passaged every 3–4 days. CAV1 mRNA levels were down-regulated in indicated cells using shRNA technology as previously described [11,20,21]. For transient transfection of cells, human TRIAP1 cDNA with a C-terminal GFP-tag cloned into pCMV6-AC-GFP was used [15]. For selection of transfected cells, 500 µg/mL G418/Neomycin (Merck/Millipore, Darmstadt, Germany) was used.

2.3. *Irradiation of Cell Cultures*

Radiation was performed using the Isovolt-320-X-ray machine (Seifert-Pantak) at 320 kV, 10 mA with a 1.65 mm aluminium filter and a distance of about 500 mm to the object being irradiated [21]. The X-ray tube operated at 90 kV (~45 keV X-rays) and the dose rate was about 3 Gy/min [22].

2.4. *Colony Formation Assay*

The long-term survival assay was carried out by seeding 250 cells/well to 15.000 cells/well in a 6-well plate and irradiation at 0, 2, 4 and 6 Gy [11,21]. The plates were left to grow for 10 days into single colonies before they were fixed in 3.7% Formaldehyde (in PBS) and 70% Ethanol. Colonies were stained with 0.05% Coomassie Brilliant Blue for 1.5–3 h. Colonies ($\geq$50 cells) were counted at fivefold magnification under the microscope.

2.5. *Flow Cytometry Analysis*

To measure and quantify the DNA-fragmentation (apoptotic sub-G1 population), as well as to quantify the cell cycle phases, cells were incubated for 30 min at RT with a staining solution containing 0.1% (*w*/*v*) sodium citrate, 50 µg/mL PI and 0.05% (*v*/*v*) Triton X-100 (*v*/*v*). Afterwards they were analysed by flow cytometry (FACS Calibur, Becton Dickinson, Heidelberg, Germany; FL-2).

2.6. *Western Blotting*

Generation of whole cell lysates was carried out by scraping cells off into ice-cold RIPA buffer (150 mmol/L NaCl, 1% NP40, 0.5% sodium-deoxycholate, 0.1% sodium-dodecylsulfate, 50 mmol/L Tris/HCl, pH 8, 10 mmol/L NaF, 1 mmol/L Na3VO4) supplemented with Protease-Inhibitor cocktail (Roche). After 2–3 freeze and thaw cycles the protein content of the lysates was measured by using DC™ Protein Assay (Bio-Rad). 50 µg to 100 µg of protein were loaded onto SDS-PAGE electrophoresis. Western blots were done as previously described [11,21] and the indicated antibodies were used to detect protein expression.

2.7. *Real-Time Reverse Transcription PCR (qRT-PCR)*

RNA was isolated using RNeasy Mini Kit (74106, Qiagen, Hilden, Germany) according to the manufacturer's instruction and as previously described [11,22]. Expression levels were normalized to the reference gene (β-actin; set as 1) and were shown as relative quantification. Specific primers were designed using Primer 3 [23] based on available NCBI nucleotide CDS sequences. Cross-reaction of primers was excluded by comparison of the sequence of interest with the NCBI database (Blast 2.2, U.S. National Centre for Biotechnology Information, Bethesda, MD) and all primers used were intron-spanning. PCR products are 200-300 bp in size. qRT-PCR was carried out using specific oligonucleotide primers (s sense, as antisense; TRIAP1s AGGATTTCGCAAGTCCAGAA, TRIAP1as GCTGATTCCACCCAAGTAT; TAGLNs TCCAGACTGTTGACCTCTTTGA, TAGLNas CCTCTCCGC TCTAACTGATGAT; ACTA2s GCCGAGATCTCACTGACTACCT, ACTA2as TGATGCTGTTGTAGGTG GTTTC; TGFB1s CCCACAACGAAATCTATGACAA, TGFB1as AACTCCGGTGACATCAAAAGAT; LAMP1s CCTGCCTTTAAAGCTGCCAA; LAMP1as CACCTTCCACCTTGAAAGCC; LAMP2s ACC ACTGTGCCATCTCCTAC, LAMP2as TGCCTGTGGAGTGAGTTGTA; ACTINs GGCACCACACTTT CTACAATGA, ACTINas TCTCTTTAATGTCACGCACGAT) as previously described [11,22].

2.8. Conditioned Media

Cells were cultured in normal growth media until confluence. Cells were left non-irradiated or irradiated with 10 Gy, media were replaced and cells cultured in the presence of 0.5% foetal bovine serum for 48 h before collection of media. Control media were generated by incubating the same medium (containing 0.5% foetal bovine serum) without cells. Conditioned media were used as 1/1 mixture with normal growth medium [11,22].

2.9. Mouse Tumour Model

Mouse xenograft tumours were generated by subcutaneous injection of 0.5×10^6 PC3 cells (+/−CAV1) either alone or mixed with 0.5×10^6 WMPY-1 cells (+/−TRIAP1) onto the hind limb of male NMRI nude mice (total volume 50 µL) as previously described [11,21]. Animals of each experimental group received a single subcutaneous injection. For radiation therapy mice were anesthetized (2% isoflurane) and tumours were exposed to a single dose of 10 Gy ± 5% in 5 mm tissue depth (~1.53 Gy/min, 300 kV, filter: 0.5 mm Cu, 10 mA, focus distance: 60 cm) using a collimated beam with an XStrahl RS 320 cabinet irradiator (XStrahl Limited, Camberly, Surrey, Great Britain). Mouse experiments were carried out in strict accordance with the recommendations of the Guide for the Care and Use of Laboratory Animals of the German Government and they were approved by the Committee on the Ethics of Animal Experiments of the responsible authorities [Landesamt für Natur, Umwelt und Verbraucherschutz (LANUV), Regierungspräsidium Düsseldorf Az.8.87-50.10.37.09.187; Az.8.87-51.04.20.09.390; Az.84-02.04.2015.A586].

2.10. Human Tumour Tissue

Tissues from human prostate carcinomas were obtained during surgery according to local ethical and biohazard regulations. All experiments were performed in strict accordance with local guidelines and regulations. Resected tissue specimens were processed for pathological diagnostic routine in agreement with institutional standards and diagnoses were made based on current WHO and updated ISUP criteria [11,21]. All studies including human tissue samples were approved by the local ethics committee (Ethik-Kommission) of the University Hospital Essen (Nr. 10-4363 and 10-4051). Human tissue samples were analysed anonymously.

2.11. Immunohistochemistry and Immunofluorescence

Immunohistochemistry was performed on 4 µm slides of formalin-fixed and paraffin-embedded prostate tissues after performing a descending alcohol-series and incubation for 10 min to 20 min in target retrieval solution (Dako, Agilent, Santa Clara, CA, USA) [11]. After blocking of the slides with 2% NGS/PBS sections were incubated with primary antibodies o/n at 4 °C. Antigen were detected with horseradish-peroxidase conjugated secondary antibodies (1:250) and developed with DAB (Dako). Nuclei were counterstained using haematoxylin.

2.12. Statistical Analysis

If not otherwise indicated, data were obtained from 3 independent experiments with at least 2–3 mice each. Total mice numbers were stated in the figure legends. Statistical significance was evaluated by 1- or 2-way ANOVA followed by Tukey's or Bonferroni multiple comparisons post-hoc test and set at the level of $p \leq 0.05$. Data analysis was performed with Prism 5.0 software (GraphPad, La Jolla, CA, USA).

3. Results

3.1. Radioresistant (CAV1-Silenced) Fibroblasts Express and Secrete Anti-Apoptotic TRIAP1

We previously reported that CAV1-deficient fibroblasts foster radiation resistance of malignant prostate epithelial cells resulting in decreased apoptosis rates *in vitro* and *in vivo*, most likely via a paracrine mechanism of action [11]. Because we hypothesized that fibroblasts could allocate CAV1-dependent apoptosis inhibiting proteins to the tumour cells, we investigated the presence and expression levels of well-known resistance-associated anti-apoptotic proteins in stromal HS5 fibroblasts being either proficient [HS5(+)] for CAV1 or CAV1-deficient [HS5(-)] achieved by a shRNA-mediated knock-down (Figure 1). Of note, CAV1-silenced HS5 fibroblasts expressed significantly higher levels of TRIAP1 at both protein (Figure 1A) and mRNA level (Figure 1B). In addition, an increased CAV1-dependent TRIAP1 secretion was confirmed in cell culture supernatants of CAV1-silenced HS5 fibroblasts, which was accompanied by increased levels of lysosomal enzymes (acid sphingomyelinase, ASM and arylsulfatase A, ASA), which might be indicative for lysosomal exocytosis (Figure 1C).

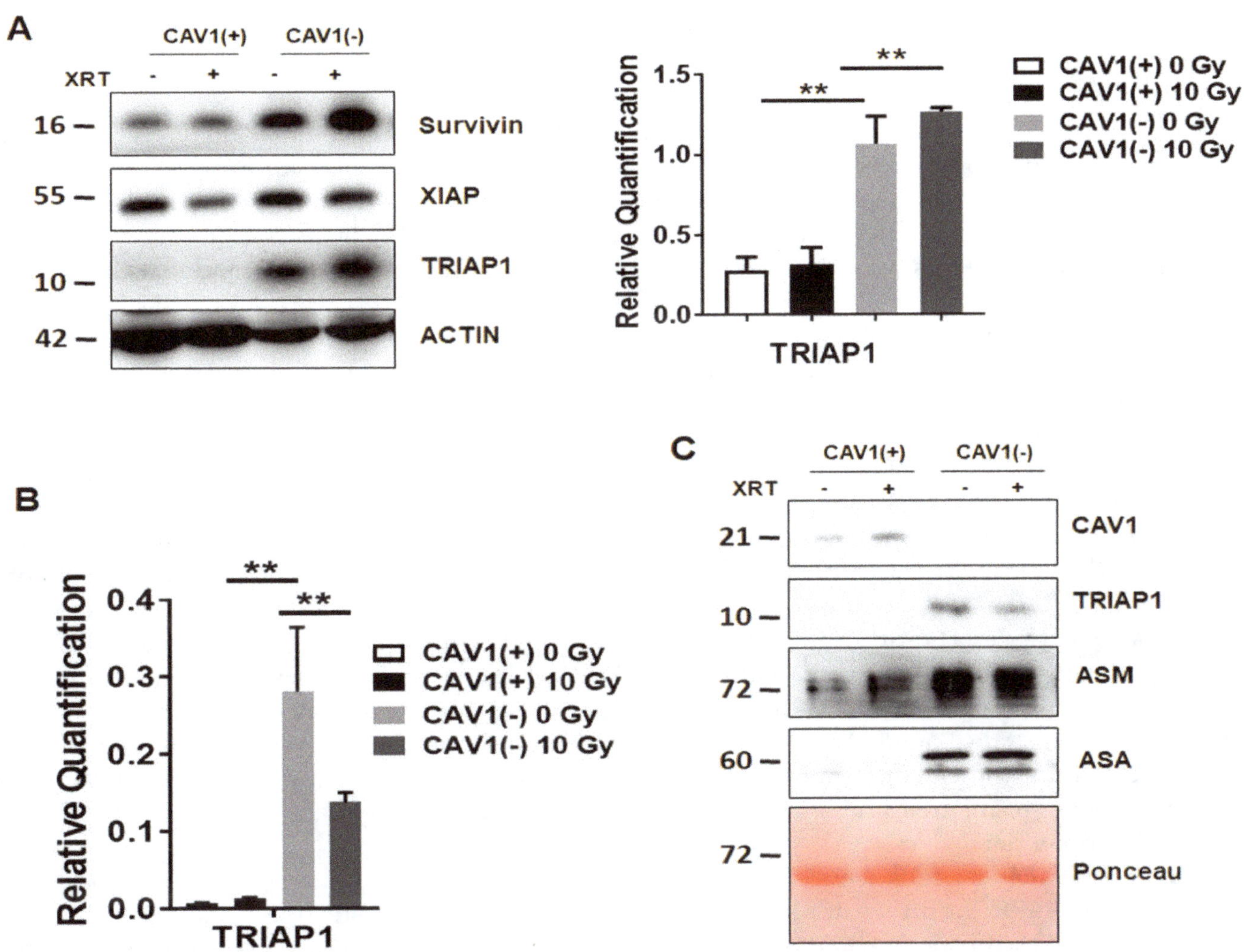

Figure 1. Radiation-resistant Caveolin-1 (CAV1)-silenced fibroblasts differentially express and secrete the apoptosis inhibiting protein TP53-regulated inhibitor of apoptosis 1 (TRIAP1).

Thus, increased expression and secretion of TRIAP1 by CAV1-silenced fibroblasts suggests that secreted TRIAP1 and then internalized by neighbouring prostate cancer cells, might account for the induced radiation resistance of these cells.

(A) Protein expression levels of apoptosis inhibiting proteins survivin, XIAP (X-linked inhibitor of apoptosis protein) and TRIAP1 were determined in CAV1-proficient [Cav1(+)] and CAV1-silenced [Cav1(-)] HS5 fibroblasts. Indicated proteins were analysed in whole protein lysates 96 h after radiation with 10 Gy by western blot analysis. Representative blots are shown. For TRIAP1 quantification, blots were analysed by densitometry and the respective signal was normalized to that from β-actin (n = 3–4 for each group). p-values were indicated: * $p \leq 0.05$, ** $p < 0.01$ by one-way ANOVA followed by post-hoc Tukey's test.

(B) qRT-PCR quantifications of TRIAP1 mRNA levels were performed 96 h post irradiation and shown as relative expression to β-actin mRNA. Data shown represent mean values ± SEM from 4 independent samples per group, each measured in duplicate. * $p \leq 0.05$, ** $p \leq 0.01$, by one-way ANOVA followed by post-hoc Tukey's test.

(C) TRIAP1 and lysosomal enzymes (ASM, acid sphingomyelinase and ASA, arylsulfatase A) secretion were further determined in cell culture supernatants derived from CAV1-silenced HS5(-) or control transfected CAV1-expressing HS5(+) fibroblasts with or without radiation treatment (10 Gy) using western blot analysis. Equal protein amounts (100 µg) were loaded. Ponceau S staining of transferred proteins was included as loading control.

3.2. Ectopic TRIAP1 Expression in Prostate Carcinoma Cells Induces Radiation Resistance

We previously have shown that cell culture supernatants of CAV1-silenced HS5 fibroblasts were able to induce radiation resistance of PC3 and LNCaP cells by decreased apoptosis [11]. We then investigated if the induced resistance of prostate cancer cells, after treatment with supernatants derived from CAV1-proficient or -deficient fibroblasts, led to higher TRIAP1 levels (not shown). However, no increased TRIAP1 levels were detectable in PC3, DU145 or LNCaP prostate carcinoma cells upon supernatants treatment most likely because the amount of tumour cell internalized TRIAP1 which was secreted from fibroblasts did not pass the threshold level of detection by western blot analysis. To provide the proof of principle that TRIAP1 mediates radiation resistance, the prostate cancer cells PC3 (p53 null), DU145 (p53 mutant) and LNCaP (p53 wild type) were transiently transfected with an expression vector encoding for human GFP-tagged TRIAP1 (Figure 2A). Empty vector transfected cells served as a control. Ectopic TRIAP1 expression resulted in decreased subG1 levels in PC3 and LNCaP cells 48 h after radiation with 10 Gy and thus increased resistance to radiation treatment. However, DU145 cells were not affected. Increased TRIAP1-levels were confirmed by western blot analysis (Figure 2B). Cell cycle analysis further revealed that ectopic TRIAP1 expression resulted in a slightly diminished G0/G1 subpopulation in PC3 cells upon radiation, while the proportion of cells in the G2/M phase increased (Figure 2C). The cell cycle of DU145 prostate carcinoma cells after TRIAP1 transfection was not affected upon radiation. Similar to PC3 cells, more TRIAP1-transfected LNCaP cells were in the G2/M phase after radiation as compared to control transfected cells. The proportions of respective cells in the S and <4n phase were rather low and not affected (not shown).

These results indicate that ectopic TRIAP1 expression mediates radiation resistance in a cell-type dependent manner and suggest that resistant prostate cancer cells will have an increased proliferation potential.

(A) Prostate cancer cells were transiently transfected with an expression vector encoding for human TRIAP1-GFP. Empty vector served as control. 24 h after transfection cells were irradiated with 0 or 10 Gy. The degree of apoptosis was quantified measuring the SubG1 fraction after radiation by flow cytometry analysis after additional 48 h of culture. Data shown represent mean values ± SEM from 4–5 independent samples per group measured in duplicates each. * $p \leq 0.05$, by two-tailed students t-test.

(B) Efficiency of TRIAP1-GFP expressions as analysed by Western blots. Representative blots from 3-4 independent experiments are shown. β-actin is used as a loading control. As additional control (Ctrl) mock transfected cells, which underwent the transfection procedure without an expression vector were shown.

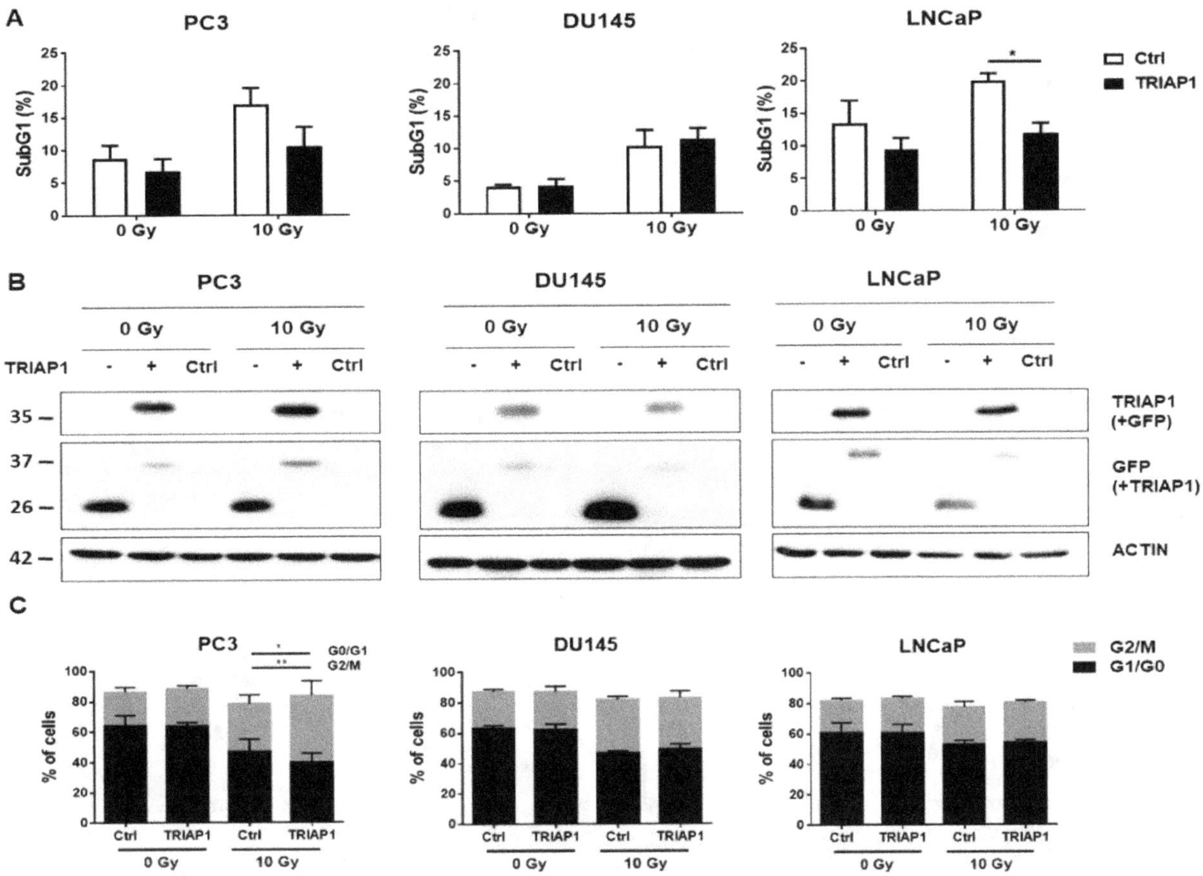

Figure 2. Ectopic TRIAP1 expression in prostate carcinoma cells results in radiation resistance.

(C) Cell cycle analysis of TRIAP1-GFP transfected prostate cancer cell lines was performed using Nicoletti/PI staining and flow cytometry. Empty vector transfected cells served as control. Data represent mean values ± SEM from 3–5 independent samples per group measured in duplicates each. ** $p \leq 0.01$, by two-way ANOVA followed by post-hoc Tukey's test.

3.3. Generation of Stromal Prostate Fibroblasts with Stable TRIAP1 Expression

Prior to investigating whether TRIAP1 derived from a reactive tumour stroma might account for the radiation resistance observed in PC3 xenografts *in vivo* [11], we assessed the suitability of another fibroblast cell type, prostate fibroblasts (WPMY-1) derived from healthy donors, to more closely mimic the human situation in future *in vivo* experiments (Figure 3).

Compared to normal HS5 fibroblasts, WPMY-1 prostate fibroblasts expressed less endogenous CAV1-expression levels (Figure 3A). Quantitative Real Time RT-PCR analysis of TRIAP1 expression levels as well as of reactive fibroblasts markers (ACTA2 and TAGLN) and tumour-promoting EMT factor transforming growth factor β (TGFB1) in WPMY-1 fibroblasts (+/− XRT) confirmed the more reactive phenotype of WPMY-1 with a less pronounced CAV1-content and furthermore of irradiated WPMY-1 fibroblasts (Figure 3B). In line with previous findings [11], colony formation assays indicated that WPMY-1 fibroblasts with a reduced CAV1 content were more resistant to radiation (Figure 3C). To further investigate a potential TRIAP1-mediated radiation resistance of prostate carcinoma cells caused by the stromal compartment, we generated TRIAP1-overexpressing WPMY-1 fibroblasts via transfection of WPMY-1 with an expression vector encoding for human TRIAP1 tagged with GFP (Figure 3D). Stably transfected and TRIAP1-GFP-sorted cells (via flow cytometry) were successfully generated. Increased TRIAP1 expression was confirmed by western blot analysis. It is worth noting that TRIAP1-overexpression did not alter CAV1 expression levels (Figure 3D). Ectopic TRIAP1 expression resulted in a significant reduced subG1 population upon radiation, which confirmed the resistant phenotype of TRIAP1-GFP expressing prostate fibroblasts (Figure 3E). TRIAP1 secretion from TRIAP1-GFP-expressing cells was confirmed by western blot analysis from cell culture supernatants and revealed an increased secretion upon radiation (Figure 3F).

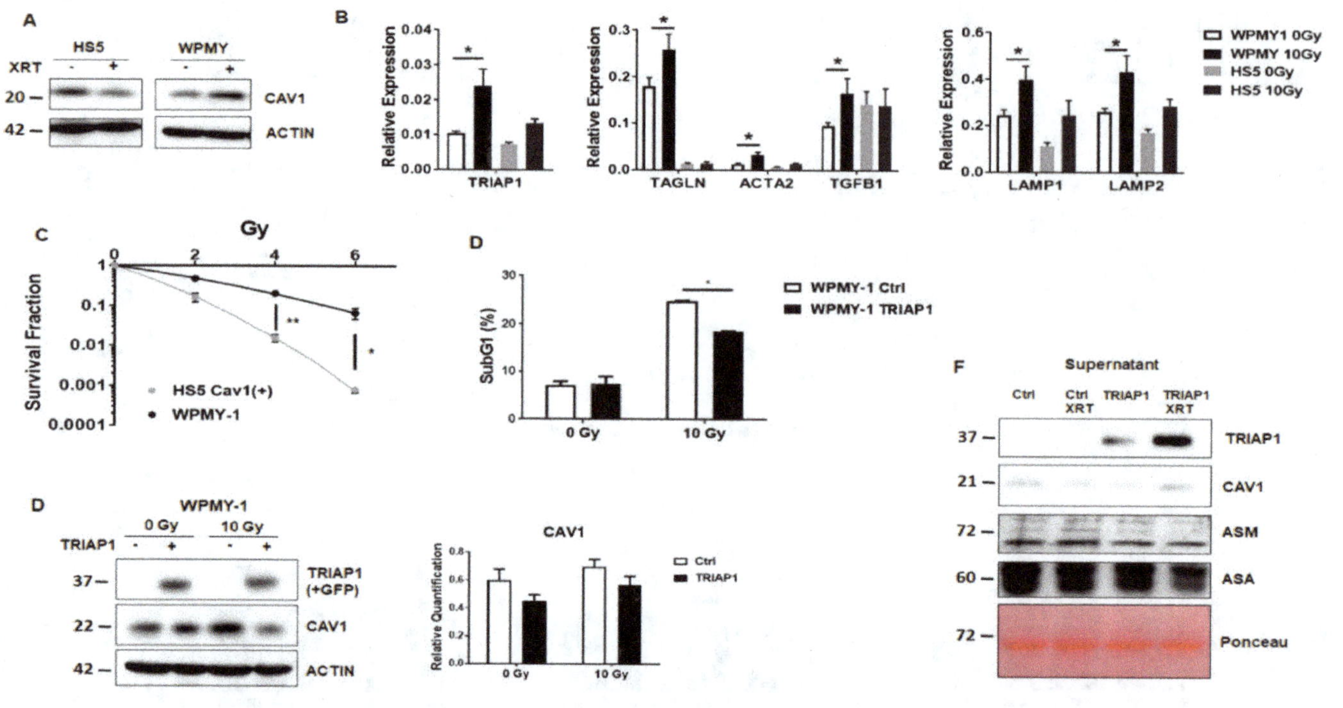

Figure 3. Characterization of the human prostate fibroblast cell line WPMY-1.

(A) CAV1 expression levels analysed by western blot in normal HS5 and prostate WPMY-1 fibroblasts, with or without radiation treatment with 10 Gy (96 h post irradiation). β-actin was included as loading control. Representative blots of at least three different experiments are shown.

(B) qRT-PCR quantifications of TRIAP1 mRNA levels, as well as reactive fibroblast markers, were performed 96 h post irradiation and shown as relative expression to β-actin mRNA. Data shown represent mean values ± SEM from 4-6 independent samples per group measured in duplicate each. * $p \leq 0.05$, ** $p \leq 0.01$, by two-way ANOVA followed by post-hoc Tukey's test.

(C) Colony formation assay of HS5 and WPMY-1 cells. Following irradiation (0–8 Gy) cells were further incubated for 10 days. Data show the surviving fractions from three independent experiments measured in triplicates each (means ± SD). *** $p \leq 0.005$, **** $p \leq 0.001$ by two-tailed students t-test.

(D) The degree of apoptosis was quantified measuring the SubG1 fraction 48 h after radiation by flow cytometry. Data shown indicate mean values ± SEM from 3 independent samples per group measured in duplicates each. * $p \leq 0.05$, by two-way ANOVA followed by post-hoc Tukey's test.

(E) WPMY-1 prostate fibroblasts were transfected with a TRIAP1-GFP encoding plasmid or empty vector in the control, selected with G418 and sorted via flow cytometry to select GFP-expressing cells. Expression levels of TRIAP1-GFP and CAV1 were confirmed by western blot analyses, with or without 10 Gy irradiations. Band signal intensity was quantified by densitometry and normalized to that from β-actin. Data represent mean ± SEM from three independent experiments. p-values were indicated: ** $p < 0.01$; *** $p \leq 0.005$ by two-way ANOVA followed by post-hoc Tukey's test.

(F) TRIAP1-GFP secretion in cell culture supernatants derived from TRIAP1-GFP or control, transfected WPMY-1 fibroblasts with or without radiation treatment (10 Gy) determined by western blot analysis. CAV1, ASM and ASA secretion levels were also investigated. Equal protein amounts (100 µg) were loaded. Ponceau S staining of transferred proteins was included as loading control.

3.4. TRIAP1-Expressing Stromal Fibroblasts Mediate Radiation Resistance

Next, we asked whether fibroblastic tumour stroma-derived TRIAP1 accounts for an increased radiation resistance in PC3 xenograft tumours [11]. To mimic the human situation we performed subcutaneous transplantations onto the hind limb of NMRI nude mice by injecting CAV1-silenced PC3(-) tumour cells in combination with control-transfected or TRIAP1-GFP-expressing WPMY-1 prostate fibroblasts (Figure 4). Prostate xenografts were implanted onto the hind limb of NMRI nude mice and were irradiated locally with a single dose of 10 Gy when the tumour reached a size of about 100 mm3 (around day 3). Tumour growth was determined by measuring the tumour volume 3 times a week (Figure 4A). Either co-implantation with WPMY-1 cells, control or TRIAP1-transfected, did not change tumour growth. The tumour growth delay after radiation was significantly decreased in PC3(-)-derived tumours co-implanted with TRIAP1-expressing WPMY-1. These tumours showed a significantly increased growth after radiation treatment when compared to PC3(-)-derived tumours co-implanted with control-transfected WPMY-1 as demonstrated by the reduced time to reach a four-fold tumour volume (Figure 2B). Immunohistochemistry using the proliferation marker PCNA (proliferating cell nuclear antigen) antibody further confirmed an increase in the proliferation rate of PC3(-) xenografts when co-implanted with TRIAP1-GFP expressing fibroblasts and a significantly decreased sensitivity to radiation treatment (Figure 4C). Thus, in line with the *in vitro* results, TRIAP1 derived from stromal fibroblasts is able to induce radiation resistance of prostate tumours.

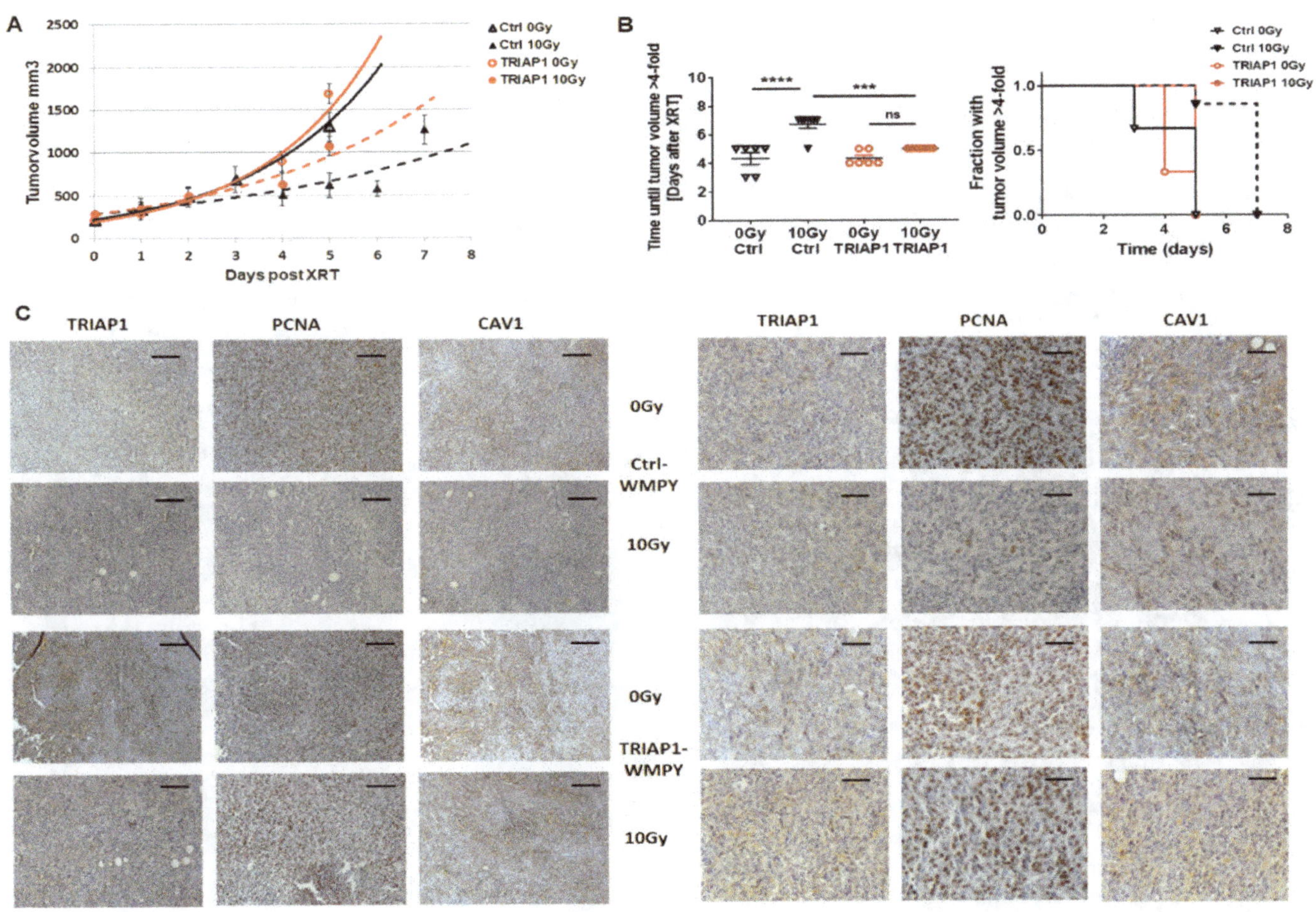

Figure 4. TRIAP1 expression in fibroblasts fosters radiation resistance in tumours derived from PC3 CAV1(-) cells.

(A) PC3 CAV1(-) cells were subcutaneously injected with either TRIAP1-expressing WPMY-1 or control-transfected fibroblasts (0.5 × 10^6 cells in total, ratio1/1) into the hind limb of NMRI nude mice. One set of animals from each group received a single tumour radiation dose of 10 Gy once its growth was easily detected (around day 3). Tumour volume was determined at indicated time points using a sliding calliper. Data are presented as mean +/− SEM from 3 independent experiments (26 mice in total: Ctrl 0 Gy n = 6; Ctrl 10 Gy n = 7; TRIAP1 0 Gy n = 6; TRIAP1 10 Gy n = 7).

(B) Tumour growth (*left panel*) and respective computed median growth delay (*right panel*) were determined as time (days) until a four-fold tumour volume was reached. *** $p < 0.005$; **** $p < 0.001$ by one-way ANOVA followed by post-hoc Tukey's test.

(C) Immunohistochemical analysis of TRIAP1, PCNA and CAV1 in isolated PC3 xenograft tumours. Sections were counterstained using haematoxylin. Representative images are shown. Magnification 200×, scale bar 50 μm (left panel), magnification 400×, scale bar 20 μm (right panel).

3.5. Human Advanced Prostate Cancer Specimens Were Characterized by an Increased TRIAP1-Immunoreactivity Indicating Radiation Resistance

As loss of stromal CAV1 is paralleled by a radiation-resistance promoting reactive tumour stroma in human prostate tissue specimens [11,21], we decided to investigate TRIAP1 expression levels, as well as the respective stromal-epithelial TRIAP1 distribution, in human prostate tissue specimens by immunohistochemistry. TRIAP1 expression in prostate epithelial cells increased with higher Gleason scores, that is, lower tumour differentiation (Figure 5). Furthermore, stromal cells of tumour samples tended to be more intensively stained in cases with higher Gleason grade (Figure 5). These results indicate that increased, potentially fibroblast-derived, TRIAP1 has implications for prostate carcinoma progression and therapy resistance.

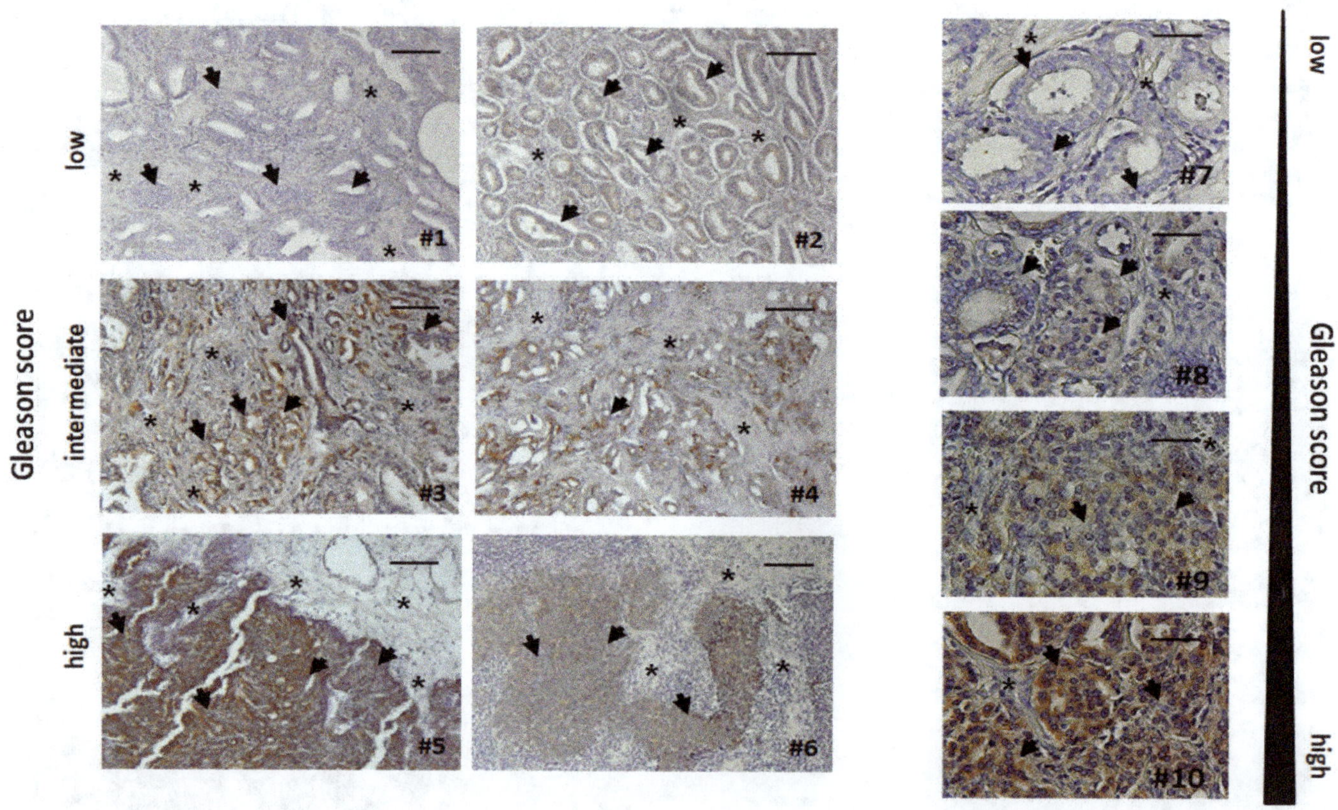

Figure 5. Immunohistochemical analysis of TRIAP1 expression levels in human prostate cancer tissues.

Paraffin-sections of human prostate cancers were stained for TRIAP1. Gleason grading scores were divided into low (Gleason Score ≥ 6, Grade group 1), intermediate (Gleason Score 7 (a/b), Grade groups 2 & 3) and high scores (Gleason Score ≥ 8, Grade groups 4 & 5). Asterisks mark stromal compartments and bold arrows point to epithelial structures. Sections were counterstained using haematoxylin. Representative images are shown. Magnification 200x. Right panel: higher magnification images: 400x.

4. Discussion

Tumours are able to diversify their microenvironment and consequently the altered, more reactive tumour microenvironment can modulate the response of tumours to therapy treatment [24–28]. Herein, activated stromal cells and in particular activated fibroblasts/CAF can mediate therapy resistance of malignant epithelial cells in a CAV1-dependent fashion [4,5,29,30].

CAV1-dependent stromal-epithelial crosstalk in tumours with the potential to induce resistance includes processes such as autophagy or the 'reverse Warburg effect' [31,32]. For example, a more reactive stromal phenotype following a decrease of CAV1 expression by lysosomal degradation in fibroblasts was observed when cancer cells induced oxidative stress in the tumour-microenvironment [33]. In turn, downregulation of CAV1 in fibroblasts leads to increased oxidative metabolism in cancer cells, fostering cell resistance [29]. Importantly, extrinsic factors from the microenvironment and in particular from activated fibroblasts/CAF, may drive resistance in a non-tumour cell autonomous mechanism [34,35]. In line with these findings, we have recently shown that CAV1-deficient fibroblasts mediate radiation resistance of human prostate carcinoma cells *in vitro* and *in vivo* and that the decrease in cell death after radiation treatment is mediated though a paracrine mechanism of action [11]. However, the exact resistance-promoting effectors, as well as the role of CAV1-dependent fibroblast-derived factors, remained elusive. We therefore hypothesized that fibroblast-derived inhibitors of apoptosis proteins could mediate cell death resistance upon radiation. Here we show that TRIAP1 is highly expressed in stromal fibroblasts in a CAV1-dependent manner. *In vitro*, an ectopic expression of TRIAP1 leads to a cell specific increased radiation resistance in p53-deficient PC3 and p53-wildtype LNCaP prostate cancer cells, whereas p53-mutant DU145 cells do not gain any radiation resistance. Conformingly and mimicking the human situation more precisely, induced over-expression of TRIAP1 in human prostate fibroblasts leads to induced radiation resistance. Further on, TRIAP1-expressing stromal fibroblasts mediate radiation resistance *in vivo* when respective cells are co-implanted with CAV1-deficient PC3 tumour cells.

The underlying mechanism by which fibroblast-derived TRIAP1 is secreted and subsequently taken up by adjacent cancer cells and/or shuttled between the stromal and the tumour cells needs to be investigated further. TRIAP1 secretion in fibroblasts with a reduced CAV1-content is paralleled by the presence of lysosomal exocytosis related proteins and enzymes, such as ASM, ASA and LAMP proteins. This indicates that fibroblasts with a reduced CAV1 content bear a higher lysosomal exocytosis activity compared to fibroblasts containing normally high amounts of CAV1. It is known that the process and regulation of lysosomal exocytosis is largely changed upon tumour progression and in transformed cells [36]. Released lysosomal hydrolases, such as cathepsins D and B, play a role in tumour growth invasion and angiogenesis [37]. LAMP2 contributes to resistance, as the so called lysosomal cell death induced by anti-cancer drugs is decreased when LAMP2 is overexpressed in fibroblasts [38]. In addition, ASM is down-regulated in several carcinomas, for example, head and neck cancer and gastrointestinal carcinoma cancer cells, leading to a destabilized lysosomal environment in combination with an anti-apoptotic adaptation by decreased ceramide production [36]. Lysosomal exocytosis in cancer cells has been suggested to facilitate the entrapment and clearance of chemotherapeutics and provide an additional line of resistance [39].

As intrinsic drug resistance might be caused, at least in part, by factors secreted by the tumour microenvironment, it is thus imperative to dissect the tumour-microenvironment interactions which may reveal important mechanisms underlying drug resistance [35,39].

Interestingly, immunohistological analysis of TRIAP1 in advanced human prostate cancer reveals increased TRIAP1 immunoreactivity in the malignant epithelial cells of the more radioresistant higher Gleason grade adenocarcinomas. This highlights fibroblast-derived TRIAP1 as a potential candidate for future CAV1-mediated radiation response modulation. TRIAP1 is also involved in prostate cancer bone metastasis [40] and sensitivity to doxorubicin in breast cancer cells [15]. In ovarian cancer cells, increased TRIAP1 levels correlate with increased proliferation, a decrease in apoptosis and overall tumour progression [41]. TRIAP1 is also found to be upregulated in multiple myeloma [42], and, in patients with nasopharyngeal carcinoma, TRIAP1 overexpression correlates with a poor survival rate [43]. Experimental knockdown of TRIAP1, by expression of micro RNA miR-320b, is able to induce apoptosis by mitochondrial deregulating mechanisms, such as cytochrome C release and membrane potential alterations [15,43].

In summary, we have specified the role of CAV1 alterations potentially induced by CAV1-deficient and more reactive, stroma in radio sensitivity of prostate carcinoma at molecular level. We have identified apoptosis inhibitor TRIAP1 as a stromal-derived factor with the potential to induce cancer cell resistance. We suggest that blocking TRIAP1 activity and avoiding drug resistance may offer a promising drug development strategy to inhibit resistance-promoting CAV1-dependent signals.

Author Contributions: J.K., A.P., A.W. and D.K. performed experiments; J.K., D.K. analysed results and made the figures; C.H., P.M., E.Y. and H.R. provided materials; H.R. performed the Gleasing scoring; D.K. and V.J. designed research, J.K. and D.K. wrote the paper and E.Y. performed language corrections. All authors reviewed and approved the manuscript. This work was supported by grants of the DFG (GRK1739/1; GRK1739/2) and the BMBF (02NUK024-D).

Acknowledgments: We thank Mohammed Benchellal and Eva Gau for their excellent technical assistance. We acknowledge support by the Open Access Publication Fund of the University of Duisburg-Essen.

References

1. Bissell, M.J. Thinking in three dimensions: Discovering reciprocal signaling between the extracellular matrix and nucleus and the wisdom of microenvironment and tissue architecture. *Mol. Biol. Cell* **2016**, *27*, 3205–3209. [CrossRef] [PubMed]
2. Bissell, M.J.; Hines, W.C. Why don't we get more cancer? A proposed role of the microenvironment in restraining cancer progression. *Nat. Med.* **2011**, *17*, 320–329. [CrossRef] [PubMed]
3. Hanahan, D.; Weinberg, R.A. Hallmarks of cancer: The next generation. *Cell* **2011**, *144*, 646–674. [CrossRef] [PubMed]
4. Chen, D.; Che, G. Value of caveolin-1 in cancer progression and prognosis: Emphasis on cancer-associated fibroblasts, human cancer cells and mechanism of caveolin-1 expression (Review). *Oncol. Lett.* **2014**, *8*, 1409–1421. [CrossRef]
5. Ketteler, J.; Klein, D. Caveolin-1, cancer and therapy resistance. *Int. J. Cancer* **2018**. [CrossRef]
6. Di Vizio, D.; Morello, M.; Sotgia, F.; Pestell, R.G.; Freeman, M.R.; Lisanti, M.P. An absence of stromal caveolin-1 is associated with advanced prostate cancer, metastatic disease and epithelial Akt activation. *Cell Cycle* **2009**, *8*, 2420–2424. [CrossRef]
7. Witkiewicz, A.K.; Dasgupta, A.; Sotgia, F.; Mercier, I.; Pestell, R.G.; Sabel, M.; Kleer, C.G.; Brody, J.R.; Lisanti, M.P. An absence of stromal caveolin-1 expression predicts early tumor recurrence and poor clinical outcome in human breast cancers. *Am. J. Pathol.* **2009**, *174*, 2023–2034. [CrossRef]
8. Ayala, G.; Morello, M.; Frolov, A.; You, S.; Li, R.; Rosati, F.; Bartolucci, G.; Danza, G.; Adam, R.M.; Thompson, T.C.; et al. Loss of caveolin-1 in prostate cancer stroma correlates with reduced relapse-free survival and is functionally relevant to tumour progression. *J. Pathol.* **2013**, *231*, 77–87. [CrossRef]
9. Eliyatkin, N.; Aktas, S.; Diniz, G.; Ozgur, H.H.; Ekin, Z.Y.; Kupelioglu, A. Expression of Stromal Caveolin- 1 May Be a Predictor for Aggressive Behaviour of Breast Cancer. *Pathol. Oncol. Res.* **2018**, *24*, 59–65. [CrossRef]
10. Shan-Wei, W.; Kan-Lun, X.; Shu-Qin, R.; Li-Li, Z.; Li-Rong, C. Overexpression of caveolin-1 in cancer-associated fibroblasts predicts good outcome in breast cancer. *Breast Care* **2012**, *7*, 477–483. [CrossRef]
11. Panic, A.; Ketteler, J.; Reis, H.; Sak, A.; Herskind, C.; Maier, P.; Rubben, H.; Jendrossek, V.; Klein, D. Progression-related loss of stromal Caveolin 1 levels fosters the growth of human PC3 xenografts and mediates radiation resistance. *Sci. Rep.* **2017**, *7*, 41138. [CrossRef] [PubMed]

12. Hammarsten, P.; Dahl Scherdin, T.; Hagglof, C.; Andersson, P.; Wikstrom, P.; Stattin, P.; Egevad, L.; Granfors, T.; Bergh, A. High Caveolin-1 Expression in Tumor Stroma Is Associated with a Favourable Outcome in Prostate Cancer Patients Managed by Watchful Waiting. *PLoS ONE* **2016**, *11*, e0164016. [CrossRef]
13. Antognelli, C.; Ferri, I.; Bellezza, G.; Siccu, P.; Love, H.D.; Talesa, V.N.; Sidoni, A. Glyoxalase 2 drives tumorigenesis in human prostate cells in a mechanism involving androgen receptor and p53-p21 axis. *Mol. Carcinog.* **2017**, *56*, 2112–2126. [CrossRef]
14. Cotter, T.G. Apoptosis and cancer: The genesis of a research field. *Nat. Rev. Cancer* **2009**, *9*, 501–507. [CrossRef] [PubMed]
15. Adams, C.; Cazzanelli, G.; Rasul, S.; Hitchinson, B.; Hu, Y.; Coombes, R.C.; Raguz, S.; Yague, E. Apoptosis inhibitor TRIAP1 is a novel effector of drug resistance. *Oncol. Rep.* **2015**, *34*, 415–422. [CrossRef] [PubMed]
16. Park, W.R.; Nakamura, Y. p53CSV, a novel p53-inducible gene involved in the p53-dependent cell-survival pathway. *Cancer Res.* **2005**, *65*, 1197–1206. [CrossRef] [PubMed]
17. Andrysik, Z.; Kim, J.; Tan, A.C.; Espinosa, J.M. A genetic screen identifies TCF3/E2A and TRIAP1 as pathway-specific regulators of the cellular response to p53 activation. *Cell Rep.* **2013**, *3*, 1346–1354. [CrossRef]
18. Klein, D.; Schmandt, T.; Muth-Kohne, E.; Perez-Bouza, A.; Segschneider, M.; Gieselmann, V.; Brustle, O. Embryonic stem cell-based reduction of central nervous system sulfatide storage in an animal model of metachromatic leukodystrophy. *Gene Ther.* **2006**, *13*, 1686–1695. [CrossRef]
19. Lansmann, S.; Ferlinz, K.; Hurwitz, R.; Bartelsen, O.; Glombitza, G.; Sandhoff, K. Purification of acid sphingomyelinase from human placenta: Characterization and N-terminal sequence. *FEBS Lett.* **1996**, *399*, 227–231. [CrossRef]
20. Barzan, D.; Maier, P.; Zeller, W.J.; Wenz, F.; Herskind, C. Overexpression of caveolin-1 in lymphoblastoid TK6 cells enhances proliferation after irradiation with clinically relevant doses. *Strahlenther. Onkol.* **2010**, *186*, 99–106. [CrossRef]
21. Klein, D.; Schmitz, T.; Verhelst, V.; Panic, A.; Schenck, M.; Reis, H.; Drab, M.; Sak, A.; Herskind, C.; Maier, P.; et al. Endothelial Caveolin-1 regulates the radiation response of epithelial prostate tumors. *Oncogenesis* **2015**, *4*, e148. [CrossRef] [PubMed]
22. Klein, D.; Steens, J.; Wiesemann, A.; Schulz, F.C.; Kaschani, F.; Roeck, K.; Yamaguchi, M.; Wirsdorfer, F.; Kaiser, M.; Fischer, J.; et al. Mesenchymal stem cell therapy protects lungs from radiation-induced endothelial cell loss by restoring superoxide dismutase 1 expression. *Antioxid. Redox Signal.* **2016**. [CrossRef] [PubMed]
23. Koressaar, T.; Remm, M. Enhancements and modifications of primer design program Primer3. *Bioinformatics* **2007**, *23*, 1289–1291. [CrossRef]
24. Singh, S.R.; Rameshwar, P.; Siegel, P. Targeting tumor microenvironment in cancer therapy. *Cancer Lett.* **2016**. [CrossRef]
25. Sun, Y. Tumor microenvironment and cancer therapy resistance. *Cancer Lett.* **2015**. [CrossRef] [PubMed]
26. Cheng, C.J.; Bahal, R.; Babar, I.A.; Pincus, Z.; Barrera, F.; Liu, C.; Svoronos, A.; Braddock, D.T.; Glazer, P.M.; Engelman, D.M.; et al. MicroRNA silencing for cancer therapy targeted to the tumour microenvironment. *Nature* **2015**, *518*, 107–110. [CrossRef] [PubMed]
27. Masuda, S.; Izpisua Belmonte, J.C. The microenvironment and resistance to personalized cancer therapy. *Nat. Rev. Clin. Oncol.* **2013**, *10*. [CrossRef]
28. Swartz, M.A.; Iida, N.; Roberts, E.W.; Sangaletti, S.; Wong, M.H.; Yull, F.E.; Coussens, L.M.; DeClerck, Y.A. Tumor microenvironment complexity: Emerging roles in cancer therapy. *Cancer Res.* **2012**, *72*, 2473–2480. [CrossRef]
29. Wang, S.; Wang, N.; Zheng, Y.; Zhang, J.; Zhang, F.; Wang, Z. Caveolin-1: An Oxidative Stress-Related Target for Cancer Prevention. *Oxidative Med. Cell. Longev.* **2017**, *2017*, 7454031. [CrossRef]
30. Kamposioras, K.; Tsimplouli, C.; Verbeke, C.; Anthoney, A.; Daoukopoulou, A.; Papandreou, C.N.; Sakellaridis, N.; Vassilopoulos, G.; Potamianos, S.P.; Liakouli, V.; et al. Silencing of caveolin-1 in fibroblasts as opposed to epithelial tumor cells results in increased tumor growth rate and chemoresistance in a human pancreatic cancer model. *Int. J. Oncol.* **2018**. [CrossRef]
31. Martinez-Outschoorn, U.E.; Balliet, R.M.; Rivadeneira, D.B.; Chiavarina, B.; Pavlides, S.; Wang, C.; Whitaker-Menezes, D.; Daumer, K.M.; Lin, Z.; Witkiewicz, A.K.; et al. Oxidative stress in cancer associated fibroblasts drives tumor-stroma co-evolution: A new paradigm for understanding tumor metabolism, the field effect and genomic instability in cancer cells. *Cell Cycle* **2010**, *9*, 3256–3276. [CrossRef] [PubMed]
32. White, E. The role for autophagy in cancer. *J. Clin. Investig.* **2015**, *125*, 42–46. [CrossRef] [PubMed]

33. Martinez-Outschoorn, U.E.; Pavlides, S.; Whitaker-Menezes, D.; Daumer, K.M.; Milliman, J.N.; Chiavarina, B.; Migneco, G.; Witkiewicz, A.K.; Martinez-Cantarin, M.P.; Flomenberg, N.; et al. Tumor cells induce the cancer associated fibroblast phenotype via caveolin-1 degradation: Implications for breast cancer and DCIS therapy with autophagy inhibitors. *Cell Cycle* **2010**, *9*, 2423–2433. [CrossRef]
34. Hirata, E.; Girotti, M.R.; Viros, A.; Hooper, S.; Spencer-Dene, B.; Matsuda, M.; Larkin, J.; Marais, R.; Sahai, E. Intravital imaging reveals how BRAF inhibition generates drug-tolerant microenvironments with high integrin beta1/FAK signaling. *Cancer Cell* **2015**, *27*, 574–588. [CrossRef]
35. Straussman, R.; Morikawa, T.; Shee, K.; Barzily-Rokni, M.; Qian, Z.R.; Du, J.; Davis, A.; Mongare, M.M.; Gould, J.; Frederick, D.T.; et al. Tumour micro-environment elicits innate resistance to RAF inhibitors through HGF secretion. *Nature* **2012**, *487*, 500–504. [CrossRef] [PubMed]
36. Kallunki, T.; Olsen, O.D.; Jaattela, M. Cancer-associated lysosomal changes: Friends or foes? *Oncogene* **2013**, *32*, 1995–2004. [CrossRef]
37. Gocheva, V.; Zeng, W.; Ke, D.; Klimstra, D.; Reinheckel, T.; Peters, C.; Hanahan, D.; Joyce, J.A. Distinct roles for cysteine cathepsin genes in multistage tumorigenesis. *Genes Dev.* **2006**, *20*, 543–556. [CrossRef]
38. Fehrenbacher, N.; Bastholm, L.; Kirkegaard-Sorensen, T.; Rafn, B.; Bottzauw, T.; Nielsen, C.; Weber, E.; Shirasawa, S.; Kallunki, T.; Jaattela, M. Sensitization to the lysosomal cell death pathway by oncogene-induced down-regulation of lysosome-associated membrane proteins 1 and 2. *Cancer Res.* **2008**, *68*, 6623–6633. [CrossRef]
39. Zhitomirsky, B.; Assaraf, Y.G. Lysosomes as mediators of drug resistance in cancer. *Drug Resist. Updates Rev. Comment. Antimicrob. Anticancer Chemother.* **2016**, *24*, 23–33. [CrossRef]
40. Siu, M.K.; Abou-Kheir, W.; Yin, J.J.; Chang, Y.S.; Barrett, B.; Suau, F.; Casey, O.; Chen, W.Y.; Fang, L.; Hynes, P.; et al. Correction: Loss of EGFR signaling-regulated miR-203 promotes prostate cancer bone metastasis and tyrosine kinase inhibitors resistance. *Oncotarget* **2018**, *9*, 32403. [CrossRef]
41. Liu, P.; Qi, X.; Bian, C.; Yang, F.; Lin, X.; Zhou, S.; Xie, C.; Zhao, X.; Yi, T. MicroRNA-18a inhibits ovarian cancer growth via directly targeting TRIAP1 and IPMK. *Oncol. Lett.* **2017**, *13*, 4039–4046. [CrossRef] [PubMed]
42. Fook-Alves, V.L.; de Oliveira, M.B.; Zanatta, D.B.; Strauss, B.E.; Colleoni, G.W. TP53 Regulated Inhibitor of Apoptosis 1 (TRIAP1) stable silencing increases late apoptosis by upregulation of caspase 9 and APAF1 in RPMI8226 multiple myeloma cell line. *Biochim. Biophys. Acta* **2016**, *1862*, 1105–1110. [CrossRef] [PubMed]
43. Li, Y.; Tang, X.; He, Q.; Yang, X.; Ren, X.; Wen, X.; Zhang, J.; Wang, Y.; Liu, N.; Ma, J. Overexpression of Mitochondria Mediator Gene TRIAP1 by miR-320b Loss Is Associated with Progression in Nasopharyngeal Carcinoma. *PLoS Genet.* **2016**, *12*, e1006183. [CrossRef] [PubMed]

6

Suppressive Role of Androgen/Androgen Receptor Signaling via Chemokines on Prostate Cancer Cells

Kouji Izumi * and Atsushi Mizokami

Department of Integrative Cancer Therapy and Urology, Kanazawa University Graduate School of Medical Science, 13-1 Takara-machi, Kanazawa, Ishikawa 920-8641, Japan; mizokami@staff.kanazawa-u.ac.jp
* Correspondence: azuizu2003@yahoo.co.jp

Abstract: Androgen/androgen receptor (AR) signaling is a significant driver of prostate cancer progression, therefore androgen-deprivation therapy (ADT) is often used as a standard form of treatment for advanced and metastatic prostate cancer patients. However, after several years of ADT, prostate cancer progresses to castration-resistant prostate cancer (CRPC). Androgen/AR signaling is still considered an important factor for prostate cancer cell survival following CRPC progression, while recent studies have reported dichotomic roles for androgen/AR signaling. Androgen/AR signaling increases prostate cancer cell proliferation, while simultaneously inhibiting migration. As a result, ADT can induce prostate cancer metastasis. Several C-C motif ligand (CCL)-receptor (CCR) axes are involved in cancer cell migration related to blockade of androgen/AR signaling. The CCL2-CCR2 axis is negatively regulated by androgen/AR signaling, with the CCL22-CCR4 axis acting as a further downstream mediator, both of which promote prostate cancer cell migration. Furthermore, the CCL5-CCR5 axis inhibits androgen/AR signaling as an upstream mediator. CCL4 is involved in prostate carcinogenesis through macrophage AR signaling, while the CCL21-CCR7 axis in prostate cancer cells is activated by tumor necrotic factor, which is secreted when androgen/AR signaling is inhibited. Finally, the CCL2-CCR2 axis has recently been demonstrated to be a key contributor to cabazitaxel resistance in CRPC.

Keywords: prostate cancer; androgen receptor; castration-resistant prostate cancer; CCL2; CCL22; CCL5; migration

---

1. Introduction

Prostate cancer is among the most frequently diagnosed malignancies worldwide in men [1]. The five-year survival rate for localized prostate cancer is close to 100%, and the prognosis for localized prostate cancer is the best among all types of cancers; however, metastatic prostate cancer is associated with a very poor prognosis, with no curative treatments currently available [1,2]. Androgen/androgen receptor (AR) signaling is known to be a significant driver of prostate cancer progression, therefore androgen-deprivation therapy (ADT)—with or without anti-androgens—is often used as a standard form of care for patients with advanced and metastatic prostate cancer [3,4]. ADT has been demonstrated to improve not only serum prostate-specific antigen levels, but also patient survival, however prostate cancer generally progresses to castration-resistant prostate cancer (CRPC) following several years of ADT [5]. Several potential mechanisms underpinning CRPC progression that relate to AR function have been identified, including androgen hypersensitivity, AR mutation, ligand promiscuity, and AR variants. Nonetheless, no radical treatments exist at present and all AR-targeting agents for CRPC eventually fail to suppress cancer cell activity [6]. Recently, some studies have reported suppressive effects of androgen/AR signaling in prostate cancer cells, therefore suppression of AR function itself may cause CRPC [7,8]. Previously, we demonstrated that androgen/AR signaling increases prostate cancer cell proliferation, while simultaneously inhibiting cancer cell migration, which is induced by the activation of several C-C motif ligand (CCL)-receptor (CCR) axes downstream or upstream of androgen/AR signaling [9–12]. This review focuses on such suppressive effects of androgen/AR signaling on prostate cancer cells through CCL-CCR axes.

2. The Role of CCL2 as a Downstream Mediator of Androgen/AR Signaling

Therapeutic approaches that solely target androgen/AR signaling are insufficient to control prostate cancer cell activity [13–15]. Genetic ablation of AR in prostate epithelial cells promotes the development of invasive prostate cancer [7], suggesting that therapeutic suppression of androgen/AR function induces unwanted signals that may promote the progression of surviving prostate cancer cells to an advanced metastatic stage. When AR function of C4-2 (a human prostate cancer cell line) cells were silenced with AR-siRNA (siAR), using scramble RNA (scr) as a control, siAR cells were observed to possess an increased migratory capacity [8]. Cytokine array analysis of conditioned media from siAR and scr cells revealed increased CCL2 expression in siAR cells, supporting a potential role for prostate cancer cell-derived CCL2 in mediating local inflammatory responses during suppression of AR [8]. CCL2 is reported to play a potential role in stimulating capillary network formation of human microvascular endothelial cells in the microenvironment of prostate cancer [16]. C4-2 siAR cells were also observed to express increased levels of epithelial-mesenchymal transition (EMT) markers and pSTAT3 via the CCL2-CCR2 axis in an autocrine manner. In addition, C4-2 siAR cells were observed to possess significantly reduced levels of PIAS3 (the endogenous protein inhibitor of activated STAT3), which is controlled by androgen/AR signaling [17]. Notably, STAT3 activation was also observed to increase CCL2 expression levels in C4-2 siAR cells. These results suggest that androgen/AR signaling in prostate cancer cells may inhibit CCL2 and pSTAT3 expression through upregulation of PIAS3 [8,9]. EMT is believed to be an essential cancer cell characteristic for invasion and metastasis to distant sites [18]; pSTAT3 activation has been reported to play an important role in EMT induction, as well as inflammation and cancer progression [19,20]. Furthermore, ADT is known to be linked to EMT induction [21]. In summary, prostatic epithelial AR silencing via siAR promotes STAT3 activation and EMT in prostate cancer cells via CCL2 induction, which may be associated with a secretory phenotype and pro-invasive characteristics of prostate cancer cells [8,9].

3. The Role of CCL22 as a Further Downstream Mediator of CCL2

CCL2 is a powerful chemotactic protein for macrophages and tumor-associated macrophages (TAMs), which infiltrate into tumors and contribute to cancer progression via immune suppression [22,23]. CCL17 and CCL22, which are high-affinity ligands for CCR4, have both been reported to be secreted by TAMs, with immunosuppressive functions [24]. Correlations have previously been reported between CCR4 expression levels and metastasis in cancer cells [25,26]. Therefore, we aimed to elucidate the relationship between the CCL2-CCR2 axis and CCL17/22-CCR4 axis in prostate cancer progression. Both CCR2 and CCR4 were observed to be expressed in human prostate cancer cell lines and prostate cancer tissues; furthermore, in vitro co-culture of prostate cancer cells and macrophages resulted in increased CCL2 and CCR2 levels in prostate cancer cells [11]. Notably, addition of CCL2 induced both CCL22 and CCR4 expression in prostate cancer cells; CCL22 subsequently promoted the migration and invasion of prostate cancer cells in an autocrine manner, via enhanced phosphorylation of Akt [11]. The CCL22-CCR4 axis is known to chemo-attract regulatory T cells (Tregs) into tumor tissues; Tregs recognize self-antigens, including tumor antigens present in tumor tissues, and efficiently suppress the activation of tumor antigen-specific effector T cells [27]. In summary, CCL2 and CCL22 secretion in the prostate cancer tumor microenvironment may induce not only direct metastasis of prostate cancer cells, but also promote the activation of TAMs and Tregs, which facilitate a suitable environment for cancer progression.

4. The Role of CCL5 as an Upstream Mediator of Androgen/AR Signaling

Skeletal metastases occur in approximately 80% of patients with advanced prostate cancer, for which no curative treatment is available [28]. We previously reported that bone stromal cells and SaOS-2 osteoblast-like cells promote prostate cancer metastasis via activation of transforming growth factor-β1 (TGF β1) [29], which in turn induces the development of an immune suppressive microenvironment [30]. CCL2 is reported to increase bone metastasis through recruitment of TAMs and osteoclasts to the tumor site and blood vessel formation through vascular endothelial growth factor-A [31,32]. Therefore, we investigated whether further chemokines could be involved in the activation of prostate cancer cells within prostate cancer bone metastases. Migration of LNCaP cells (an AR-positive prostate cancer cell line) increased significantly when co-cultured with bone stromal cells isolated from prostate cancer bone metastases. Cytokine array analysis of conditioned media from bone stromal cell cultures subsequently identified CCL5, a high-affinity ligand of CCR5, as a concentration-dependent promoter of LNCaP cell migration [12]. LNCaP cell migration was observed to be suppressed by the addition of a CCL5-neutralizing antibody to cocultures with bone stromal cells, while AR knockdown using siRNA was observed to increase LNCaP cell migration compared with control cells [12]. As CCL5 was unable to promote migration of LNCaP siAR cells, it was concluded that elevated CCL5 secretion by bone stromal cells from metastatic lesions induced prostate cancer cell migration in a CCL5-dependent manner, upstream of AR signaling [12]. Upregulation of CCL5 has previously been reported to increase the aggressive potential of breast cancer cells and the invasiveness of prostate cancer cells [33–35]. In addition, Luo et al. found that CCL5 upregulation in bone marrow mesenchymal stem cells increased the metastatic potential of prostate cancer cells, and subsequently downregulated AR signaling, due to inhibition of AR nuclear translocation [36]. Furthermore, CCL5 has been found to suppress prolyl hydroxylase expression, leading to suppression of VHL-mediated HIF2α ubiquitination and suppression of AR signaling [37]. Results obtained using LNCaP siAR cells indicate that CCL5 activity is located upstream of AR signaling. Moreover, SaOS-2 did not promote the migration of PC-3 AR-negative prostate cancer cells [12]. These results suggest that the migratory potential of AR-positive prostate cancer cells in bone metastases is increased by CCL5, secreted by bone stromal cells via the suppression of androgen/AR signaling. CCL5 is also secreted by prostate cancer-associated fibroblasts and recruited macrophages into the prostate cancer microenvironment [38]. Estrogen receptor α could reduce prostate cancer cell invasion through reduction of CCL5 secretion from fibroblasts and macrophage infiltration prostate cancer [38].

5. Treatment Strategies Targeting CCL-CCR Axes and Androgen/AR Signaling

5.1. CCL2-CCR2 Axis

A new role for AR silencing in the mediation of EMT induction via activation of the CCL2-CCR2 axis in the tumor microenvironment provides new therapeutic targets for preventing potential prostate cancer metastasis at later stages. A previous study reported on treatment of forty-six CRPC patients with the human CCL2 monoclonal antibody, carlumab, in a phase 2 trial. Unfortunately, this single-arm study was not able to meet its primary objective to demonstrate potential therapeutic benefits of carlumab alone in patients with metastatic CRPC who had failed prior docetaxel-based treatment [39]. Carlumab was unable to sustain durable free CCL2 suppression, permitting rapid rebound and increases in CCL2 to baseline or higher concentrations; this insufficient suppression meant meaningful clinical responses could not be achieved [39]. To entirely suppress the CCL2-CCR2 axis, a receptor antagonist may provide a more suitable treatment method as receptor blockade efficiency is irrespective of serum CCL2 concentrations. Several CCR2 antagonists has been reported in the literature [40,41]. As stated in Section 2, CCL2-CCR2 axis and STAT3 activate each other in prostate cancer cells, therefore STAT3 is also regarded as a potential treatment target. We confirmed inhibition of STAT3 activity by a STAT3 inhibitor, AG490, resulted in down regulation of EMT gene expression in C4-2 siAR cells [8].

5.2. CCL22-CCR4 Axis

Tissue microarray analysis revealed a correlation between staining intensity of CCR4 and prostate cancer progression, however no such correlation existed with CCR2, despite the fact that CCR2 and CCR4 intensities were correlated with one another. Therefore, the CCL22-CCR4 axis may prove a more significant driver of prostate cancer migration and invasion than the CCL2-CCR2 axis [11]. Phosphorylation of Akt proteins is more effectively inhibited by CCR4 antagonists than CCR2 antagonists, further indicating the efficiency of CCR4 antagonist therapy against prostate cancer migration and invasion [11]. Akt activation is controlled by phosphorylation of the two key residues threonine 308 (Thr308) and serine 473 (Ser473) [42], and their phosphorylation promotes prostate cancer cell growth, proliferation, motility, and survival [43–45]. Our previous results indicated that the CCL22-CCR4 axis controls phosphorylation of Ser473. It has been previously been demonstrated that CCL2 promotes prostate cancer cell proliferation, migration, and survival via Akt-activation-dependent mechanisms [46–48]. Taken together, these results indicate that the CCL22-CCR4 axis may prove to be a better therapeutic target than the CCL2–CCR2 axis for prostate cancer patients. CCR4 expression is observed on tumor cells derived from the majority of adult T-cell leukemia-lymphoma (ATL) patients, therefore mogamulizumab, an anti-CCR4 antibody, has been approved in Japan for the treatment of relapsed/refractory ATL [49,50]. In addition, potent inhibition of Akt signaling using an Akt inhibitor was associated with a tolerable safety profile and meaningful disease control in a subgroup of patients with solid tumors during phase 1 and 2 trials [51,52]. Application of such CCL22-CCR4 targeting agents for therapy of CRPC patients is expected in the near future.

5.3. CCL5-CCR5 Axis and Others

In a recent study of ours that focused on the effects of coffee compounds on prostate cancer cells, the coffee diterpenes kahweol acetate and cafestol were observed to synergistically inhibit prostate cancer cell proliferation and migration [53]. These diterpenes were capable of inhibiting androgen/AR signaling without inducing prostate cancer cell secretion on CCL2 and CCL5 [53]. It is noteworthy that expression of CCR2 and CCR5, receptors for CCL2 and CCL5, respectively, visibly decreased following diterpene administration in prostate cancer cells [53]. Kahweol acetate and cafestol may, therefore, represent potential therapeutic candidates, especially in combination therapy for the treatment of both castration-sensitive prostate cancer and CRPC [53].

6. CCL Involvement Various Pathways of Prostate Cancer Progression

6.1. Carcinogenesis

Co-culturing of immortalized prostate epithelial cells with macrophages has been observed to induce prostate tumorigenesis, involving the signaling alteration of macrophage AR-inflammatory chemokine CCL4-pSTAT3 activation, EMT, and p53/PTEN tumor suppressor down-regulation [54]. Furthermore, in vivo studies have demonstrated that PTEN(+/−) mice lacking macrophage AR develop fewer prostatic intraepithelial neoplasia lesions, supporting an important role for macrophage AR signaling during prostate tumorigenesis [54]. CCL4-neutralizing antibodies effectively inhibited macrophage-induced prostate tumorigenic signaling, while CCL4 upregulation was associated with increased Snail expression and p53/PTEN down-regulation in high-grade prostatic intraepithelial neoplasia and prostate cancer [54]. This study identified the AR-CCL4-pSTAT3 axis is a key regulator during prostate tumorigenesis and highlighted the important roles of infiltrating macrophages and inflammatory cytokines during prostate tumorigenesis [54,55].

6.2. Lymph Node Metastasis

Tumor necrotic factor (TNF) is negatively regulated by androgen/AR signaling, and androgen/AR signaling blockade induces TNF mRNA in prostatic stroma [56]. We observed that human prostate cancer cells (PC-3, DU145, LNCaP, and LNCaP-SF) express both TNF-α and CCR7 and that low concentrations of TNF-α can induce CCR7 expression in prostate cancer cells through phosphorylation of extracellular signal-regulated kinases in an autocrine manner [57]. CCL21, a ligand of CCR7 that is secreted by fibroblastic reticular cells in lymph nodes and is abundant in the T-cell zone of the lymph node [58], was found to promote prostate cancer cell migration via protein kinase p38 phosphorylation [57]. These results suggest that TNF-α induces CCR7 expression and that the CCL21-CCR7 axis is capable of increasing the metastatic potential of prostate cancer cells during lymph node metastasis. In combination with ADT, the CCL21-CCR7 axis may, therefore, prove a superior target compared with single androgen/AR signaling-targeted therapy for treatment of patients with prostate cancer and lymph node metastasis.

6.3. Resistance to Taxanes

Understanding the underlying mechanisms behind chemoresistance and disease progression in patients with prostate cancer is important in order to develop novel treatment strategies. In particular, cabazitaxel resistance is a considerable challenge in CRPC patients as cabazitaxel is often administered as a last resort [59]. The mechanism through which cabazitaxel resistance develops is still unclear, however. We previously established a cabazitaxel-resistant prostate cancer cell line, DU145-TxR/CxR, from a paclitaxel-resistant cell line, DU145-TxR [60]. The cDNA microarray analysis revealed that CCL2 expression was upregulated in both DU145-TxR and DU145-TxR/CxR cells, compared with control DU145 cells. Furthermore, the secreted CCL2 protein level in DU145-TxR and DU145-TxR/CxR cells was observed to be higher than in the parental DU145 cells [61]. Stimulation of DU145 cells with CCL2 increased proliferation during cabazitaxel treatment, while CCR2 antagonist suppressed the proliferation of DU145-TxR and DU145-TxR/CxR cells during cabazitaxel treatment [61]. The CCL2-CCR2 axis was found to reduce apoptosis through inhibition of caspase-3 and poly(ADP-ribose) polymerase (PARP), indicating that CCL2 is potentially a key contributor to cabazitaxel resistance in prostate cancer cells [61]. Inhibition of the CCL2-CCR2 axis may provide a potential therapeutic strategy against both chemosensitive CRPC and chemoresistant CRPC.

7. Concluding Remarks

Our recent studies have elucidated several CCL-CCR axes involved in prostate cancer progression, some of which are negatively regulated by androgen/AR signaling and vice versa (Figure 1). Other CCL-CCR axes may play significant roles in prostate cancer progression, while CCL-CCR axes form chemokine-networks in which CCL-CCR axes are capable of activating and/or inactivating one another. In summary, the complete elucidation of this chemokine-network, including the exact function of each chemokine, is required to control prostate cancer cells across every stage.

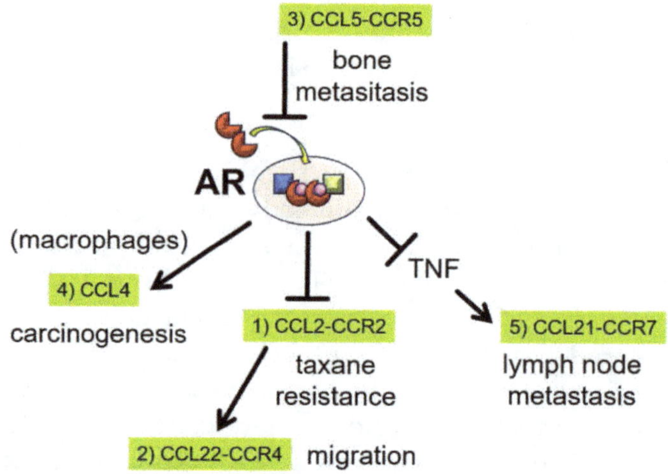

Figure 1. Androgen receptor (AR) and C-C motif ligand (CCL)-receptor (CCR) axes. (1) Androgen/AR signaling negatively regulates CCL2 secretion and inhibits prostate cancer cell migration. The CCL2-CCR2 axis contributes to chemoresistance to taxanes. (2) The CCL22-CCR4 axis is located downstream of the CCL2-CCR2 axis and increases the migratory capacity of prostate cancer cells. (3) CCL5 activity occurs upstream of AR signaling and increases the migratory capacity of prostate cancer cells via inhibition of androgen/AR signaling. (4) CCL4 is positively regulated by androgen/AR signaling in macrophages, and is a key regulator during prostate tumorigenesis. (5) TNF, which is negatively regulated by androgen/AR signaling, induces CCR7 expression in prostate cancer cells. CCR7 subsequently binds CCL21 from fibroblastic reticular cells in lymph node, resulting in increased migratory capacity.

Author Contributions: K.I. wrote the manuscript. A.M. revised the manuscript.

Abbreviations

AR	androgen receptor
ADT	androgen-deprivation therapy
CRPC	castration-resistant prostate cancer
CCL	C-C motif ligand
CCR	CCL-receptor
EMT	epithelial-mesenchymal transition
siAR	AR-siRNA
scr	scramble RNA
TAMs	tumor-associated macrophages
Tregs	regulatory T cells
TGF β1	transforming growth factor-β1
Thr308	threonine 308
Ser473	serine 473
ATL	adult T-cell leukemia-lymphoma
TNF	tumor necrotic factor
PARP	poly(ADP-ribose) polymerase

References

1. Siegel, R.L.; Miller, K.D.; Jemal, A. Cancer statistics, 2019. *CA Cancer J. Clin.* **2019**, *69*, 7–34. [CrossRef]
2. Ferlay, J.; Parkin, D.M.; Steliarova-Foucher, E. Estimates of cancer incidence and mortality in Europe in 2008. *Eur. J. Cancer* **2010**, *46*, 765–781. [CrossRef] [PubMed]

3. Maximum androgen blockade in advanced prostate cancer: An overview of the randomised trials. Prostate Cancer Trialists' Collaborative Group. *Lancet* **2000**, *355*, 1491–1498.
4. Samson, D.J.; Seidenfeld, J.; Schmitt, B.; Hasselblad, V.; Albertsen, P.C.; Bennett, C.L.; Wilt, T.J.; Aronson, N. Systematic review and meta-analysis of monotherapy compared with combined androgen blockade for patients with advanced prostate carcinoma. *Cancer* **2002**, *95*, 361–376. [CrossRef]
5. Scher, H.I.; Sawyers, C.L. Biology of progressive, castration-resistant prostate cancer: Directed therapies targeting the androgen-receptor signaling axis. *J. Clin. Oncol.* **2005**, *23*, 8253–8261. [CrossRef]
6. Chandrasekar, T.; Yang, J.C.; Gao, A.C.; Evans, C.P. Mechanisms of resistance in castration-resistant prostate cancer (CRPC). *Transl. Androl. Urol.* **2015**, *4*, 365–380. [CrossRef] [PubMed]
7. Niu, Y.; Altuwaijri, S.; Lai, K.P.; Wu, C.T.; Ricke, W.A.; Messing, E.M.; Yao, J.; Yeh, S.; Chang, C. Androgen receptor is a tumor suppressor and proliferator in prostate cancer. *Proc. Natl. Acad. Sci. USA* **2008**, *105*, 12182–12187. [CrossRef]
8. Izumi, K.; Fang, L.Y.; Mizokami, A.; Namiki, M.; Li, L.; Lin, W.J.; Chang, C. Targeting the androgen receptor with siRNA promotes prostate cancer metastasis through enhanced macrophage recruitment via CCL2/CCR2-induced STAT3 activation. *EMBO. Mol. Med.* **2013**, *5*, 1383–1401. [CrossRef] [PubMed]
9. Lin, T.H.; Izumi, K.; Lee, S.O.; Lin, W.J.; Yeh, S.; Chang, C. Anti-androgen receptor ASC-J9 versus anti-androgens MDV3100 (Enzalutamide) or Casodex (Bicalutamide) leads to opposite effects on prostate cancer metastasis via differential modulation of macrophage infiltration and STAT3-CCL2 signaling. *Cell Death Dis.* **2013**, *4*, e764. [CrossRef]
10. Izumi, K.; Mizokami, A.; Lin, H.P.; Ho, H.M.; Iwamoto, H.; Maolake, A.; Natsagdorj, A.; Kitagawa, Y.; Kadono, Y.; Miyamoto, H.; et al. Serum chemokine (CC motif) ligand 2 level as a diagnostic, predictive, and prognostic biomarker for prostate cancer. *Oncotarget* **2016**, *7*, 8389–8398. [CrossRef] [PubMed]
11. Maolake, A.; Izumi, K.; Shigehara, K.; Natsagdorj, A.; Iwamoto, H.; Kadomoto, S.; Takezawa, Y.; Machioka, K.; Narimoto, K.; Namiki, M.; et al. Tumor-associated macrophages promote prostate cancer migration through activation of the CCL22-CCR4 axis. *Oncotarget* **2017**, *8*, 9739–9751. [CrossRef] [PubMed]
12. Urata, S.; Izumi, K.; Hiratsuka, K.; Maolake, A.; Natsagdorj, A.; Shigehara, K.; Iwamoto, H.; Kadomoto, S.; Makino, T.; Naito, R.; et al. C-C motif ligand 5 promotes migration of prostate cancer cells in the prostate cancer bone metastasis microenvironment. *Cancer Sci.* **2018**, *109*, 724–731. [CrossRef]
13. Harris, W.P.; Mostaghel, E.A.; Nelson, P.S.; Montgomery, B. Androgen deprivation therapy: Progress in understanding mechanisms of resistance and optimizing androgen depletion. *Nat. Clin. Pract. Urol.* **2009**, *6*, 76–85. [CrossRef]
14. Kasper, S.; Cookson, M.S. Mechanisms leading to the development of hormone-resistant prostate cancer. *Urol. Clin. North. Am.* **2006**, *33*, 201–210. [CrossRef] [PubMed]
15. Yamaoka, M.; Hara, T.; Kusaka, M. Overcoming persistent dependency on androgen signaling after progression to castration-resistant prostate cancer. *Clin. Cancer Res.* **2010**, *16*, 4319–4324. [CrossRef] [PubMed]
16. Nalla, A.K.; Estes, N.; Patel, J.; Rao, J.S. N-cadherin mediates angiogenesis by regulating monocyte chemoattractant protein-1 expression via PI3K/Akt signaling in prostate cancer cells. *Exp. Cell Res.* **2011**, *317*, 2512–2521. [CrossRef] [PubMed]
17. Junicho, A.; Matsuda, T.; Yamamoto, T.; Kishi, H.; Korkmaz, K.; Saatcioglu, F.; Fuse, H.; Muraguchi, A. Protein inhibitor of activated STAT3 regulates androgen receptor signaling in prostate carcinoma cells. *Biochem. Biophys. Res. Commun.* **2000**, *278*, 9–13. [CrossRef] [PubMed]
18. Friedl, P.; Alexander, S. Cancer invasion and the microenvironment: Plasticity and reciprocity. *Cell* **2011**, *147*, 992–1009. [CrossRef] [PubMed]
19. Abdulghani, J.; Gu, L.; Dagvadorj, A.; Lutz, J.; Leiby, B.; Bonuccelli, G.; Lisanti, M.P.; Zellweger, T.; Alanen, K.; Mirtti, T.; et al. Stat3 promotes metastatic progression of prostate cancer. *Am. J. Pathol.* **2008**, *172*, 1717–1728. [CrossRef]
20. Azare, J.; Leslie, K.; Al-Ahmadie, H.; Gerald, W.; Weinreb, P.H.; Violette, S.M.; Bromberg, J. Constitutively activated Stat3 induces tumorigenesis and enhances cell motility of prostate epithelial cells through integrin beta 6. *Mol. Cell Biol.* **2007**, *27*, 4444–4453. [CrossRef] [PubMed]
21. Sun, Y.; Wang, B.E.; Leong, K.G.; Yue, P.; Li, L.; Jhunjhunwala, S.; Chen, D.; Seo, K.; Modrusan, Z.; Gao, W.Q.; Settleman, J.; et al. Androgen deprivation causes epithelial-mesenchymal transition in the prostate: Implications for androgen-deprivation therapy. *Cancer Res.* **2012**, *72*, 527–536. [CrossRef] [PubMed]

22. Matsushima, K.; Larsen, C.G.; DuBois, G.C.; Oppenheim, J.J. Purification and characterization of a novel monocyte chemotactic and activating factor produced by a human myelomonocytic cell line. *J. Exp. Med.* **1989**, *169*, 1485–1490. [CrossRef] [PubMed]
23. Loberg, R.D.; Ying, C.; Craig, M.; Yan, L.; Snyder, L.A.; Pienta, K.J. CCL2 as an important mediator of prostate cancer growth in vivo through the regulation of macrophage infiltration. *Neoplasia* **2007**, *9*, 556–562. [CrossRef] [PubMed]
24. Hefetz-Sela, S.; Stein, I.; Klieger, Y.; Porat, R.; Sade-Feldman, M.; Zreik, F.; Nagler, A.; Pappo, O.; Quagliata, L.; Dazert, E.; et al. Acquisition of an immunosuppressive protumorigenic macrophage phenotype depending on c-Jun phosphorylation. *Proc. Natl. Acad. Sci. USA* **2014**, *111*, 17582–17587. [CrossRef]
25. Olkhanud, P.B.; Baatar, D.; Bodogai, M.; Hakim, F.; Gress, R.; Anderson, R.L.; Deng, J.; Xu, M.; Briest, S.; Biragyn, A. Breast cancer lung metastasis requires expression of chemokine receptor CCR4 and regulatory T cells. *Cancer Res.* **2009**, *69*, 5996–6004. [CrossRef]
26. Mantovani, A.; Sozzani, S.; Locati, M.; Allavena, P.; Sica, A. Macrophage polarization: Tumor-associated macrophages as a paradigm for polarized M2 mononuclear phagocytes. *Trends. Immunol.* **2002**, *23*, 549–555. [CrossRef]
27. Nishikawa, H.; Sakaguchi, S. Regulatory T cells in cancer immunotherapy. *Curr. Opin. Immunol.* **2014**, *27*, 1–7. [CrossRef]
28. Rubin, M.A.; Putzi, M.; Mucci, N.; Smith, D.C.; Wojno, K.; Korenchuk, S.; Pienta, K.J. Rapid ("warm") autopsy study for procurement of metastatic prostate cancer. *Clin. Cancer Res.* **2000**, *6*, 1038–1045.
29. Izumi, K.; Mizokami, A.; Li, Y.Q.; Narimoto, K.; Sugimoto, K.; Kadono, Y.; Kitagawa, Y.; Konaka, H.; Koh, E.; Keller, E.T.; et al. Tranilast inhibits hormone refractory prostate cancer cell proliferation and suppresses transforming growth factor beta1-associated osteoblastic changes. *Prostate* **2009**, *69*, 1222–1234. [CrossRef]
30. Yang, L.; Pang, Y.; Moses, H.L. TGF-beta and immune cells: An important regulatory axis in the tumor microenvironment and progression. *Trends. Immunol.* **2010**, *31*, 220–227. [CrossRef]
31. Mizutani, K.; Sud, S.; McGregor, N.A.; Martinovski, G.; Rice, B.T.; Craig, M.J.; Varsos, Z.S.; Roca, H.; Pienta, K.J. The chemokine CCL2 increases prostate tumor growth and bone metastasis through macrophage and osteoclast recruitment. *Neoplasia* **2009**, *11*, 1235–1242. [CrossRef]
32. Li, X.; Loberg, R.; Liao, J.; Ying, C.; Snyder, L.A.; Pienta, K.J.; McCauley, L.K. A destructive cascade mediated by CCL2 facilitates prostate cancer growth in bone. *Cancer Res.* **2009**, *69*, 1685–1692. [CrossRef]
33. Yasuhara, R.; Irie, T.; Suzuki, K.; Sawada, T.; Miwa, N.; Sasaki, A.; Tsunoda, Y.; Nakamura, S.; Mishima, K. The beta-catenin signaling pathway induces aggressive potential in breast cancer by up-regulating the chemokine CCL5. *Exp. Cell Res.* **2015**, *338*, 22–31. [CrossRef]
34. Vaday, G.G.; Peehl, D.M.; Kadam, P.A.; Lawrence, D.M. Expression of CCL5 (RANTES) and CCR5 in prostate cancer. *Prostate* **2006**, *66*, 124–134. [CrossRef]
35. Kato, T.; Fujita, Y.; Nakane, K.; Mizutani, K.; Terazawa, R.; Ehara, H.; Kanimoto, Y.; Kojima, T.; Nozawa, Y.; Deguchi, T.; et al. CCR1/CCL5 interaction promotes invasion of taxane-resistant PC3 prostate cancer cells by increasing secretion of MMPs 2/9 and by activating ERK and Rac signaling. *Cytokine* **2013**, *64*, 251–257. [CrossRef]
36. Luo, J.; Ok Lee, S.; Liang, L.; Huang, C.K.; Li, L.; Wen, S.; Chang, C. Infiltrating bone marrow mesenchymal stem cells increase prostate cancer stem cell population and metastatic ability via secreting cytokines to suppress androgen receptor signaling. *Oncogene* **2014**, *33*, 2768–2778. [CrossRef]
37. Luo, J.; Lee, S.O.; Cui, Y.; Yang, R.; Li, L.; Chang, C. Infiltrating bone marrow mesenchymal stem cells (BM-MSCs) increase prostate cancer cell invasion via altering the CCL5/HIF2alpha/androgen receptor signals. *Oncotarget* **2015**, *6*, 27555–27565. [CrossRef]
38. Yeh, C.R.; Slavin, S.; Da, J.; Hsu, I.; Luo, J.; Xiao, G.Q.; Ding, J.; Chou, F.J.; Yeh, S. Estrogen receptor α in cancer associated fibroblasts suppresses prostate cancer invasion via reducing CCL5, IL6 and macrophage infiltration in the tumor microenvironment. *Mol. Cancer* **2016**, *15*, 7. [CrossRef]
39. Pienta, K.J.; Machiels, J.P.; Schrijvers, D.; Alekseev, B.; Shkolnik, M.; Crabb, S.J.; Li, S.; Seetharam, S.; Puchalski, T.A.; Takimoto, C.; et al. Phase 2 study of carlumab (CNTO 888), a human monoclonal antibody against CC-chemokine ligand 2 (CCL2), in metastatic castration-resistant prostate cancer. *Investig. New. Drugs.* **2013**, *31*, 760–768. [CrossRef]

40. Vilums, M.; Zweemer, A.J.; Dekkers, S.; Askar, Y.; de Vries, H.; Saunders, J.; Stamos, D.; Brussee, J.; Heitman, L.H.; IJzerman, A.P. Design and synthesis of novel small molecule CCR2 antagonists: Evaluation of 4-aminopiperidine derivatives. *Bioorg. Med. Chem. Lett.* **2014**, *24*, 5377–5380. [CrossRef]
41. Junker, A.; Kokornaczyk, A.K.; Zweemer, A.J.; Frehland, B.; Schepmann, D.; Yamaguchi, J.; Itami, K.; Faust, A.; Hermann, S.; Wagner, S.; et al. Synthesis, binding affinity and structure-activity relationships of novel, selective and dual targeting CCR2 and CCR5 receptor antagonists. *Org. Biomol. Chem.* **2015**, *13*, 2407–2422. [CrossRef]
42. Manning, B.D.; Cantley, L.C. AKT/PKB signaling: Navigating downstream. *Cell* **2007**, *129*, 1261–1274. [CrossRef]
43. Assinder, S.J.; Dong, Q.; Kovacevic, Z.; Richardson, D.R. The TGF-beta, PI3K/Akt and PTEN pathways: Established and proposed biochemical integration in prostate cancer. *Biochem. J.* **2009**, *417*, 411–421. [CrossRef]
44. Majumder, P.K.; Sellers, W.R. Akt-regulated pathways in prostate cancer. *Oncogene* **2005**, *24*, 7465–7474. [CrossRef]
45. New, D.C.; Wu, K.; Kwok, A.W.; Wong, Y.H. G protein-coupled receptor-induced Akt activity in cellular proliferation and apoptosis. *FEBS. J.* **2007**, *274*, 6025–6036. [CrossRef]
46. Loberg, R.D.; Day, L.L.; Harwood, J.; Ying, C.; St John, L.N.; Giles, R.; Neeley, C.K.; Pienta, K.J. CCL2 is a potent regulator of prostate cancer cell migration and proliferation. *Neoplasia* **2006**, *8*, 578–586. [CrossRef]
47. Loberg, R.D.; Ying, C.; Craig, M.; Day, L.L.; Sargent, E.; Neeley, C.; Wojno, K.; Snyder, L.A.; Yan, L.; Pienta, K.J. Targeting CCL2 with systemic delivery of neutralizing antibodies induces prostate cancer tumor regression in vivo. *Cancer Res.* **2007**, *67*, 9417–9424. [CrossRef]
48. Roca, H.; Varsos, Z.; Pienta, K.J. CCL2 protects prostate cancer PC3 cells from autophagic death via phosphatidylinositol 3-kinase/AKT-dependent survivin up-regulation. *J. Biol. Chem.* **2008**, *283*, 25057–25073. [CrossRef]
49. Ishida, T.; Utsunomiya, A.; Jo, T.; Yamamoto, K.; Kato, K.; Yoshida, S.; Takemoto, S.; Suzushima, H.; Kobayashi, Y.; Imaizumi, Y.; et al. Mogamulizumab for relapsed adult T-cell leukemia-lymphoma: Updated follow-up analysis of phase I and II studies. *Cancer Sci.* **2017**, *108*, 2022–2029. [CrossRef]
50. Yamamoto, K.; Utsunomiya, A.; Tobinai, K.; Tsukasaki, K.; Uike, N.; Uozumi, K.; Yamaguchi, K.; Yamada, Y.; Hanada, S.; Tamura, K.; et al. Phase I study of KW-0761, a defucosylated humanized anti-CCR4 antibody, in relapsed patients with adult T-cell leukemia-lymphoma and peripheral T-cell lymphoma. *J. Clin. Oncol.* **2010**, *28*, 1591–1598. [CrossRef]
51. Saura, C.; Roda, D.; Rosello, S.; Oliveira, M.; Macarulla, T.; Pérez-Fidalgo, J.A.; Morales-Barrera, R.; Sanchis-García, J.M.; Musib, L.; Budha, N.; et al. A First-in-Human Phase I Study of the ATP-Competitive AKT Inhibitor Ipatasertib Demonstrates Robust and Safe Targeting of AKT in Patients with Solid Tumors. *Cancer Discov.* **2017**, *7*, 102–113. [CrossRef]
52. Kim, S.B.; Dent, R.; Im, S.A.; Espié, M.; Blau, S.; Tan, A.R.; Isakoff, S.J.; Oliveira, M.; Saura, C.; Wongchenko, M.J.; et al. Ipatasertib plus paclitaxel versus placebo plus paclitaxel as first-line therapy for metastatic triple-negative breast cancer (LOTUS): A multicentre, randomised, double-blind, placebo-controlled, phase 2 trial. *Lancet Oncol.* **2017**, *18*, 1360–1372. [CrossRef]
53. Iwamoto, H.; Izumi, K.; Natsagdorj, A.; Naito, R.; Makino, T.; Kadomoto, S.; Hiratsuka, K.; Shigehara, K.; Kadono, Y.; Narimoto, K.; et al. Coffee diterpenes kahweol acetate and cafestol synergistically inhibit the proliferation and migration of prostate cancer cells. *Prostate* **2019**, *79*, 468–479. [CrossRef]
54. Fang, L.Y.; Izumi, K.; Lai, K.P.; Liang, L.; Li, L.; Miyamoto, H.; Lin, W.J.; Chang, C. Infiltrating macrophages promote prostate tumorigenesis via modulating androgen receptor-mediated CCL4-STAT3 signaling. *Cancer. Res.* **2013**, *73*, 5633–5646. [CrossRef]
55. Izumi, K.; Li, L.; Chang, C. Androgen receptor and immune inflammation in benign prostatic hyperplasia and prostate cancer. *Clin. Investig.* **2014**, *4*, 935–950. [CrossRef]
56. Davis, J.S.; Nastiuk, K.L.; Krolewski, J.J. TNF is necessary for castration-induced prostate regression, whereas TRAIL and FasL are dispensable. *Mol. Endocrinol.* **2011**, *25*, 611–620. [CrossRef]
57. Maolake, A.; Izumi, K.; Natsagdorj, A.; Iwamoto, H.; Kadomoto, S.; Makino, T.; Naito, R.; Shigehara, K.; Kadono, Y.; Hiratsuka, K.; et al. Tumor necrosis factor-alpha induces prostate cancer cell migration in lymphatic metastasis through CCR7 upregulation. *Cancer Sci.* **2018**, *109*, 1524–1531. [CrossRef]

58. Barone, F.; Gardner, D.H.; Nayar, S.; Steinthal, N.; Buckley, C.D.; Luther, S.A. Stromal Fibroblasts in Tertiary Lymphoid Structures: A Novel Target in Chronic Inflammation. *Front. Immunol.* **2016**, *7*, 477. [CrossRef]
59. de Bono, J.S.; Oudard, S.; Ozguroglu, M.; Hansen, S.; Machiels, J.P.; Kocak, I.; Gravis, G.; Bodrogi, I.; Mackenzie, M.J.; Shen, L.; et al. Prednisone plus cabazitaxel or mitoxantrone for metastatic castration-resistant prostate cancer progressing after docetaxel treatment: A randomised open-label trial. *Lancet* **2010**, *376*, 1147–1154. [CrossRef]
60. Machioka, K.; Izumi, K.; Kadono, Y.; Iwamoto, H.; Naito, R.; Makino, T.; Kadomoto, S.; Natsagdorj, A.; Keller, E.T.; Zhang, J.; et al. Establishment and characterization of two cabazitaxel-resistant prostate cancer cell lines. *Oncotarget* **2018**, *9*, 16185–16196. [CrossRef]
61. Natsagdorj, A.; Izumi, K.; Hiratsuka, K.; Machioka, K.; Iwamoto, H.; Naito, R.; Makino, T.; Kadomoto, S.; Shigehara, K.; Kadono, Y.; et al. CCL2 induces resistance to the antiproliferative effect of cabazitaxel in prostate cancer cells. *Cancer Sci.* **2019**, *110*, 279–288. [CrossRef]

7

Pirfenidone, an Anti-Fibrotic Drug, Suppresses the Growth of Human Prostate Cancer Cells by Inducing G_1 Cell Cycle Arrest

Kenichiro Ishii [1,2,*], Takeshi Sasaki [1], Kazuhiro Iguchi [3], Manabu Kato [1], Hideki Kanda [1], Yoshifumi Hirokawa [2], Kiminobu Arima [1], Masatoshi Watanabe [2] and Yoshiki Sugimura [1]

1 Department of Nephro-Urologic Surgery and Andrology, Mie University Graduate School of Medicine, Tsu, Mie 514-8507, Japan; t-sasaki@clin.medic.mie-u.ac.jp (T.S.); katouuro@clin.medic.mie-u.ac.jp (M.K.); kanda331@clin.medic.mie-u.ac.jp (H.K.); kiminobu@clin.medic.mie-u.ac.jp (K.A.); sugimura@clin.medic.mie-u.ac.jp (Y.S.)
2 Department of Oncologic Pathology, Mie University Graduate School of Medicine, Tsu, Mie 514-8507, Japan; ultray2k@doc.medic.mie-u.ac.jp (Y.H.); mawata@doc.medic.mie-u.ac.jp (M.W.)
3 Laboratory of Community Pharmacy, Gifu Pharmaceutical University, Gifu, Gifu 501-1196, Japan; iguchi@gifu-pu.ac.jp
* Correspondence: kenishii@clin.medic.mie-u.ac.jp

Abstract: Pirfenidone (PFD) is an anti-fibrotic drug used to treat idiopathic pulmonary fibrosis by inducing G_1 cell cycle arrest in fibroblasts. We hypothesize that PFD can induce G_1 cell cycle arrest in different types of cells, including cancer cells. To investigate the effects of PFD treatment on the growth of human prostate cancer (PCa) cells, we used an androgen-sensitive human PCa cell line (LNCaP) and its sublines (androgen-low-sensitive E9 and F10 cells and androgen-insensitive AIDL cells), as well as an androgen-insensitive human PCa cell line (PC-3). PFD treatment suppressed the growth of all PCa cells. Transforming growth factor β1 secretion was significantly increased in PFD-treated PCa cells. In both LNCaP and PC-3 cells, PFD treatment increased the population of cells in the G_0/G_1 phase, which was accompanied by a decrease in the S/G_2 cell population. CDK2 protein expression was clearly decreased in PFD-treated LNCaP and PC-3 cells, whereas p21 protein expression was increased in only PFD-treated LNCaP cells. In conclusion, PFD may serve as a novel therapeutic drug that induces G_1 cell cycle arrest in human PCa cells independently of androgen sensitivity. Thus, in the tumor microenvironment, PFD might target not only fibroblasts, but also heterogeneous PCa cells of varying androgen-sensitivity levels.

Keywords: prostate cancer; androgen sensitivity; pirfenidone; TGFβ1; G_1 cell cycle arrest

---

1. Introduction

The number of males diagnosed with prostate cancer (PCa) is increasing worldwide [1]. Most patients with early-stage PCa can be treated with therapies such as radical prostatectomy or irradiation, whereas androgen deprivation therapy (ADT) is the standard systemic therapy given to patients with advanced PCa. Even though ADT induces temporary remission, the majority of patients (approximately 60%) eventually progress to castration-resistant PCa (CRPC), which is associated with a high mortality rate [2,3].

PCa is characterized by multifocal and heterogeneous progression of the primary tumor. In PCa progression, a decrease or loss of androgen sensitivity in PCa cells is a significant clinical concern. CRPC, a heterogeneous disease, exhibits varying degrees of androgen sensitivity. Once PCa cells lose sensitivity to ADT, effective therapies are limited [4]. In the past few years, however, several new options for the treatment of CRPC have been approved, including CYP17 inhibitors, androgen receptor

(AR) antagonists, and taxane [5]. Despite progress in the development of new drugs, there is a strong medical need to optimize the sequence and combination of approved drugs.

Drug repositioning or repurposing is the process of finding new uses for existing drugs [6], provided that additional clinical trials are relatively easy to perform, and the drug safety profiles have been established. In PCa, there have been a number of drug repositioning studies of non-cancer drugs, including the antidiabetic drug troglitazone, which is a ligand for peroxisome proliferator-activated receptor gamma [7]; the antihypertensive drug candesartan, which is an angiotensin II receptor blocker [8]; naftopidil, which is a selective α_1-adrenoceptor antagonist used to treat benign prostatic hyperplasia [9]; and the antiallergy drug, tranilast [10]. A drug repositioning approach helps identify new pharmaceutical processes to transform existing drugs into useful sources of new anticancer drugs [11].

Pirfenidone (PFD) is an established anti-fibrotic and anti-inflammatory drug used to treat idiopathic pulmonary fibrosis, an interstitial lung disease characterized by accumulation of fibroblasts/ myofibroblasts, excessive extracellular matrix production, and altered transforming growth factor β (TGFβ)/bone morphogenetic protein signaling [12,13]. A number of studies have reported that PFD treatment suppresses the growth of and induces G_1 cell cycle arrest in stromal cells, rat hepatic stellate cells [14], and human Tenon fibroblasts [15,16]. Interestingly, PFD treatment has also been reported to suppress the growth of epithelial cells/cancer cells, including human lens epithelial cells [17] and human hepatocellular carcinoma cells [18]. Epidemiologically, Miura et al. reported a reduced incidence of lung cancer in patients with idiopathic pulmonary fibrosis treated with PFD [19]; however, the mechanism of PFD-induced cancer cell suppression is not well characterized.

Many studies on CRPC have used androgen-insensitive PCa cell lines, such as PC-3 and DU145 cells, which do not express AR [20]. These cell lines were derived from highly anaplastic tumors from different metastatic sites in the bone and brain [21,22]. The PC-3 and DU145 cell lines both differ strongly in aggressiveness compared with the androgen-sensitive, AR-positive LNCaP cell line, which was derived from a lymph node metastasis [23]. Comparisons between androgen-sensitive LNCaP cells and androgen-insensitive PC-3 and DU145 cell lines may not be relevant to the acquisition of androgen insensitivity in clinical PCa, because many clinical androgen-insensitive PCa cases express AR. A more accurate model of clinical cancer requires, at the very least, an androgen-insensitive, AR-positive cancer cell line. To compare the biochemical characteristics of androgen-insensitive and sensitive PCa cells, we generated three sublines from androgen-sensitive LNCaP cells: E9 and F10 (androgen-low-sensitive) and AIDL (androgen-insensitive) cells [24–26]. The parental LNCaP cell line and its derivative E9, F10, and AIDL sublines express similar levels of the AR protein, but androgen-dependent secretion of the prostate-specific antigen (PSA) is only detected in LNCaP cells [27]. In this study, we used the LNCaP cell line and its sublines to investigate the effects of PFD treatment on the growth of human PCa cells, focusing on androgen sensitivity.

2. Materials and Methods

2.1. Materials

PFD was purchased from Tokyo Chemical Industry Co., Ltd. (Tokyo, Japan). Rabbit monoclonal anti-p21 and anti-CDK2 antibodies were purchased from Cell Signaling Technology (Danvers, MA, USA). Rabbit polyclonal anti-phospho-Akt (Ser473) and anti-Akt antibodies were purchased from Cell Signaling Technology. Mouse monoclonal anti-β-actin (clone AC-15) antibody was purchased from Sigma-Aldrich Co. (St. Louis, MO, USA). Rabbit polyclonal anti-PSA antibody was purchased from Dako Cytomation (Copenhagen, Denmark). Rabbit polyclonal anti-AR (N-20) antibody was purchased from Santa Cruz Biotechnology (Santa Cruz, CA, USA).

2.2. Cell Culture

The androgen-sensitive, AR-positive human PCa cell line LNCaP, and the androgen-insensitive, AR-negative human PCa cell line PC-3 were obtained from the American Type Culture Collection (Manassas, VA, USA). LNCaP and PC-3 cells were authenticated by the short tandem repeat method and cultured in RPMI 1640 medium (Nacalai Tesque, Kyoto, Japan) supplemented with 10% fetal bovine serum (Sigma-Aldrich Co.). Androgen-low-sensitive E9 and F10 cells were established from androgen-sensitive LNCaP cells using a limiting dilution method under regular culture conditions [24,25]. In contrast, androgen-insensitive AIDL cells were established from LNCaP cells by continuous passaging under hormone-depleted conditions [26]. The androgen sensitivity of the parental LNCaP cells and the E9, F10, and AIDL cells were confirmed by the change in *KLK3* (PSA) mRNA expression after treatment with the synthetic androgen R1881 [20].

2.3. Cell Viability Assay

To assess cell viability after the PFD treatments, LNCaP, E9, F10, AIDL, and PC-3 cells were plated in 12-well plates at 5×10^4 to 1×10^5 cells/well. PFD (0.1 and 0.3 mg/mL) or vehicle-only (0.1% dimethyl sulfoxide [DMSO]) was added on day two, and the cells were cultured for an additional three days. The cells were detached by trypsinization and counted using the Countess II Automated Cell Counter (Thermo Fisher Scientific Inc., Waltham, MA, USA). Cell viability was assessed by trypan blue exclusion assay.

2.4. Cell Cycle Analysis

LNCaP or PC-3 cells (1.5×10^5 cells) were seeded into 100-mm culture dishes (Sumitomo Bakelite Co., Ltd., Tokyo, Japan). Twenty-four hours after seeding, the cells were treated with 0.1 or 0.3 mg/mL PFD or vehicle (0.1% DMSO) for 24 h. After treatment, the cells were isolated, and the nuclei were stained using the BD Cycletest Plus DNA Reagent Kit (BD Biosciences, San Jose, CA, USA). To determine the cell cycle distribution, the DNA content of the stained cells was analyzed using the BD FACS Canto II flow cytometer (BD Biosciences), as described previously [28].

2.5. Apoptosis Assay

LNCaP cells (6×10^5 cells) and PC-3 cells (4×10^5 cells) were seeded in 100 mm culture dishes (Sumitomo Bakelite Co., Ltd.). 24 h after seeding, the cells were treated with 0.1 or 0.3 mg/mL PFD, or vehicle (0.1% DMSO), for 48 h (LNCaP cells) or 72 h (PC-3 cells). After treatment, the cells were trypsinized, collected, and stained with annexin V–fluorescein isothiocyanate and propidium iodide simultaneously using the Annexin V-FITC Apoptosis Detection kit (BD Biosciences). The cell suspensions were analyzed using the BD FACS Canto II flow cytometer (BD Biosciences) to determine the percentage of apoptotic (annexin V–fluorescein isothiocyanate staining) and necrotic (propidium iodide staining) cells, as described previously [28]. A minimum of 20,000 cells were collected for all samples.

2.6. ELISA

For quantitative determination of TGFβ1 and PSA proteins, aliquots of conditioned medium from PCa cells were collected and subjected to ELISA using the Quantikine® human TGF-β1 immunoassay kit (R&D Systems, Inc., Minneapolis, MN, USA) and PSA Enzyme Immunoassay Test Kit (Hope Laboratories, Belmont, CA, USA), respectively.

2.7. Preparation of Cell Lysates

LNCaP or PC-3 cells (1×10^6) were seeded in 100 mm culture dishes (Sumitomo Bakelite Co., Ltd.). 24 h after seeding, the cells were treated with PFD (0.1 or 0.3 mg/mL) or vehicle (0.1% DMSO) for 48 h. The cells were harvested by scraping, and whole cell lysates were prepared as described previously [27]. Briefly, the cells were washed with ice-cold phosphate-buffered saline and lysed with CelLytic™ (Sigma-Aldrich Co.) containing 1% Nonidet P-40, 10 mM 4-(2-aminoethyl) benzensulfonyl fluoride, 0.8 mM aprotinin, 50 mM bestatin, 15 mM E-64, 20 mM leupeptin, and 10 mM pepstatin. After 60 min on ice, the lysates were centrifuged at 10,000 *g* for 10 min, and the supernatants were collected. The protein concentration was measured using the NanoDrop 2000 instrument (Thermo Fisher Scientific Inc.).

2.8. Western Blot Analysis

Extracted proteins were separated by gel electrophoresis and transferred to Immobilon polyvinylidene difluoride membranes (Merck Millipore, Darmstadt, Germany) following our previously reported protocol [27]. The anti-AR, anti-PSA, anti-phospho-Akt (Ser473), anti-Akt, and anti-β-actin antibodies were used at dilutions of 1:2500, 1:5000, 1:1000, 1:1000, and 1:5000, respectively. Specific protein bands were visualized using the SuperSignal™ West Pico Chemiluminescent Substrate (Thermo Fisher Scientific Inc.) with the LAS-4000 Mini (Fuji Photo Film, Tokyo, Japan).

2.9. Statistical Analysis

Results are expressed as means ± standard deviation. Differences between two groups were determined using Student's *t*-test. Values of $p < 0.05$ were considered statistically significant.

3. Results

3.1. Effects of Pirfenidone Treatment on the Growth of Prostate Cancer Cells (LNCaP, LNCaP Sublines, and PC-3)

First, we confirmed that PFD treatment suppresses the growth of fibroblasts. PFD treatment (0.3 mg/mL) for 72 h suppressed the growth of commercially available prostate stromal cells (data not shown). Using these experimental conditions, we treated the PCa cells (LNCaP, E9, F10, AIDL, and PC-3) with PFD and found that PFD treatment suppressed the growth of all cell lines (Figure 1A). Among the LNCaP cells and sublines, growth suppression was more pronounced in LNCaP and E9 cells than in F10 and AIDL cells. We also assessed TGFβ1 secretion from PCa cells because of its relationship to cell cycle and apoptosis. TGFβ1 levels were measured in the culture medium of the PCa cells using ELISA. TGFβ1 secretion was significantly increased by PFD treatment in all PCa cells evaluated (Figure 1B). Among the LNCaP cells and sublines, the increase in TGFβ1 secretion was greater in LNCaP and E9 cells than in F10 and AIDL cells.

(A)
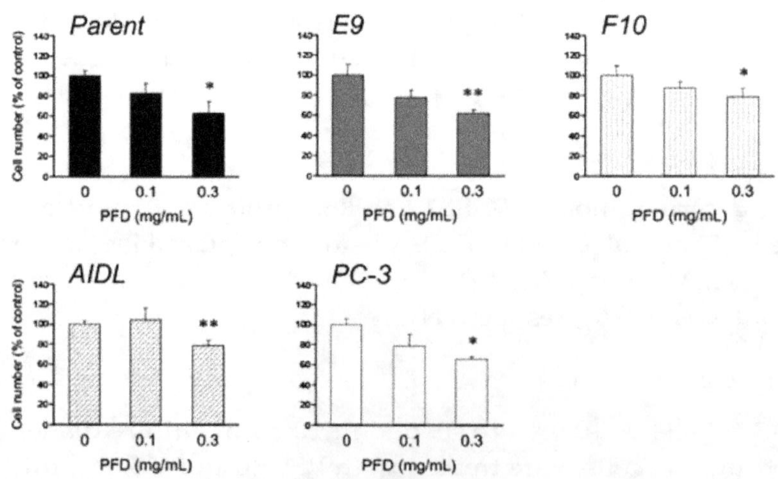

Figure 1. *Cont.*

(B)

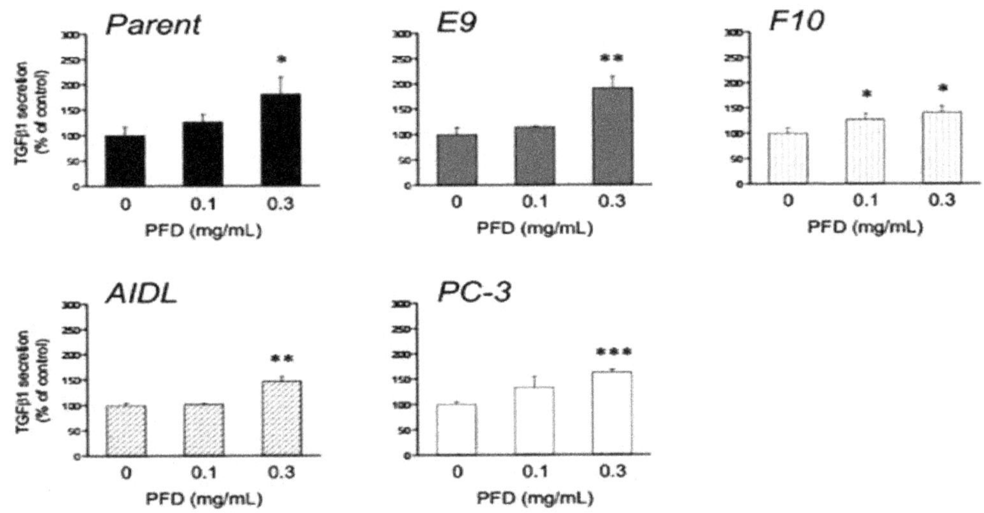

Figure 1. Effects of pirfenidone (PFD) treatment on the growth and secretion of transforming growth factor β1 (TGFβ1) of human prostate cancer (PCa) cells. Human PCa cells (parental LNCaP cell line and the LNCaP sublines E9, F10, and AIDL; and the PC-3 cell line) were plated in 12-well plates and treated with PFD for three days. Effects of PFD treatment on the (**A**) growth and (**B**) TGFβ1 secretion of human PCa cells. Data are representative of three independent experiments, and the values represent the means ± standard deviation. *, $p < 0.05$; **, $p < 0.01$; ***, $p < 0.001$ versus the vehicle-treated control.

3.2. Pirfenidone Antiproliferative Mechanisms in LNCaP and PC-3 Cells

To investigate whether PFD treatment affects the cell cycle, we performed flow cytometric and Western blot analyses of cell-cycle regulatory proteins. In both LNCaP and PC-3 cells, PFD treatment increased the population of cells in the G_0/G_1 phase, which was accompanied by a decrease in S/G_2 phase cells (Figure 2, Tables 1 and 2). p21 protein expression was increased by PFD treatment in LNCaP cells, but was not detected in PC-3 cells (Figure 3). Of note, PFD-increased p21 protein expression was the highest in E9 cells (Figure S1). In contrast, CDK2 protein expression was clearly decreased in both PFD-treated LNCaP and PC-3 cells. Of note, PFD treatment did not induce early apoptosis in either LNCaP or PC-3 cells.

Table 1. Effects of pirfenidone (PFD) treatment on cell cycle progression in LNCaP cells.

PFD (mg/mL)	Phase (%)		
	G_0/G_1	S	G_2/M
0	61.8 ± 0.9	19.2 ± 0.3	17.8 ± 0.5
0.1	64.9 ± 0.7 **	16.8 ± 0.7 *	17.2 ± 0.5
0.3	72.3 ± 0.7 ***	12.0 ± 0.2 ***	14.7 ± 0.5 **

*, $p < 0.05$; **, $p < 0.01$; ***, $p < 0.001$ versus vehicle-treated control.

Table 2. Effects of pirfenidone (PFD) treatment on cell cycle progression in PC-3 cells.

PFD (mg/mL)	Phase (%)		
	G_0/G_1	S	G_2/M
0	45.5 ± 1.3	20.5 ± 0.4	26.7 ± 0.7
0.1	51.1 ± 0.7 **	16.1 ± 0.1 **	25.5 ± 0.3
0.3	55.6 ± 0.6 **	15.2 ± 0.9 **	23.0 ± 0.9 **

**, $p < 0.01$ versus vehicle-treated control.

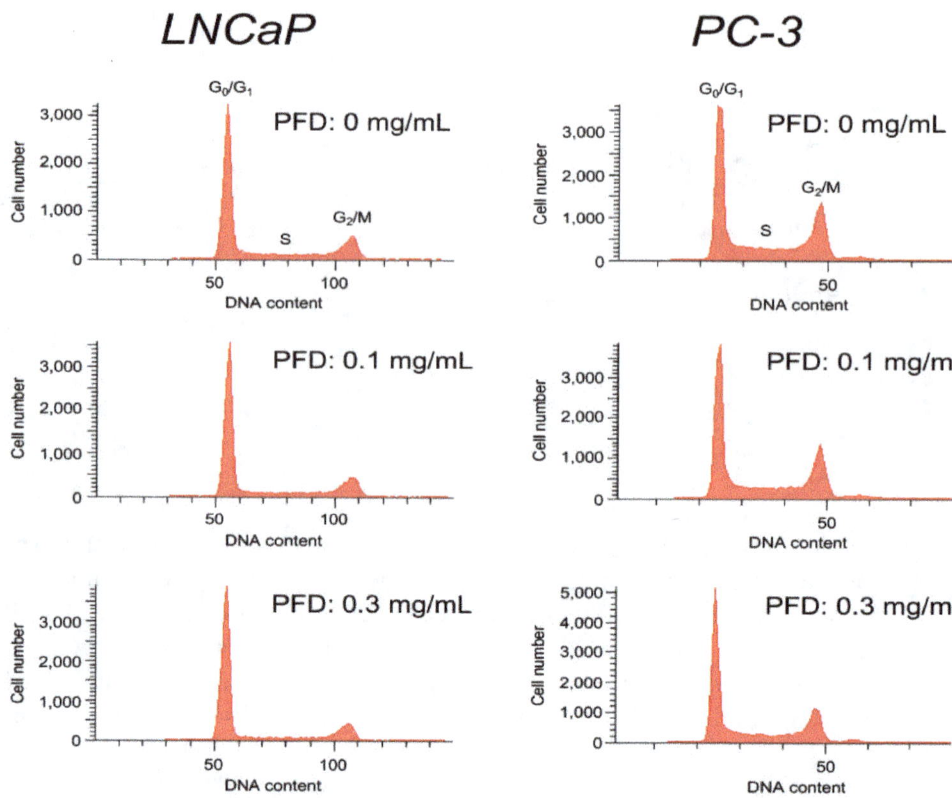

Figure 2. Cell cycle analysis by flow cytometry of human prostate cancer cells treated with pirfenidone (PFD). The cell cycle was determined by propidium iodide (PI) staining, as detailed in the "Material and Methods" section. The proportions of cells in the G_0/G_1, S, and G_2/M phase were calculated from one representative experiment (n = 3).

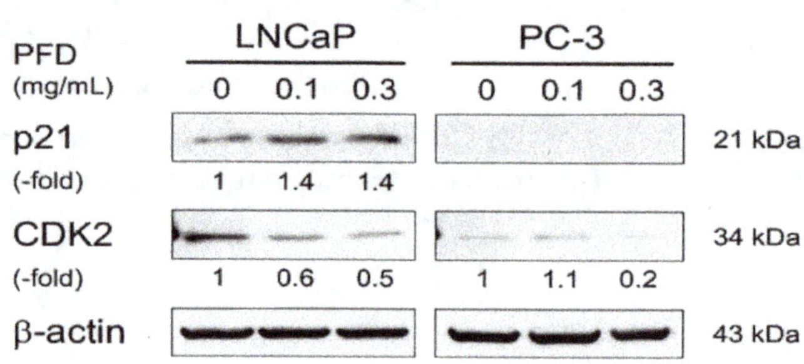

Figure 3. Effects of pirfenidone (PFD) treatment on the expression of cell cycle-related proteins in human prostate cancer cells. Both LNCaP and PC-3 cells were plated in 100 mm dishes and treated with PFD for two days. Cell lysates (50 μg) were separated by electrophoresis using a 12.5% SDS–polyacrylamide gel. After separation, the proteins in the gel were transferred to a polyvinylidene difluoride membrane by electroblotting. p21 and CDK2 protein levels were determined by Western blot analysis using specific antibodies. Equal loading of the samples was confirmed by measuring β-actin protein levels.

3.3. Effects of Pirfenidone Treatment on Androgen Receptor Signaling-Related Protein Levels in LNCaP and PC-3 Cells

To confirm specific inhibition of PFD treatment on the AR signaling pathway, we evaluated the protein levels of AR and PSA in LNCaP cells by Western blot analysis. AR protein expression was not changed, but PSA protein expression was decreased by PFD treatment (Figure 4). In both LNCaP and PC-3 cells, PFD treatment slightly decreased the level of phospho-Akt (Ser473), suggesting slight inhibition of Akt phosphorylation. Of note, AR and PSA protein expression was not detected in PC-3 cells as reported previously [20].

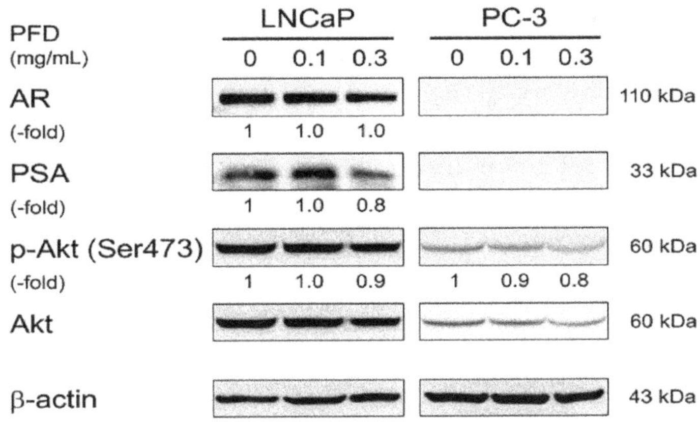

Figure 4. Effects of pirfenidone (PFD) treatment on androgen receptor signaling-related protein levels in human prostate cancer cells. Both LNCaP and PC-3 cells were plated in 100 mm dishes and treated with PFD for an additional two days. Cell lysates (50 μg) were separated by electrophoresis using a 12.5% SDS–polyacrylamide gel. After separation, proteins in the gel were transferred to a polyvinylidene difluoride membrane by electroblotting. Androgen receptor, prostate-specific antigen, phospho-Akt (Ser473), and total Akt protein levels were determined by Western blot analysis using specific antibodies. Equal loading of the samples was confirmed by measuring β-actin levels.

We further evaluated the effects of PFD treatment on PSA secretion by measuring PSA protein levels in conditioned medium from PFD-treated LNCaP cell cultures using ELISA. The PSA protein level was significantly reduced in LNCaP cell culture medium, suggesting that PSA secretion was inhibited by PFD treatment (Figure 5).

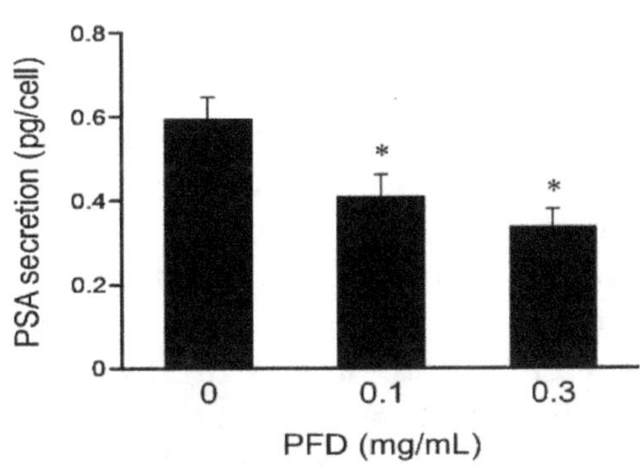

Figure 5. Effects of pirfenidone (PFD) treatment on prostate-specific antigen (PSA) secretion from human prostate cancer cells. The level of PSA secreted from LNCaP cells was determined by measuring the PSA level in LNCaP conditioned medium by ELISA. Values represent the means ± standard deviation. *, $p < 0.05$ versus the vehicle-treated control.

4. Discussion

Drug repositioning of the anti-fibrotic compound PFD to cancer treatment is not a novel idea, but investigation of the effects of PFD on PCa progression, considering the extracellular-matrix-rich microenvironment of PCa, would provide meaningful information for this potential application of PFD. In this study, we demonstrated that PFD treatment suppressed the growth and induced G_1 cell cycle arrest in various PCa cell lines that differed in androgen sensitivity, suggesting that PFD may target not only fibroblasts but also heterogeneous PCa cells within the tumor microenvironment.

In animal models of fibrosis, PFD induced anti-fibrotic effects mainly via inhibition of TGFβ signaling in fibroblasts. TGFβ is a multifunctional cytokine that regulates cell proliferation, extracellular matrix production and degradation, cell differentiation, and apoptosis [29]. In animal models of fibrosis, PFD treatment inhibited fibrosis, which was associated with down-regulation of TGFβ, platelet-derived growth factor, and collagen synthesis in various types of cells, including human lung fibroblasts [30], rat hepatic stellate cells [14], human pancreatic stellate cells [31], rat renal fibroblasts [32], human Tenon fibroblasts [15], and rat cardiac fibroblasts [33].

In contrast, our results showed that PFD treatment significantly increased TGFβ secretion from all PCa cells evaluated, regardless of androgen sensitivity. Our previous study reported that TGFβ1 secretion from PCa cells was quite low compared with that from fibroblasts, especially carcinoma-associated fibroblasts [27]. TGFβ participates in cell proliferation and differentiation not only in normal processes such as embryonic development and wound healing, but also abnormal processes such as cancer progression and angiogenesis [34]. Although a number of studies have investigated the role of TGFβ, the results are still controversial. Importantly, the TGFβ signaling pathway is involved in both tumor-suppressive and tumor-promoting roles. The presence of TGFβ in the tumor microenvironment may promote tumor growth by enhancing stromal support and angiogenesis and by impairing immune surveillance [35]. In contrast, TGFβ plays a tumor-suppressive role by inducing G_1 cell cycle arrest in various cell types, such as epithelial, endothelial, and hematopoietic cells and fibroblasts [36]. Cell-cycle inhibition by TGFβ is mediated in part by the up-regulation of antiproliferative proteins such as $p15^{INK4b}$, $p21^{CIP1}$, and $p27^{KIP1}$. In this study, the $p21^{CIP1}$ protein level was increased in PFD-treated LNCaP cells. Increased expression of $p21^{CIP1}$, a cell-cycle-inhibitory protein, is not only associated with cell cycle inhibition, but also cell differentiation and senescence [37].

Lin et al. reported that the Akt pathway is associated with AR activation in LNCaP cells [38]. Iguchi et al. demonstrated that inhibition of Akt phosphorylation by the PI3K inhibitor LY294002 reduced PSA expression in LNCaP cells [39]. Previous studies reported that PFD treatment inhibits phosphorylation of Akt in rat hepatocytes [40], human lung fibroblasts [30], and human Tenon fibroblasts [16]. Similarly, in this study, PFD treatment slightly inhibited the phosphorylation of Akt in both LNCaP and PC-3 cells. In addition, PFD treatment reduced PSA protein expression and secretion in LNCaP cells.

Serum PSA levels are influenced by a number of drugs, such as non-steroidal anti-inflammatory drugs and statins [41,42]; for example, the serum PSA level was found to be lower in aspirin users than non-users [41]. In contrast, Iguchi et al. reported that betamethasone, an agonist of the glucocorticoid receptor, increased PSA mRNA expression in LNCaP cells [43]. Similar to PFD, the antidiabetic drug troglitazone, which is a ligand of peroxisome proliferator-activated receptor gamma, reduced PSA expression in LNCaP cells [7]. Previous studies and our results suggest that PFD treatment reduces PSA expression, which is associated with inhibition of Akt phosphorylation in LNCaP cells.

The tumor microenvironment of the prostate is highly complex and heterogeneous, and is composed of carcinoma-associated fibroblasts as well as epithelial cancer cells that infiltrate into the surrounding tumor stroma, referred to as reactive stroma [44]. This heterogenous stromal component of the prostate contains multiple populations of fibroblasts that are associated with tumorigenesis [45,46]. In this study, we demonstrated that the anti-fibrotic drug PFD suppressed the growth of human PCa cells by inducing G_1 cell cycle arrest. In addition, our data suggest that PFD-induced growth suppression occurs independently of androgen sensitivity. Therefore, PFD may provide a novel therapeutic option for targeting not only fibroblasts surrounding cancer cells, but also heterogeneous PCa cells of varying androgen sensitivities within patients with CRPC.

5. Conclusions

In our studies of drug repositioning, we demonstrate that PFD may serve as a novel therapeutic drug that induces G_1 cell cycle arrest in human PCa cells independently of androgen sensitivity.

Supplementary Materials:
Figure S1: Effects of pirfenidone (PFD) treatment on the expression of cell cycle-related proteins in human prostate cancer cells. E9, F10, and AIDL cells were plated in 100-mm dishes and treated with PFD for 2 days. Cell lysates (50 μg) were separated by electrophoresis using a 12.5% SDS–polyacrylamide gel. After separation, the proteins in the gel were transferred to a polyvinylidene difluoride membrane by electroblotting. p21 and CDK2 protein levels were determined by Western blot analysis using specific antibodies. Equal loading of the samples was confirmed by measuring β-actin protein levels.

Author Contributions: conceptualization, M.W. and Y.S.; investigation, T.S., M.K., K.I. (Kazuhiro Iguchi) and H.K.; writing—original draft preparation, K.I. (Kenichiro Ishii); writing—review and editing, Y.H. and K.A.

Acknowledgments: We would like to thank Yumi Yoshikawa and Izumi Matsuoka for their technical support.

References

1. Gronberg, H. Prostate cancer epidemiology. *Lancet* **2003**, *361*, 859–864. [CrossRef]
2. Huggins, C.; Hodges, C.V. Studies on prostatic cancer: I. The effect of castration, of estrogen and of androgen injection on serum phosphatases in metastatic carcinoma of the prostate. *J. Urol.* **2002**, *168*, 9–12. [CrossRef]
3. Fizazi, K.; Higano, C.S.; Nelson, J.B.; Gleave, M.; Miller, K.; Morris, T.; Nathan, F.E.; McIntosh, S.; Pemberton, K.; Moul, J.W. Phase III, randomized, placebo-controlled study of docetaxel in combination with zibotentan in patients with metastatic castration-resistant prostate cancer. *J. Clin. Oncol.* **2013**, *31*, 1740–1747. [CrossRef] [PubMed]
4. Mukherji, D.; Omlin, A.; Pezaro, C.; Shamseddine, A.; de Bono, J. Metastatic castration-resistant prostate cancer (CRPC): Preclinical and clinical evidence for the sequential use of novel therapeutics. *Cancer Metast. Rev.* **2014**, *33*, 555–566. [CrossRef]
5. Nevedomskaya, E.; Baumgart, S.J.; Haendler, B. Recent Advances in Prostate Cancer Treatment and Drug Discovery. *Int. J. Mol. Sci.* **2018**, *19*, 1359. [CrossRef]
6. Shim, J.S.; Liu, J.O. Recent advances in drug repositioning for the discovery of new anticancer drugs. *Int. J. Biol. Sci.* **2014**, *10*, 654–663. [CrossRef] [PubMed]
7. Hisatake, J.I.; Ikezoe, T.; Carey, M.; Holden, S.; Tomoyasu, S.; Koeffler, H.P. Down-Regulation of prostate-specific antigen expression by ligands for peroxisome proliferator-activated receptor gamma in human prostate cancer. *Cancer Res.* **2000**, *60*, 5494–5498.
8. Uemura, H.; Ishiguro, H.; Nakaigawa, N.; Nagashima, Y.; Miyoshi, Y.; Fujinami, K.; Sakaguchi, A.; Kubota, Y. Angiotensin II receptor blocker shows antiproliferative activity in prostate cancer cells: A possibility of tyrosine kinase inhibitor of growth factor. *Mol. Cancer Ther.* **2003**, *2*, 1139–1147. [PubMed]
9. Kanda, H.; Ishii, K.; Ogura, Y.; Imamura, T.; Kanai, M.; Arima, K.; Sugimura, Y. Naftopidil, a selective alpha-1 adrenoceptor antagonist, inhibits growth of human prostate cancer cells by G1 cell cycle arrest. *Int. J. Cancer* **2008**, *122*, 444–451. [CrossRef] [PubMed]
10. Izumi, K.; Mizokami, A.; Li, Y.Q.; Narimoto, K.; Sugimoto, K.; Kadono, Y.; Kitagawa, Y.; Konaka, H.; Koh, E.; Keller, E.T.; et al. Tranilast inhibits hormone refractory prostate cancer cell proliferation and suppresses transforming growth factor beta1-associated osteoblastic changes. *Prostate* **2009**, *69*, 1222–1234. [CrossRef]
11. Olgen, S.; Kotra, L. Drug Repurposing in the Development of Anticancer Agents. *Curr. Med. Chem.* **2018**. [CrossRef] [PubMed]
12. Raghu, G.; Selman, M. Nintedanib and pirfenidone. New antifibrotic treatments indicated for idiopathic pulmonary fibrosis offer hopes and raises questions. *Am. J. Respir. Crit. Care Med.* **2015**, *191*, 252–254. [CrossRef] [PubMed]
13. Koli, K.; Myllarniemi, M.; Vuorinen, K.; Salmenkivi, K.; Ryynanen, M.J.; Kinnula, V.L.; Keski-Oja, J. Bone morphogenetic protein-4 inhibitor gremlin is overexpressed in idiopathic pulmonary fibrosis. *Am. J. Pathol.* **2006**, *169*, 61–71. [CrossRef]
14. Xiang, X.H.; Jiang, T.P.; Zhang, S.; Song, J.; Li, X.; Yang, J.Y.; Zhou, S. Pirfenidone inhibits proliferation, arrests the cell cycle, and downregulates heat shock protein-47 and collagen type I in rat hepatic stellate cells in vitro. *Mol. Med. Rep.* **2015**, *12*, 309–314. [CrossRef] [PubMed]
15. Lin, X.; Yu, M.; Wu, K.; Yuan, H.; Zhong, H. Effects of pirfenidone on proliferation, migration, and collagen contraction of human Tenon's fibroblasts in vitro. *Invest. Ophthalmol. Vis. Sci.* **2009**, *50*, 3763–3770. [CrossRef]
16. Guo, X.; Yang, Y.; Liu, L.; Liu, X.; Xu, J.; Wu, K.; Yu, M. Pirfenidone Induces G1 Arrest in Human Tenon's Fibroblasts In Vitro Involving AKT and MAPK Signaling Pathways. *J. Ocul. Pharmacol. Ther.* **2017**, *33*, 366–374. [CrossRef]

17. Yang, Y.; Ye, Y.; Lin, X.; Wu, K.; Yu, M. Inhibition of pirfenidone on TGF-β2 induced proliferation, migration and epithlial-mesenchymal transition of human lens epithelial cells line SRA01/04. *PLoS ONE* **2013**, *8*, e56837. [CrossRef]
18. Zou, W.J.; Huang, Z.; Jiang, T.P.; Shen, Y.P.; Zhao, A.S.; Zhou, S.; Zhang, S. Pirfenidone Inhibits Proliferation and Promotes Apoptosis of Hepatocellular Carcinoma Cells by Inhibiting the Wnt/beta-Catenin Signaling Pathway. *Med. Sci. Monit.* **2017**, *23*, 6107–6113. [CrossRef]
19. Miura, Y.; Saito, T.; Tanaka, T.; Takoi, H.; Yatagai, Y.; Inomata, M.; Nei, T.; Saito, Y.; Gemma, A.; Azuma, A. Reduced incidence of lung cancer in patients with idiopathic pulmonary fibrosis treated with pirfenidone. *Respir. Investig.* **2018**, *56*, 72–79. [CrossRef]
20. Ishii, K.; Imamura, T.; Iguchi, K.; Arase, S.; Yoshio, Y.; Arima, K.; Hirano, K.; Sugimura, Y. Evidence that androgen-independent stromal growth factor signals promote androgen-insensitive prostate cancer cell growth in vivo. *Endocr. Relat. Cancer* **2009**, *16*, 415–428. [CrossRef]
21. Kaighn, M.E.; Narayan, K.S.; Ohnuki, Y.; Lechner, J.F.; Jones, L.W. Establishment and characterization of a human prostatic carcinoma cell line (PC-3). *Investig. Urol.* **1979**, *17*, 16–23.
22. Stone, K.R.; Mickey, D.D.; Wunderli, H.; Mickey, G.H.; Paulson, D.F. Isolation of a human prostate carcinoma cell line (DU 145). *Int. J. Cancer* **1978**, *21*, 274–281. [CrossRef] [PubMed]
23. Horoszewicz, J.S.; Leong, S.S.; Chu, T.M.; Wajsman, Z.L.; Friedman, M.; Papsidero, L.; Kim, U.; Chai, L.S.; Kakati, S.; Arya, S.K.; et al. The LNCaP cell line—A new model for studies on human prostatic carcinoma. *Prog. Clin. Biol. Res.* **1980**, *37*, 115–132. [PubMed]
24. Iguchi, K.; Ishii, K.; Nakano, T.; Otsuka, T.; Usui, S.; Sugimura, Y.; Hirano, K. Isolation and characterization of LNCaP sublines differing in hormone sensitivity. *J. Androl.* **2007**, *28*, 670–678. [CrossRef] [PubMed]
25. Iguchi, K.; Hayakawa, Y.; Ishii, K.; Matsumoto, K.; Usui, S.; Sugimura, Y.; Hirano, K. Characterization of the low pH/low nutrient-resistant LNCaP cell subline LNCaP-F10. *Oncol. Rep.* **2012**, *28*, 2009–2015. [CrossRef]
26. Onishi, T.; Yamakawa, K.; Franco, O.E.; Kawamura, J.; Watanabe, M.; Shiraishi, T.; Kitazawa, S. Mitogen-activated protein kinase pathway is involved in alpha6 integrin gene expression in androgen-independent prostate cancer cells: Role of proximal Sp1 consensus sequence. *Biochim. Biophys. Acta* **2001**, *1538*, 218–227. [CrossRef]
27. Ishii, K.; Sasaki, T.; Iguchi, K.; Kajiwara, S.; Kato, M.; Kanda, H.; Hirokawa, Y.; Arima, K.; Mizokami, A.; Sugimura, Y. Interleukin-6 induces VEGF secretion from prostate cancer cells in a manner independent of androgen receptor activation. *Prostate* **2018**, *78*, 849–856. [CrossRef]
28. Ishii, K.; Matsuoka, I.; Kajiwara, S.; Sasaki, T.; Miki, M.; Kato, M.; Kanda, H.; Arima, K.; Shiraishi, T.; Sugimura, Y. Additive naftopidil treatment synergizes docetaxel-induced apoptosis in human prostate cancer cells. *J. Cancer Res. Clin. Oncol.* **2018**, *144*, 89–98. [CrossRef]
29. Thomas, B.J.; Kan, O.K.; Loveland, K.L.; Elias, J.A.; Bardin, P.G. In the Shadow of Fibrosis: Innate Immune Suppression Mediated by Transforming Growth Factor-beta. *Am. J. Respir. Cell Mol. Biol.* **2016**, *55*, 759–766. [CrossRef]
30. Conte, E.; Gili, E.; Fagone, E.; Fruciano, M.; Iemmolo, M.; Vancheri, C. Effect of pirfenidone on proliferation, TGF-beta-induced myofibroblast differentiation and fibrogenic activity of primary human lung fibroblasts. *Eur. J. Pharm. Sci.* **2014**, *58*, 13–19. [CrossRef] [PubMed]
31. Kozono, S.; Ohuchida, K.; Eguchi, D.; Ikenaga, N.; Fujiwara, K.; Cui, L.; Mizumoto, K.; Tanaka, M. Pirfenidone inhibits pancreatic cancer desmoplasia by regulating stellate cells. *Cancer Res.* **2013**, *73*, 2345–2356. [CrossRef] [PubMed]
32. Hewitson, T.D.; Kelynack, K.J.; Tait, M.G.; Martic, M.; Jones, C.L.; Margolin, S.B.; Becker, G.J. Pirfenidone reduces in vitro rat renal fibroblast activation and mitogenesis. *J. Nephrol.* **2001**, *14*, 453–460. [PubMed]
33. Shi, Q.; Liu, X.; Bai, Y.; Cui, C.; Li, J.; Li, Y.; Hu, S.; Wei, Y. In vitro effects of pirfenidone on cardiac fibroblasts: Proliferation, myofibroblast differentiation, migration and cytokine secretion. *PLoS ONE* **2011**, *6*, e28134. [CrossRef] [PubMed]
34. Bhowmick, N.A.; Neilson, E.G.; Moses, H.L. Stromal fibroblasts in cancer initiation and progression. *Nature* **2004**, *432*, 332–337. [CrossRef] [PubMed]
35. Miyazono, K.; Katsuno, Y.; Koinuma, D.; Ehata, S.; Morikawa, M. Intracellular and extracellular TGF-beta signaling in cancer: Some recent topics. *Front. Med.* **2018**. [CrossRef] [PubMed]
36. Siegel, P.M.; Massague, J. Cytostatic and apoptotic actions of TGF-beta in homeostasis and cancer. *Nat. Rev. Cancer* **2003**, *3*, 807–821. [CrossRef]

37. Georgakilas, A.G.; Martin, O.A.; Bonner, W.M. p21: A Two-Faced Genome Guardian. *Trends Mol. Med.* **2017**, *23*, 310–319. [CrossRef]
38. Lin, H.K.; Hu, Y.C.; Yang, L.; Altuwaijri, S.; Chen, Y.T.; Kang, H.Y.; Chang, C. Suppression versus induction of androgen receptor functions by the phosphatidylinositol 3-kinase/Akt pathway in prostate cancer LNCaP cells with different passage numbers. *J. Biol. Chem.* **2003**, *278*, 50902–50907. [CrossRef]
39. Iguchi, K.; Fukami, K.; Ishii, K.; Otsuka, T.; Usui, S.; Sugimura, Y.; Hirano, K. Low androgen sensitivity is associated with low levels of Akt phosphorylation in LNCaP-E9 cells. *J. Androl.* **2012**, *33*, 660–666. [CrossRef]
40. Nakanishi, H.; Kaibori, M.; Teshima, S.; Yoshida, H.; Kwon, A.H.; Kamiyama, Y.; Nishizawa, M.; Ito, S.; Okumura, T. Pirfenidone inhibits the induction of iNOS stimulated by interleukin-1beta at a step of NF-kappaB DNA binding in hepatocytes. *J. Hepatol.* **2004**, *41*, 730–736. [CrossRef] [PubMed]
41. Murad, A.S.; Down, L.; Davey Smith, G.; Donovan, J.L.; Athene Lane, J.; Hamdy, F.C.; Neal, D.E.; Martin, R.M. Associations of aspirin, nonsteroidal anti-inflammatory drug and paracetamol use with PSA-detected prostate cancer: Findings from a large, population-based, case-control study (the ProtecT study). *Int. J. Cancer* **2011**, *128*, 1442–1448. [CrossRef] [PubMed]
42. Yokomizo, A.; Shiota, M.; Kashiwagi, E.; Kuroiwa, K.; Tatsugami, K.; Inokuchi, J.; Takeuchi, A.; Naito, S. Statins reduce the androgen sensitivity and cell proliferation by decreasing the androgen receptor protein in prostate cancer cells. *Prostate* **2011**, *71*, 298–304. [CrossRef] [PubMed]
43. Iguchi, K.; Hashimoto, M.; Kubota, M.; Yamashita, S.; Nakamura, M.; Usui, S.; Sugiyama, T.; Hirano, K. Effects of 14 frequently used drugs on prostate-specific antigen expression in prostate cancer LNCaP cells. *Oncol. Lett.* **2014**, *7*, 1665–1668. [CrossRef] [PubMed]
44. Ishii, K.; Takahashi, S.; Sugimura, Y.; Watanabe, M. Role of Stromal Paracrine Signals in Proliferative Diseases of the Aging Human Prostate. *J. Clin. Med.* **2018**, *7*. [CrossRef]
45. Franco, O.E.; Jiang, M.; Strand, D.W.; Peacock, J.; Fernandez, S.; Jackson, R.S., 2nd; Revelo, M.P.; Bhowmick, N.A.; Hayward, S.W. Altered TGF-beta signaling in a subpopulation of human stromal cells promotes prostatic carcinogenesis. *Cancer Res.* **2011**, *71*, 1272–1281. [CrossRef] [PubMed]
46. Kiskowski, M.A.; Jackson, R.S., 2nd; Banerjee, J.; Li, X.; Kang, M.; Iturregui, J.M.; Franco, O.E.; Hayward, S.W.; Bhowmick, N.A. Role for stromal heterogeneity in prostate tumorigenesis. *Cancer Res.* **2011**. [CrossRef]

8

Loss of Fibroblast-Dependent Androgen Receptor Activation in Prostate Cancer Cells is Involved in the Mechanism of Acquired Resistance to Castration

Kenichiro Ishii [1,2,*], Izumi Matsuoka [1], Takeshi Sasaki [1], Kohei Nishikawa [1], Hideki Kanda [1], Hiroshi Imai [3], Yoshifumi Hirokawa [2], Kazuhiro Iguchi [4], Kiminobu Arima [1] and Yoshiki Sugimura [1]

1 Department of Nephro-Urologic Surgery and Andrology, Mie University Graduate School of Medicine, Tsu, Mie 514-8507, Japan
2 Department of Oncologic Pathology, Mie University Graduate School of Medicine, Tsu, Mie 514-8507, Japan
3 Pathology Division, Mie University Hospital, Tsu, Mie 514-8507, Japan
4 Laboratory of Community Pharmacy, Gifu Pharmaceutical University, Gifu, Gifu 501-1196, Japan
* Correspondence: kenishii@clin.medic.mie-u.ac.jp

Abstract: Loss of androgen receptor (AR) dependency in prostate cancer (PCa) cells is associated with progression to castration-resistant prostate cancer (CRPC). The tumor stroma is enriched in fibroblasts that secrete AR-activating factors. To investigate the roles of fibroblasts in AR activation under androgen deprivation, we used three sublines of androgen-sensitive LNCaP cells (E9 and F10 cells: low androgen sensitivity; and AIDL cells: androgen insensitivity) and original fibroblasts derived from patients with PCa. We performed in vivo experiments using three sublines of LNCaP cells and original fibroblasts to form homotypic tumors. The volume of tumors derived from E9 cells plus fibroblasts was reduced following androgen deprivation therapy (ADT), whereas that of F10 or AIDL cells plus fibroblasts was increased even after ADT. In tumors derived from E9 cells plus fibroblasts, serum prostate-specific antigen (PSA) decreased rapidly after ADT, but was still detectable. In contrast, serum PSA was increased even in F10 cells inoculated alone. In indirect cocultures with fibroblasts, PSA production was increased in E9 cells. Epidermal growth factor treatment stimulated Akt and p44/42 mitogen-activated protein kinase phosphorylation in E9 cells. Notably, AR splice variant 7 was detected in F10 cells. Overall, we found that fibroblast-secreted AR-activating factors modulated AR signaling in E9 cells after ADT and loss of fibroblast-dependent AR activation in F10 cells may be responsible for CRPC progression.

Keywords: prostate cancer; androgen deprivation therapy; androgen sensitivity; androgen receptor dependency; fibroblast-dependent androgen receptor activation

---

1. Introduction

The number of men diagnosed with prostate cancer (PCa) is increasing worldwide [1]. Most patients with early-stage PCa can be treated with therapies such as radical prostatectomy or irradiation. In contrast, patients with advanced PCa are treated with androgen deprivation therapy (ADT), the standard systemic therapy. Although ADT induces temporary remission, the majority of patients (approximately 60%) eventually develop and progress to castration-resistant prostate cancer (CRPC), which is associated with a high mortality rate [2,3]. In the development and progression of CRPC, a decrease or loss of androgen sensitivity in PCa cells often occurs. Low androgen sensitivity of PCa cells is associated with a more malignant phenotype and is difficult to cure. Changes in the androgen sensitivity of PCa cells are often caused artificially as negative effects of ADT and by spontaneously arising variants of androgen receptor (AR) even before ADT is started [4,5].

ADT for patients with advanced PCa aims to decrease the concentration of circulating androgen and block AR signaling in PCa cells [6]. Well-differentiated PCa cells are generally androgen and AR dependent, i.e., AR signaling regulates cell cycle progression and differentiation. Loss of AR signaling after ADT triggers AR-independent outgrowth, generating poorly differentiated uncontrollable PCa cells [7]. To prevent the development and progression of CRPC, we hypothesize that preservation of AR signaling after ADT is essential. Nelson et al. described four molecular-state frameworks for activation of AR signaling after ADT as follows: state 1, endocrine androgen and AR dependent; state 2, intracrine androgen and AR dependent; state 3, androgen independent and AR dependent; and state 4, androgen and AR independent [8]. Several molecular mechanisms responsible for changes in the AR dependency of PCa cells have been suggested, e.g., androgen-independent activation of AR signaling by mutations in the AR or altered levels of coactivators, and activation of alternative growth factor pathways [4,5]. In patients with CRPC, a number of growth factors and cytokines contribute to malignancy of PCa cells through activation of AR signaling in an androgen-independent manner, which is often called the "outlaw pathway" [8]. Previous studies have suggested that epidermal growth factor (EGF), fibroblast growth factor (FGF)-7 (also known as keratinocyte growth factor (KGF)), insulin-like growth factor (IGF)-1, and interleukin (IL)-6 can activate AR signaling in the absence of androgen [9–11] via various signaling pathways, including Akt, signal transducer and activator of transcription (STAT) 3, and p44/42 mitogen-activated protein kinase (MAPK) pathways [12]. In the tumor microenvironment, tumor stroma surrounding PCa cells is enriched in fibroblasts that secrete AR-activating factors, such as EGF, FGF-7/KGF, IGF-1, and IL-6 [13,14].

Most fibroblasts do not express AR and can survive in the absence of androgen [15–17]. Several studies have reported that androgen-independent interactions between PCa cells and fibroblasts determine how PCa cells respond to ADT [18,19]. In androgen-insensitive PCa cells, we demonstrated that fibroblast-derived FGF-7/KGF may bypass the functionally inactive AR and may promote cell proliferation after ADT [19]. In androgen-sensitive PCa cells, however, we have recently reported that fibroblast-derived EGF, IGF-1, and IL-6 can activate AR signaling, leading to preservation of AR signaling after ADT [17]. Thus, we hypothesize that fibroblast-dependent AR activation may preserve AR signaling after ADT and may play a critical role in the prevention of CRPC development and progression. Clinically, PCa is a heterogeneous disease with various biological behaviors, such as androgen sensitivity and response to ADT. To investigate the relationship between fibroblast-dependent AR activation and androgen sensitivity in PCa cells, well-established PCa cell lines with a variety of androgen sensitivities are strongly required.

LNCaP human PCa cell lines are androgen-sensitive PCa cell lines that are useful for investigating the molecular mechanisms responsible for changes in the androgen sensitivity and AR dependency of PCa cells. Notably, LNCaP cells are a heterogeneous cell population containing various clones with naturally occurring differences in androgen sensitivity caused by spontaneously arising changes [20,21]. Accordingly, we generated two sublines of LNCaP cells (E9 and F10) showing low androgen sensitivity by using a limiting dilution method with regular culture conditions [22,23]. In addition, we established androgen-insensitive AIDL cells from parental LNCaP cells by continuous passaging under hormone-depleted conditions [24]. The parental LNCaP cell line and its derivative sublines (E9, F10, and AIDL) expressed similar levels of AR protein, and AR-dependent secretion of prostate-specific antigen (PSA) was detected in LNCaP and E9 cells [14]. As compared with parental LNCaP cells, we have previously reported that combination of E9 or AIDL cells with embryonic rat urogenital sinus mesenchyme promoted tumor progression in vivo even under androgen ablation [19]. Comparing the characteristic features of paternal LNCaP cells and its sublines allows us to investigate the molecular mechanisms responsible for changes in the androgen sensitivity and AR dependency of PCa cells.

In this study, we defined "androgen sensitivity" in PCa cells as the degree of androgen-dependent AR activation in vitro. In contrast, we defined "AR dependency" in PCa cells as the degree of androgen- and growth factor/cytokine-dependent AR activation. The objective of this study was to investigate the role of fibroblast-dependent AR activation in the tumorigenesis of three LNCaP sublines differing in androgen sensitivity under androgen deprivation conditions.

2. Materials and Methods

2.1. Materials

Dihydrotestosterone (DHT) and anti-androgen bicalutamide were purchased from Sigma-Aldrich Co., LLC. (St. Louis, MO, USA). Recombinant human EGF, FGF-2, FGF-7, FGF-10, hepatocyte growth factor (HGF), IGF-1, transforming growth factor (TGF) β1, vascular endothelial growth factor (VEGF), and IL-6 were purchased from PeproTech, Inc. (Rocky Hill, NJ, USA). Rabbit polyclonal anti-PSA and mouse monoclonal anti-neuron-specific enolase (NSE; BBS/NC/V1-H14) antibodies were purchased from Dako Cytomation (Copenhagen, Denmark). Rabbit polyclonal anti-AR (N-20) and anti-EGFR (1005) antibodies were purchased from Santa Cruz Biotechnology (Santa Cruz, CA, USA). Rabbit polyclonal anti-CD31 and anti-Ki-67 and rabbit monoclonal anti-AR splice variant 7 (AR-V7; EPR15656) antibodies were purchased from Abcam Inc. (Cambridge, MA, USA). Rabbit monoclonal anti-phospho-STAT3 (Tyr705) (D3A7), anti-STAT3 (D3Z2G), anti-phospho-Akt (Ser473) (D9E), and anti-Akt (pan) (C67E7) antibodies were purchased from Cell Signaling Technology, Inc. (Beverly, MA, USA). Rabbit polyclonal anti-p44/42 MAPK and mouse monoclonal anti-phospho-p44/42 MAPK (Thr202/Tyr204) (E10) antibodies were purchased from Cell Signaling Technology, Inc. Mouse monoclonal anti-β-actin (AC-15) antibodies were purchased from Sigma-Aldrich Co., LLC.

2.2. Cell Culture

The androgen-sensitive, AR-positive human PCa cell lines LNCaP and 22Rv1 were obtained from the American Type Culture Collection (Manassas, VA, USA). Both LNCaP and 22Rv1 cells were authenticated by the short tandem repeat-PCR method and were cultured in phenol red (+) RPMI-1640 supplemented with 10% fetal bovine serum (FBS; Sigma-Aldrich Co., LLC.). E9 and F10 cells (showing low sensitivity to androgen) were obtained from the parental androgen-sensitive LNCaP cells using a limiting dilution method under regular culture conditions [22,23]. In contrast, androgen-insensitive AIDL cells were established from LNCaP cells by continuous passaging under hormone-depleted conditions [24]. AIDL cells were cultured in phenol red (-) RPMI-1640 supplemented with 10% charcoal-stripped (CS)-FBS. The androgen sensitivity of parental LNCaP, E9, F10, and AIDL cells was confirmed by changes in *KLK3* (PSA) mRNA expression in cell cultures treated with synthetic androgen R1881 [19]. The nontumorigenic human prostate epithelial cell line BPH-1 was obtained from Dr. Simon W. Hayward (Northshore University HealthSystem, Chicago, IL, USA) and was cultured in phenol red (+) RPMI-1640 supplemented with 10% FBS.

Commercially available human prostate stromal cells (PrSC) were purchased from Lonza Group Ltd. (Basel, Switzerland). pcPrFs (pcPrF-M5, pcPrF-M6, and pcPrF-M7) were primary cultured from PCa specimens collected from biopsies of patients with advanced PCa [17]. PrSC and pcPrFs were cultured in medium prepared from an SCBM Bullet Kit (Lonza Group Ltd.). The four fibroblast lines (PrSC, pcPrF-M5, pcPrF-M6, and pcPrF-M7) do not express AR protein and do not respond to DHT stimulation on cell proliferation as previously reported [17].

2.3. Indirect Coculture of Prostate Cancer Cell Lines (E9, F10, and AIDL Cells) with Fibroblasts

E9, F10, and AIDL cells were co-cultured with each of the four fibroblast lines (PrSC, pcPrF-M5, pcPrF-M6, and pcPrF-M7) in six-well plates using cell culture inserts (BD Falcon, Franklin Lakes, NJ, USA) as previously described [17]. E9, F10, and AIDL cells (4×10^4 cells/well) were seeded into six-well plates in their respective recommended medium, whereas fibroblasts (PrSC, pcPrF-M5, pcPrF-M6, and pcPrF-M7; 2×10^4 cells/well) were seeded in SCBM media into cell culture inserts for 2 days. The culture medium for PCa cells and fibroblasts was replaced with phenol red (-) RPMI-1640 supplemented with 1% CS-FBS containing DHT (0.1 nM), and the inserts with fibroblasts were then placed into six-well plates for an additional 4 days. DHT concentrations in the incubation medium were chosen based on previous studies of tissue DHT levels in recurrent PCa [25].

2.4. *Stimulation of Cell Growth by Treatment with Growth Factors and Cytokines*

Examination of the effects of growth factor and cytokine stimulation was performed as previously described [17], with minor modifications. E9 (5×10^3 cells/well), F10 (4×10^3 cells/well), and AIDL cells (6×10^3 cells/well) were cultured in phenol red (+) RPMI-1640 supplemented with 10% FBS for 2 days, and the culture medium was then replaced with phenol red (-) RPMI-1640 supplemented with 1% CS-FBS containing DHT (0.1 nM). One day later, the culture medium was replaced with phenol red (-) RPMI-1640 containing 10 ng/mL each of recombinant EGF, FGF-2, FGF-7, FGF-10, HGF, IGF-1, TGFβ1, VEGF, and IL-6. Cells were then incubated for 4 days before analysis.

2.5. *Enzyme-Linked Immunosorbent Assay*

Serum PSA levels in mice were assayed using a PSA Enzyme Immunoassay test kit (Hope Laboratories, Belmont, CA, USA).

2.6. *Preparation of Cell Lysates*

LNCaP, E9, F10, AIDL, and 22Rv1 cells were harvested by scraping, and whole cell lysates were prepared as previously described [14]. Briefly, the cell surface was washed with ice-cold phosphate-buffered saline and then lysed with CelLytic (Sigma-Aldrich Co., LLC.) containing 1% Nonidet P-40, 10 mM 4-(2-aminoethyl) benzensulfonyl fluoride, 0.8 mM aprotinin, 50 mM bestatin, 15 mM E-64, 20 mM leupeptin, and 10 mM pepstatin. After 60 min on ice, the lysates were centrifuged at 10,000× *g* for 10 min, and the supernatants were collected. The protein concentration was measured using a NanoDrop 2000 (Thermo Fisher Scientific Inc., Waltham, MA, USA).

2.7. *Western Blot Analyses*

Extracted proteins were separated by gel electrophoresis and transferred to Immobilon polyvinylidene difluoride membranes following our previously reported protocol [26]. Anti-AR, anti-PSA, anti-NSE, anti-phospho-STAT3, anti-STAT3, anti-phospho-Akt, anti-Akt, anti-phospho-p44/42 MAPK, anti-p44/42 MAPK, and anti-β-actin antibodies were used at dilutions of 1:2500, 1:5000, 1:5000, 1:2000, 1:2000, 1:1000, 1:1000, 1:2000, 1:1000, and 1:5000, respectively. Specific protein bands were assessed with a LAS-4000 Mini (Fuji Photo Film, Tokyo, Japan) using SuperSignal West Pico Chemiluminescent Substrate (Thermo Fisher Scientific Inc.).

2.8. *Animal Studies*

All animals were maintained in a specific pathogen-free environment. Mie University's Committee on Animal Investigations approved the experimental protocol. Male athymic nude mice (BALB/c, nu/nu, 6–8 weeks old) were purchased from CLEA Japan, Inc. (Tokyo, Japan) and used for all experiments.

2.9. *In Vivo Xenograft Model*

Examination of the effects of ADT on tumorigenesis of E9, F10, and AIDL cells was performed as previously described [17]. Subconfluent cultures of E9, F10, AIDL, and pcPrF-M5 cells were trypsinized and counted. Xenografts without pcPrF-M5 cells contained 5×10^5 PCa cells. Xenografts with pcPrF-M5 cells were prepared by mixing 2.5×10^5 PCa cells and 2.5×10^5 pcPrF-M5 cells in suspension. Pelleted cells were resuspended in 50 μL neutralized type I rat tail collagen gels and then grafted beneath the renal capsule of male athymic mice (6–8 weeks old). In total, 1×10^6 PCa cells were grafted in each mouse. For the androgen deprivation experiments, mice were randomized on day 14 after transplantation. Mice treated with ADT were castrated and orally administered a bicalutamide (5 mg/kg/day) suspension with 0.5% carboxymethylcellulose in a 5-days-on/2-days-off schedule through a 22-gauge gavage needle; the control group underwent sham operation and received the diluent. Mice were killed on days 14 and 21 after ADT, and tumor weights and serum PSA levels were then measured. The tumor volume was determined by direct measurement with calipers (volume = long axis × short axis ×short axis × 0.5236), as previously described [27].

2.10. *Histopathology and Immunohistochemistry*

For histopathological analysis, tumors were fixed in 10% neutral buffered formalin and embedded in paraffin. General tissue morphology was visualized by standard hematoxylin and eosin staining. Next, immunohistochemical staining was performed with an ImmPRESS Reagent Kit (Vector Laboratories, Inc., Burlingame, CA, USA). Antigen retrieval was performed using 10 mM sodium citrate buffer (pH 6.0) for AR, NSE, Ki-67, and CD31. Antigen Unmasking Solution (Vector Laboratories, Inc.) was used for PSA. The antigen-antibody reaction was visualized using 3′,3-diaminobenzidine tertahydrochloride as a substrate. Sections were counterstained with hematoxylin and examined by light microscopy.

Cell proliferation in tumors was determined by the percentage of Ki67-positive nuclei in 10 different areas at 400× magnification from each tissue specimen. A 'microvessel' was defined as mouse-specific CD31-positive endothelial cells that formed a vascular lumen. The number of microvessels was counted in 10 different areas at 400× magnification from each tissue specimen, as previously described [27]. The results were independently reviewed by 2 blinded investigators.

2.11. *Statistical Analysis*

Results are expressed as means ± standard deviations. Differences between the two groups were determined using Student's *t*-tests. Results with *p* values of less than 0.05 were considered statistically significant.

3. Results

3.1. *Effects of ADT on Tumor Growth and Serum PSA Kinetics of Xenograft Derived from Co-Inoculation of E9, F10, and AIDL Cells with pcPrF-M5 Cells In Vivo*

Regardless of the presence of pcPrF-M5 cells, tumor volumes in mice inoculated with E9 cells rapidly decreased following ADT (Figure 1A) and were not altered between days 14 and 21 after ADT, i.e., the effects of ADT on E9 tumors were maintained until at least day 21 after ADT. In contrast, in the presence or absence of pcPrF-M5 cells, tumor volumes in mice inoculated with F10 cells temporally decreased following ADT, but increased between days 14 and 21 after ADT (Figure 1A). Moreover, in the presence or absence of pcPrF-M5 cells, tumor volumes in mice inoculated with AIDL cells gradually increased following ADT, similar to the results in the sham group (Figure 1A).

Ki-67 labeling indices in mice inoculated with E9 cells, with or without pcPrF-M5 cells, rapidly decreased following ADT (Figure 1B) and were not significantly increased between days 14 and 21 after ADT; that is, the effects of ADT on E9 tumors were maintained until at least day 21 after ADT. In contrast, Ki-67 labeling indices in mice inoculated with F10 or AIDL cells, with or without pcPrF-M5 cells, were not altered following ADT as compared with that in the sham group (Figure 1B).

Serum PSA titers in mice inoculated with E9 cells alone rapidly decreased after ADT (Figure 1C). Importantly, those in mice inoculated with E9 cells plus pcPrF-M5 cells rapidly decreased following ADT and became detectable on day 21 after ADT. Serum PSA titers in mice inoculated with F10 cells, with or without pcPrF-M5 cells, increased gradually (Figure 1C), whereas those in mice inoculated with AIDL cells, with or without pcPrF-M5 cells, were not detected (Figure 1C).

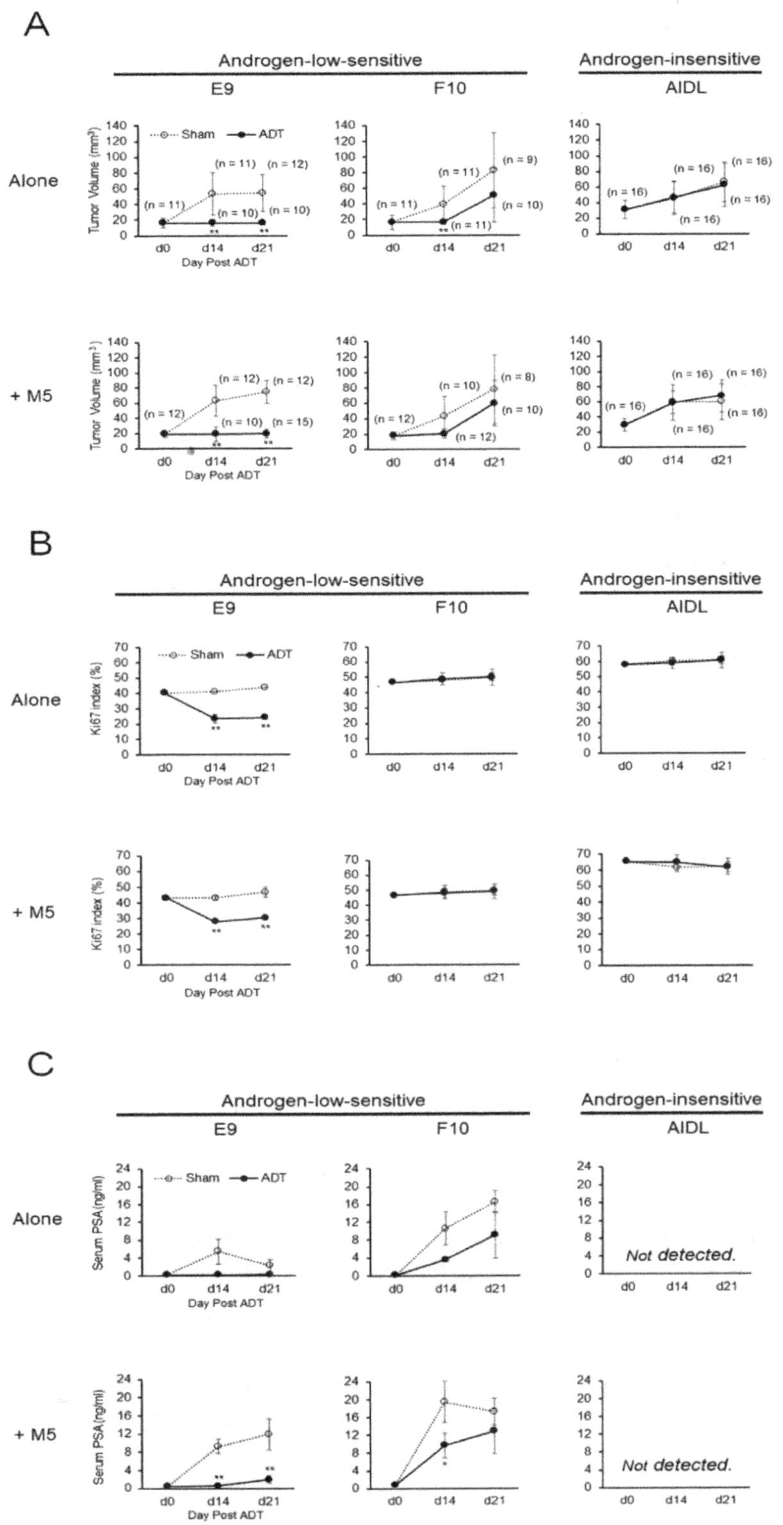

Figure 1. Effects of androgen deprivation therapy (ADT) on tumor growth and serum PSA kinetics of xenografts derived from co-inoculation of LNCaP sublines with pcPrF-M5 cells in vivo. Changes in tumor volume (**A**), Ki67 index (**B**), and serum PSA (**C**) of xenografts were compared in untreated (sham-operated) or ADT-treated mice inoculated with PCa cells with or without M5 cells on days 0, 14, and 21 after ADT. ** $P < 0.01$ versus sham-operated control. ADT, androgen deprivation therapy; PSA, prostate-specific antigen; M5, pcPrF-M5.

E9-derived tumors, with or without pcPrF-M5 cells, grown in mice treated with ADT showed reduced tumorigenesis as compared with those in untreated (sham-operated) mice (Figure S1). In contrast, both F10- and AIDL-derived tumors, with or without pcPrF-M5 cells, grown in mice treated with ADT showed no changes in tumorigenesis as compared with those in untreated mice (Figures S2 and S3). Additionally, AR and PSA proteins were expressed in both E9- and F10-derived tumors, with or without pcPrF-M5 cells, on day 21 after ADT, although serum PSA levels were very low in mice inoculated with E9 cells alone (Figures S1 and S2). Moreover, AR protein was expressed in AIDL-derived tumors, with or without pcPrF-M5 cells, on day 21 after ADT, whereas PSA protein was not detected because of mutated AR in AIDL cells (Figure S3). NSE staining was diffuse among ADT-treated and untreated hosts on day 21 after ADT (Figures S1–S3). Microvessel density (MVD) in all tumors, regardless of the presence of pcPrF-M5 cells, was not altered among ADT-treated and untreated hosts on days 14 and 21 after ADT (Tables S1–S3).

3.2. *Effects of Indirect Coculture with Fibroblasts on E9, F10, and AIDL Cells In Vitro*

Cell proliferation of E9 and AIDL cells was significantly increased when cells were cocultured with PrSCs or pcPrFs, whereas that of F10 cells was not affected (Figure 2A). Expression levels of PSA proteins were increased in E9 cells cocultured with pcPrFs but not PrSC and were not affected in F10 cells (Figure 2B). In contrast, expression levels of PSA proteins were decreased in F10 cells cocultured with PrSC but not pcPrFs. PSA protein expression was not detected in AIDL cells cocultured with PrSC or pcPrFs (Figure 2B). NSE protein expression and STAT3 phosphorylation were increased in E9 cells cocultured with PrSC or pcPrFs (Figure 2B). NSE proteins were not changed in F10 cells cocultured with PrSC or pcPrFs, whereas STAT3 phosphorylation was increased in F10 cells cocultured with pcPrF-M6 cells (Figure 2B). NSE protein expression was decreased in AIDL cells cocultured with pcPrFs but not PrSC, whereas phosphorylation of STAT3 was increased in AIDL cells cocultured with pcPrFs (Figure 2B).

3.3. *Effects of Growth Factors and Cytokines on E9, F10, and AIDL Cells In Vitro*

Cell proliferation of E9 and AIDL cells was significantly increased by treatment with growth factors and cytokines, such as EGF and IL-6, whereas that of F10 cells was significantly decreased by treatment with FGF-10, HGF, IGF-1, and TGFβ1 (Figure 3A). Phosphorylation of Akt and p44/42 MAPK in E9 and AIDL cells was strongly increased by treatment with EGF, whereas that in F10 cells was not affected (Figure 3B). Phosphorylation of Akt and p44/42 MAPK in E9, F10, and AIDL cells was not affected by treatment with HGF (data not shown). EGFR protein was detectable in all PCa cells, including LNCaP, E9, F10, AIDL cells, whereas expression of EGFR protein in AIDL cells was considerably higher than that in E9 and F10 cells (Figure S4). Detection of EGFR protein in BPH-1 cells was used as a positive control for anti-EGFR antibodies. In addition, PSA secretion from E9 cells was significantly increased by treatment with EGF but not HGF, whereas that from F10 cells was significantly decreased by treatment with both EGF and HGF (Figure S5). Notably, PSA secretion from AIDL cells was not detected.

Full-length AR protein was detectable in all PCa cells, including LNCaP, E9, F10, AIDL, and 22Rv1 cells, whereas AR-V7 protein was detectable only in F10 cells (Figure 4). Detection of AR-V7 protein in 22Rv1 cells was used as a positive control for anti-AR-V7 antibodies.

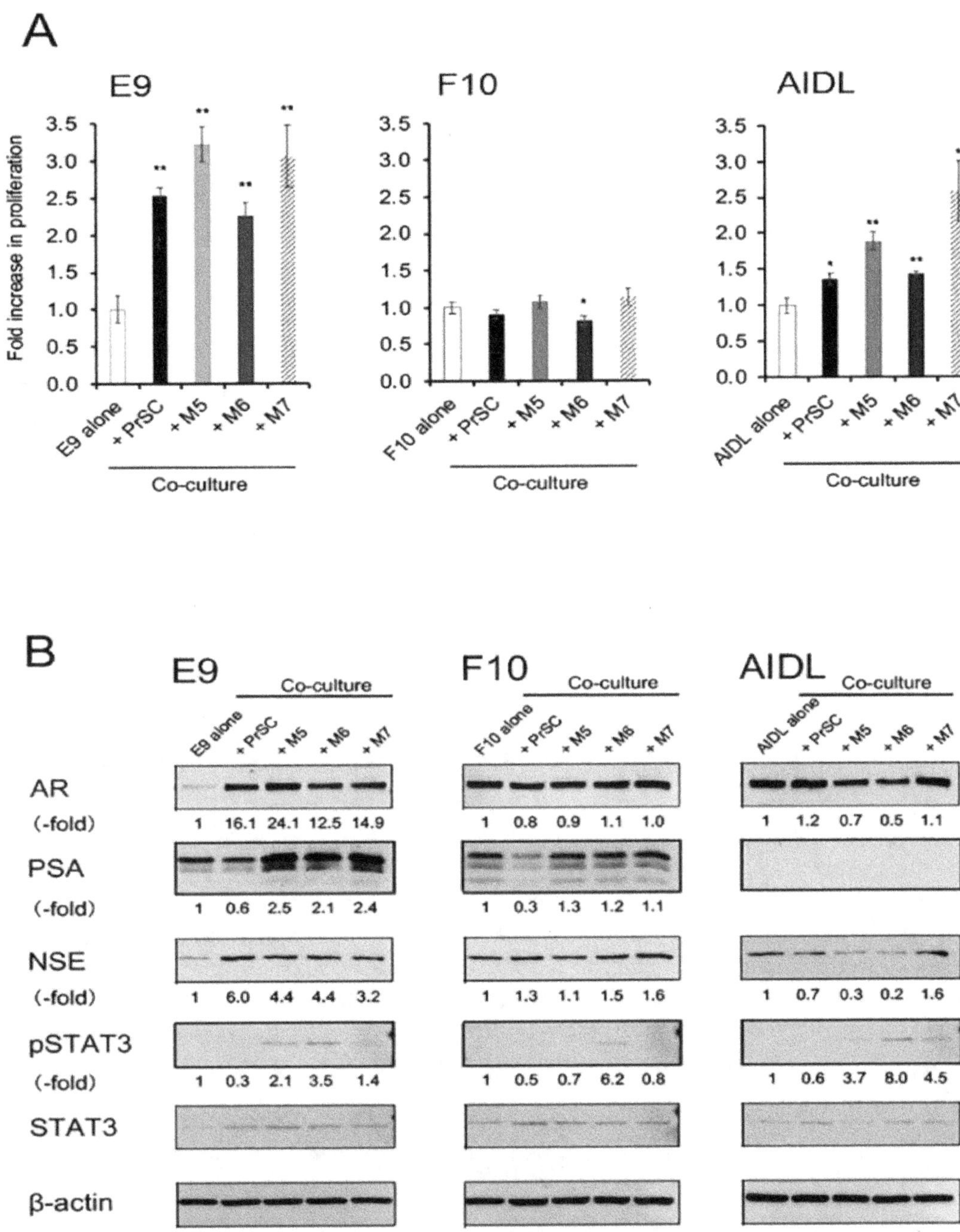

Figure 2. Effects of indirect coculture with fibroblasts on cell proliferation and PSA expression in LNCaP sublines in vitro. (**A** and **B**) LNCaP sublines were co-cultured with fibroblasts using cell culture inserts for 4 days in phenol red (−) RPMI-1640 with 1% CS-FBS containing DHT (0.1 nM). (**A**) Cell proliferation. * $P < 0.05$, ** $P < 0.01$ versus LNCaP sublines alone. (**B**) Cell lysates from co-cultures were subjected to western blotting and probed with antibodies against each protein. Protein levels were compared using actin expression as a loading control. AR, androgen receptor; DHT, dihydrotestosterone; NSE, neuron-specific enolase; PSA, prostate-specific antigen; M5, pcPrF-M5; M6, pcPrF-M6; M7, pcPrF-M7.

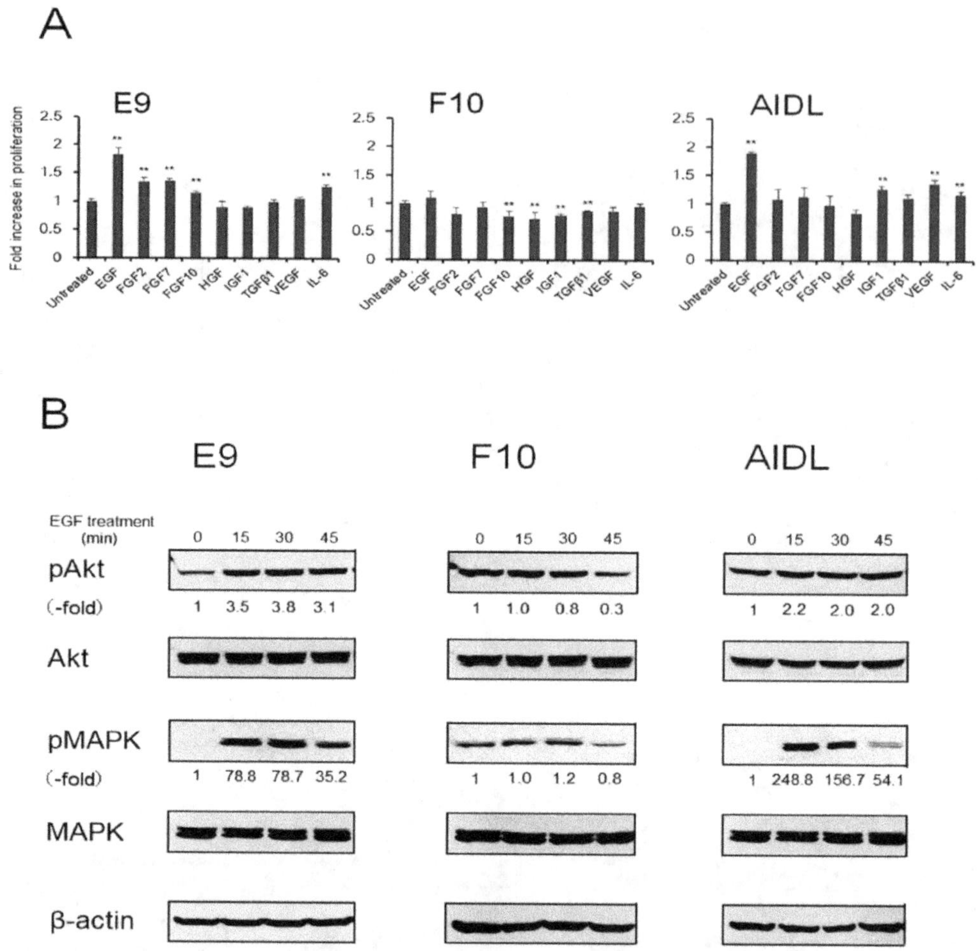

Figure 3. Effects of growth factors and cytokines on cell proliferation and cellular signaling in LNCaP sublines in vitro. (**A** and **B**) LNCaP sublines were treated with 10 ng/mL of each growth factor and cytokine for 4 days in phenol red (−) RPMI-1640 with 1% CS-FBS containing DHT (0.1 nM). (**A**) Cell proliferation. ** $P < 0.01$ versus untreated control. (**B**) Cell lysates from cultures of LNCaP sublines were subjected to western blotting and probed with antibodies against each target protein. Protein levels were compared with actin expression as a loading control. DHT, dihydrotestosterone.

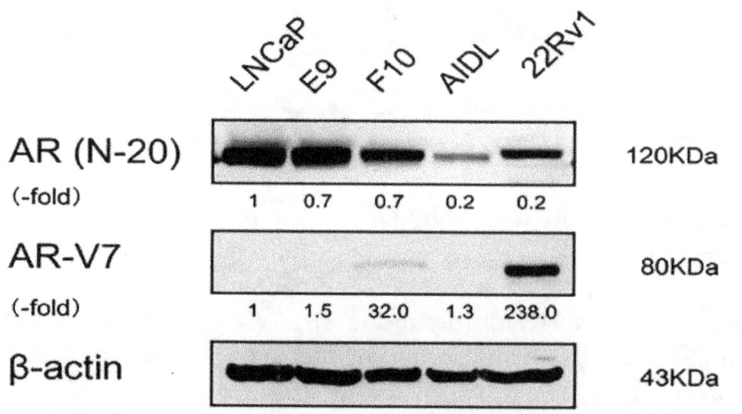

Figure 4. Expression of AR-V7 protein in human PCa cell lines. Cell lysates from growing cultures of parental LNCaP cells, LNCaP sublines (E9, F10, and AIDL cells), and 22Rv1 cells were subjected to western blotting and probed with antibodies against each protein. Protein levels were compared using actin as a loading control. 22Rv1 cells were used as a positive control for detection of AR-V7 protein. AR, androgen receptor; AR-V7, androgen receptor splice variant 7.

4. Discussion

The reduced AR dependency of PCa cells is an important clinical development because of its association with progression to CRPC. In this study, we found that fibroblast-secreted AR-activating factors preserved AR signaling in E9 cells after ADT, indicating that these PCa cells could be controlled by ADT. In contrast, loss of fibroblast-dependent AR activation in F10 cells may be responsible for the development and progression of CRPC.

Development and progression of CRPC after ADT are mediated by multiple molecular mechanisms, classified as adaptation to a low-concentration androgen environment caused by ADT, or clonal selection [28–31]. Androgen-insensitive PCa cells can be generated by adaptation of androgen-sensitive PCa cells to a low androgen environment. In contrast, ADT results in the expansion of androgen-insensitive PCa cells, which coexist with androgen-sensitive PCa cell populations in PCa tissue, i.e., clonal selection of androgen-insensitive PCa cells [28]. PCa tissue consists of heterogeneous cell populations. Tumor heterogeneity is reflected in the increased subclonal populations in PCa tissue [32]. These subclones may interact in complex ways with each other or with the surrounding microenvironment. Thus, we hypothesize that tumor heterogeneity in PCa tissue is an extremely important phenomenon not only for understanding tumor progression but also for developing truly personalized treatment regimens for patients with PCa.

To compare the biochemical characteristics of androgen-sensitive and -insensitive PCa cells, we generated three sublines from androgen-sensitive AR-positive LNCaP cells: E9 and F10 cells (showing low androgen sensitivity) and AIDL cells (showing androgen insensitivity) [22–24]. E9 cells are less sensitive to androgen-related responses, such as growth stimulation and PSA production, than parental LNCaP cells [22]. Moreover, E9 cells have a more aggressive tumorigenic phenotype than parental LNCaP cells in vivo. We have previously investigated the mechanisms underlying the low androgen sensitivity of E9 cells and found that decreased phosphorylation of Akt was associated with low androgen sensitivity of E9 cells [33]. The Akt and p44/42 MAPK pathways are known to be associated with the regulation of androgen responses [34,35]. We also demonstrated that PSA production was significantly decreased in parental LNCaP cells when Akt phosphorylation was suppressed by phosphatidylinositol 3-kinase or Akt inhibitors [33]. Thus, E9 cells may be a useful model to reflect high-grade Gleason tumors with low phosphorylation of Akt. Similar to E9 cells, F10 cells are also less sensitive to androgen-related responses than parental LNCaP cells and have a more aggressive tumorigenic phenotype than parental LNCaP cells in vivo [23]. Interestingly, F10 cells can survive under low-pH/low-nutrient conditions, whereas parental LNCaP cells show significant cell death under such conditions. The intratumor environment is characterized by low-pH, low-nutrient, and chronic hypoxic conditions owing to poor vascular development [36,37]. Thus, we suggest that F10 cells may be a useful model to determine the mechanisms underlying their adaptation to a low-pH/low-nutrient environment. In contrast to E9 and F10 cells, AIDL cells are insensitive to androgen-related responses [24]. Parental LNCaP cells harbor an AR mutation at codon 877 (T877A). In addition to the T877A mutation, we found that AIDL cells harbored a point mutation at codon 741 (W741C), suggesting that the T877A/W741C double mutation may be responsible for the androgen insensitivity of AIDL cells [38]. Thus, AIDL cells may be a useful model to investigate the mechanisms underlying the mutated AR in PCa cells.

In PCa, the tumor microenvironment is highly complex and heterogeneous and is composed of carcinoma-associated fibroblasts (CAFs) as well as epithelial cancer cells that infiltrate into the surrounding tumor stroma, referred to as the reactive stroma [39]. This heterogeneous stromal component of PCa tissues contains multiple populations of fibroblasts that are associated with tumorigenesis [40,41]. CAFs contribute to the malignancy of PCa cells by enhancing the proliferation and invasion of cancer cells and promoting angiogenesis in tumors [42]. Thus, inhibition of CAF generation and function in PCa tissue could be a new target for controlling primary cancer progression. Importantly, most fibroblasts in the prostate stroma are negative for AR [15–17], and the phenotypes of human PCa fibroblasts are strongly heterogeneous [13]. CAFs secrete abundant growth factors and cytokines, which enhance the proliferation of PCa cells. However, the proliferation of PCa cells is regulated by AR signaling, suggesting that stromal paracrine factors derived from CAFs can activate AR signaling in PCa cells. In patients with CRPC, PCa cells can grow in the absence of androgen, indicating that AR signaling in PCa cells is activated by CAF-derived growth factors and cytokines instead of androgen. Therefore, CAFs could be an important target to prevent androgen-independent outgrowth.

In the clinical setting, serum PSA is the most useful biomarker to detect PCa. However, increased levels of serum PSA are also observed in cases of benign prostatic hyperplasia or inflammation of the prostate. PSA is a serine protease and member of the tissue kallikrein family of proteases and is produced in both normal luminal epithelial cells and well-differentiated PCa cells [43]. Transcription of the *PSA* gene is normally regulated by androgens through activation of AR signaling. In addition to androgens, PSA expression is induced through activation of AR signaling by CAF-derived growth factors and cytokines. In our laboratory, Sasaki et al. reported that fibroblasts directly affected PSA expression in LNCaP cells cocultured in vitro [17]. Among various CAF-derived growth factors and cytokines, we confirmed that EGF, IGF-1, and IL-6 directly increased PSA expression in LNCaP cells, suggesting that soluble factors derived from fibroblasts may function as AR-activating factors in the absence of androgen.

In our previous work, Sasaki et al. found that the PSA kinetics after ADT were not an accurate prognostic marker when considering serum PSA levels after ADT to determine the number of viable PCa cells [44,45]. Compared with rapid decreases in PSA after ADT, prolonged PSA decreases after ADT can predict favorable progression-free survival and overall survival. In this study, AR signaling in E9 cells, but not in F10 cells, was activated by paracrine signals derived from fibroblasts, suggesting that the androgen sensitivity of PCa cells may not reflect the AR dependency of PCa cells. Preservation of AR signaling after ADT may have an important role in maintaining the AR dependency of PCa cells. Thus, fibroblast-dependent AR activation after ADT may cause persistent activation of AR signaling in PCa cells, preventing loss of AR dependency after ADT. Notably, Sasaki et al. demonstrated that fibroblasts could enhance the treatment efficacy of ADT during in vivo tumorigenesis, resulting in a more favorable prognosis, e.g., prolonged serum PSA decline and maintenance of the efficacy of ADT [17]. Other researchers also demonstrated that normal human fibroblasts could inhibit the proliferation of tumor cells [46,47]. We still know very little about the tumor-promoting CAFs and the factors that distinguish CAFs from other fibroblasts found in the same tissue. Future studies are needed to identify the specific profiles of fibroblast-derived factors responsible for disease progression. Additionally, several AR variants (ARVs), derived from alternative splicing of the AR transcript, have been identified [48–51]. ARVs may emerge as more common mediators of androgen-independent and AR-dependent tumor progression, although their functions are still unclear. AR-V7 is a major splice variant expressed in human PCa that is associated with the development and progression of CRPC [52]. In human PCa cell lines, expression of AR-V7 mediates resistance to a new generation of AR-targeted therapies, such as enzalutamide and abiraterone [53]. In addition, Shimizu et al. recently reported that knockdown of AR-V7 in LNCaP95-DR cells did not restore sensitivity to docetaxel and cabazitaxel, suggesting that AR-V7 may be not involved in taxane resistance [54]. Thus, expression of AR-V7 has been proposed for the assessment of suitability for taxane chemotherapy [55].

With regard to the androgen sensitivity of PCa cells, our data showed the following results, in the presence or absence of fibroblasts: (1) tumor growth of E9 cells was significantly diminished after ADT compared with that in the sham group; (2) tumor growth of F10 cells was temporally reduced after ADT, but was restarted under androgen deprivation; and (3) tumor growth of AIDL cells was not decreased after ADT. These results established some important clinical concepts. For example, to treat certain PCa cells (e.g., E9 cells), fibroblast-target therapy should be avoided because of the preservation of AR signaling after ADT. Additionally, for certain PCa cells expressing AR-V7 (e.g., F10 cells), ADT may not be effective for treating CRPC. In these cells, the responsiveness to fibroblasts may not be associated with tumor growth, and the efficacy of ADT may be limited. Finally, in androgen-insensitive PCa cells (e.g., AIDL cells), ADT is completely useless because of the AR independence of PCa cells.

Identification of improved, personalized treatments will be supported by recent major progress in the molecular characterization of early- and late-stage PCa. Indeed, such advancements have already led to novel classifications of prostate tumors based on gene expression profiles and mutation status and should greatly facilitate the choice of novel targeted therapies best tailored to the needs of patients [56,57], particularly for individual subgroups of patients, representing a major step towards personalized medicine adapted to the individual needs of patients with PCa. Selecting the optimal drug or drug sequence and combination for PCa treatment will be improved by the identification of molecular biomarkers predictive of response and progression. Similar to breast cancer, subdivisions of luminal A, luminal B, and basal subtypes, which exhibit different clinical prognoses and responses to ADT, have been proposed for PCa [58]. Such a classification may greatly support treatment choices for early- and late-stage disease and ultimately improve the overall survival rates and quality of life in patients with PCa. As we learn more about the genetic heterogeneity of PCa cells and mechanisms of treatment resistance, we expect that markers for treatment choice and response will be developed and validated to better guide treatment decisions [59,60].

Using three sublines of androgen-sensitive LNCaP cells, we demonstrated that loss of fibroblast-dependent AR activation in PCa cells (e.g., F10 cells) may be responsible for the development and progression of CRPC. In the absence of androgen, the AR dependence of PCa cells interacting with fibroblasts reflected the efficacy of ADT; for example, E9 cells could be controlled by ADT. To choose appropriate patients with advanced PCa for ADT, it is necessary to evaluate the degree of AR dependence in PCa cells interacting with fibroblasts before ADT is started. Further investigations are needed to identify clear molecular markers using biopsy tissue samples or bodily fluid samples derived from patients with advanced PCa.

5. Conclusions

To prevent the development and progression of CRPC, the preservation of AR signaling after ADT is essential. In this study, we demonstrated that loss of fibroblast-dependent AR activation and expression of AR-V7 protein in PCa cells may be involved in the mechanism of acquired resistance to castration. In near future, clear molecular markers are needed to identify the degree of AR dependence in PCa cells interacting with fibroblasts before ADT is started, e.g., expression of AR-V7 proteins is one of the candidates.

Supplementary Materials:
Figure S1: Effects of ADT on histopathological characteristics of xenografts derived from co-inoculation of E9 cells with fibroblasts in vivo; Figure S2: Effects of ADT on histopathological characteristics of xenografts derived from co-inoculation of F10 cells with fibroblasts in vivo; Figure S3: Effects of ADT on histopathological characteristics of xenografts derived from co-inoculation of AIDL cells with fibroblasts in vivo; Figure S4: Expression of EGFR protein in human PCa cell lines; Figure S5: Effects of growth factors on PSA secretion from LNCaP sublines in vitro; Table S1: MVD changes in E9 tumors after ADT; Table S2: MVD changes in F10 tumors after ADT; Table S3: MVD changes in AIDL tumors after ADT.

Author Contributions: Conceptualization, Ke.I. (Kenichiro Ishii) and Y.S.; investigation, I.M., T.S., Ka.I. (Kazuhiro Iguchi), K.N. and H.K.; writing—original draft preparation, Ke.I. (Kenichiro Ishii); writing—review and editing, H.I., Y.H. and K.A.

Acknowledgments: We would like to thank Yumi Yoshikawa for their technical support.

References

1. Gronberg, H. Prostate cancer epidemiology. *Lancet* **2003**, *361*, 859–864. [CrossRef]
2. Huggins, C.; Hodges, C.V. Studies on prostatic cancer: I. The effect of castration, of estrogen and of androgen injection on serum phosphatases in metastatic carcinoma of the prostate. 1941. *J. Urol.* **2002**, *168*, 9–12. [CrossRef]

3. Fizazi, K.; Higano, C.S.; Nelson, J.B.; Gleave, M.; Miller, K.; Morris, T.; Nathan, F.E.; McIntosh, S.; Pemberton, K.; Moul, J.W. Phase III, randomized, placebo-controlled study of docetaxel in combination with zibotentan in patients with metastatic castration-resistant prostate cancer. *J. Clin. Oncol* **2013**, *31*, 1740–1747. [CrossRef] [PubMed]
4. Taplin, M.E.; Ho, S.M. Clinical review 134: The endocrinology of prostate cancer. *J. Clin. Endocrinol. Metab.* **2001**, *86*, 3467–3477. [CrossRef] [PubMed]
5. So, A.; Gleave, M.; Hurtado-Col, A.; Nelson, C. Mechanisms of the development of androgen independence in prostate cancer. *World J. Urol.* **2005**, *23*, 1–9. [CrossRef] [PubMed]
6. Culig, Z.; Klocker, H.; Bartsch, G.; Hobisch, A. Androgen receptors in prostate cancer. *Endocr. Relat. Cancer* **2002**, *9*, 155–170. [CrossRef] [PubMed]
7. Jennbacken, K.; Tesan, T.; Wang, W.; Gustavsson, H.; Damber, J.E.; Welen, K. N-cadherin increases after androgen deprivation and is associated with metastasis in prostate cancer. *Endocr. Relat. Cancer* **2010**, *17*, 469–479. [CrossRef]
8. Nelson, P.S. Molecular states underlying androgen receptor activation: A framework for therapeutics targeting androgen signaling in prostate cancer. *J. Clin. Oncol.* **2012**, *30*, 644–646. [CrossRef]
9. Culig, Z.; Hobisch, A.; Cronauer, M.V.; Radmayr, C.; Trapman, J.; Hittmair, A.; Bartsch, G.; Klocker, H. Androgen receptor activation in prostatic tumor cell lines by insulin-like growth factor-I, keratinocyte growth factor, and epidermal growth factor. *Cancer Res.* **1994**, *54*, 5474–5478. [CrossRef]
10. Ueda, T.; Mawji, N.R.; Bruchovsky, N.; Sadar, M.D. Ligand-independent activation of the androgen receptor by interleukin-6 and the role of steroid receptor coactivator-1 in prostate cancer cells. *J. Biol. Chem.* **2002**, *277*, 38087–38094. [CrossRef]
11. Kim, J.; Coetzee, G.A. Prostate specific antigen gene regulation by androgen receptor. *J. Cell Biochem.* **2004**, *93*, 233–241. [CrossRef] [PubMed]
12. Zhu, M.L.; Kyprianou, N. Androgen receptor and growth factor signaling cross-talk in prostate cancer cells. *Endocr. Relat. Cancer* **2008**, *15*, 841–849. [CrossRef] [PubMed]
13. Ishii, K.; Mizokami, A.; Tsunoda, T.; Iguchi, K.; Kato, M.; Hori, Y.; Arima, K.; Namiki, M.; Sugimura, Y. Heterogenous induction of carcinoma-associated fibroblast-like differentiation in normal human prostatic fibroblasts by co-culturing with prostate cancer cells. *J. Cell Biochem.* **2011**, *112*, 3604–3611. [CrossRef] [PubMed]
14. Ishii, K.; Sasaki, T.; Iguchi, K.; Kajiwara, S.; Kato, M.; Kanda, H.; Hirokawa, Y.; Arima, K.; Mizokami, A.; Sugimura, Y. Interleukin-6 induces VEGF secretion from prostate cancer cells in a manner independent of androgen receptor activation. *Prostate* **2018**, *78*, 849–856. [CrossRef] [PubMed]
15. Tanner, M.J.; Welliver, R.C., Jr.; Chen, M.; Shtutman, M.; Godoy, A.; Smith, G.; Mian, B.M.; Buttyan, R. Effects of androgen receptor and androgen on gene expression in prostate stromal fibroblasts and paracrine signaling to prostate cancer cells. *PLoS ONE* **2011**, *6*, e16027. [CrossRef] [PubMed]
16. Gravina, G.L.; Mancini, A.; Ranieri, G.; Di Pasquale, B.; Marampon, F.; Di Clemente, L.; Ricevuto, E.; Festuccia, C. Phenotypic characterization of human prostatic stromal cells in primary cultures derived from human tissue samples. *Int. J. Oncol.* **2013**, *42*, 2116–2122. [CrossRef]
17. Sasaki, T.; Ishii, K.; Iwamoto, Y.; Kato, M.; Miki, M.; Kanda, H.; Arima, K.; Shiraishi, T.; Sugimura, Y. Fibroblasts prolong serum prostate-specific antigen decline after androgen deprivation therapy in prostate cancer. *Lab. Invest.* **2016**, *96*, 338–349. [CrossRef]
18. Halin, S.; Hammarsten, P.; Wikstrom, P.; Bergh, A. Androgen-insensitive prostate cancer cells transiently respond to castration treatment when growing in an androgen-dependent prostate environment. *Prostate* **2007**, *67*, 370–377. [CrossRef]
19. Ishii, K.; Imamura, T.; Iguchi, K.; Arase, S.; Yoshio, Y.; Arima, K.; Hirano, K.; Sugimura, Y. Evidence that androgen-independent stromal growth factor signals promote androgen-insensitive prostate cancer cell growth in vivo. *Endocr. Relat. Cancer* **2009**, *16*, 415–428. [CrossRef]
20. Horoszewicz, J.S.; Leong, S.S.; Chu, T.M.; Wajsman, Z.L.; Friedman, M.; Papsidero, L.; Kim, U.; Chai, L.S.; Kakati, S.; Arya, S.K.; et al. The LNCaP cell line–a new model for studies on human prostatic carcinoma. *Prog. Clin. Biol. Res.* **1980**, *37*, 115–132.
21. Wan, X.S.; Zhou, Z.; Steele, V.; Kopelovich, L.; Kennedy, A.R. Establishment and characterization of sublines of LNCaP human prostate cancer cells. *Oncol. Rep.* **2003**, *10*, 1569–1575. [CrossRef] [PubMed]

22. Iguchi, K.; Ishii, K.; Nakano, T.; Otsuka, T.; Usui, S.; Sugimura, Y.; Hirano, K. Isolation and characterization of LNCaP sublines differing in hormone sensitivity. *J. Androl.* **2007**, *28*, 670–678. [CrossRef] [PubMed]
23. Iguchi, K.; Hayakawa, Y.; Ishii, K.; Matsumoto, K.; Usui, S.; Sugimura, Y.; Hirano, K. Characterization of the low pH/low nutrient-resistant LNCaP cell subline LNCaP-F10. *Oncol. Rep.* **2012**, *28*, 2009–2015. [CrossRef] [PubMed]
24. Onishi, T.; Yamakawa, K.; Franco, O.E.; Kawamura, J.; Watanabe, M.; Shiraishi, T.; Kitazawa, S. Mitogen-activated protein kinase pathway is involved in alpha6 integrin gene expression in androgen-independent prostate cancer cells: Role of proximal Sp1 consensus sequence. *Biochim. Biophys. Acta* **2001**, *1538*, 218–227. [CrossRef]
25. Titus, M.A.; Schell, M.J.; Lih, F.B.; Tomer, K.B.; Mohler, J.L. Testosterone and dihydrotestosterone tissue levels in recurrent prostate cancer. *Clin. Cancer Res.* **2005**, *11*, 4653–4657. [CrossRef] [PubMed]
26. Ishii, K.; Sugimura, Y. Identification of a new pharmacological activity of the phenylpiperazine derivative naftopidil: Tubulin-binding drug. *J. Chem. Biol.* **2015**, *8*, 5–9. [CrossRef] [PubMed]
27. Ishii, K.; Matsuoka, I.; Kajiwara, S.; Sasaki, T.; Miki, M.; Kato, M.; Kanda, H.; Arima, K.; Shiraishi, T.; Sugimura, Y. Additive naftopidil treatment synergizes docetaxel-induced apoptosis in human prostate cancer cells. *J. Cancer Res. Clin. Oncol.* **2018**, *144*, 89–98. [CrossRef] [PubMed]
28. Craft, N.; Chhor, C.; Tran, C.; Belldegrun, A.; DeKernion, J.; Witte, O.N.; Said, J.; Reiter, R.E.; Sawyers, C.L. Evidence for clonal outgrowth of androgen-independent prostate cancer cells from androgen-dependent tumors through a two-step process. *Cancer Res.* **1999**, *59*, 5030–5036.
29. Feldman, B.J.; Feldman, D. The development of androgen-independent prostate cancer. *Nat. Rev. Cancer* **2001**, *1*, 34–45. [CrossRef]
30. Dutt, S.S.; Gao, A.C. Molecular mechanisms of castration-resistant prostate cancer progression. *Future Oncol.* **2009**, *5*, 1403–1413. [CrossRef]
31. Tombal, B. What is the pathophysiology of a hormone-resistant prostate tumour? *Eur. J. Cancer* **2011**, *47*, S179–S188. [CrossRef]
32. Yadav, S.S.; Stockert, J.A.; Hackert, V.; Yadav, K.K.; Tewari, A.K. Intratumor heterogeneity in prostate cancer. *Urol. Oncol.* **2018**, *36*, 349–360. [CrossRef] [PubMed]
33. Iguchi, K.; Fukami, K.; Ishii, K.; Otsuka, T.; Usui, S.; Sugimura, Y.; Hirano, K. Low androgen sensitivity is associated with low levels of Akt phosphorylation in LNCaP-E9 cells. *J. Androl.* **2012**, *33*, 660–666. [CrossRef] [PubMed]
34. Lin, H.K.; Hu, Y.C.; Yang, L.; Altuwaijri, S.; Chen, Y.T.; Kang, H.Y.; Chang, C. Suppression versus induction of androgen receptor functions by the phosphatidylinositol 3-kinase/Akt pathway in prostate cancer LNCaP cells with different passage numbers. *J. Biol. Chem.* **2003**, *278*, 50902–50907. [CrossRef] [PubMed]
35. Jaworski, T. Degradation and beyond: Control of androgen receptor activity by the proteasome system. *Cell Mol. Biol. Lett.* **2006**, *11*, 109–131. [CrossRef] [PubMed]
36. Kizaka-Kondoh, S.; Inoue, M.; Harada, H.; Hiraoka, M. Tumor hypoxia: A target for selective cancer therapy. *Cancer Sci.* **2003**, *94*, 1021–1028. [CrossRef] [PubMed]
37. Stewart, G.D.; Ross, J.A.; McLaren, D.B.; Parker, C.C.; Habib, F.K.; Riddick, A.C. The relevance of a hypoxic tumour microenvironment in prostate cancer. *BJU Int.* **2010**, *105*, 8–13. [CrossRef] [PubMed]
38. Otsuka, T.; Iguchi, K.; Fukami, K.; Ishii, K.; Usui, S.; Sugimura, Y.; Hirano, K. Androgen receptor W741C and T877A mutations in AIDL cells, an androgen-independent subline of prostate cancer LNCaP cells. *Tumour Biol.* **2011**, *32*, 1097–1102. [CrossRef]
39. Ishii, K.; Takahashi, S.; Sugimura, Y.; Watanabe, M. Role of Stromal Paracrine Signals in Proliferative Diseases of the Aging Human Prostate. *J. Clin. Med.* **2018**, *7*, 68. [CrossRef]
40. Franco, O.E.; Jiang, M.; Strand, D.W.; Peacock, J.; Fernandez, S.; Jackson, R.S., 2nd; Revelo, M.P.; Bhowmick, N.A.; Hayward, S.W. Altered TGF-beta signaling in a subpopulation of human stromal cells promotes prostatic carcinogenesis. *Cancer Res.* **2011**, *71*, 1272–1281. [CrossRef]
41. Kiskowski, M.A.; Jackson, R.S., 2nd; Banerjee, J.; Li, X.; Kang, M.; Iturregui, J.M.; Franco, O.E.; Hayward, S.W.; Bhowmick, N.A. Role for stromal heterogeneity in prostate tumorigenesis. *Cancer Res.* **2011**, *71*, 3459–3470. [CrossRef] [PubMed]
42. Cunha, G.R.; Hayward, S.W.; Wang, Y.Z.; Ricke, W.A. Role of the stromal microenvironment in carcinogenesis of the prostate. *Int. J. Cancer* **2003**, *107*, 1–10. [CrossRef] [PubMed]

43. Yousef, G.M.; Diamandis, E.P. The new human tissue kallikrein gene family: Structure, function, and association to disease. *Endocr. Rev.* **2001**, *22*, 184–204. [CrossRef] [PubMed]
44. Sasaki, T.; Onishi, T.; Hoshina, A. Nadir PSA level and time to PSA nadir following primary androgen deprivation therapy are the early survival predictors for prostate cancer patients with bone metastasis. *Prostate Cancer Prostatic Dis.* **2011**, *14*, 248–252. [CrossRef] [PubMed]
45. Sasaki, T.; Onishi, T.; Hoshina, A. Cutoff value of time to prostate-specific antigen nadir is inversely correlated with disease progression in advanced prostate cancer. *Endocr. Relat. Cancer* **2012**, *19*, 725–730. [CrossRef] [PubMed]
46. Flaberg, E.; Markasz, L.; Petranyi, G.; Stuber, G.; Dicso, F.; Alchihabi, N.; Olah, E.; Csizy, I.; Jozsa, T.; Andren, O.; et al. High-throughput live-cell imaging reveals differential inhibition of tumor cell proliferation by human fibroblasts. *Int. J. Cancer* **2011**, *128*, 2793–2802. [CrossRef] [PubMed]
47. Alkasalias, T.; Flaberg, E.; Kashuba, V.; Alexeyenko, A.; Pavlova, T.; Savchenko, A.; Szekely, L.; Klein, G.; Guven, H. Inhibition of tumor cell proliferation and motility by fibroblasts is both contact and soluble factor dependent. *Proc. Natl. Acad. Sci. USA* **2014**, *111*, 17188–17193. [CrossRef] [PubMed]
48. Dehm, S.M.; Schmidt, L.J.; Heemers, H.V.; Vessella, R.L.; Tindall, D.J. Splicing of a novel androgen receptor exon generates a constitutively active androgen receptor that mediates prostate cancer therapy resistance. *Cancer Res.* **2008**, *68*, 5469–5477. [CrossRef] [PubMed]
49. Guo, Z.; Yang, X.; Sun, F.; Jiang, R.; Linn, D.E.; Chen, H.; Chen, H.; Kong, X.; Melamed, J.; Tepper, C.G.; et al. A novel androgen receptor splice variant is up-regulated during prostate cancer progression and promotes androgen depletion-resistant growth. *Cancer Res.* **2009**, *69*, 2305–2313. [CrossRef]
50. Hu, R.; Dunn, T.A.; Wei, S.; Isharwal, S.; Veltri, R.W.; Humphreys, E.; Han, M.; Partin, A.W.; Vessella, R.L.; Isaacs, W.B.; et al. Ligand-independent androgen receptor variants derived from splicing of cryptic exons signify hormone-refractory prostate cancer. *Cancer Res.* **2009**, *69*, 16–22. [CrossRef]
51. Sun, S.; Sprenger, C.C.; Vessella, R.L.; Haugk, K.; Soriano, K.; Mostaghel, E.A.; Page, S.T.; Coleman, I.M.; Nguyen, H.M.; Sun, H.; et al. Castration resistance in human prostate cancer is conferred by a frequently occurring androgen receptor splice variant. *J. Clin. Invest.* **2010**, *120*, 2715–2730. [CrossRef] [PubMed]
52. Hu, R.; Lu, C.; Mostaghel, E.A.; Yegnasubramanian, S.; Gurel, M.; Tannahill, C.; Edwards, J.; Isaacs, W.B.; Nelson, P.S.; Bluemn, E.; et al. Distinct transcriptional programs mediated by the ligand-dependent full-length androgen receptor and its splice variants in castration-resistant prostate cancer. *Cancer Res.* **2012**, *72*, 3457–3462. [CrossRef] [PubMed]
53. Antonarakis, E.S.; Lu, C.; Wang, H.; Luber, B.; Nakazawa, M.; Roeser, J.C.; Chen, Y.; Mohammad, T.A.; Chen, Y.; Fedor, H.L.; et al. AR-V7 and resistance to enzalutamide and abiraterone in prostate cancer. *N. Engl. J. Med.* **2014**, *371*, 1028–1038. [CrossRef] [PubMed]
54. Shimizu, Y.; Tamada, S.; Kato, M.; Hirayama, Y.; Takeyama, Y.; Iguchi, T.; Sadar, M.D.; Nakatani, T. Androgen Receptor Splice Variant 7 Drives the Growth of Castration Resistant Prostate Cancer without Being Involved in the Efficacy of Taxane Chemotherapy. *J. Clin. Med.* **2018**, *7*, 444. [CrossRef] [PubMed]
55. Scher, H.I.; Graf, R.P.; Schreiber, N.A.; Jayaram, A.; Winquist, E.; McLaughlin, B.; Lu, D.; Fleisher, M.; Orr, S.; Lowes, L.; et al. Assessment of the Validity of Nuclear-Localized Androgen Receptor Splice Variant 7 in Circulating Tumor Cells as a Predictive Biomarker for Castration-Resistant Prostate Cancer. *JAMA Oncol.* **2018**, *4*, 1179–1186. [CrossRef] [PubMed]
56. Nevedomskaya, E.; Baumgart, S.J.; Haendler, B. Recent Advances in Prostate Cancer Treatment and Drug Discovery. *Int. J. Mol. Sci.* **2018**, *19*, 1359. [CrossRef] [PubMed]
57. Nuhn, P.; De Bono, J.S.; Fizazi, K.; Freedland, S.J.; Grilli, M.; Kantoff, P.W.; Sonpavde, G.; Sternberg, C.N.; Yegnasubramanian, S.; Antonarakis, E.S. Update on Systemic Prostate Cancer Therapies: Management of Metastatic Castration-resistant Prostate Cancer in the Era of Precision Oncology. *Eur. Urol.* **2018**. [CrossRef]
58. Zhao, S.G.; Chang, S.L.; Erho, N.; Yu, M.; Lehrer, J.; Alshalalfa, M.; Speers, C.; Cooperberg, M.R.; Kim, W.; Ryan, C.J.; et al. Associations of Luminal and Basal Subtyping of Prostate Cancer With Prognosis and Response to Androgen Deprivation Therapy. *JAMA Oncol.* **2017**, *3*, 1663–1672. [CrossRef] [PubMed]
59. Mukherji, D.; Omlin, A.; Pezaro, C.; Shamseddine, A.; de Bono, J. Metastatic castration-resistant prostate cancer (CRPC): Preclinical and clinical evidence for the sequential use of novel therapeutics. *Cancer Metastasis Rev.* **2014**, *33*, 555–566. [CrossRef]
60. Kita, Y.; Goto, T.; Akamatsu, S.; Yamasaki, T.; Inoue, T.; Ogawa, O.; Kobayashi, T. Castration-Resistant Prostate Cancer Refractory to Second-Generation Androgen Receptor Axis-Targeted Agents: Opportunities and Challenges. *Cancers* **2018**, *10*, 345. [CrossRef]

9

The Importance of Time to Prostate-Specific Antigen (PSA) Nadir after Primary Androgen Deprivation Therapy in Hormone-Naïve Prostate Cancer Patients

Takeshi Sasaki and Yoshiki Sugimura *

Department of Nephro-Urologic Surgery and Andrology, Mie University Graduate School of Medicine, Tsu, Mie 514-8507, Japan; t-sasaki@clin.medic.mie-u.ac.jp

* Correspondence: sugimura@clin.medic.mie-u.ac.jp

Abstract: Prostate-specific antigen (PSA) is currently the most useful biomarker for detection of prostate cancer (PCa). The ability to measure serum PSA levels has affected all aspects of PCa management over the past two decades. The standard initial systemic therapy for advanced PCa is androgen-deprivation therapy (ADT). Although PCa patients with metastatic disease initially respond well to ADT, they often progress to castration-resistant prostate cancer (CRPC), which has a high mortality rate. We have demonstrated that time to PSA nadir (TTN) after primary ADT is an important early predictor of overall survival and progression-free survival for advanced PCa patients. In in vivo experiments, we demonstrated that the presence of fibroblasts in the PCa tumor microenvironment can prolong the period for serum PSA decline after ADT, and enhance the efficacy of ADT. Clarification of the mechanisms that affect TTN after ADT could be useful to guide selection of optimal PCa treatment strategies. In this review, we discuss recent in vitro and in vivo findings concerning the involvement of stromal–epithelial interactions in the biological mechanism of TTN after ADT to support the novel concept of "tumor regulating fibroblasts".

Keywords: prostate-specific antigen; androgen deprivation therapy; time to PSA nadir; fibroblasts

---

1. Introduction

Androgen-deprivation therapy (ADT) is the standard initial systemic therapy for advanced prostate cancer (PCa). Generally, ADT is performed with surgical castration or pharmacologic castration (luteinizing hormone-releasing hormone agonist or antagonist) accompanied with/without an antiandrogen (combined androgen blockade: CAB). Even PCa patients with metastatic disease initially respond well to ADT, but most eventually progress to castration-resistant prostate cancer (CRPC), which has a high mortality rate. Thus, there is an urgent need for predictors of which patients are more likely to develop CRPC. Several prognostic factors identified in clinical and laboratory studies could be used to predict survival including performance status, T stage, and extent of bone metastases, as well as serum alkaline phosphatase, hemoglobin, and testosterone levels [1,2]. The J-CAPRA score is a novel, validated score encompassing initial prostate-specific antigen (PSA) levels, Gleason score, and clinical stage that can be used to predict outcomes among patients undergoing primary ADT who represent the full spectrum of risk and stage, including advanced disease [3].

PSA is currently the most useful biomarker for detection of PCa. The ability to measure serum PSA levels affected all aspects of PCa management over the past two decades. Serum PSA levels are generally proportional to tumor volume and clinical stage of the disease. Thus, despite recognized limitations, measurement of PSA is essential for screening and monitoring of treatment response, prognosis, and progression in patients with PCa [4]. We demonstrated that time to PSA nadir (TTN) after primary ADT is an important early predictor for overall survival and progression-free survival for advanced PCa patients [5,6]. Moreover, we gathered evidence from in vivo experiments to support the role of fibroblasts in the PCa tumor microenvironment in prolonging the period of serum PSA decline after ADT and enhancing the efficacy of ADT [7]. In this review, we briefly summarize the importance of TTN after primary ADT for hormone-naïve advanced PCa patients with a focus on results from both in vitro and in vivo experiments.

2. Prostate-Specific Antigen (PSA)

PSA is an androgen-regulated serine protease and member of the tissue kallikrein family of proteases [8]. In humans, the prostate gland consists of a single layer of secretory epithelial cells, which are surrounded by a continuous layer of basal cells and a basement membrane [9]. PSA is produced by secretory epithelial cells in the prostate gland, and is directly secreted into the prostate lumen. One characteristic of PCa is disruption of the basal cell layer and basement membrane, and this loss of normal glandular architecture results in increases in serum PSA [10]. Transcription of the *PSA* gene is normally regulated by androgens via the androgen receptor (AR) [11]. The AR is a steroid hormone receptor that binds as a homodimer to a specific DNA sequence termed the androgen-responsive element (ARE). Consensus AREs are located at −156 to −170 from the transcriptional start site of the *PSA* gene [12]. Meanwhile, the *PSA* distal enhancer is located approximately 4.2 kb upstream of the transcription start site in a region that contains a single strong consensus ARE (ARE III). The presence of multiple additional weak non-consensus AREs has also been demonstrated by binding studies showing that the cooperative binding of multiple ARs in this region likely accounts for its strong androgen-dependent activity [13–16].

3. PSA Expression after Androgen Deprivation Therapy (ADT)

The goal of ADT treatment for PCa patients is to downregulate concentrations of circulating androgen or block transcriptional activation of the AR [17]. The decrease in PSA levels after primary ADT most likely results from tumor cell death and/or decreased expression of AR-stimulated PSA in surviving tumor cells (Figure 1). As a result, in some cases, primary ADT may have larger effects on PSA production than on tumor survival.

In the absence of androgen, AR is activated by the protein kinase A and/or protein kinase C pathway. In LNCaP cells, androgen-independent induction of *PSA* gene expression is regulated by the AR-dependent pathway [18–20]. Mitogen-activated protein kinase signaling may also regulate *PSA* transcription in an androgen-independent manner [21]. A number of growth factors and cytokines, including insulin-like growth factor (IGF) 1, keratinocyte growth factor (KGF; known as fibroblast growth factor (FGF) 7), epidermal growth factor (EGF), and interleukin (IL) 6, stimulate AR signaling and PSA expression in the context of androgen deficiency [22].

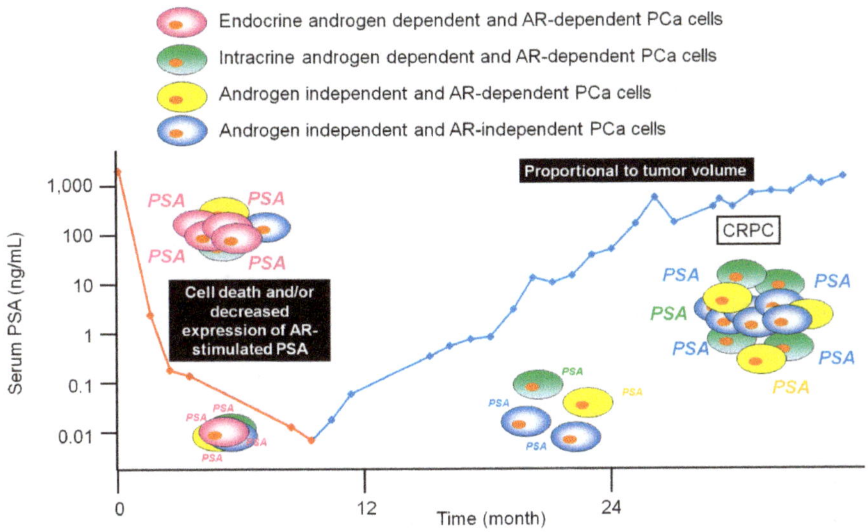

Figure 1. Serum PSA kinetics after primary ADT. Decreases in PSA levels following primary ADT are due to tumor cell death and/or simply decreased expression of AR-stimulated PSA in response to the absence of androgen. Figure 1 refers to [23]. AR: androgen receptor; PCa: prostate cancer; CRPC: castration-resistant prostate cancer; PSA: prostate-specific antigen; ADT: androgen-deprivation therapy.

4. Time to PSA Nadir (TTN) after Primary ADT

For PCa patients undergoing ADT, PSA kinetics are an important indicator of ADT response. However, the prognostic significance of PSA kinetics remains controversial [24]. In particular, the prognostic importance of various PSA indexes after treatment, such as PSA level at initial diagnosis, pattern of PSA decrease after hormone therapy, PSA half-life, nadir PSA level after treatment, TTN and percentage of PSA decrease, is unclear. Furthermore, few studies have examined whether these PSA indexes can accurately predict the likelihood of progression to CRPC [25]. Intuitively, most urologists expect that a more rapid PSA decline in response to primary ADT would be positively associated with extended survival. Indeed, clinical studies performed in the 1990s indicated that rapid PSA declines were associated with longer remission periods [26]. However, we have recently reported clinical evidence that a prolonged period for serum PSA decline after primary ADT was strongly indicative of disease progression in patients with advanced PCa [5,6]. Several recent studies describing results from large, multicenter investigations also demonstrated that longer TTN periods after primary ADT can predict favorable progression-free survival and overall survival in various hormone-naïve patient populations [24,27,28]. Akbay et al. evaluated PSA decline pattern after primary ADT in advanced PCa patients [29]. They showed that rapid PSA decline patients (fast decline slope) patients had higher rates of PSA progression, while prolonged PSA decline patients (slow decline slope) patients had lower rates of PSA progression. Choueiri et al. also demonstrated that higher PSA decline ($\geq$52 ng/mL/year) was associated with shorter survival in univariate analysis [27]. These findings may seem counterintuitive in that they suggest that a more rapid response to primary ADT indicates more aggressive disease.

Here, we summarize several lines of clinical evidence showing the prognostic importance of TTN after primary ADT in hormone-naïve PCa patients (Table 1). With respect to disease progression, Morote et al. reported that a PSA nadir $\leq$ 0.2 ng/mL and TTN $\geq$ 12 months in metastatic PCa patients was associated with a low risk of PSA progression [30]. Hori et al. reported that PSA nadir < 1 ng/mL and TTN > 12 months was associated with a low risk of biochemical relapse in PCa patients with bone metastasis, whereas patients without bone metastasis with a PSA nadir < 0.1 ng/mL and TTN > 24 months had a low risk of biochemical relapse [31]. These studies could be limited by patient classification in that the study participants were stratified into two groups according to PSA nadir level or TTN. Inadequate PSA responders, with shorter TTN due to small declines in PSA from an initial high PSA level, which is obviously associated with poor progression, might have been included in the group having a shorter TTN. Interestingly, Huang et al. stratified their study participants into four groups by combining PSA nadir level and TTN. In their study, they demonstrated that patients with advanced or metastatic disease and a PSA nadir $\geq$ 0.2 ng/mL and TTN < 10 months had a significantly reduced disease-free progression time [32]. However, for adequate PSA responders (PSA nadir $\leq$ 0.2 ng/mL), they found that prolonged TTN did not correlate with longer progression-free survival due to the inclusion of patients having various clinical stages, as well as those who underwent different pretreatments, such as radical prostatectomy or radiotherapy. We propose that the optimal cut-off value for TTN be >11 months for advanced PCa patients without bone metastasis and >8 months for those with bone metastasis in both groups with $\leq$0.2 PSA nadir and >0.2 with and without metastasis, respectively, for progression-free survival [6]. With respect to survival, Hussain et al. reported that a PSA $\leq$ 4 ng/mL after 7 months of androgen deprivation is a strong and specific predictor for risk of death [33]. Choueiri et al. recently reported that a PSA nadir < 0.2 ng/mL and a TTN of > 6 months were an optimal predictor for a longer overall survival in patients with metastatic disease [27]. Our analysis showed that a PSA nadir < 0.2 ng/mL and a TTN of > 9 months were an optimal predictor of longer overall survival in patients with bone metastatic disease [5].

Table 1. Examination of TTN after primary ADT in hormone-naïve PCa patients as a prognostic marker for disease outcome.

Study	Patients (*N*)	Treatment	TTN Cutoff Threshold (Months)	Outcome
Morote et al. 2004 [30]	283 (98 locally advanced, 185 metastatic)	Orchidectomy or maximal androgen blockade	12	Progression-free survival
Morote et al. 2005 [34]	185 (metastatic)	Orchidectomy or LHRH agonist with antiandrogen	9	Progression-free survival
Choueiri et al. 2009 [27]	179 (metastatic, 47.5% had prior RP or RT)	LHRH agonist or orchidectomy with or without antiandrogen	6	Overall survival
Hori et al. 2011 [31]	155 (46 with bone metastasis, 109 without bone metastasis)	LHRH agonist or orchidectomy with or without antiandrogen	24 (without bone metastasis)	Progression-free survival
Huang et al. 2011 [32]	650 (advanced or metastatic, 35% had RP or RT)	LHRH agonist or orchidectomy with or without antiandrogen	10	Progression-free survival
Huang et al. 2011 [35]	650 (advanced or metastatic, 35% had RP or RT)	LHRH agonist or orchidectomy with or without antiandrogen	10	Overall survival
Sasaki et al. 2011 [5]	87 (with bone metastasis)	LHRH agonist or orchidectomy with antiandrogen	9	Overall survival
Sasaki et al. 2012 [6]	184 (advanced, 101 with bone metastasis, 83 without bone metastasis)	LHRH agonist or orchidectomy with antiandrogen	8 with bone metastasis, 11 without bone metastasis	Progression-free survival
Hong et al. 2012 [36]	131 (metastatic)	LHRH agonist or orchidectomy with antiandrogen	8	Progression-free survival
Zhang et al. 2013 [37]	332 (advanced or metastatic)	LHRH antagonist or orchidectomy with flutamide	10	Overall survival Progression-free survival
Kitagawa et al. 2014 [28]	10,958 (all stage)	LHRH agonist or orchidectomy with or without antiandrogen	9	Overall survival Progression-free survival
Tomioka et al. 2014 [38]	286 (metastatic)	LHRH agonist or orchidectomy with or without antiandrogen	<6, 6–12, $\geq$12	Overall survival Progression-free survival
Teoh et al. 2017 [39]	419 (metastatic)	LHRH agonist or orchidectomy	<3, 3–17, >17	Overall survival Progression-free survival
Akbay et al. 2017 [29]	97 (advanced)	LHRH agonist or orchidectomy with or without antiandrogen	12	Progression-free survival

LHRH: luteinizing hormone releasing hormone; RP: radical prostatectomy; RT: radiation therapy; TTN: PSA nadir; ADT: androgen-deprivation therapy; PCa: prostate cancer.

5. Biological Mechanism of TTN after ADT

The above findings indicate that a rapid decline in PSA expression after primary ADT appears to be a strong indicator of more aggressive disease. However, the mechanisms that mediate this effect remain unknown.

Generally, well-differentiated PCa cells are AR-dependent and PSA positive, whereas poorly differentiated PCa cells are AR-independent and PSA negative. Nelson et al. provided four molecular-state frameworks for AR activation in PCa after ADT—State 1: Endocrine androgen-dependent and AR-dependent; State 2: Intracrine androgen-dependent and AR-dependent; State 3: Androgen-independent and AR-dependent; State 4: Androgen-independent and AR-independent [40]. State 4 is considered to be a fatal stage in which AR signaling is abolished. The transition of PCa cells to an androgen-independent phenotype is a complex process that involves selection and outgrowth of pre-existing clones of androgen-independent cells (clonal selection) as well as adaptive upregulation of genes that promote cancer cell survival and growth after ADT (adaptation) [41]. These two mechanisms share an important prerequisite characteristic: PCa are heterogeneous tumors comprising various subpopulations of cells that respond differently to ADT [41]. Acute loss of AR function after ADT is associated not only with apoptosis and reduced PSA secretion by PCa cells, but also with triggering of AR-independent growth. Disruption of androgen signaling by ADT may inhibit cell cycle control, which could contribute to carcinogenesis [42]. Further in vitro and in vivo experiments to characterize the interaction between AR-dependent and AR-independent PCa cells will be required to confirm these clinical findings.

We demonstrated a critical role for fibroblasts in tumor stroma in the regulation of androgen dependency of PCa cells and PSA expression after ADT [7]. The tumor stroma surrounding cancer cells is enriched in fibroblasts that secrete AR-stimulating factors, vascular endothelial growth factor (VEGF), and transforming growth factor (TGF) β [43]. Stromal–epithelial cell interactions play a crucial role in carcinogenesis and tumor progression [44]. We previously reported that stromal remodeling after castration is accompanied by changes in the expression levels of these growth factors in the prostate [45]. Importantly, most fibroblastic cells in the prostate stroma are negative for AR [46,47], and the phenotypes of human PCa fibroblastic stromal cells are broadly heterogeneous [48]. Several studies showed that that androgen-sensitive and -insensitive interactions between stromal and epithelial cells determine how prostate epithelial cells respond to androgen ablation [49,50]. In our in vitro and in vivo experiments, we found that the AR-independent and heterogeneous characteristics of fibroblasts in PCa tissue could regulate ADT efficacy as measured by tumor volume and Ki67 index, which is related to the decline in serum PSA after ADT [7]. Even though there is a limitation of animal experiments, we found that fibroblasts had two mechanisms for regulating declines in serum PSA after ADT: 1) maintenance of tumor microvessels and 2) secretion of soluble AR-stimulating factors. Fibroblasts have a diverse capacity for neovascularization and varying expression patterns of these soluble factors. In a low androgen environment, stromal–epithelial interactions may be an important mechanism to control AR activity and AR-regulated PSA expression. Thus, we advocate for the adoption of a new concept, "tumor regulating fibroblasts", which describes the possible action of fibroblasts after ADT in PCa patients (Figure 2). Tumor-promoting "aggressive" fibroblasts, also called cancer-associated fibroblasts (CAFs), are well known [51]. CAFs surround cancer cells to support the survival and proliferation of cancer cells in a paracrine fashion. These "aggressive" fibroblasts can increase the selective pressure to promote preferential selection of more aggressive epithelial phenotypes. On the other hand, Hayashi et al. reported that rat urogenital mesenchyme, which shares similar features with CAFs, but has a physiological role that involves facilitating the development, differentiation, and, ultimately, growth quiescence of the prostate, elicited a reduction in tumorigenic potential of Dunning prostatic adenocarcinoma [52]. Recent studies also demonstrated that normal human fibroblasts can inhibit the proliferation of tumor cells [53,54]. "Protective" fibroblasts that contribute to a prolonged serum PSA decline period could enhance treatment efficacy, resulting in a more favorable prognosis.

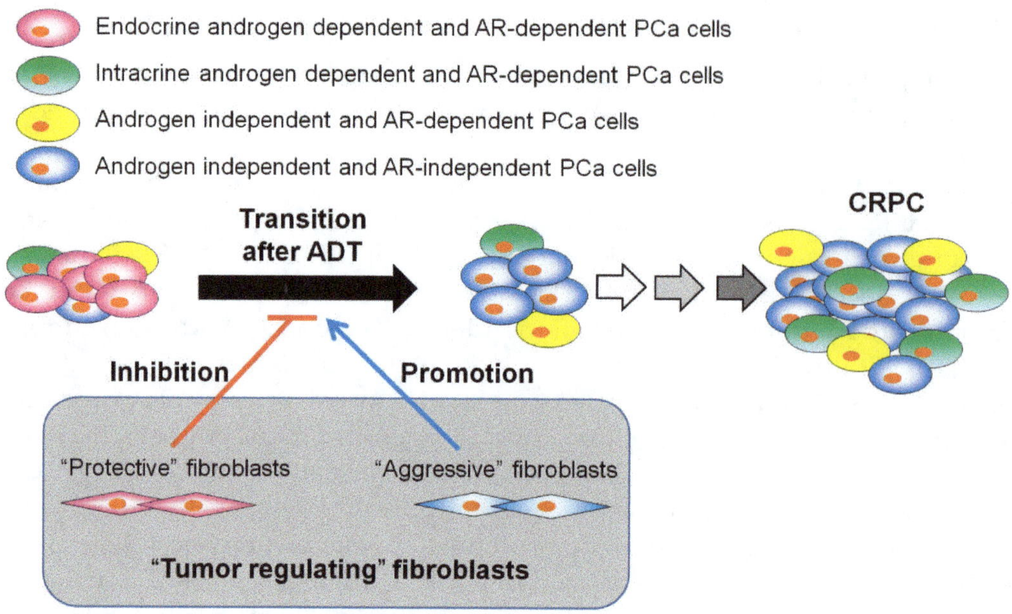

Figure 2. "Tumor regulating fibroblasts" and the role of fibroblasts after ADT in PCa patients. "Protective" fibroblasts inhibit the transition of PCa cells from AR-dependent to AR-independent, whereas "aggressive" fibroblasts promote the transition from AR-dependent to AR-independent [7].

6. Concluding Remarks

TTN after primary ADT for advanced PCa patients is a powerful tool for predicting disease progression and overall survival. Clarification of the mechanisms associated with TTN after primary ADT could help inform treatment decision-making to determine optimal strategies for PCa treatment. In this review, we focused on stromal–epithelial interactions to develop a clearer picture of the biological mechanism of TTN after ADT, based on findings from in vitro and in vivo experiments to provide a novel concept of "tumor regulating fibroblasts".

Author Contributions: T.S. wrote the first draft of the manuscript; Y.S. revised the manuscript and added additional text.

Acknowledgments: We are grateful to Kenichiro Ishii of the Department of Oncologic Pathology, Mie University Graduate School of Medicine, for valuable discussions.

References

1. De Voogt, H.J.; Suciu, S.; Sylvester, R.; Pavone-Macaluso, M.; Smith, P.H.; de Pauw, M. Multivariate analysis of prognostic factors in patients with advanced prostatic cancer: Results from 2 european organization for research on treatment of cancer trials. *J. Urol.* **1989**, *141*, 883–888. [CrossRef]
2. Ishikawa, S.; Soloway, M.S.; Van der Zwaag, R.; Todd, B. Prognostic factors in survival free of progression after androgen deprivation therapy for treatment of prostate cancer. *J. Urol.* **1989**, *141*, 1139–1142. [CrossRef]
3. Cooperberg, M.R.; Hinotsu, S.; Namiki, M.; Ito, K.; Broering, J.; Carroll, P.R.; Akaza, H. Risk assessment among prostate cancer patients receiving primary androgen deprivation therapy. *J. Clin. Oncol.* **2009**, *27*, 4306–4313. [CrossRef] [PubMed]
4. Hernandez, J.; Thompson, I.M. Prostate-specific antigen: A review of the validation of the most commonly used cancer biomarker. *Cancer* **2004**, *101*, 894–904. [CrossRef] [PubMed]
5. Sasaki, T.; Onishi, T.; Hoshina, A. Nadir PSA level and time to PSA nadir following primary androgen deprivation therapy are the early survival predictors for prostate cancer patients with bone metastasis. *Prostate Cancer Prostatic Dis.* **2011**, *14*, 248–252. [CrossRef] [PubMed]

6. Sasaki, T.; Onishi, T.; Hoshina, A. Cutoff value of time to prostate-specific antigen nadir is inversely correlated with disease progression in advanced prostate cancer. *Endocr. Relat. Cancer* **2012**, *19*, 725–730. [CrossRef] [PubMed]
7. Sasaki, T.; Ishii, K.; Iwamoto, Y.; Kato, M.; Miki, M.; Kanda, H.; Arima, K.; Shiraishi, T.; Sugimura, Y. Fibroblasts prolong serum prostate-specific antigen decline after androgen deprivation therapy in prostate cancer. *Lab. Invest.* **2016**, *96*, 338–349. [CrossRef] [PubMed]
8. Yousef, G.M.; Diamandis, E.P. The new human tissue kallikrein gene family: Structure, function, and association to disease. *Endocr. Rev.* **2001**, *22*, 184–204. [CrossRef] [PubMed]
9. Sasaki, T.; Franco, O.E.; Hayward, S.W. Male reproductive tract: Prostate overview. In *Encyclopedia of Reproduction*, 2nd ed.; Skinner, M., Ed.; Elsevier: Cambridge, MA, USA, 2018; Volume 1, pp. 309–314, ISBN 9780128012383.
10. Balk, S.P.; Ko, Y.J.; Bubley, G.J. Biology of prostate-specific antigen. *J. Clin. Oncol.* **2003**, *21*, 383–391. [CrossRef] [PubMed]
11. Young, C.Y.; Montgomery, B.T.; Andrews, P.E.; Qui, S.D.; Bilhartz, D.L.; Tindall, D.J. Hormonal regulation of prostate-specific antigen messenger RNA in human prostatic adenocarcinoma cell line lncap. *Cancer Res.* **1991**, *51*, 3748–3752. [PubMed]
12. Riegman, P.H.; Vlietstra, R.J.; van der Korput, J.A.; Brinkmann, A.O.; Trapman, J. The promoter of the prostate-specific antigen gene contains a functional androgen responsive element. *Mol. Endocrinol.* **1991**, *5*, 1921–1930. [CrossRef] [PubMed]
13. Schuur, E.R.; Henderson, G.A.; Kmetec, L.A.; Miller, J.D.; Lamparski, H.G.; Henderson, D.R. Prostate-specific antigen expression is regulated by an upstream enhancer. *J. Biol. Chem.* **1996**, *271*, 7043–7051. [CrossRef] [PubMed]
14. Cleutjens, K.B.; van der Korput, H.A.; van Eekelen, C.C.; van Rooij, H.C.; Faber, P.W.; Trapman, J. An androgen response element in a far upstream enhancer region is essential for high, androgen-regulated activity of the prostate-specific antigen promoter. *Mol. Endocrinol.* **1997**, *11*, 148–161. [CrossRef] [PubMed]
15. Zhang, S.; Murtha, P.E.; Young, C.Y. Defining a functional androgen responsive element in the 5′ far upstream flanking region of the prostate-specific antigen gene. *Biochem. Biophys. Res. Commun.* **1997**, *231*, 784–788. [CrossRef] [PubMed]
16. Huang, W.; Shostak, Y.; Tarr, P.; Sawyers, C.; Carey, M. Cooperative assembly of androgen receptor into a nucleoprotein complex that regulates the prostate-specific antigen enhancer. *J. Biol. Chem.* **1999**, *274*, 25756–25768. [CrossRef] [PubMed]
17. Culig, Z.; Klocker, H.; Bartsch, G.; Hobisch, A. Androgen receptors in prostate cancer. *Endocr. Relat. Cancer* **2002**, *9*, 155–170. [CrossRef] [PubMed]
18. Nazareth, L.V.; Weigel, N.L. Activation of the human androgen receptor through a protein kinase a signaling pathway. *J. Biol. Chem.* **1996**, *271*, 19900–19907. [CrossRef] [PubMed]
19. Sadar, M.D. Androgen-independent induction of prostate-specific antigen gene expression via cross-talk between the androgen receptor and protein kinase a signal transduction pathways. *J. Biol. Chem.* **1999**, *274*, 7777–7783. [CrossRef] [PubMed]
20. Ikonen, T.; Palvimo, J.J.; Kallio, P.J.; Reinikainen, P.; Janne, O.A. Stimulation of androgen-regulated transactivation by modulators of protein phosphorylation. *Endocrinology* **1994**, *135*, 1359–1366. [CrossRef] [PubMed]
21. Franco, O.E.; Onishi, T.; Yamakawa, K.; Arima, K.; Yanagawa, M.; Sugimura, Y.; Kawamura, J. Mitogen-activated protein kinase pathway is involved in androgen-independent *PSA* gene expression in lncap cells. *Prostate* **2003**, *56*, 319–325. [CrossRef] [PubMed]
22. Kim, J.; Coetzee, G.A. Prostate specific antigen gene regulation by androgen receptor. *J. Cell. Biochem.* **2004**, *93*, 233–241. [CrossRef] [PubMed]
23. Mizokami, A.; Namiki, M. Reconsideration of progression to CRPC during androgen deprivation therapy. *J. Steroid Biochem. Mol. Biol.* **2015**, *145*, 164–171. [CrossRef] [PubMed]
24. Crawford, E.D.; Bennett, C.L.; Andriole, G.L.; Garnick, M.B.; Petrylak, D.P. The utility of prostate-specific antigen in the management of advanced prostate cancer. *BJU Int.* **2013**, *112*, 548–560. [CrossRef] [PubMed]
25. Kwak, C.; Jeong, S.J.; Park, M.S.; Lee, E.; Lee, S.E. Prognostic significance of the nadir prostate specific antigen level after hormone therapy for prostate cancer. *J. Urol.* **2002**, *168*, 995–1000. [CrossRef]

26. Arai, Y.; Yoshiki, T.; Yoshida, O. Prognostic significance of prostate specific antigen in endocrine treatment for prostatic cancer. *J. Urol.* **1990**, *144*, 1415–1419. [CrossRef]
27. Choueiri, T.K.; Xie, W.; D'Amico, A.V.; Ross, R.W.; Hu, J.C.; Pomerantz, M.; Regan, M.M.; Taplin, M.E.; Kantoff, P.W.; Sartor, O.; et al. Time to prostate-specific antigen nadir independently predicts overall survival in patients who have metastatic hormone-sensitive prostate cancer treated with androgen-deprivation therapy. *Cancer* **2009**, *115*, 981–987. [CrossRef] [PubMed]
28. Kitagawa, Y.; Ueno, S.; Izumi, K.; Mizokami, A.; Hinotsu, S.; Akaza, H.; Namiki, M. Nadir prostate-specific antigen (PSA) level and time to PSA nadir following primary androgen deprivation therapy as independent prognostic factors in a Japanese large-scale prospective cohort study (J-CaP). *J. Cancer Res. Clin. Oncol.* **2014**, *140*, 673–679. [CrossRef] [PubMed]
29. Akbay, E.; Bozlu, M.; Cayan, S.; Kara, P.O.; Tek, M.; Aytekin, C. Prostate-specific antigen decline pattern in advanced prostate cancer receiving androgen deprivation therapy and relationship with prostate-specific antigen progression. *Aging Male* **2017**, *20*, 175–183. [CrossRef] [PubMed]
30. Morote, J.; Trilla, E.; Esquena, S.; Abascal, J.M.; Reventos, J. Nadir prostate-specific antigen best predicts the progression to androgen-independent prostate cancer. *Int. J. Cancer* **2004**, *108*, 877–881. [CrossRef] [PubMed]
31. Hori, S.; Jabbar, T.; Kachroo, N.; Vasconcelos, J.C.; Robson, C.N.; Gnanapragasam, V.J. Outcomes and predictive factors for biochemical relapse following primary androgen deprivation therapy in men with bone scan negative prostate cancer. *J. Cancer Res. Clin. Oncol.* **2011**, *137*, 235–241. [CrossRef] [PubMed]
32. Huang, S.P.; Bao, B.Y.; Wu, M.T.; Choueiri, T.K.; Goggins, W.B.; Huang, C.Y.; Pu, Y.S.; Yu, C.C.; Huang, C.H. Impact of prostate-specific antigen (PSA) nadir and time to PSA nadir on disease progression in prostate cancer treated with androgen-deprivation therapy. *Prostate* **2011**, *71*, 1189–1197. [CrossRef] [PubMed]
33. Hussain, M.; Tangen, C.M.; Higano, C.; Schelhammer, P.F.; Faulkner, J.; Crawford, E.D.; Wilding, G.; Akdas, A.; Small, E.J.; Donnelly, B.; et al. Absolute prostate-specific antigen value after androgen deprivation is a strong independent predictor of survival in new metastatic prostate cancer: Data from southwest oncology group trial 9346 (INT-0162). *J. Clin. Oncol.* **2006**, *24*, 3984–3990. [CrossRef] [PubMed]
34. Morote, J.; Esquena, S.; Abascal, J.M.; Trilla, E.; Cecchini, L.; Raventos, C.X.; Orsola, A.; Planas, J.; Catalan, R.; Reventos, J. Usefulness of prostate-specific antigen nadir as predictor of androgen-independent progression of metastatic prostate cancer. *Int. J. Biol. Markers* **2005**, *20*, 209–216. [CrossRef] [PubMed]
35. Huang, S.P.; Bao, B.Y.; Wu, M.T.; Choueiri, T.K.; Goggins, W.B.; Liu, C.C.; Huang, C.Y.; Pu, Y.S.; Yu, C.C.; Wu, T.T.; et al. Significant associations of prostate-specific antigen nadir and time to prostate-specific antigen nadir with survival in prostate cancer patients treated with androgen-deprivation therapy. *Aging Male* **2012**, *15*, 34–41. [CrossRef] [PubMed]
36. Hong, S.Y.; Cho, D.S.; Kim, S.I.; Ahn, H.S.; Kim, S.J. Prostate-specific antigen nadir and time to prostate-specific antigen nadir following maximal androgen blockade independently predict prognosis in patients with metastatic prostate cancer. *Korean J. Urol.* **2012**, *53*, 607–613. [CrossRef] [PubMed]
37. Zhang, L.M.; Jiang, H.W.; Tong, S.J.; Zhu, H.Q.; Liu, J.; Ding, Q. Prostate-specific antigen kinetics under androgen deprivation therapy and prostate cancer prognosis. *Urol. Int.* **2013**, *91*, 38–48. [CrossRef] [PubMed]
38. Tomioka, A.; Tanaka, N.; Yoshikawa, M.; Miyake, M.; Anai, S.; Chihara, Y.; Okajima, E.; Hirayama, A.; Hirao, Y.; Fujimoto, K. Nadir PSA level and time to nadir PSA are prognostic factors in patients with metastatic prostate cancer. *BMC Urol.* **2014**, *14*, 33. [CrossRef] [PubMed]
39. Teoh, J.Y.; Tsu, J.H.; Yuen, S.K.; Chiu, P.K.; Chan, S.Y.; Wong, K.W.; Ho, K.L.; Hou, S.S.; Ng, C.F.; Yiu, M.K. Association of time to prostate-specific antigen nadir and logarithm of prostate-specific antigen velocity after progression in metastatic prostate cancer with prior primary androgen deprivation therapy. *Asian J. Androl.* **2017**, *19*, 98–102. [PubMed]
40. Nelson, P.S. Molecular states underlying androgen receptor activation: A framework for therapeutics targeting androgen signaling in prostate cancer. *J. Clin. Oncol.* **2012**, *30*, 644–646. [CrossRef] [PubMed]
41. So, A.; Gleave, M.; Hurtado-Col, A.; Nelson, C. Mechanisms of the development of androgen independence in prostate cancer. *World J. Urol.* **2005**, *23*, 1–9. [CrossRef] [PubMed]
42. Jennbacken, K.; Tesan, T.; Wang, W.; Gustavsson, H.; Damber, J.E.; Welen, K. N-cadherin increases after androgen deprivation and is associated with metastasis in prostate cancer. *Endocr. Relat. Cancer* **2010**, *17*, 469–479. [CrossRef] [PubMed]
43. Zhu, M.L.; Kyprianou, N. Androgen receptor and growth factor signaling cross-talk in prostate cancer cells. *Endocr. Relat. Cancer* **2008**, *15*, 841–849. [CrossRef] [PubMed]

44. Sasaki, T.; Franco, O.E.; Hayward, S.W. Interaction of prostate carcinoma-associated fibroblasts with human epithelial cell lines in vivo. *Differentiation* **2017**, *96*, 40–48. [CrossRef] [PubMed]
45. Kato, M.; Ishii, K.; Iwamoto, Y.; Sasaki, T.; Kanda, H.; Yamada, Y.; Arima, K.; Shiraishi, T.; Sugimura, Y. Activation of fgf2-fgfr signaling in the castrated mouse prostate stimulates the proliferation of basal epithelial cells. *Biol. Reprod.* **2013**, *89*, 81. [CrossRef] [PubMed]
46. Gravina, G.L.; Mancini, A.; Ranieri, G.; Di Pasquale, B.; Marampon, F.; Di Clemente, L.; Ricevuto, E.; Festuccia, C. Phenotypic characterization of human prostatic stromal cells in primary cultures derived from human tissue samples. *Int. J. Oncol.* **2013**, *42*, 2116–2122. [CrossRef] [PubMed]
47. Tanner, M.J.; Welliver, R.C., Jr.; Chen, M.; Shtutman, M.; Godoy, A.; Smith, G.; Mian, B.M.; Buttyan, R. Effects of androgen receptor and androgen on gene expression in prostate stromal fibroblasts and paracrine signaling to prostate cancer cells. *PLoS ONE* **2011**, *6*, e16027. [CrossRef] [PubMed]
48. Ishii, K.; Mizokami, A.; Tsunoda, T.; Iguchi, K.; Kato, M.; Hori, Y.; Arima, K.; Namiki, M.; Sugimura, Y. Heterogenous induction of carcinoma-associated fibroblast-like differentiation in normal human prostatic fibroblasts by co-culturing with prostate cancer cells. *J. Cell. Biochem.* **2011**, *112*, 3604–3611. [CrossRef] [PubMed]
49. Ishii, K.; Imamura, T.; Iguchi, K.; Arase, S.; Yoshio, Y.; Arima, K.; Hirano, K.; Sugimura, Y. Evidence that androgen-independent stromal growth factor signals promote androgen-insensitive prostate cancer cell growth in vivo. *Endocr. Relat. Cancer* **2009**, *16*, 415–428. [CrossRef] [PubMed]
50. Halin, S.; Hammarsten, P.; Wikstrom, P.; Bergh, A. Androgen-insensitive prostate cancer cells transiently respond to castration treatment when growing in an androgen-dependent prostate environment. *Prostate* **2007**, *67*, 370–377. [CrossRef] [PubMed]
51. Olumi, A.F.; Grossfeld, G.D.; Hayward, S.W.; Carroll, P.R.; Tlsty, T.D.; Cunha, G.R. Carcinoma-associated fibroblasts direct tumor progression of initiated human prostatic epithelium. *Cancer Res.* **1999**, *59*, 5002–5011. [PubMed]
52. Hayashi, N.; Cunha, G.R. Mesenchyme-induced changes in the neoplastic characteristics of the dunning prostatic adenocarcinoma. *Cancer Res.* **1991**, *51*, 4924–4930. [PubMed]
53. Flaberg, E.; Markasz, L.; Petranyi, G.; Stuber, G.; Dicso, F.; Alchihabi, N.; Olah, E.; Csizy, I.; Jozsa, T.; Andren, O.; et al. High-throughput live-cell imaging reveals differential inhibition of tumor cell proliferation by human fibroblasts. *Int. J. Cancer* **2011**, *128*, 2793–2802. [CrossRef] [PubMed]
54. Alkasalias, T.; Flaberg, E.; Kashuba, V.; Alexeyenko, A.; Pavlova, T.; Savchenko, A.; Szekely, L.; Klein, G.; Guven, H. Inhibition of tumor cell proliferation and motility by fibroblasts is both contact and soluble factor dependent. *Proc. Natl. Acad. Sci. USA* **2014**, *111*, 17188–17193. [CrossRef] [PubMed]

10

Androgen Receptor Splice Variant 7 Drives the Growth of Castration Resistant Prostate Cancer without being Involved in the Efficacy of Taxane Chemotherapy

Yasuomi Shimizu [1], Satoshi Tamada [1], Minoru Kato [1,*], Yukiyoshi Hirayama [2], Yuji Takeyama [1], Taro Iguchi [1], Marianne D. Sadar [2] and Tatsuya Nakatani [1]

1 Department of Urology, Graduate School of Medicine, Osaka City University, Osaka 545-8585, Japan; yasuomis0922@gmail.com (Y.S.); s-tamada@med.osaka-cu.ac.jp (S.T.); yuji101@outlook.jp (Y.T.); taro@msic.med.osaka-cu.ac.jp (T.I.); nakatani@med.osaka-cu.ac.jp (T.N.)

2 Genome Sciences Centre, BC Cancer, Vancouver, BC V5Z 1L3, Canada; ukiyoc@hotmail.com (Y.H.); msadar@bcgsc.ca (M.D.S.)

* Correspondence: kato.minoru@med.osaka-cu.ac.jp

Abstract: Expression of androgen receptor (AR) splice variant 7 (AR-V7) has been identified as the mechanism associated with the development of castration-resistant prostate cancer (CRPC). However, a potential link between AR-V7 expression and resistance to taxanes, such as docetaxel or cabazitaxel, has not been unequivocally demonstrated. To address this, we used LNCaP95-DR cells, which express AR-V7 and exhibit resistance to enzalutamide and docetaxel. Interestingly, LNCaP95-DR cells showed cross-resistance to cabazitaxel. Furthermore, these cells had increased levels of P-glycoprotein (P-gp) and their sensitivity to both docetaxel and cabazitaxel was restored through treatment with tariquidar, a P-gp antagonist. Results generated demonstrated that P-gp mediated cross-resistance between docetaxel and cabazitaxel. Although the LNCaP95-DR cells had increased expression of AR-V7 and its target genes (UBE2C, CDC20), the knockdown of AR-V7 did not restore sensitivity to docetaxel or cabazitaxel. However, despite resistance to docetaxel and carbazitaxel, EPI-002, an antagonist of the AR amino-terminal domain (NTD), had an inhibitory effect on the proliferation of LNCaP95-DR cells, which was similar to that achieved with the parental LNCaP95 cells. On the other hand, enzalutamide had no effect on the proliferation of either cell line. In conclusion, our results suggested that EPI-002 may be an option for the treatment of AR-V7-driven CRPC, which is resistant to taxanes.

Keywords: androgen receptor; docetaxel; cabazitaxel; castration-resistant prostate cancer; chemotherapy; P-glycoprotein; EPI-002; splice variant

---

1. Introduction

The primary effective treatment for most recurring prostate cancer (PC) is androgen deprivation therapy (ADT). Although initially effective, the malignancy will eventually form castration-resistant prostate cancer (CRPC) [1]. Current treatment options for CRPC are androgen receptor- (AR-) targeted therapies, such as enzalutamide and abiraterone, as well as taxanes, such as docetaxel and cabazitaxel. However, no curative CRPC therapy is available for the presentation of treatment resistance [2]. The mechanisms of CRPC development include overexpression of AR [3,4], gain-of-function mutations in the AR ligand-binding domain (LBD) [5], intratumoral androgen synthesis [6], altered expression and function of the AR coactivators [7,8], aberrant post-translational modification of the AR [9], and the AR splice variants (AR-Vs) lacking the LBD [10]. Expression of AR-V7 in human prostate cancer cell lines mediates resistance to enzalutamide and abiraterone [11,12]. EPI compounds are AR amino-terminal domain- (NTD-) targeting drugs that block the transcriptional activities of full-length (FL)-AR and AR-Vs in vitro and exhibit antitumor activity in CRPC xenografts [13–15]. Resistance to taxane-based chemotherapy is frequently attributed to the overexpression of the transporter protein, P-glycoprotein (P-gp), which is also known as ATP-binding cassette subfamily B member 1 (ABCB1) or multidrug resistance protein 1 (MDR-1) [16,17]. Prostate cancer specimens from CRPC patients have increased levels of P-gp [18]. Cabazitaxel is highly cytotoxic with a low affinity for P-gp [19], and therefore, it is not considered to be clinically cross-resistant with docetaxel, thereby providing a survival benefit for docetaxel-pretreated patients [20]. However, a recent report indicated that P-gp could mediate cabazitaxel–docetaxel cross-resistance in advanced prostate cancer [21]. Alternative mechanisms of resistance to taxanes may involve tubulin mutations [22], although the precise association has yet to be elucidated. The presence of AR-V7 in circulating tumor cells from men with metastatic CRPC is not associated with primary resistance to taxane chemotherapy [23,24], which is contrary to pre-clinical data suggesting that the expression of AR-V7 mediates resistance to docetaxel in LuCap23.1 human prostate cancer xenografts [25]. In this study, we focused specifically on the role of AR-V7 in taxane resistance in pre-clinical models of prostate cancer to address this clinically important question. We employed the CRPC cell line, LNCaP95, which endogenously expresses AR-V7, to examine the status of cross-resistance between docetaxel and cabazitaxel, and to assess the involvement of AR-V7 with taxane resistance. We further evaluated the effect of EPI-002, an NTD-targeting drug, on enzalutamide resistant LNCaP95 cells with acquired resistance to taxanes.

2. Materials and Methods

2.1. Cell Lines and Culture Conditions

Prostate cancer cell lines, DU145, PC3, and LNCaP were purchased from the American Type Culture Collection (ATCC). LNCaP95 was a generous gift from Dr. Jun O. Luo (Johns Hopkins University, Baltimore, MD, USA). Cells were authenticated by short tandem repeat analysis (Takara Bio Inc., Shiga, Japan) and then tested by DDC Medical (Thermo Fisher Scientific, Waltham, MA, USA) in April 2018, to ensure that the cells were mycoplasma-free. Cells were maintained as monolayer cultures at 37 °C and 5% CO_2. The cell line DU145 was cultured in DMEM supplemented with 10% FBS, PC-3 in RPMI 1640 with 10% FBS, LNCaP in phenol red-free RPMI 1640 with 10% FBS, and LNCaP95 in phenol red-free RPMI 1640 with 10% charcoal stripped serum (CSS). A docetaxel-resistant cell line variant, LNCaP95-DR, was developed over a period of 6 months by exposure to gradually increased concentrations of docetaxel (Sigma-Aldrich, St. Louis, MO, USA). A time-matched parental cell line, LNCaP95-C, was developed in a medium containing vehicle (DMSO) at the corresponding concentration. Finally, LNCaP95-DR cells were maintained in medium containing 15 nM docetaxel.

2.2. Cell Proliferation Assay

The effect of drugs on cell proliferation was assessed using the Premix WST-1 Cell Proliferation Assay System (Takara Bio) according to the manufacturer's protocol. Cell viability was normalized to the viability of vehicle-treated control cells (DMSO).

BrdU ELISA was performed to evaluate the inhibitory effect of EPI-002 on the proliferation of cells. LNCaP95-P, LNCaP95-C, and LNCaP95-DR cells were treated with vehicle (DMSO) or EPI-002 for 48 h, and BrdU incorporation was measured using the BrdU ELISA kit (Roche Diagnostics, Basel, Switzerland).

2.3. Western Blot Analysis

Western blots were performed as previously described in Reference [26]. The primary antibodies used were: AR (1:1000; Santa Cruz Biotechnology), AR-V7 (1:400; Precision), GR (1:1000; BD transduction laboratories), PSA (1:1000; Santa Cruz Biotechnology), FKBP5 (1:1000; Santa Cruz Biotechnology), UBE2C (1:1000; Boston Biochem), NSE (1:1000; Merck), Mdr-1 (1:1000; Santa

Cruz), Aurora A (1:1000), BRN-2 (1:1000), total-STAT3 (1:1000), p-STAT3Tyr705 (1:1000), total-AKT (1:1000), p-AktSer473 (1:1000), total-S6 (1:1000), p-S6 (1:2500), total-p44/42MAPKErk1/2 (1:1000), p-p44/42MAPKErk1/2 (1:1000), 110α (1:1000), 110β (1:1000), 110γ (1:1000), PI3KClass III (1:1000), p85 (1:1000), 4EBP1 (1:1000), and p-4EBP1 (1:1000), from Cell Signaling Technology. Beta-actin (1:1000, Abcam and Cell Signaling Technology) was used as a loading control.

2.4. Real-Time Quantitative Reverse Transcription PCR (Real-Time RT-qPCR)

Total RNA was isolated using the RNAqueous Total RNA Isolation Kit (Life Technologies, Waltham, MA, USA) and it was reverse transcribed to cDNA using the High Capacity cDNA Reverse Transcription Kit (Applied Biosystems, Waltham, MA, USA). Real-time RT-qPCR was performed in triplicate for each biological sample. Transcript levels for each gene were normalized to levels of the GAPDH transcript. Primers were purchased from Applied Biosystems: AR (Hs00171172_m1), KLK3 (Hs02576345_m1), FKBP5 (Hs01561006_m1), UBE2C (Hs00964100_g1), CDC20 (Hs00426680_mH), GAPDH (Hs00266705_g1), AR-V7 (forward, 5′-CCATCTTGTCGTCTTCGGAAATGTTA-3′; reverse, 5′-TTTGAATGAGGCAAGTCAGCCTTTCT-3′).

2.5. AR-Driven PSA(6.1kb)-Luciferase Reporter Gene Assay

PSA (6.1kb)-luciferase reporter plasmid encodes nucleotides −6000/+12 relative to the transcription start site of the human PSA/KLK3 gene and it includes the PSA promoter, with AREII (−395 to 376) and AREI (−170 to −156), and enhancer regions with AREIII (−4148 to −4134), as described in References [27,28]. LNCaP95-C and LNCaP95-DR cells seeded in 24-well plates were transfected using FuGENE HD Transfection Reagent (Promega, Madison, WI, USA), with a plasmid encoding the prostate-specific antigen (PSA) (6.1 kb)-luciferase reporter gene construct. The next day, the cells were pre-treated with vehicle (DMSO), enzalutamide (10 μM), docetaxel (5 nM), or cabazitaxel (10 nM) for 1 h before adding R1881 or EtOH (vehicle) under serum-free, phenol red-free conditions. After 48 h of incubation, the cells were harvested and lysed using the lysis buffer that was provided with the Luciferase Assay System (Promega). PSA-luciferase activity was measured using with the Wallac 1420 ARVOsx multi-label plate reader (PerkinElmer, Waltham, MA, USA) and normalized to protein concentration by the Bradford method as explained in Reference [29].

2.6. Knockdown of AR-V7

AR-V7 expression was transiently knocked down in LNCaP95-DR cells using Lipofectamine™ RNAiMAX (Invitrogen, Carlsbad, CA, USA) according to the manufacturer's instructions. AR-V7 siRNAs (Silencer® Select siRNAs) were obtained from Life Technologies™ (Grand Island, NY, USA). The sense sequence of siRNA for AR-V7 was 5′-GUAGUUGUGAGUAUCAUGATT-3′.

2.7. Statistical Analysis

Statistical analyses were performed using GraphPad Prism 7 (GraphPad Software, Inc., La Jolla CA, USA). All experiments were performed in triplicates for each biological sample. Data for cell proliferation assays, real-time RT-qPCR, and luciferase assays were depicted as mean ± SD from 3 to 4 independent experiments. IC50 values and 95% confidence intervals (CI) were calculated using the nonlinear regression analysis of percentage inhibition. The comparison of LogIC50 was calculated using the extra sum-of-squares F-test. One-way ANOVA followed by a Sidak's post hoc test was used to assess the difference between the data of real-time RT-qPCR and luciferase assay. $p < 0.05$ was considered to indicate a statistically significant difference.

3. Results

3.1. LNCaP95-DR Cells Were Cross-Resistant to Cabazitaxel

To evaluate the inhibitory effect of docetaxel and cabazitaxel on prostate cancer cell lines, the MTT assay was performed (Figure 1A,B). LNCaP cells were highly sensitive to docetaxel and cabazitaxel, whereas LNCaP95 cells were less sensitive than LNCaP cells. A docetaxel resistant LNCaP95 cell line, LNCaP95-DR, was obtained by exposing parental cells to gradually increasing concentrations of docetaxel. As shown in Figure 1C, LNCaP95-DR cells were significantly less sensitive to docetaxel than LNCaP95-C cells. Furthermore, LNCaP95-DR cells were less sensitive to cabazitaxel than LNCaP95-C cells (Figure 1D). A table showing the IC50s of all these cell lines is provided in Figure 1E. These data suggest that the acquired resistance to docetaxel results in the cross-resistance to cabazitaxel.

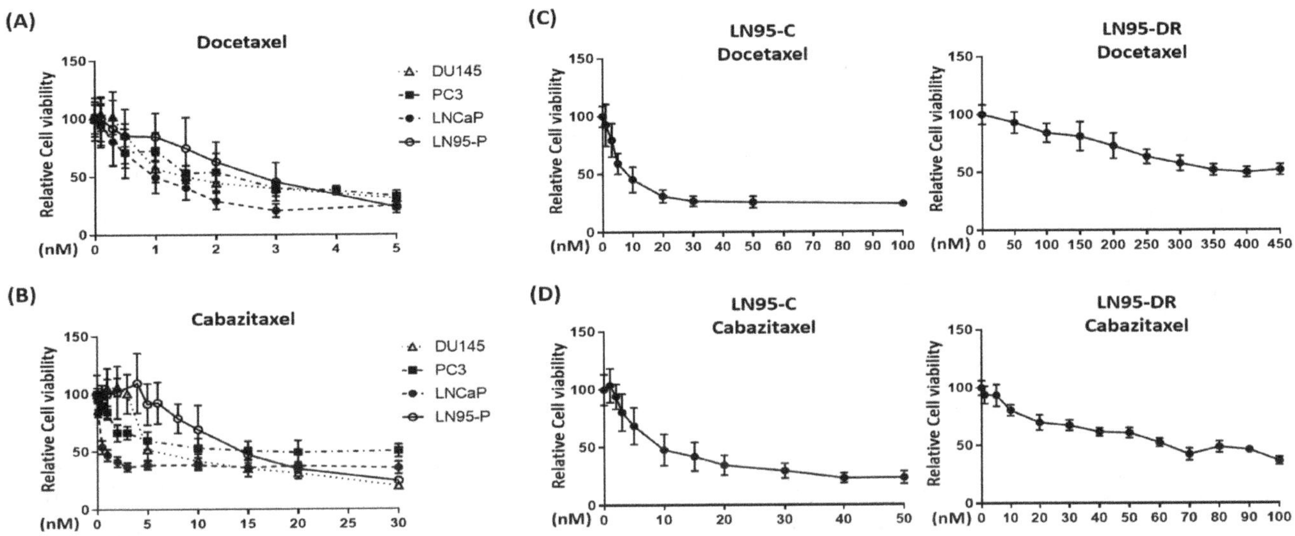

(E)

	IC50 (95%CI) <nM>					
	DU145	PC3	LNCaP	LN95-P	LN95-C	LN95-DR
Docetaxel	1.816 (1.552–2.148)	2.132 (1.771–2.545)	1.038 (0.8548–1.252)	3.199 (2.904–3.566)	10.07 (8.926–11.36)	412.1 (382.8–450.2)
Cabazitaxel	13.25 (10.52–16.75)	20.79 (16.26–27.77)	1.624 (1.042–2.375)	13.07 (11.65–15.17)	11.78 (10.53–13.17)	66.36 (62.17–71.23)

Figure 1. LN95-DR shows cross-resistance to cabazitaxel. Dose responses for docetaxel (**A**) and cabazitaxel (**B**) on the viability of prostate cancer cell lines (DU145, PC3, LNCaP, and LN95-P) assessed by the MTT assay; Dose responses for docetaxel (**C**) or cabazitaxel (**D**) on the viability of LN95-C and LN95-DR after 72 h; (**E**) A table showing IC50 values and 95% confidence intervals for docetaxel and cabazitaxel on prostate cancer cell lines. LN95-P: parental LNCaP95; LN95-C: time-matched parental LNCaP95 cells treated with DMSO as a vehicle control; LN95-DR: LNCaP95 with acquired resistance to docetaxel.

3.2. P-gp Was Overexpressed in LNCaP95-DR Cells and Tariquidar Restored Sensitivity to Docetaxel and Cabazitaxel

Consistent with a known mechanism of acquired resistance to taxanes, P-gp was overexpressed in LNCaP95-DR cells as measured by the Western blot analysis (Figure 2A). To test whether this high level of P-gp protein in LNCaP95-DR cells played a direct role in the resistance to docetaxel and cabazitaxel, a P-gp inhibitor was tested. Tariquidar is a potent P-gp antagonist that inhibits P-gp mediated drug efflux [30–33]. We found that the monotherapy with tariquidar showed no effect on the proliferation of LNCaP95-DR (data not shown), whilst tariquidar restored the sensitivity of LNCaP95-DR cells to both docetaxel and cabazitaxel (Figure 2B–D). These data indicated that the cross-resistance between docetaxel and cabazitaxel in LNCaP95-DR cells was mainly mediated by P-gp.

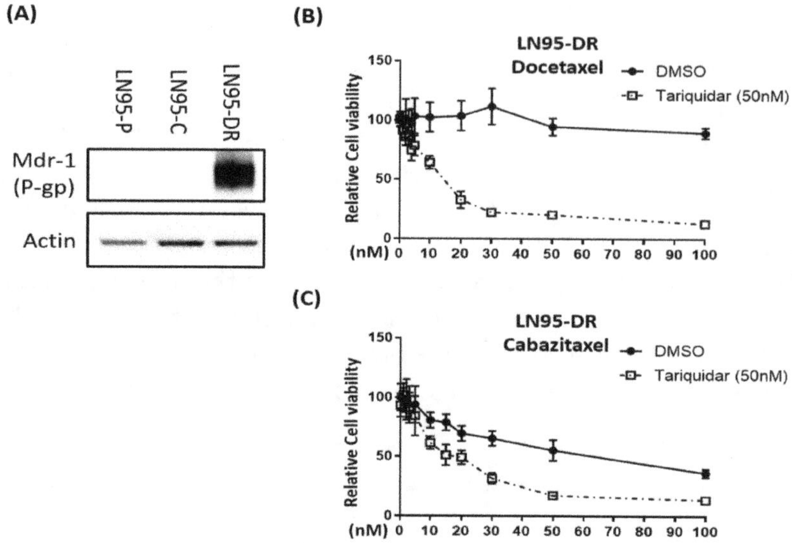

(D)

IC50 (95%CI) <nM>	
LN95-DR (Tariquidar 50nM)	
Docetaxel	15.43(12.94–18.55)
Cabazitaxel	16.86(14.99–18.98)

Figure 2. Tariquidar restored the sensitivity of LNCaP95-DR to docetaxel and cabazitaxel. (**A**) Levels of P-gp protein in LN95-P, LN95-C, and LN95-DR cell lysates using b-actin as a loading control; Effects of inhibition of p-gp on the viability of LN95-C and LN95-DR cells incubated with DMSO or a combination of tariquidar (50 nM, inhibitor of P-gp) and increasing concentrations of docetaxel (**B**) or cabazitaxel (**C**); (**D**) Table showing the IC50s of docetaxel and cabazitaxel in LN95-DR cells incubated with a combination of 50 nM tariquidar.

3.3. *Expression of AR-V7-Regulated Genes Was Increased in LNCaP95-DR*

To elucidate other potential contributing factors involved in the mechanism of taxane resistance and provide clues for possible intervention, we compared the levels of expression of several key genes in LNCaP95-DR cells using Western blot analysis and real-time RT-qPCR. LNCaP95-DR cells had higher levels of glucocorticoid receptor (GR), UBE2C, and phosphorylated S6 (pS6), but lower levels of BRN-2 proteins as compared to levels in LNCaP95-C (Figures 3A,B and A1C).

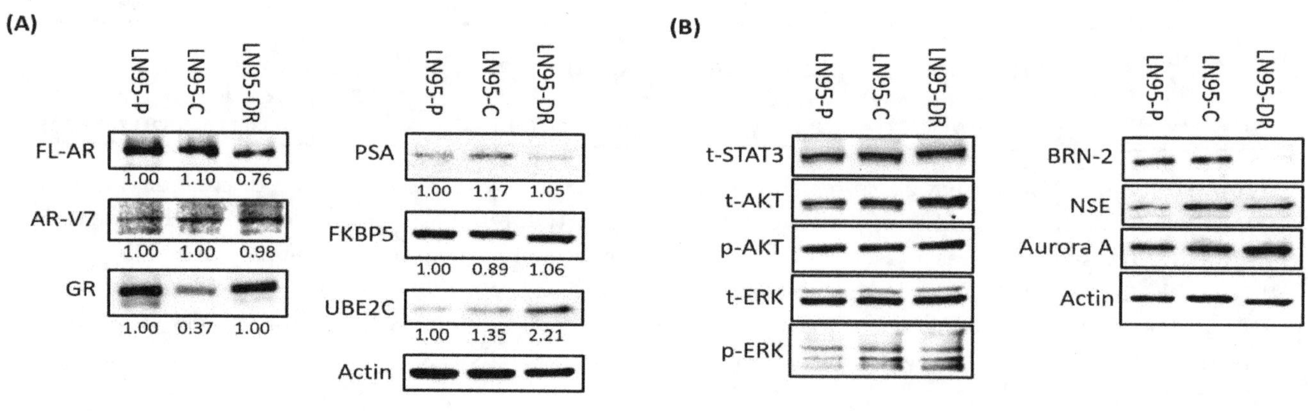

Figure 3. Levels of proteins suspected to play a role in the resistance to therapies for castration-resistant prostate cancer (CRPC). Western blot analyses using whole cell lysates from cell lines. Expression of proteins in androgen receptor (AR) and AR-V7 signaling (**A**), Jak/STAT, PI3K/AKT/mTOR, Ras/MAPK pathway and neuroendocrine markers (**B**). The ratio among the three cell lines was described, which was normalized to that in LN95-P, comparing the protein expression values of FL-AR and AR-V7 regulated molecules normalized to beta-actin as an internal control (**A**).

Real-time RT-qPCR revealed that the transcript levels of FL-AR and its target gene KLK3 in LNCaP95-DR cells did not differ from those in LNCaP-C cells, whereas the transcript levels of AR-V7 and its target genes, UBE2C and CDC20, were all increased in LNCaP95-DR cells as compared to levels in LNCaP-C cells (Figure 4). Interestingly, neither docetaxel nor cabazitaxel suppressed the expression of genes regulated by FL-AR (Figure 4A,B). FL-AR was functional in LNCaP95-DR cells as indicated by the induction of PSA-luciferase activity, as well as KLK3 and FKBP5 gene expression in response to the synthetic androgen, R1881 (Figure 4). Neither docetaxel nor cabazitaxel reduced the transcriptional activity of FL-AR in LNCaP95-C or LNCaP95-DR when measuring a PSA-luciferase reporter or endogenous expression of KLK3 and FKBP5 in response to R1881 (Figure 4).

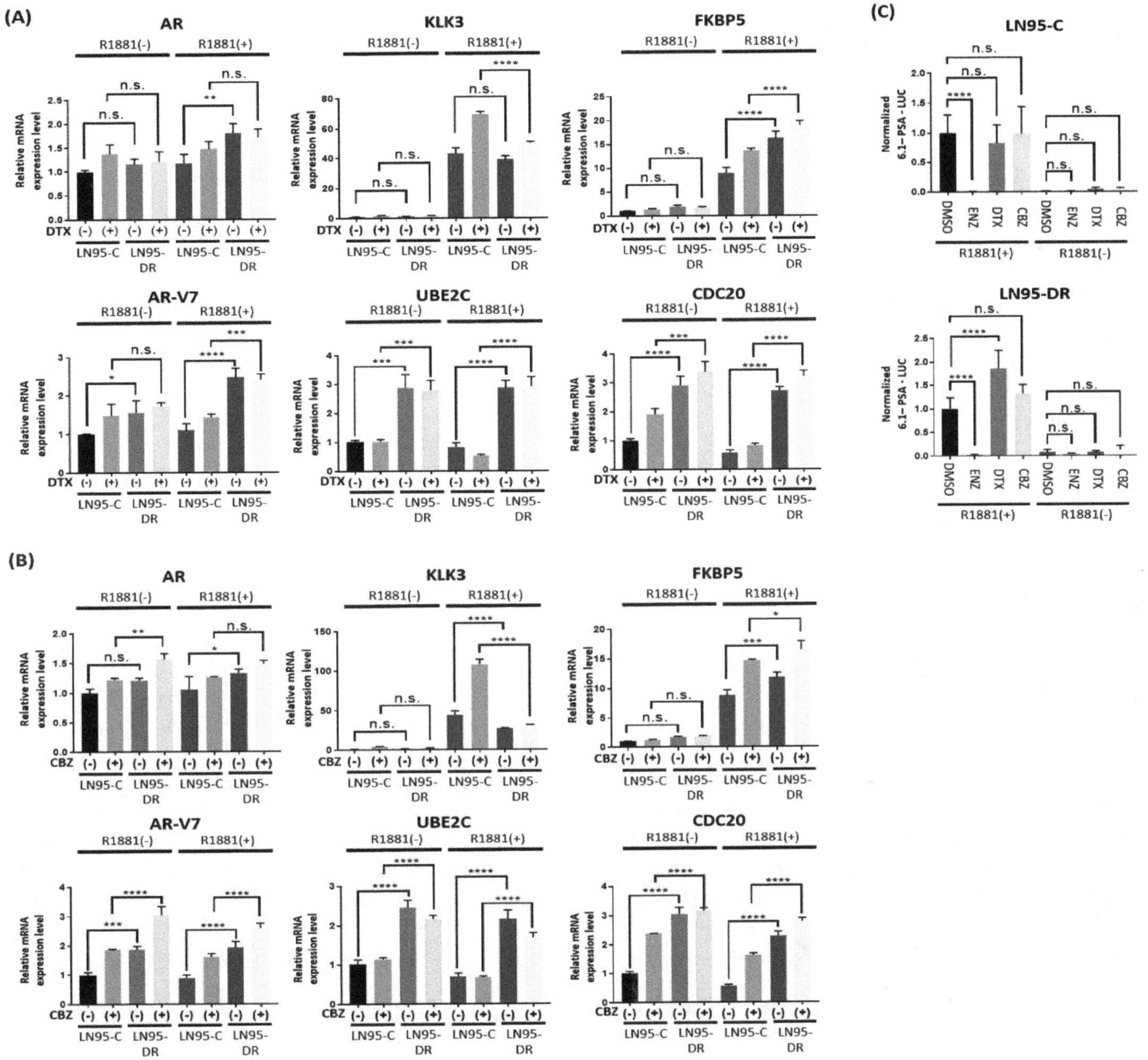

Figure 4. Increased expression of AR-V7 target genes in LNCaP95-DR. Levels of mRNA for FL-AR, AR-V7 and their target genes plus PSA (6.1 kb)-luciferase activities in LN95-C and LN95-DR in response to the synthetic androgen R1881 and taxanes. Transcript levels of FL-AR, KLK3, FKBP5, AR-V7, UBE2C, and CDC20 normalized to transcript levels of GAPDH. LN95-C and LN95-DR were treated with DMSO, docetaxel (5 nM) (**A**) or cabazitaxel (10 nM) (**B**) for 1 h prior to the addition of R1881 (1 nM) or EtOH for 48 h. For the luciferase assay (**C**), LN95-C and LN95-DR were treated with DMSO, enzalutamide (10 μM), docetaxel (5 nM), or cabazitaxel (10 nM) for 1 h prior to treatment with R1881 (1 nM) or EtOH for 48 h. n.s.: not significant; * $p < 0.05$; ** $p < 0.01$; *** $p < 0.001$; **** $p < 0.0001$.

BRN2 is a transcription factor that is proposed to play a role in enzalutamide-induced neuroendocrine transdifferentiation [34]. LNCaP95-DR cells had reduced expression of this transcription factor (Figure 3). Therefore, we tested whether the altered expression of BRN2 might correlate to increased sensitivity to enzalutamide. Unfortunately, no difference in cell viability was measured in response to enzalutamide in LNCaP95-DR cells (Figure A1A). All cell lines remained resistant to enzalutamide.

Glucocorticoid receptor (GR), is proposed to play a role in CRPC as an alternative steroid receptor for AR [35]. Although levels of GR protein were elevated in the LNCaP95 cells (Figure 3), the GR agonist, dexamethasone, did not affect the proliferation of the LNCaP95 cells (Figure A1B). This suggested that the GR is not driving proliferation in this model.

Alterations in expression of components of the PI3K/Akt/mTOR occur in 42% of primary prostate tumors and 100% of metastatic tumors [36]. Thus, targeting the PI3K/Akt/mTOR pathway is considered a promising approach for the treatment of CRPC [15,37,38]. To determine a possible role of this pathway, western blot analysis was performed to assess the status of this pathway in LNCaP95-DR cells. We found no significant change in the expression levels of proteins related to the PI3K/Akt/mTOR pathway, with the exception of pS6 (Figures 3B and A1C). However, the mTOR inhibitor, everolimus, did not mediate differential effects in the different cell lines (Figure A1D), which could suggest that an increased pS6 expression was not important for the acquired resistance to docetaxel.

3.4. Knockdown of AR-V7 Has No Effect on Sensitivity to Docetaxel and Cabazitaxel

Given the expression of AR-V7, and its target genes UBE2C and CDC20 being increased in LNCaP95-DR, we examined whether AR-V7 may contribute to the acquisition of resistance to taxanes and whether the targeting of AR-V7 might be a good intervention strategy. To test these we used two approaches, knockdown of AR-V7 and an inhibitor of the AR-Vs transcriptional activities. When AR-V7 expression was transiently knocked down in LNCaP95-DR cells using small interfering RNA (siRNA) (Figure 5A), proliferation of LNCaP95-DR cells was decreased by 37% (Figure 5B). AR-V7 knockdown did not restore the sensitivity of LNCaP95-DR cells to docetaxel or cabazitaxel (Figure 5C).

AR-NTD targeting drugs are a potential treatment strategy for CRPC represented by LNCaP95-DR cells which have acquired resistance to enzalutamide, docetaxel, and cabazitaxel. This is because AR-NTD is essential for the transcriptional activities exerted by both the FL-AR and AR-Vs. Thus, antagonists of AR-NTD, such as EPI-002, could have a therapeutic effect on LNCaP95 driven by AR-V7. Importantly, EPI-002 had an inhibitory effect on the proliferation of LNCaP95-DR cells that was similar to the effect measured with the parental LNCaP95 cells (Figure 5D). Together, the data revealed that AR-NTD-targeting drugs are a feasible intervention for taxane-resistant prostate cancers that are driven by AR-Vs.

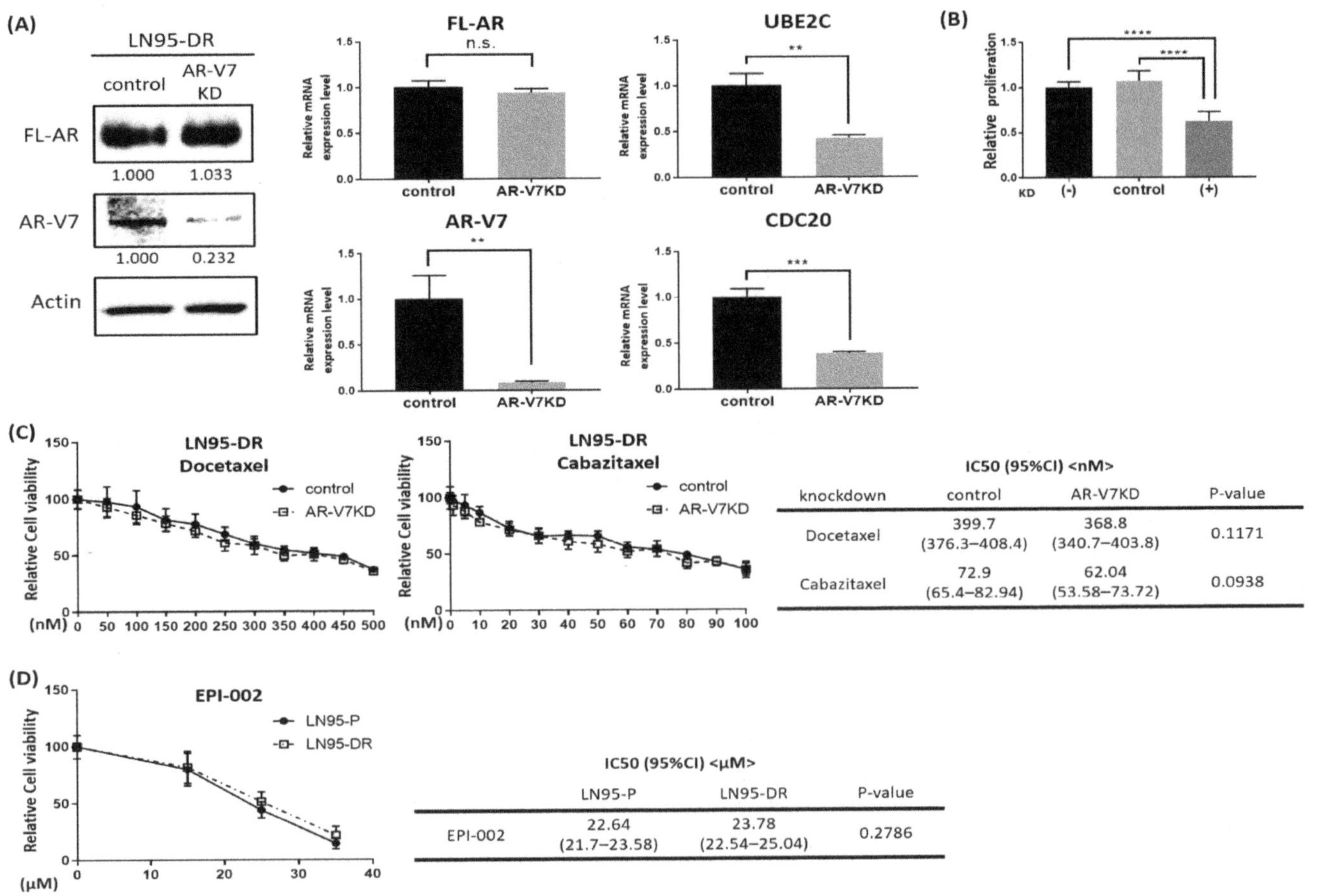

knockdown	IC50 (95%CI) <nM> control	AR-V7KD	P-value
Docetaxel	399.7 (376.3–408.4)	368.8 (340.7–403.8)	0.1171
Cabazitaxel	72.9 (65.4–82.94)	62.04 (53.58–73.72)	0.0938

	IC50 (95%CI) <μM> LN95-P	LN95-DR	P-value
EPI-002	22.64 (21.7–23.58)	23.78 (22.54–25.04)	0.2786

Figure 5. EPI-002 inhibits proliferation of LN95-DR. (**A**) Levels of proteins of FL-AR and AR-V7 and transcripts of FL-AR, AR-V7, UBE2C, and CDC20 in LN95-DR that were transfected with AR-V7 siRNA. After 48 h of transfection with 5 nM AR-V7 siRNA, LN95-DR cells were incubated in serum-free conditions for 48 h prior to collecting the proteins and for 96 h prior to collecting RNA. Transfection with AR-V7 siRNA sufficiently decreased the expression of AR-V7 and the target genes (UBE2C and CDC20) without affecting the level of the FL-AR; (**B**) Knockdown of AR-V7 decreased the proliferation of LN95-DR by 37%. After 48 h of transfection with 5 nM AR-V7 siRNA, LN95-DR cells were incubated in serum-free conditions for 72 h prior to measuring proliferation; (**C**) Dose response curves for docetaxel or cabazitaxel on the viability of LN95-DR cells treated 48 h after transfection with 5 nM AR-V7 siRNA (AR-V7KD). The table shows IC50s of docetaxel and cabazitaxel in LN95-DR cells after knockdown of AR-V7; (**D**) Dose response curve for EPI-002 on proliferation of both LN95-P and LN95-DR cells as measured by the BrdU ELISA assay. The Table shows IC50 values of EPI-002 in LN95-P and LN95-DR cells. n.s.: not significant; ** $p < 0.01$; *** $p < 0.001$; **** $p < 0.0001$.

4. Discussion

AR-V7 is a major splice variant expressed in human prostate cancer that is associated with the development and progression of CRPC [39,40]. AR-V7 potentially contributes to the resistance to enzalutamide and abiraterone in CRPC [11,41,42]. The involvement of AR-V7 in taxane resistance is not well understood. The LNCaP95 cell line was derived from the LNCaP cell line and it has acquired resistance to androgen depletion conditions. LNCaP95 cells express full-length AR and AR-V7, but the level of AR-V567es is negligible [43]. Proliferation of LNCaP95 cells is driven by AR-V7, despite the endogenous expression of functional FL-AR [39,44]. These cells are resistant to enzalutamide [45]. To elucidate whether AR-V7 plays a role in the acquired resistance to taxanes, the LNCaP95-DR cell line was developed and used as a model for CRPC.

Resistance to taxanes can be associated with the overexpression of P-gp [16,17], which has been confirmed in CRPC patients [18]. Levels of P-gp expression are higher in docetaxel-resistant TaxR and DU145-DTXR cells as compared to that in docetaxel-sensitive, parental C4-2B and DU145 cells, respectively [46,47]. Similarly, P-gp is overexpressed in docetaxel-resistant DU145R and CWR22rv1R cells derived from the parental DU145 and CWR22rv1 cells, respectively [48]. Consistent with these reports, we showed that P-gp was overexpressed in LNCaP95-DR cells, and that tariquidar treatment restored sensitivity to docetaxel. Therefore, overexpression of P-gp played a major role in the acquired resistance of LNCaP95 to docetaxel. Tariquidar is an anthranilic acid-derived third-generation P-gp inhibitor. Its efficacy has been evaluated in several clinical trials on different types of cancer including lung cancer, but there are no reports on its use for prostate cancer. Our study was the first report using tariquidar for prostate cancer cell lines.

Cabazitaxel is a next-generation semisynthetic taxane chemotherapeutic agent that is effective in patients with docetaxel-resistant CRPC [49]. In the TROPIC clinical trial, cabazitaxel significantly improved the overall survival in CRPC patients during or after docetaxel treatment, but the survival benefit was limited to 2.4 months [20]. In our in vitro study, LNCaP95-DR cells were resistant to high doses of docetaxel with an IC50 of >400 nM; however, these cells maintained some sensitivity to cabazitaxel with an IC50 of approximately 70 nM (Figure 1C,D). The data was consistent with that observed clinically with the resistance to docetaxel. Importantly, tariquidar restored sensitivity to docetaxel, as well as to cabazitaxel, thereby indicating that the cross-resistance between docetaxel and cabazitaxel was mediated by the overexpression of P-gp.

Expression of AR-V7 is clinically important and has been proposed for the assessment of which patients should receive inhibitors of the androgen receptor or taxanes [24]. In this study, we showed that the expression of AR-V7-regulated genes was increased in LNCaP95-DR, but that knockdown of AR-V7 did not restore sensitivity to docetaxel and cabazitaxel. These data support the idea that AR-V7 was not involved in taxane resistance. Contrary to these data, Tadani-Mulero M et al. reported that the expression of AR-V7 resulted in taxane resistance in a mouse model of CRPC due to the absence of the AR hinge region, which appears to be critical for microtubule binding [25]. This report compared FL-AR with AR-V567es-expressing LuCaP86.2 tumor xenografts and FL-AR with AR-V7-expressing LuCaP23.1 tumor xenografts. That report concluded that AR-V7, but not AR-V567es, was important for resistance to docetaxel which was not supported by the work presented here in the LNCaP95-DR cells. Consistent with data presented in this study, are those obtained from clinical studies that have shown that the detection of AR-V7 in circulating tumor cells from men with metastatic CRPC was not associated with primary resistance to taxane chemotherapy [23,24,50].

Some recent reports suggest that taxanes can inhibit AR signaling in prostate cancer cells [51–53]. However, in our study, neither docetaxel or cabazitaxel decreased the expression of genes regulated by FL-AR nor were there any effects on the transcriptional activity of the AR. Zhu ML and Darshan MS examined the effect of taxanes on the androgen/AR axis using very high concentrations of paclitaxel (1 μM and 100 nM, respectively), which were over the clinically effective range of paclitaxel [54,55]. In the present study, we tested docetaxel and cabazitaxel at the concentrations close to their IC50 (5 nM and 10 nM, respectively), which was clinically feasible [56,57]. Our data suggested that taxane chemotherapy did not affect the androgen/AR axis in LNCaP95-C and LNCaP95-DR when used at the clinically feasible concentration.

AR-NTD is essential for the transcriptional activities of both FL-AR and AR-Vs. Therefore, AR-NTD-targeting therapy has benefits over the drugs targeting the AR-LBD. EPI-002 targets the NTD of the AR and can block the signaling induced by the FL-AR and AR-Vs [58]. In this study, we showed that the inhibitory effect by EPI-002 on the proliferation of LNCaP95-DR cells was similar to that achieved with the parental LNCaP95 cells. LNCaP95-DR proliferation remained driven by AR-V7, suggesting that AR-NTD could be a therapeutic target for cancers such as LNCaP95-DR, with acquired resistance to taxanes and enzalutamide.

5. Conclusions

In summary, we have demonstrated that docetaxel-resistant LNCaP95 cells are cross-resistant to cabazitaxel. We showed that resistance to docetaxel and cabazitaxel depended on the increased expression of P-gp and the inhibition of P-gp with tariquidar restored to docetaxel and cabazitaxel. Furthermore, expression of AR-V7-regulated genes was increased in LNCaP95-DR cells, although AR-V7 did not contribute to taxane resistance. Finally, EPI-002, an antagonist of AR-NTD, inhibited proliferation of LNCaP95-DR. In conclusion, the present study described a potential option for the treatment of docetaxel-resistant, AR-V7-driven CRPC.

Author Contributions: Conceptualization, M.K. and M.D.S.; Methodology, M.K.; Software, Y.S.; Validation, Y.S., Y.T., Y.H., M.K. and M.D.S.; Formal Analysis, M.K.; Investigation, Y.S.; Resources, M.D.S.; Data Curation, S.T.; Writing—Original Draft Preparation, Y.S.; Writing—Review & Editing, M.K.; Visualization, T.I.; Supervision, T.N.; Project Administration, M.K.; Funding Acquisition, M.K. and M.D.S.

Acknowledgments: This work was supported by the US National Cancer Institute (#R01 CA105304) awarded to MD Sadar and Grants–in–Aid for Scientific Research (JP16K20156) awarded to M.K.

Appendix A

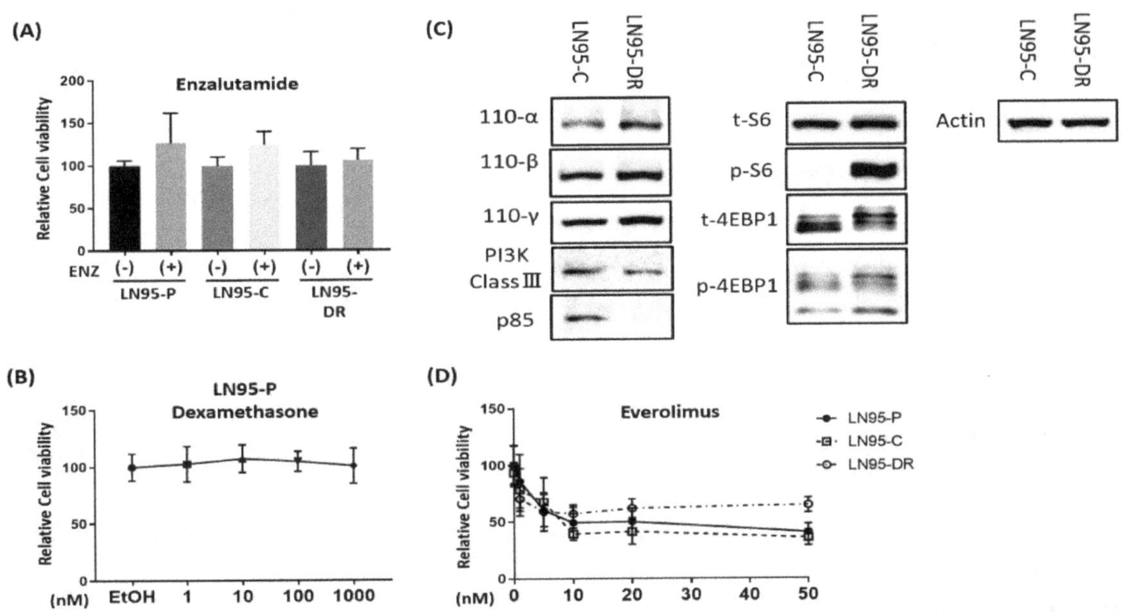

Figure A1. Effect of enzalutamide, dexamethasone, or everolimus on taxane-resistant cells. (**A**) Resistance of LN95-P, LN95-C, and LN95-DR cell viability to 10 μM enzalutamide for 72 h; (**B**) Dose response of dexamethasone on the growth and viability of LN95-P cells; (**C**) Levels of proteins related to the PI3K/Akt/mTOR axis in whole cell lysates; (**D**) Dose response of everolimus on the viability of LN95-P, LN95-C, and LN95-DR cells.

References

1. Rini, B.I.; Small, E.J. Hormone-refractory Prostate Cancer. *Curr. Treat. Options Oncol.* **2002**, *3*, 437–446. [CrossRef] [PubMed]
2. Knudsen, K.E.; Scher, H.I. Starving the addiction: New opportunities for durable suppression of AR signaling in prostate cancer. *Clin. Cancer Res.* **2009**, *15*, 4792–4798. [CrossRef] [PubMed]

3. Visakorpi, T.; Hyytinen, E.; Koivisto, P.; Tanner, M.; Keinanen, R.; Palmberg, C.; Palotie, A.; Tammela, T.; Isola, J.; Kallioniemi, O.P. In vivo amplification of the androgen receptor gene and progression of human prostate cancer. *Nat. Genet.* **1995**, *9*, 401–406. [CrossRef] [PubMed]
4. Koivisto, P.; Visakorpi, T.; Kallioniemi, O.P. Androgen receptor gene amplification: A novel molecular mechanism for endocrine therapy resistance in human prostate cancer. *Scand. J. Clin. Lab. Investig. Suppl.* **1996**, *226*, 57–63. [CrossRef]
5. Culig, Z.; Hobisch, A.; Cronauer, M.V.; Cato, A.C.; Hittmair, A.; Radmayr, C.; Eberle, J.; Bartsch, G.; Klocker, H. Mutant androgen receptor detected in an advanced-stage prostatic carcinoma is activated by adrenal androgens and progesterone. *Mol. Endocrinol.* **1993**, *7*, 1541–1550. [CrossRef] [PubMed]
6. Cai, C.; Chen, S.; Ng, P.; Bubley, G.J.; Nelson, P.S.; Mostaghel, E.A.; Marck, B.; Matsumoto, A.M.; Simon, N.I.; Wang, H.; et al. Intratumoral de novo steroid synthesis activates androgen receptor in castration-resistant prostate cancer and is upregulated by treatment with CYP17A1 inhibitors. *Cancer Res.* **2011**, *71*, 6503–6513. [CrossRef] [PubMed]
7. Ueda, T.; Mawji, N.R.; Bruchovsky, N.; Sadar, M.D. Ligand-independent activation of the androgen receptor by interleukin-6 and the role of steroid receptor coactivator-1 in prostate cancer cells. *J. Biol. Chem.* **2002**, *277*, 38087–38094. [CrossRef] [PubMed]
8. Xu, J.; Wu, R.C.; O'Malley, B.W. Normal and cancer-related functions of the p160 steroid receptor co-activator (SRC) family. *Nat. Rev. Cancer* **2009**, *9*, 615–630. [CrossRef] [PubMed]
9. Gioeli, D.; Paschal, B.M. Post-translational modification of the androgen receptor. *Mol. Cell. Endocrinol.* **2012**, *352*, 70–78. [CrossRef] [PubMed]
10. Karantanos, T.; Corn, P.G.; Thompson, T.C. Prostate cancer progression after androgen deprivation therapy: Mechanisms of castrate resistance and novel therapeutic approaches. *Oncogene* **2013**, *32*, 5501–5511. [CrossRef] [PubMed]
11. Li, Y.; Chan, S.C.; Brand, L.J.; Hwang, T.H.; Silverstein, K.A.; Dehm, S.M. Androgen receptor splice variants mediate enzalutamide resistance in castration-resistant prostate cancer cell lines. *Cancer Res.* **2013**, *73*, 483–489. [CrossRef] [PubMed]
12. Antonarakis, E.S.; Lu, C.; Wang, H.; Luber, B.; Nakazawa, M.; Roeser, J.C.; Chen, Y.; Mohammad, T.A.; Chen, Y.; Fedor, H.L.; et al. AR-V7 and resistance to enzalutamide and abiraterone in prostate cancer. *N. Engl. J. Med.* **2014**, *371*, 1028–1038. [CrossRef] [PubMed]
13. Myung, J.K.; Banuelos, C.A.; Fernandez, J.G.; Mawji, N.R.; Wang, J.; Tien, A.H.; Yang, Y.C.; Tavakoli, I.; Haile, S.; Watt, K.; et al. An androgen receptor N-terminal domain antagonist for treating prostate cancer. *J. Clin. Investig.* **2013**, *123*, 2948–2960. [CrossRef] [PubMed]
14. Martin, S.K.; Banuelos, C.A.; Sadar, M.D.; Kyprianou, N. N-terminal targeting of androgen receptor variant enhances response of castration resistant prostate cancer to taxane chemotherapy. *Mol. Oncol.* **2014**, *9*, 628–639. [CrossRef] [PubMed]
15. Kato, M.; Banuelos, C.A.; Imamura, Y.; Leung, J.K.; Caley, D.P.; Wang, J.; Mawji, N.R.; Sadar, M.D. Cotargeting Androgen Receptor Splice Variants and mTOR Signaling Pathway for the Treatment of Castration-Resistant Prostate Cancer. *Clin. Cancer Res.* **2016**, *22*, 2744–2754. [CrossRef] [PubMed]
16. David-Beabes, G.L.; Overman, M.J.; Petrofski, J.A.; Campbell, P.A.; de Marzo, A.M.; Nelson, W.G. Doxorubicin-resistant variants of human prostate cancer cell lines DU 145, PC-3, PPC-1, and TSU-PR1: Characterization of biochemical determinants of antineoplastic drug sensitivity. *Int. J. Oncol.* **2000**, *17*, 1077–1086. [CrossRef] [PubMed]
17. Takeda, M.; Mizokami, A.; Mamiya, K.; Li, Y.Q.; Zhang, J.; Keller, E.T.; Namiki, M. The establishment of two paclitaxel-resistant prostate cancer cell lines and the mechanisms of paclitaxel resistance with two cell lines. *Prostate* **2007**, *67*, 955–967. [CrossRef] [PubMed]
18. Mahon, K.L.; Henshall, S.M.; Sutherland, R.L.; Horvath, L.G. Pathways of chemotherapy resistance in castration-resistant prostate cancer. *Endocr.-Relat. Cancer* **2011**, *18*, R103–R123. [CrossRef] [PubMed]
19. Di Lorenzo, G.; Buonerba, C.; Autorino, R.; De Placido, S.; Sternberg, C.N. Castration-resistant prostate cancer: Current and emerging treatment strategies. *Drugs* **2010**, *70*, 983–1000. [CrossRef] [PubMed]
20. de Bono, J.S.; Oudard, S.; Ozguroglu, M.; Hansen, S.; Machiels, J.P.; Kocak, I.; Gravis, G.; Bodrogi, I.; Mackenzie, M.J.; Shen, L.; et al. Prednisone plus cabazitaxel or mitoxantrone for metastatic castration-resistant prostate cancer progressing after docetaxel treatment: A randomised open-label trial. *Lancet* **2010**, *376*, 1147–1154. [CrossRef]

21. Lombard, A.P.; Liu, C.; Armstrong, C.M.; Cucchiara, V.; Gu, X.; Lou, W.; Evans, C.P.; Gao, A.C. ABCB1 Mediates Cabazitaxel-Docetaxel Cross-Resistance in Advanced Prostate Cancer. *Mol. Cancer Ther.* **2017**, *16*, 2257–2266. [CrossRef] [PubMed]
22. Ploussard, G.; Terry, S.; Maille, P.; Allory, Y.; Sirab, N.; Kheuang, L.; Soyeux, P.; Nicolaiew, N.; Coppolani, E.; Paule, B.; et al. Class III beta-tubulin expression predicts prostate tumor aggressiveness and patient response to docetaxel-based chemotherapy. *Cancer Res.* **2010**, *70*, 9253–9264. [CrossRef] [PubMed]
23. Antonarakis, E.S.; Lu, C.; Luber, B.; Wang, H.; Chen, Y.; Nakazawa, M.; Nadal, R.; Paller, C.J.; Denmeade, S.R.; Carducci, M.A.; et al. Androgen Receptor Splice Variant 7 and Efficacy of Taxane Chemotherapy in Patients With Metastatic Castration-Resistant Prostate Cancer. *JAMA Oncol.* **2015**, *1*, 582–591. [CrossRef] [PubMed]
24. Scher, H.I.; Graf, R.P.; Schreiber, N.A.; Jayaram, A.; Winquist, E.; McLaughlin, B.; Lu, D.; Fleisher, M.; Orr, S.; Lowes, L.; et al. Assessment of the Validity of Nuclear-Localized Androgen Receptor Splice Variant 7 in Circulating Tumor Cells as a Predictive Biomarker for Castration-Resistant Prostate Cancer. *JAMA Oncol.* **2018**, *4*, 1179–1186. [CrossRef] [PubMed]
25. Thadani-Mulero, M.; Portella, L.; Sun, S.; Sung, M.; Matov, A.; Vessella, R.L.; Corey, E.; Nanus, D.M.; Plymate, S.R.; Giannakakou, P. Androgen receptor splice variants determine taxane sensitivity in prostate cancer. *Cancer Res.* **2014**, *74*, 2270–2282. [CrossRef] [PubMed]
26. Kato, M.; Wei, M.; Yamano, S.; Kakehashi, A.; Tamada, S.; Nakatani, T.; Wanibuchi, H. DDX39 acts as a suppressor of invasion for bladder cancer. *Cancer Sci.* **2012**, *103*, 1363–1369. [CrossRef] [PubMed]
27. Sadar, M.D. Androgen-independent induction of prostate-specific antigen gene expression via cross-talk between the androgen receptor and protein kinase A signal transduction pathways. *J. Biol. Chem.* **1999**, *274*, 7777–7783. [CrossRef] [PubMed]
28. Cleutjens, K.B.; van der Korput, H.A.; van Eekelen, C.C.; van Rooij, H.C.; Faber, P.W.; Trapman, J. An androgen response element in a far upstream enhancer region is essential for high, androgen-regulated activity of the prostate-specific antigen promoter. *Mol. Endocrinol.* **1997**, *11*, 148–161. [CrossRef] [PubMed]
29. Bradford, M.M. A rapid and sensitive method for the quantitation of microgram quantities of protein utilizing the principle of protein-dye binding. *Anal. Biochem.* **1976**, *72*, 248–254. [CrossRef]
30. Martin, C.; Berridge, G.; Mistry, P.; Higgins, C.; Charlton, P.; Callaghan, R. The molecular interaction of the high affinity reversal agent XR9576 with P-glycoprotein. *Br. J. Pharmacol.* **1999**, *128*, 403–411. [CrossRef] [PubMed]
31. Stewart, A.; Steiner, J.; Mellows, G.; Laguda, B.; Norris, D.; Bevan, P. Phase I trial of XR9576 in healthy volunteers demonstrates modulation of P-glycoprotein in CD56+ lymphocytes after oral and intravenous administration. *Clin. Cancer Res.* **2000**, *6*, 4186–4191. [PubMed]
32. Agrawal, M.; Abraham, J.; Balis, F.M.; Edgerly, M.; Stein, W.D.; Bates, S.; Fojo, T.; Chen, C.C. Increased 99mTc-sestamibi accumulation in normal liver and drug-resistant tumors after the administration of the glycoprotein inhibitor, XR9576. *Clin. Cancer Res.* **2003**, *9*, 650–656. [PubMed]
33. Abraham, J.; Edgerly, M.; Wilson, R.; Chen, C.; Rutt, A.; Bakke, S.; Robey, R.; Dwyer, A.; Goldspiel, B.; Balis, F.; et al. A phase I study of the P-glycoprotein antagonist tariquidar in combination with vinorelbine. *Clin. Cancer Res.* **2009**, *15*, 3574–3582. [CrossRef] [PubMed]
34. Luo, J.; Attard, G.; Balk, S.P.; Bevan, C.; Burnstein, K.; Cato, L.; Cherkasov, A.; De Bono, J.S.; Dong, Y.; Gao, A.C.; et al. Role of Androgen Receptor Variants in Prostate Cancer: Report from the 2017 Mission Androgen Receptor Variants Meeting. *Eur. Urol.* **2018**, *73*, 715–723. [CrossRef] [PubMed]
35. Arora, V.K.; Schenkein, E.; Murali, R.; Subudhi, S.K.; Wongvipat, J.; Balbas, M.D.; Shah, N.; Cai, L.; Efstathiou, E.; Logothetis, C.; et al. Glucocorticoid receptor confers resistance to antiandrogens by bypassing androgen receptor blockade. *Cell* **2013**, *155*, 1309–1322. [CrossRef] [PubMed]
36. Taylor, B.S.; Schultz, N.; Hieronymus, H.; Gopalan, A.; Xiao, Y.; Carver, B.S.; Arora, V.K.; Kaushik, P.; Cerami, E.; Reva, B.; et al. Integrative genomic profiling of human prostate cancer. *Cancer Cell* **2010**, *18*, 11–22. [CrossRef] [PubMed]
37. Bitting, R.L.; Armstrong, A.J. Targeting the PI3K/Akt/mTOR pathway in castration-resistant prostate cancer. *Endocr.-Relat. Cancer* **2013**, *20*, R83–99. [CrossRef] [PubMed]
38. Sarker, D.; Reid, A.H.; Yap, T.A.; de Bono, J.S. Targeting the PI3K/AKT pathway for the treatment of prostate cancer. *Clin. Cancer Res.* **2009**, *15*, 4799–4805. [CrossRef] [PubMed]

39. Hu, R.; Lu, C.; Mostaghel, E.A.; Yegnasubramanian, S.; Gurel, M.; Tannahill, C.; Edwards, J.; Isaacs, W.B.; Nelson, P.S.; Bluemn, E.; et al. Distinct transcriptional programs mediated by the ligand-dependent full-length androgen receptor and its splice variants in castration-resistant prostate cancer. *Cancer Res.* **2012**, *72*, 3457–3462. [CrossRef] [PubMed]
40. Hornberg, E.; Ylitalo, E.B.; Crnalic, S.; Antti, H.; Stattin, P.; Widmark, A.; Bergh, A.; Wikstrom, P. Expression of androgen receptor splice variants in prostate cancer bone metastases is associated with castration-resistance and short survival. *PLoS ONE* **2011**, *6*, e19059. [CrossRef] [PubMed]
41. Yu, Z.; Chen, S.; Sowalsky, A.G.; Voznesensky, O.S.; Mostaghel, E.A.; Nelson, P.S.; Cai, C.; Balk, S.P. Rapid induction of androgen receptor splice variants by androgen deprivation in prostate cancer. *Clin. Cancer Res.* **2014**, *20*, 1590–1600. [CrossRef] [PubMed]
42. Liu, L.L.; Xie, N.; Sun, S.; Plymate, S.; Mostaghel, E.; Dong, X. Mechanisms of the androgen receptor splicing in prostate cancer cells. *Oncogene* **2014**, *33*, 3140–3150. [CrossRef] [PubMed]
43. Yang, Y.C. Development of Novel Small Molecule Inhibitor of Androgen Receptor to Treat Castration-Resistant Prostate Cancer. Ph.D Thesis, The University of British Columbia, Vancouver, BC, Canada, August 2015.
44. Yang, Y.C.; Meimetis, L.G.; Tien, A.H.; Mawji, N.R.; Carr, G.; Wang, J.; Andersen, R.J.; Sadar, M.D. Spongian diterpenoids inhibit androgen receptor activity. *Mol. Cancer Ther.* **2013**, *12*, 621–631. [CrossRef] [PubMed]
45. Yang, Y.C.; Banuelos, C.A.; Mawji, N.R.; Wang, J.; Kato, M.; Haile, S.; McEwan, I.J.; Plymate, S.; Sadar, M.D. Targeting Androgen Receptor Activation Function-1 with EPI to Overcome Resistance Mechanisms in Castration-Resistant Prostate Cancer. *Clin. Cancer Res.* **2016**, *22*, 4466–4477. [CrossRef] [PubMed]
46. Zhu, Y.; Liu, C.; Nadiminty, N.; Lou, W.; Tummala, R.; Evans, C.P.; Gao, A.C. Inhibition of ABCB1 expression overcomes acquired docetaxel resistance in prostate cancer. *Mol. Cancer Ther.* **2013**, *12*, 1829–1836. [CrossRef] [PubMed]
47. Zhu, Y.; Liu, C.; Armstrong, C.; Lou, W.; Sandher, A.; Gao, A.C. Antiandrogens Inhibit ABCB1 Efflux and ATPase Activity and Reverse Docetaxel Resistance in Advanced Prostate Cancer. *Clin. Cancer Res.* **2015**, *21*, 4133–4142. [CrossRef] [PubMed]
48. Domingo-Domenech, J.; Vidal, S.J.; Rodriguez-Bravo, V.; Castillo-Martin, M.; Quinn, S.A.; Rodriguez-Barrueco, R.; Bonal, D.M.; Charytonowicz, E.; Gladoun, N.; de la Iglesia-Vicente, J.; et al. Suppression of acquired docetaxel resistance in prostate cancer through depletion of notch- and hedgehog-dependent tumor-initiating cells. *Cancer Cell* **2012**, *22*, 373–388. [CrossRef] [PubMed]
49. Galsky, M.D.; Dritselis, A.; Kirkpatrick, P.; Oh, W.K. Cabazitaxel. *Nat. Rev. Drug Discov.* **2010**, *9*, 677–678. [CrossRef] [PubMed]
50. Onstenk, W.; Sieuwerts, A.M.; Kraan, J.; Van, M.; Nieuweboer, A.J.; Mathijssen, R.H.; Hamberg, P.; Meulenbeld, H.J.; De Laere, B.; Dirix, L.Y.; et al. Efficacy of Cabazitaxel in Castration-resistant Prostate Cancer Is Independent of the Presence of AR-V7 in Circulating Tumor Cells. *Eur. Urol.* **2015**, *68*, 939–945. [CrossRef] [PubMed]
51. Zhu, M.L.; Horbinski, C.M.; Garzotto, M.; Qian, D.Z.; Beer, T.M.; Kyprianou, N. Tubulin-targeting chemotherapy impairs androgen receptor activity in prostate cancer. *Cancer Res.* **2010**, *70*, 7992–8002. [CrossRef] [PubMed]
52. Kuroda, K.; Liu, H.; Kim, S.; Guo, M.; Navarro, V.; Bander, N.H. Docetaxel down-regulates the expression of androgen receptor and prostate-specific antigen but not prostate-specific membrane antigen in prostate cancer cell lines: Implications for PSA surrogacy. *Prostate* **2009**, *69*, 1579–1585. [CrossRef] [PubMed]
53. Gan, L.; Chen, S.; Wang, Y.; Watahiki, A.; Bohrer, L.; Sun, Z.; Wang, Y.; Huang, H. Inhibition of the androgen receptor as a novel mechanism of taxol chemotherapy in prostate cancer. *Cancer Res.* **2009**, *69*, 8386–8394. [CrossRef] [PubMed]
54. Kobayashi, M.; Sakamoto, J.; Namikawa, T.; Okamoto, K.; Okabayashi, T.; Ichikawa, K.; Araki, K. Pharmacokinetic study of paclitaxel in malignant ascites from advanced gastric cancer patients. *World J. Gastroenterol.* **2006**, *12*, 1412–1415. [CrossRef] [PubMed]
55. Gianni, L.; Kearns, C.M.; Giani, A.; Capri, G.; Vigano, L.; Lacatelli, A.; Bonadonna, G.; Egorin, M.J. Nonlinear pharmacokinetics and metabolism of paclitaxel and its pharmacokinetic/pharmacodynamic relationships in humans. *J. Clin. Oncol.* **1995**, *13*, 180–190. [CrossRef] [PubMed]

56. Brunsvig, P.F.; Andersen, A.; Aamdal, S.; Kristensen, V.; Olsen, H. Pharmacokinetic analysis of two different docetaxel dose levels in patients with non-small cell lung cancer treated with docetaxel as monotherapy or with concurrent radiotherapy. *BMC Cancer* **2007**, *7*, 197. [CrossRef] [PubMed]
57. Ferron, G.M.; Dai, Y.; Semiond, D. Population pharmacokinetics of cabazitaxel in patients with advanced solid tumors. *Cancer Chemother. Pharmacol.* **2013**, *71*, 681–692. [CrossRef] [PubMed]
58. Silberstein, J.L.; Taylor, M.N.; Antonarakis, E.S. Novel Insights into Molecular Indicators of Response and Resistance to Modern Androgen-Axis Therapies in Prostate Cancer. *Curr. Urol. Rep.* **2016**, *17*, 29. [CrossRef] [PubMed]

11

NCL1, a Highly Selective Lysine-Specific Demethylase 1 Inhibitor, Suppresses Castration-Resistant Prostate Cancer Growth via Regulation of Apoptosis and Autophagy

Toshiki Etani [1], Taku Naiki [1,2,*], Aya Naiki-Ito [2], Takayoshi Suzuki [3,4], Keitaro Iida [1], Satoshi Nozaki [1], Hiroyuki Kato [2], Yuko Nagayasu [2], Shugo Suzuki [2], Noriyasu Kawai [1], Takahiro Yasui [1] and Satoru Takahashi [2]

1 Department of Nephro-Urology, Nagoya City University, Graduate School of Medical Sciences, Nagoya 467-8601, Japan; uroetani@med.nagoya-cu.ac.jp (T.E.); ikeitarou1009@gmail.com (K.I.); snozaki@med.nagoya-cu.ac.jp (S.N.); n-kawai@med.nagoya-cu.ac.jp (N.K.); yasui@med.nagoya-cu.ac.jp (T.Y.)
2 Department of Experimental Pathology and Tumor Biology, Nagoya City University, Graduate School of Medical Sciences, Nagoya 467-8601, Japan; ayaito@med.nagoya-cu.ac.jp (A.N.-I.); h.kato@med.nagoya-cu.ac.jp (H.K.); naga-p@dk.pdx.ne.jp (Y.N.); shugo@med.nagoya-cu.ac.jp (S.S.); sattak@med.nagoya-cu.ac.jp (S.T.)
3 Department of Chemistry, Kyoto Prefectural University of Medicine, Graduate School of Medical Science, Kyoto 606-0823, Japan; suzukit@koto.kpu-m.ac.jp
4 CREST, Japan Science and Technology Agency (JST), Kawaguchi 332-0015, Japan
* Correspondence: naiki@med.nagoya-cu.ac.jp

Abstract: Recent studies have shown that epigenetic alterations lead to oncogenic activation, thus indicating that these are therapeutic targets. Herein, we analyzed the efficacy and therapeutic potential of our developed histone lysine demethylase 1 (LSD1) inhibitor, NCL1, in castration-resistant prostate cancer (CRPC). The CRPC cell lines 22Rv1, PC3, and PCai1CS were treated with NCL1, and LSD1 expression and cell viability were assessed. The epigenetic effects and mechanisms of NCL1 were also evaluated. CRPC cells showed strong LSD1 expression, and cell viability was decreased by NCL1 in a dose-dependent manner. Chromatin immunoprecipitation analysis indicated that NCL1 induced histone H3 lysine 9 dimethylation accumulation at promoters of P21. As shown by Western blot and flow cytometry analyses, NCL1 also dose-dependently induced caspase-dependent apoptosis. The stimulation of autophagy was observed in NCL1-treated 22Rv1 cells by transmission electron microscopy and LysoTracker analysis. Furthermore, WST-8 assay revealed that the anti-tumor effect of NCL1 was reinforced when autophagy was inhibited by chloroquine in 22Rv1 cells. Combination index analysis revealed that a concurrent use of these drugs had a synergistic effect. In ex vivo analysis, castrated nude mice were injected subcutaneously with PCai1 cells and intraperitoneally with NCL1. Tumor volume was found to be reduced with no adverse effects in NCL1-treated mice compared with controls. Finally, immunohistochemical analysis using consecutive human specimens in pre- and post-androgen deprivation therapy demonstrated that LSD1 expression levels in CRPC, including neuroendocrine differentiation cases, were very high, and identical to levels observed in previously examined prostate biopsy specimens. NCL1 effectively suppressed prostate cancer growth in vitro and ex vivo without adverse events via the regulation of apoptosis and autophagy, suggesting that NCL1 is a potential therapeutic agent for CRPC.

Keywords: LSD1; epigenetics; castration-resistant prostate cancer; autophagy

1. Introduction

Prostate cancer is one of the most frequently diagnosed cancers in Western countries. In advanced prostate cancer, androgen deprivation therapy (ADT) has remained a first-line therapy for decades [1]. After showing an initial response, most patients develop progressive disease, referred to as castration-resistant prostate cancer (CRPC). Intriguingly, CRPC is not androgen independent and several new drugs designed to further suppress the androgen receptor (AR) pathway have led to improved survival, including abiraterone acetate and enzalutamide [2–5]. The human AR gene encodes for a protein with an atomic mass of 110 kDa that consists of an N-terminal domain, a DNA-binding domain, and a ligand-binding domain. AR controls the growth of the prostate gland, and much evidence from preclinical and clinical studies has shown that multiple androgen/AR signaling pathways implicated throughout the various stages of prostate cancer [6]. In addition, recent reports have described the potential therapeutic implications of estrogen and related receptors in prostate cancer [7,8]. However, not all patients respond equally to these newer AR-targeting drugs. Approximately 20–40% of patients with CRPC have a poor clinical response to such agents, and nearly all patients who initially respond acquire secondary resistance. Prospective trials are ongoing to develop the best biomarker strategy for identifying patients resistant to these drugs.

Prostate cancer progresses in a multistep process in response to changes in genetic mechanisms. However, in addition to genetic mutations, epigenetic alterations have also been identified as activating oncogenes and causing a loss of function of tumor suppressor genes [9,10]. Methylation is a form of post-translational covalent modification of histones that epigenetically regulates specific gene expression patterns. Lysine-specific demethylase 1 (LSD1), a member of the flavin adenine dinucleotide dependent enzyme family, behaves like a histone demethylase. LSD1 acts by removing one or two methyl (but not three) groups from lysine residues 4 or 9 in histone H3 (H3K4 and H3K9, respectively) [11].

Growing evidence indicates that LSD1 is critical for human tumorigenesis, and its expression is increased in several malignancies, including prostate, breast, lung, ovarian, and colon cancers [12–16]. Therefore, the inhibition of LSD1 holds promise as a novel anticancer strategy. We have previously developed a novel and selective LSD1 inhibitor called NCL1 (*N*-[(1S)-3-[3-(trans-2-aminocyclopropyl) phenoxy]-1-(benzylcarbamoyl)propyl] benzamide) after using a combination of in vitro screening and protein structure similarity clustering [17]. In addition, we have reported that NCL1 impairs LSD1 demethylase activity and inhibits cell proliferation in castration-naïve prostate cancer [16].

In this study, we examined the LSD1 status in CRPC cell lines and human specimens including aggressive neuroendocrine differentiated (NED) phenotypes. In addition, we tested the therapeutic significance of NCL1 in CRPC cells in vitro and in an ex vivo subcutaneous model. Furthermore, we investigated the pharmacological mechanism of NCL1 using chromatin immunoprecipitation (ChIP), flow cytometry, and Western blot analyses. To our knowledge, we are the first laboratory to describe the inhibition of LSD1-induced cell death in CRPC through the regulation of autophagy by NCL1.

2. Results

2.1. LSD1 Is Highly Expressed in CRPC

To determine the status of LSD1 in human prostate cancer, the expression of LSD1 and Nkx3.1, a sensitive specific marker of differentiated adenocarcinoma originating from the prostate [18], was examined in human prostate biopsy specimens by immunohistochemistry and the staining intensity scored. We found that LSD1 expression levels in CRPC were very high, as previously found in prostate biopsy specimens obtained from patients (Figure 1A–F,K). Interestingly, neuroendocrine-differentiated tumors after androgen deprivation therapy, which had no expression of Nkx3.1, had high levels of LSD1 expression (Figure 1G–J).

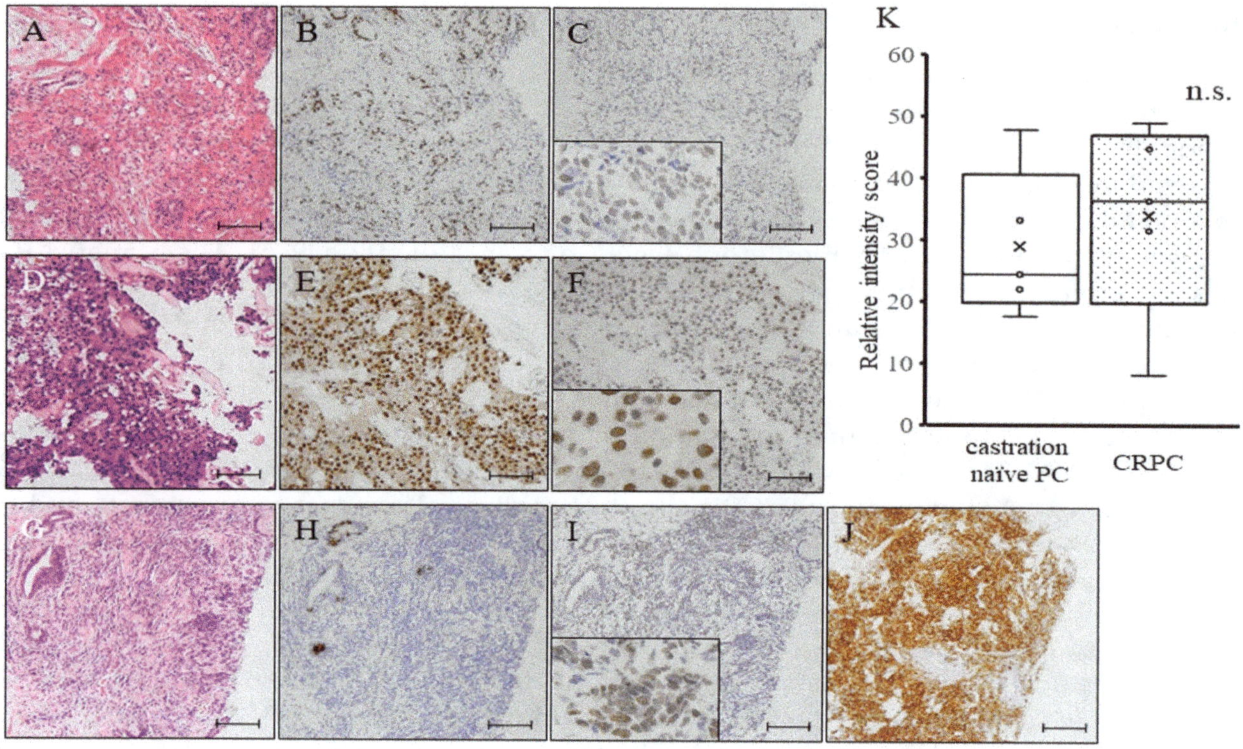

Figure 1. (**A–F**) Hematoxylin and eosin (HE) staining (**A**), and immunohistochemistry for Nkx3.1, a sensitive specific marker of differentiated adenocarcinoma originating from the prostate (**B**), and histone lysine demethylase 1 (LSD1) (**C**) of castration-naïve prostate cancer (castration-naïve PC) specimens obtained by prostate biopsy for an initial diagnosis in patients. HE staining (**D**), and immunohistochemistry for Nkx3.1 (**E**), and LSD1 (**F**) of castration-resistant prostate cancer (CRPC) specimens obtained by prostate biopsy after the acquisition of castration resistance and treatment with androgen deprivation therapy. Nuclei were counterstained with hematoxylin. Scale bar is 50 μm. (**G–J**) HE staining (**G**), and immunohistochemistry for Nkx3.1 (**H**), LSD1 (**I**), and synaptophysin (**J**) in prostate biopsy specimens obtained after the acquisition of castration resistance and neuroendocrine differentiation after treatment with androgen deprivation therapy. Scale bar is 50 μm. (**K**) A graphical comparison of intensity levels of LSD1 expression between castration-naïve PC and CRPC biopsy samples. n.s.: not significant.

2.2. LSD1 Expression in Prostate Cancer Cell Lines and Suppression of Prostate Cancer Cell Proliferation by NCL1

First, to determine whether LSD1 inhibition influences specific gene methylation status, 22Rv1 cultured prostate cancer cells treated with NCL1 were subjected to chromatin immunoprecipitation (ChIP) assay. As a result, and consistent with our previous report, NCL1 specifically impaired the demethylation of histone H3 lysine 4 (H3K4me2) at the containing promoter lesion of P21 genes (Figure 2A), reflecting the increased level of fold enrichment compared with the IgG control in ChIP assay, and increased levels of P21 protein expression in Western blot analysis, compared to the control. Next, we examined the status of LSD1 in CRPC cell lines, and found by Western blot analysis that LSD1 protein was highly expressed (Figure 2B). The proliferation of prostate cancer cells was significantly decreased by NCL1 treatment in a dose-dependent manner in the cancer cell lines tested, as determined by cell proliferation assay (Figure 2C). These findings suggest that NCL1 attenuated CRPC cell proliferation by demethylating H3K4me2 via LSD1 inhibition.

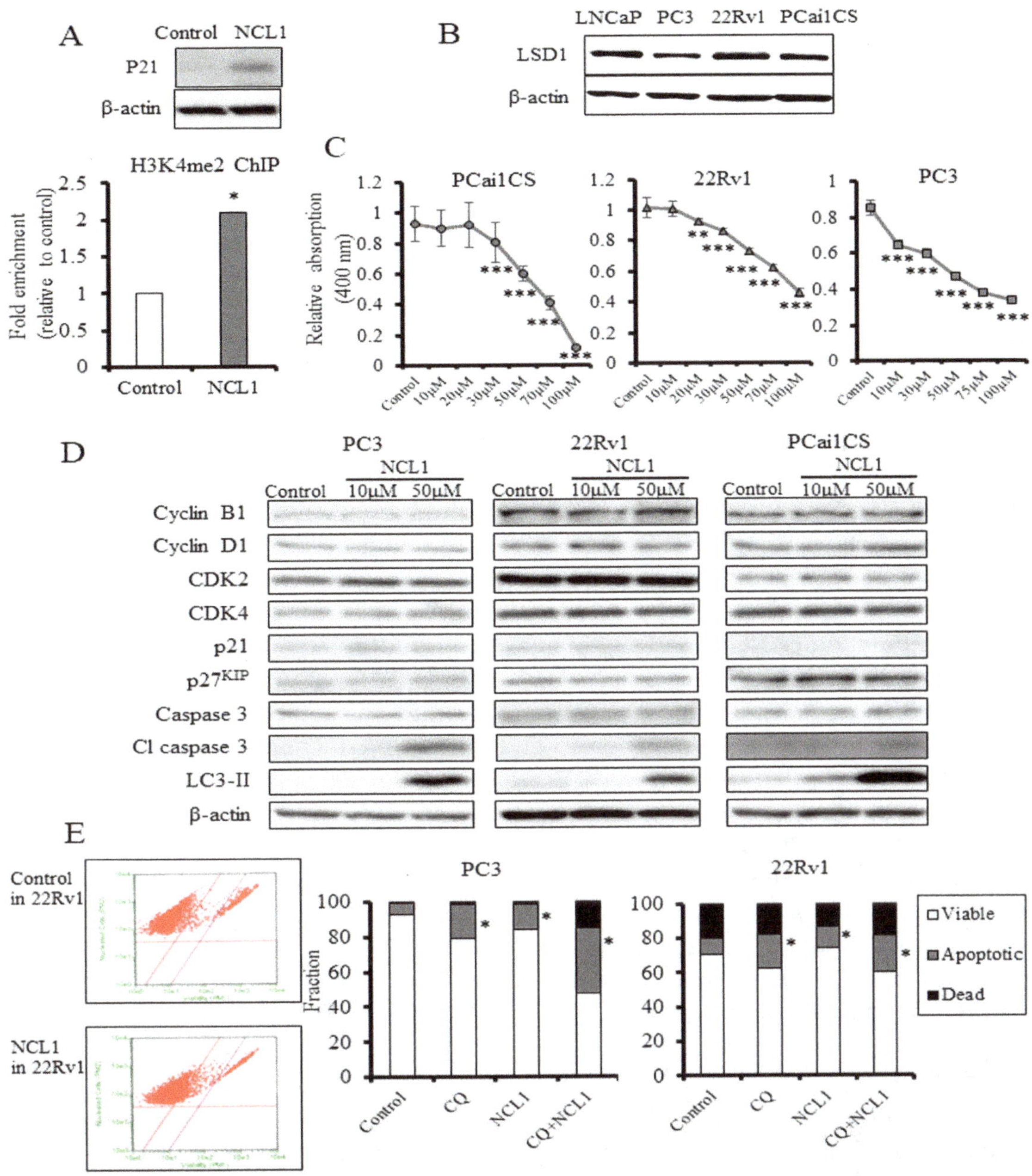

Figure 2. (**A**) Chromatin immunoprecipitation (ChIP) analysis in 22Rv1 cells using a histone H3 lysine 4 dimethylation (H3K4me2) antibody showed that NCL1 induced the attenuation of demethylation of H3K4me2 in the promoter regions of P21. Western blot analysis of P21 in 22Rv1 cells is shown. The protein expression of P21 was increased, reflecting the results of the ChIP analysis. β-actin was used as an internal loading control. (**B**) Western blot analysis of PCai1CS, 22Rv1, and PC3 cells for LSD1. All castration-resistant prostate cancer (CRPC) cell lines expressed LSD1. β-actin was used as an internal loading control. (**C**) PCai1CS and 22Rv1 cells were treated with vehicle (control) or NCL1, and subjected to WST-8 assay to measure cell proliferation. NCL1 treatment reduced the cell viability of the two CRPC cell lines in a dose-dependent manner. (**D**) Western blot analyses 48 h after NCL1 treatment of 22Rv1, PCai1CS, and PC3 cells. The cell cycle-related protein expression of cyclin B1, cyclin D1, CDK2, CDK4, and $p27^{KIP}$ were unchanged. Treatment with NCL1 resulted in a marked elevation in cleaved caspase 3 without any change in caspase 3. In addition, protein expression of microtubule-associated protein light chain 3 (LC3)-II was elevated in NCL1-treated CRPC cells. β-actin was used as an internal loading control. (**E**) Guava® apoptosis analysis of PC3 and 22Rv1 cells. NCL1, the autophagy inhibitor chloroquine (CQ), and a combination of these drugs induced apoptosis in CRPC cells. Mean ± standard deviation (SD); * $p < 0.05$, ** $p < 0.001$, *** $p < 0.0001$.

2.3. NCL1 Inhibits CRPC Cell Growth by Apoptotic Mechanisms

To determine how NCL1 induced growth inhibition, proteins involved in the cell cycle and apoptosis were examined in NCL1-treated CRPC cell lines. Reflecting the inhibition of LSD1, the expression of P21 was enhanced (Figure 2A). Cleaved caspase 3 was markedly elevated after treatment with NCL1 but caspase 3 expression remained unchanged. However, examination of cell cycle-related proteins showed that cyclin B1, cyclin D1, cyclin-dependent kinase (CDK)2, CDK4, and $p27^{KIP}$ expression were not changed by NCL1 treatment (Figure 2D). Therefore, analyses by Guava® ViaCount assay were performed. As a result, we found that NCL1 treatment of PC3 and 22Rv1 cells led to a significant induction of apoptosis (Figure 2E). These results suggested that selective attenuation of LSD1 using NCL1 inhibits cell proliferation by caspase-dependent apoptosis.

2.4. NCL1 Potentially Regulates Autophagy to Induce Cell Death in 22Rv1 Cells

The conversion of microtubule-associated protein light chain 3 (LC3)-I to LC3-II and the formation of LC3 puncta were used to determine whether NCL1 treatment induced autophagy in CRPC cells. We found that NCL1 induced an increase of LC3-II protein levels in 22Rv1, PCai1CS, and PC3 cells as determined by Western blotting (Figure 2D). Therefore, to confirm the contribution of NCL1 to autophagy, we then raised the pH of the lumen of lysosomes and/or autolysosomes to inhibit autophagic flux using chloroquine (CQ), thereby preventing autophagic degradation. Flow cytometry revealed that a combination of NCL1 and CQ increased apoptotic cell numbers (Figure 2E). In addition, it was revealed by LysoTracker analysis that NCL1 treatment led to a further accumulation of activated lysosomes (Figure 3B), and the addition of CQ caused an attenuation of the phenomenon (Figure 3D). By WST-8 cell counting assay, CQ alone was shown to have an effect on 22Rv1 cell viability, while CQ enhanced the inhibition of cell growth by NCL1 (Figure 3E). Furthermore, combination index analysis revealed that the force of combination was shown to be synergistic (Figure 3F). NCL1 treatment for 3 h led to the formation of autophagosomes, as shown in Figure 3G. The cytoplasm also showed an increase in structures (shown in the 72 h figure) from 24 h to 72 h; the results obtained with LysoTracker suggest that these structures were lysosomes. CQ treatment led to the inhibition of the degradation of structures incorporated into phagosomes. Using a combined treatment, these findings revealed colocalization (Figure 3G). These results suggest that NCL1 may induce CRPC cell death by regulating autophagy potential in addition to regulating an apoptotic anticancer pathway.

2.5. Ex Vivo Regulation of Tumorigenesis by NCL1

We examined the expression level of LSD1 after castration in a PCai1 subcutaneous tumor model. We found a high level of LSD1 expression that remained unchanged 1 week after castration (Figure 4D,E), and continued at this level for 8 weeks. We next assessed the role of NCL1 in tumor progression ex vivo. After PCai1 cells were injected subcutaneously into castrated nude mice, animals were subsequently treated with vehicle control or 1.0 mg/kg of NCL1. The NCL1-treated group showed a significant inhibition of tumor size compared to vehicle controls (Figure 4A). The size of other organs and body weights were not affected by NCL1 treatment, and differences in the relative weights of organs and blood parameters between the two groups were not found (Tables 1 and 2). Vacuolization was found to be increased in the groups treated with NCL1 compared with the vehicle control group (Figure 4F,G). Mechanisms involved in the inhibition of tumor growth by NCL1 in an animal model were examined using terminal deoxy nucleotidyl transferase-mediated dUTP nick end labeling (TUNEL) assays. Increased numbers of TUNEL-positive cells, and therefore apoptosis, were noted after treatment with NCL1 compared with vehicle (Figure 4B,H,I). In addition, we undertook immunohistochemical staining of CD31 to examine tumor vascularity. NCL1-treated tumors were found to have significantly decreased numbers of CD-31 positive blood vessels (Figure 4C,J,K). These results suggest that NCL1 regulates apoptosis to induce cell death and decrease tumor vascularity, both in vitro and ex vivo.

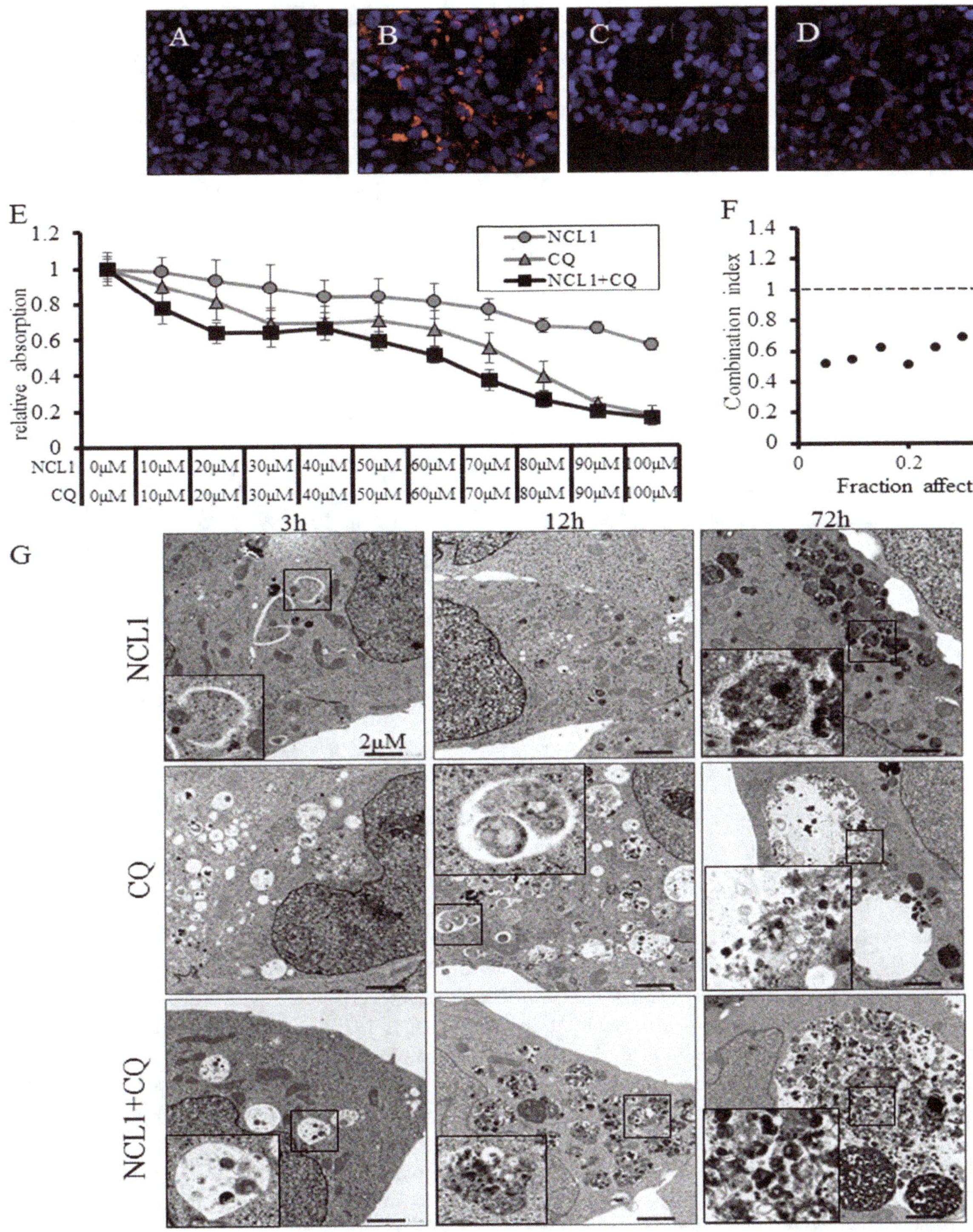

Figure 3. (**A–D**) Detection of the activation of lysosomes using LysoTracker analysis in 22Rv1 cells. Cells were treated with vehicle control (**A**), 50 μM NCL1 (**B**), 50 μM chloroquine (CQ) (**C**), or with 50 μM NCL1 and 50 μM CQ (**D**). Blue: nuclei, red: lysosomes. (**E**) 22Rv1 cells were treated with 50 μM NCL1 and/or 50 μM CQ. A WST-8 assay, in which the dye absorption rate positively correlated with cell viability, revealed that a combination of NCL1 and CQ decreased cell growth. (**F**) A combination index was calculated from the results of the WST-8 assay in Figure 3E. The combination of NCL1 and CQ showed a synergistic effect. (**G**) Cells were treated with 50 μM NCL1 for 3 h, 12 h, and 72 h. Three hours after NCL1 treatment, the formation of autophagosomes was noted by transmission electron microscopy (TEM). The cytoplasm also showed increased numbers of structures (visible in the 72 h figure) from 24 h to 72 h. Scale bar is 20 μm.

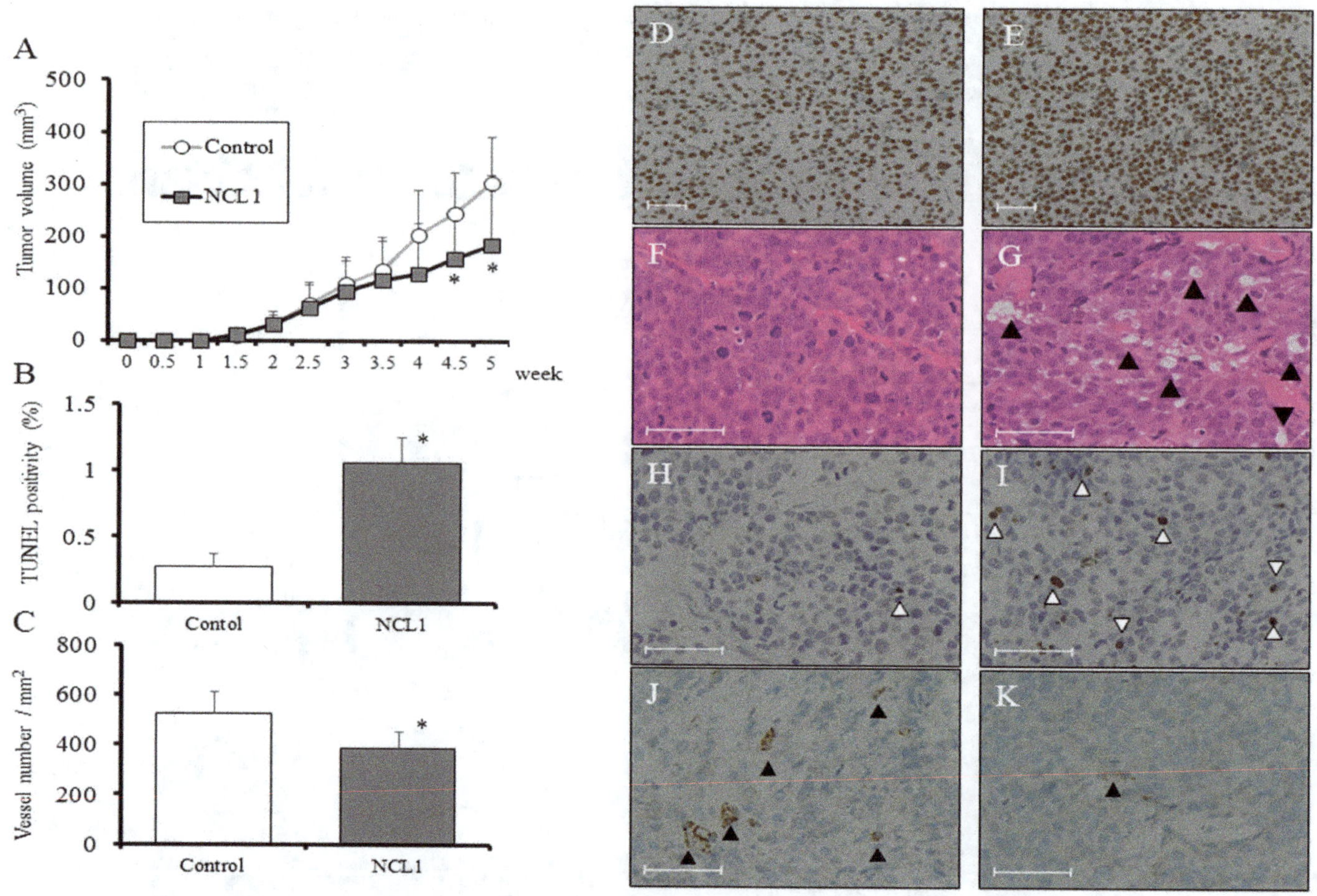

Figure 4. (**A**) Tumor growth was significantly inhibited in mice treated with 1.0 mg/kg NCL1 as compared to vehicle controls. (**B**) A terminal deoxy nucleotidyl transferase-mediated dUTP nick end labeling (TUNEL) assay was performed in NCL1-treated and control mice, and quantified as the mean TUNEL labeling percentage based on at least five randomly selected high-power microscope fields per individual. (**C**) Immunohistochemistry for CD31. Positivity was quantified as the mean number of vessels/mm^2 based on at least five randomly selected high-power microscopic fields per individual. (**D,E**) Representative immunohistochemistry of LSD1 in a subcutaneous PCai1 tumor. Uncastrated tumor (**D**), and 1 week after castration (**E**). Scale bar is 50 μm. (**F,G**) Hematoxylin and eosin (HE) staining in subcutaneous tumors from vehicle control (**F**) and 1.0 mg/kg NCL1-treated (**G**) mice. Vacuolation (black arrowheads) was increased in the NCL1-treated group compared with controls. Scale bar is 50 μm. (**H,I**) TUNEL staining for apoptosis in subcutaneous tumors from vehicle control (**H**) and 1.0 mg/kg NCL1-treated (**I**) mice. White arrowheads indicate TUNEL-positive cells. Scale bar is 50 μm. (**J,K**) Representative immunohistochemical images of CD31 in subcutaneous tumors from control (**J**) and 1.0 mg/kg NCL1-treated (**K**) mice. Black arrowheads indicate CD31-positive cells. Scale bar is 50 μm. Mean ± standard deviation (SD); * $p < 0.05$.

Table 1. Relative organ weights at the experiment's termination in a PCai1 mouse tumor model. BW: body weight; R: right; L: left.

	No. of Mice	BW (g)	Liver (%)	R-Kidney (%)	L-Kidney (%)
Control	11	23.4 ± 1.3	5.14 ± 0.22	0.78 ± 0.02	0.80 ± 0.05
NCL1 1.0 mg/kg	10	23.5 ± 1.5	5.30 ± 0.26	0.81 ± 0.03	0.82 ± 0.04

Table 2. Blood results at the experiment's termination in a PCai1mouse tumor model. AST: aspartate aminotransferase; ALT: alanine aminotransferase; ALP: alkaline phosphatase; T-Bil: total bilirubin; T-Chol: total cholesterol; Crea: creatinine; BUN: blood urea nitrogen; Na: sodium; K: potassium; Cl: chloride; Ca: calcium. Mean ± standard deviation (SD).

	Control	**NCL1 1.0 mg/kg**
AST	56.2 ± 10.3	55.9 ± 10.6
ALT	28.2 ± 6.3	27.1 ± 3.6
ALP	240.7 ± 20.6	250.2 ± 30.9
T-Bil	0.04 ± 0.01	0.04 ± 0.01
T-Chol	83.5 ± 8.3	85.1 ± 8.3
Crea	0.09 ± 0.01	0.10 ± 0.01
BUN	23.4 ± 3.2	23.5 ± 2.7
Na	146.3 ± 3.8	146.8 ± 2.7
K	6.7 ± 1.0	5.9 ± 0.7
Cl	110.4 ± 6.5	109.7 ± 5.3
Ca	8.5 ± 0.2	8.4 ± 0.3

3. Discussion

In the present study, we examined whether CRPC proliferation is repressed by LSD1 and targeted this molecule using a specific inhibitor, NCL1. The inhibition of LSD1 by NCL1 led to an increase in H3K4me2 modifications in the promoter region (Figure 2A) and induced increases in P21 protein expression (Figure 2A). NCL1 significantly inhibited growth in vitro (Figure 2C) as well as tumor growth ex vivo at low concentration levels (Figure 4A). In addition, adverse events were not noted in the general condition of the mice and in their blood analyses (Table 1-2). These results demonstrated the safety and efficacy of NCL1 in CRPC, highlighting its potential as a new treatment for this disease. We are the first to demonstrate the therapeutic potential of NCL1 ex vivo using a CRPC animal model.

Currently, CRPC patients are treated by ADT, including AR- and non-AR-targeting drugs [19]. However, the selection of more appropriate treatment and the sequencing of these drugs is increasingly being investigated. Above all, NED phenotypes have been identified in about 50% of cases of CRPC, which express NED markers such as synaptophysin and chromogranin [20]. The presence of NED markers has been shown to indicate a poorer prognosis when treated with AR-targeting drugs, including enzalutamide [21,22]. In previous reports, including our previous article, the overexpression of LSD1 in prostate cancer was shown to be a predictive marker for aggressive tumor biology and tumor recurrence during therapy [16,23,24]. However, reports describing changes in LSD1 expression using consecutive pre- and post-ADT specimens do not exist. Our immunohistochemical analyses demonstrated that the overexpression of LSD1 in aggressive cancer was maintained in castration-resistant cancer cells (Figure 1A–F). In addition, although only in one case, the overexpression of LSD1 was maintained after changes to an NED tumor (Figure 1G–J). Furthermore, using a PCai1 ex vivo model, a high level of LSD1 expression was maintained from 1 week after castration (Figure 4D,E), and cell growth of castration-resistant PCai1 cells was effectively suppressed by NCL1 both in vitro and ex vivo. Therefore, NCL1 may have therapeutic potential for CRPC, including NED phenotypes, from an early phase after the acquisition of castration resistance. Further studies are needed to clearly test its in vivo potential in combination with ADT, as well as with AR-targeting drugs.

We observed high protein expression of LSD1 in CRPC cells (Figure 2B), and the inhibition of LSD1 activity using NCL1 to reduce cell proliferation in vitro (Figure 2C). In addition, the common mechanism of cell death induced both in vitro and ex vivo by NCL1 was revealed to be caspase-dependent apoptosis (Figure 2D,E and Figure 4A,B,H,I). Apoptosis is an active cell suicide process that maintains cellular homeostasis; however, cancer cells can override apoptotic cell death by upregulating anti-apoptotic machinery and/or downregulating pro-apoptotic programs [25,26]. It is generally accepted that autophagy can function as an adaptive response to maintain cell survival and growth [27,28]. A recent report reveals that the mammalian target of rapamycin (mTOR) inhibition protects cancer cells from apoptosis during nutrient limitation [29]. Several studies have established that autophagy is associated with drug resistance in prostate cancer cells to ADT and inhibitors of PI3K/Akt/mTOR signaling [30–32]. In a previous study, we reported that LSD1 inhibition stimulated autophagy in castration-naïve prostate cancer [16]. Therefore, to confirm this phenomenon in CRPC, we examined drug-induced autophagy in 22Rv1 cells by detecting LC3-II expression, as well as LysoTracker analysis, a non-specific autophagic marker, using TEM and Western blotting. We showed that NCL1 induced autophagy in 22Rv1, PC3, and PCai1CS cells in a concentration-dependent manner (Figure 2D). WST-8 assay revealed that the anti-tumor effect of NCL1 was reinforced when autophagy was inhibited by CQ in 22Rv1 cells. In addition, combination index analysis revealed that a combination of these drugs showed a synergistic effect (Figure 3F). These results suggest that the stimulation of autophagy in CRPC protects cells against anti-tumor agents through LSD1. Also, when treated in combination with drugs that regulate autophagy, NCL1 may be more effective in the suppression of CRPC growth.

LSD1 plays a key role in many physiological functions. Previous studies have described how LSD1 inhibition reduces cell growth by affecting the expression of several genes involved in proliferation and the cell cycle [33–37]. Previous reports described that estrogen-induced demethylation of H3K9me2 by LSD1 caused reactive oxygen species-induced DNA damage and subsequently caused apoptosis via the regulation of phosphorylation with DNA damage repair enzymes in hormone-responsive cells [38,39]. In addition, a recent report described how a combination of LSD1 knockdown and cisplatin effectively suppressed the proliferation of PC3 cells, and that vascular endothelial growth factor, one of the most important promoters of angiogenesis, was downregulated by LSD1 siRNA treatment [40]. Our results suggest a similar mechanism. The inhibitory mechanisms of cancer growth by NCL1 appeared to be related not only to direct effects on cell proliferation but also to effects on angiogenesis, as shown by a reduction in CD 31-positive vessels ex vivo (Figure 4C,J,K). Recently, an abnormal mRNA splice variant of the androgen receptor, called AR-V7 [41,42], was shown to convey resistance to ADT [43,44]. In addition, in patients with metastatic CRPC, the presence of detectable AR-V7 transcripts in circulating tumor cells has been associated with a high positive predictive value for a non-response to AR-targeting agents, including enzalutamide, in several studies [45,46]. Prospective trials are ongoing to develop the best biomarker strategy for identifying treatment-resistant patients. Interestingly, in 2018, Regufe da Mota et al. [47] reported that LSD1 inhibition caused attenuation of the expression of not only wild-type AR, but also AR-V7. Further prospective trials using biomarkers to help select patients are warranted to evaluate the benefits of a strategic sequence of several drugs, including NCL1, for patients with CRPC.

4. Conclusions

In summary, NCL1 suppressed CRPC growth in vitro and ex vivo, showing strong efficacy without adverse events by regulating autophagy and apoptosis. Further, the strong expression of LSD1 was noted in human CRPC cells including those with NED phenotypes. These findings highlight how NCL1 may be considered a novel potential therapeutic agent for CRPC.

5. Materials and Methods

5.1. Human Castration-Naïve and Castration-Resistant Prostate Cancer Specimens

We obtained five castration-naïve prostate cancer specimens from five patients by needle biopsy. In addition, consecutive castration-resistant prostate cancer specimens after treatment were obtained by surgery or biopsy from the same patients from which previous biopsy specimens had been collected at Nagoya City University and affiliated hospitals between 2010 and 2016. All specimens were obtained after patients had provided written informed consent for the use of their tissues, according to an

Institutional Review Board approved protocol with approval number 1168. All cases were evaluated by a panel of experienced pathologists.

5.2. Chemicals

NCL1 was synthesized as previously described [17].

5.3. Prostate Cancer Cell Lines

The human prostate cancer cell lines PC3 and 22Rv1 were obtained from the American Type Culture Collection (Rockville, MD, USA); these cells plus a castration-resistant rat prostate cancer cell line, PCai1, were cultured as previously described [46,47]. The cell lines used included PC3, which is human CRPC cell line without AR expression; 22Rv1, which is human CRPC cell line with AR expression; and PCai1CS, which is established from CRPC originating from a transgenic rat model as previously described [48,49]. They were cultured in media with 10% charcoal-stripped serum, and then treated with dimethyl sulfoxide (DMSO) as a vehicle that was equal in concentration to that used for 1–100 μM NCL1 for 72 h. Finally, its effect on cell proliferation determined. To assess the effects of autophagy on cell proliferation, PC3 and 22Rv1 cells were treated with 50 μM NCL1 and/or 50 μM chloroquine (CQ) for 24 h, an inhibitor of autophagy. All experiments were performed in triplicate.

5.4. Cell Proliferation Assay

A WST-8 Cell Counting Kit (Wako, Osaka, Japan) was used to assess the proliferation of cells grown in 96-well microplates. Prostate cancer cells were seeded in DMEM medium containing 10% fetal bovine serum (FBS) in 96-well plates (5×10^3 cells/well), and a WST-8 assay was performed as previously described [16].

5.5. Analysis of the Cytotoxic Effect of NCL1 in Combination with Chloroquine

The effect of a drug combination was calculated according to the median effect principle. First, we constructed dose-response curves for the cytotoxic effects of NCL1 and chloroquine, both alone and in combination, in 22Rv1 cells using a WST-8 assay. The data were used to determine the 'combination index' (CI), using the equation: CI = (D)1/(Dx)1 + (D)2/(Dx)2, where (D)1 and (D)2 are the combinations doses that kill x% of cells, and (Dx)1 and (Dx)2 are the doses of each drug alone that kill x% of cells. If CI < 1, then synergism is indicated as previously described [50,51].

5.6. ChIP Assay

22Rv1 cells were incubated in the presence or absence of 50 μM NCL1 as indicated. Formaldehyde (1%) was used to cross-link cells, and the chromatin was collected and subjected to immunoprecipitation using an H3K4-me2 antibody (Cell Signaling Technology, Beverly, MA, USA). As a negative control, isotype-specific IgG was used. Extracted DNA was dissolved in TE buffer, and real-time PCR using specific primers (F-GGGGCGGTTGTATATCAGG, R-GGCTCCACAAGGAACTGACT) was used to confirm the methylation status of the promoter regions of P21 (CDKN1A).

5.7. Western Blot Analysis

Cells were lysed in SDS buffer, and 10 μL of protein lysate sample was dissolved in 12% polyacrylamide gels and transferred onto Hybond ECL membranes (GE Healthcare, Piscataway, NJ, USA). Antibodies against $P21^{WAF1}$ (Cell Signaling), cyclin B1 (Santa Cruz Biotechnology, Santa Cruz, CA, USA), cyclin D1 (Santa Cruz), CDK2 (Santa Cruz), CDK4 (Santa Cruz), caspase 3 (Cell Signaling), cleaved caspase 3 (Cell Signaling), and LC3-II (Abcam, Cambridge, UK) were used to assess protein expression levels. A monoclonal anti-beta-actin antibody (Sigma, St. Louis, MO, USA) was used to evaluate beta actin expression as a protein loading control.

5.8. Flow Cytometry Analysis

PC3 and 22Rv1 cells (1×10^5 per line) were treated with 50 μM NCL1 for 48 h, with or without 50 μM CQ, and cell suspensions were then prepared and stained with Guava® ViaCount reagent and propidium iodide according to a Guava® Assay protocol (Guava Technologies, Hayward, CA, USA). CytoSoft Software was used to analyze apoptosis and cell cycle phase distributions on a Guava® PCA Instrument.

5.9. Lysosome Localization and Activity Using LysoTracker® and LysoSensor™ Dyes

22Rv1 cells (3×10^4) were seeded in 8-well chamber slides with RPMI medium and 5% FBS. Cells were incubated for 48 h and treated with 50 μM NCL1 or 50 μM CQ, or a combination of 50 μM NCL1 and 50 μM CQ. Control cells were treated with the same amount of solvent (DMSO and distilled water). We removed the medium from the dish after 48 h of treatment, and add the prewarmed (37 °C) probe-containing medium. The cells were then incubated for 60 minutes. Lastly, we replaced the loading solution with fresh medium and observed the cells using an IN Cell Analyzer 6000 (GE Healthcare, Chicago, IL, USA).

5.10. Transmission Electron Microscopy (TEM)

22Rv1 cells were seeded in 6-well plates (1×10^5 cells/well) in DMEM containing 10% FBS. After an overnight incubation, cells were treated with or without 50 μM NCL1, and with or without 50 μM CQ for 3 or 72 h. Glutaraldehyde (2.5%) was used to pre-fix cells in 0.1 M phosphate buffer (pH 7.4) at 4 °C. Specimens were then post-fixed in 1% osmium tetroxide in 0.1 M phosphate buffer (pH 7.4) for 45 min. A graded series of ethanol was used to dehydrate specimens, which were subsequently embedded in epoxy resin. An Ultracut-S ultramicrotome (LEICA, Wetzlar, Germany) and a diamond knife were used to cut ultra-thin sections, which were then stained with 2% uranyl acetate in distilled water for 15 min and a lead staining solution for 5 min. A JEM-1011J (JEOL, Tokyo, Japan) electron microscope at 80 KV was used to observe sections.

5.11. Ex Vivo Studies Using a Subcutaneous Castration-Resistant PCai1 Model

Six-week-old male KSN/nu-nu nude mice from Nippon SLC (Hamamatsu, Japan) were maintained as previously described [48,49]. PCai1 cells cultured in T-75 flasks were grown to confluence, trypsinized, and counted. PCai1 cells (1×10^6 in 100 μL serum-free DMEM) were subcutaneously injected into the dorsal side of each mouse under isoflurane anesthesia. After 1, 3, and 4 weeks, mice (n = 5) were castrated, while other mice (n = 5) were left uncastrated as negative controls. Five weeks after implantation, all mice were sacrificed, and the LSD1 expression of PCai1 tumors was analyzed. For the next experiment, all nude mice were castrated, and 1×10^6 PCai1 cells resuspended in 100 μL serum-free DMEM were subcutaneously implanted as described above. Ten days later, an intraperitoneal injection of DMSO as a vehicle that was equal in concentration to that used for 1.0 mg/kg NCL1 (n = 10), or 1.0 mg/kg (n = 10) NCL1 was performed twice per week. Tumor size (determined by caliper measurement) and body weight were measured twice per week. Mice were sacrificed 5 weeks after the implantation of cells.

All animal experiments were performed according to protocols approved by the Institutional Animal Care and Use Committee of Nagoya City University Graduate School of Medical Sciences; the approval number was H24M-58.

5.12. Immunohistochemical Analysis

Deparaffinized tissue arrays were incubated with anti-LSD1 (1:200; Cell Signaling), anti-Nkx3.1 (1:400; Cell Signaling), or anti-synaptophysin (1:100; Cell Signaling). Deparaffinized animal tissues were incubated with anti-CD31 (1:100; Santa Cruz). Antibody binding was visualized by a conventional immunostaining method, as described previously [48,49], using an autoimmunostaining apparatus (HX System, Ventana, Tucson, AZ, USA).

LSD1 expression was evaluated using intensity scores for normal prostate glands and carcinoma cores from patients. For LSD1 immunoreactivity in nuclei, raw nuclear intensity data for tumor cells in prostate cancer cores and luminal cells in normal prostate glands were measured using a BZ-9000 multifunctional microscope and analysis software (Keyence Japan, Osaka, Japan). For each patient, evaluations were repeated five times and an average intensity score was calculated for each core.

5.13. TUNEL Assay

A TUNEL assay using an in situ Apoptosis Detection Kit from Takara (Otsu, Japan) was used according to the manufacturer's protocol to determine apoptotic cells in deparaffinized tissues. The relative ratio of TUNEL-positive cells was determined using five random microscopic fields for each group.

5.14. Statistical Analysis

Student's *t*, ANOVA, or Kruskal–Wallis tests were used to assess the association between variables. A value of $p < 0.05$ was considered statistically significant.

Author Contributions: All the authors have read and approved the manuscript and agree with its submission to this journal. Details regarding authorship, conflicts of interest, and ethics approval are given in the accompanying Author Submission Requirement Form. The contribution of each author to the manuscript sufficient enough for each to take public responsibility for appropriate portions of the content. T.E. made critical revisions of the manuscript. T.N., A.N.-I., T.S., H.K., Y.N., S.S. carried out the acquisition of data, and coordinated and helped to draft the manuscript. K.I. conducted statistical analyses concerning this study. T.Y., and S.T. supervised this manuscript. All authors read and approved the final manuscript.

Acknowledgments: The authors would like to thank Hiroshi Takase for the assistance in the Core Laboratory of the Research Equipment Sharing Center at Nagoya City University.

References

1. Frydenberg, M.; Stricker, P.D.; Kaye, K.W. Prostate cancer diagnosis and management. *Lancet* **1997**, *349*, 1681–1687. [CrossRef]
2. de Bono, J.S.; Logothetis, C.J.; Molina, A.; Fizazi, K.; North, S.; Chu, L.; Chi, K.N.; Jones, R.J.; Goodman, O.B.; Saad, F., Jr.; et al. Abiraterone and increased survival in metastatic prostate cancer. *N. Engl. J. Med.* **2011**, *364*, 1995–2005. [CrossRef] [PubMed]
3. Ryan, C.J.; Smith, M.R.; de Bono, J.S.; Molina, A.; Logothetis, C.J.; de Souza, P.; Fizazi, K.; Mainwaring, P.; Piulats, J.M.; Ng, S.; et al. Abiraterone in metastatic prostate cancer without previous chemotherapy. *N. Engl. J. Med.* **2013**, *368*, 138–148. [CrossRef] [PubMed]
4. Beer, T.M.; Armstrong, A.J.; Rathkopf, D.E.; Loriot, Y.; Sternberg, C.N.; Higano, C.S.; Iversen, P.; Bhattacharya, S.; Carles, J.; Chowdhury, S.; et al. Enzalutamide in metastatic prostate cancer before chemotherapy. *N. Engl. J. Med.* **2014**, *371*, 424–433. [CrossRef] [PubMed]
5. Scher, H.I.; Fizazi, K.; Saad, F.; Taplin, M.E.; Sternberg, C.N.; Miller, K.; de Wit, R.; Mulders, P.; Chi, K.N.; Shore, N.D.; et al. Increased survival with enzalutamide in prostate cancer after chemotherapy. *N. Engl. J. Med.* **2012**, *367*, 1187–1197. [CrossRef] [PubMed]
6. Di Zazzo, E.; Galasso, G.; Giovannelli, P.; Di Donato, M.; Di Santi, A.; Cernera, G.; Rossi, V.; Abbondanza, C.; Moncharmont, B.; Sinisi, A.A.; et al. Prostate cancer stem cells: The role of androgen and estrogen receptors. *Oncotarget* **2016**, *7*, 193–208. [CrossRef] [PubMed]
7. Di Zazzo, E.; Galasso, G.; Giovannelli, P.; Di Donato, M.; Castoria, G. Estrogen and Their Receptors in Prostate Cancer: Therapeutic Implications. *Front. Oncol.* **2018**, *8*, 2. [CrossRef] [PubMed]

8. Rossi, V.; Di Zasso, E.; Galasso, G.; De Rosa, C.; Abbondanza, C.; Sinisi, A.A.; Altucci, L.; Migliaccio, A.; Castoria, G. Estrogens Modulate Somatostatin Receptors Expression and Synergize With the Somatostatin Analog Pasireotide in Prostate Cells. *Front. Pharmacol.* **2019**, *10*, 28. [CrossRef] [PubMed]
9. Shi, Y.; Lan, F.; Matson, C.; Mulligan, P.; Whetstine, J.R.; Cole, P.A.; Casero, R.A. Histone demethylation mediated by the nuclear amine oxidase homolog LSD1. *Cell* **2004**, *119*, 941–953. [CrossRef] [PubMed]
10. Kouzarides, T. Chromatin modifications and their function. *Cell* **2007**, *128*, 693–705. [CrossRef] [PubMed]
11. Metzger, E.; Wissmann, M.; Yin, N.; Muller, J.M.; Schneider, R.; Peters, A.H.; Gunther, T.; Buettner, R.; Schule, R. LSD1 demethylates repressive histone marks to promote androgen-receptor-dependent transcription. *Nature* **2005**, *437*, 436–439. [CrossRef] [PubMed]
12. Hayami, S.; Kelly, J.D.; Cho, H.S.; Yoshimatsu, M.; Unoki, M.; Tsunoda, T.; Field, H.I.; Neal, D.E.; Yamaue, H.; Ponder, B.A.; et al. Overexpression of LSD1 contributes to human carcinogenesis through chromatin regulation in various cancers. *Int. J. Cancer* **2011**, *128*, 574–586. [CrossRef] [PubMed]
13. Lim, S.; Janzer, A.; Becker, A.; Zimmer, A.; Schule, R.; Buettner, R.; Kirfel, J. Lysine-specific demethylase 1 (LSD1) is highly expressed in ER-negative breast cancers and a biomarker predicting aggressive biology. *Carcinogenesis* **2010**, *31*, 512–520. [CrossRef] [PubMed]
14. Lv, T.; Yuan, D.; Miao, X.; Lv, Y.; Zhan, P.; Shen, X.; Song, Y. Over-expression of LSD1 promotes proliferation, migration and invasion in non-small cell lung cancer. *PLoS ONE* **2012**, *7*, e35065. [CrossRef] [PubMed]
15. Nagasawa, S.; Sedukhina, A.S.; Nakagawa, Y.; Maeda, I.; Kubota, M.; Ohnuma, S.; Tsugawa, K.; Ohta, T.; Roche-Molina, M.; Bernal, J.A.; et al. LSD1 overexpression is associated with poor prognosis in basal-like breast cancer, and sensitivity to PARP inhibition. *PLoS ONE* **2015**, *10*, e0118002. [CrossRef] [PubMed]
16. Etani, T.; Suzuki, T.; Naiki, T.; Naiki-Ito, A.; Ando, R.; Iida, K.; Kawai, N.; Tozawa, K.; Miyata, N.; Kohri, K.; et al. NCL1, a highly selective lysine-specific demethylase 1 inhibitor, suppresses prostate cancer without adverse effect. *Oncotarget* **2015**, *6*, 2865–2878. [CrossRef] [PubMed]
17. Ueda, R.; Suzuki, T.; Mino, K.; Tsumoto, H.; Nakagawa, H.; Hasegawa, M.; Sasaki, R.; Mizukami, T.; Miyata, N. Identification of cell-active lysine specific demethylase 1-selective inhibitors. *J. Am. Chem. Soc.* **2009**, *131*, 17536–17537. [CrossRef]
18. Gurel, B.; Ali, T.Z.; Montgomery, E.A.; Begum, S.; Hicks, J.; Goggins, M.; Eberhart, C.G.; Clark, D.P.; Bieberich, C.J.; Epstein, J.I.; et al. NKX3.1 as a marker of prostatic origin in metastatic tumors. *Am. J. Surg. Pathol.* **2010**, *34*, 1097–1105. [CrossRef] [PubMed]
19. Parker, C.; Nilsson, S.; Heinrich, D.; Helle, S.I.; O'Sullivan, J.M.; Fossa, S.D.; Chodacki, A.; Wiechno, P.; Logue, J.; Seke, M.; et al. Alpha emitter radium-223 and survival in metastatic prostate cancer. *N. Engl. J. Med.* **2013**, *369*, 213–223. [CrossRef] [PubMed]
20. Culine, S.; El Demery, M.; Lamy, P.J.; Iborra, F.; Avances, C.; Pinguet, F. Docetaxel and cisplatin in patients with metastatic androgen independent prostate cancer and circulating neuroendocrine markers. *J. Urol.* **2007**, *178*, 844–848, discussion 848. [CrossRef]
21. Conteduca, V.; Burgio, S.L.; Menna, C.; Carretta, E.; Rossi, L.; Bianchi, E.; Masini, C.; Amadori, D.; De Giorgi, U. Chromogranin A is a potential prognostic marker in prostate cancer patients treated with enzalutamide. *Prostate* **2014**, *74*, 1691–1696. [CrossRef] [PubMed]
22. Burgio, S.L.; Conteduca, V.; Menna, C.; Carretta, E.; Rossi, L.; Bianchi, E.; Kopf, B.; Fabbri, F.; Amadori, D.; De Giorgi, U. Chromogranin A predicts outcome in prostate cancer patients treated with abiraterone. *Endocr. Relat. Cancer* **2014**, *21*, 487–493. [CrossRef] [PubMed]
23. Kashyap, V.; Ahmad, S.; Nilsson, E.M.; Helczynski, L.; Kenna, S.; Persson, J.L.; Gudas, L.J.; Mongan, N.P. The lysine specific demethylase-1 (LSD1/KDM1A) regulates VEGF-A expression in prostate cancer. *Mol. Oncol.* **2013**, *7*, 555–566. [CrossRef] [PubMed]
24. Sehrawat, A.; Gao, L.; Wang, Y.; Bankhead, A., 3rd; McWeeney, S.K.; King, C.J.; Schwartzman, J.; Urrutia, J.; Bisson, W.H.; Coleman, D.J.; et al. LSD1 activates a lethal prostate cancer gene network independently of its demethylase function. *Proc. Natl. Acad. Sci. USA* **2008**, *115*, E4179–E4188. [CrossRef]
25. Blagosklonny, M.V. Cell death beyond apoptosis. *Leukemia* **2000**, *14*, 1502–1508. [CrossRef]
26. Demidenko, Z.N.; Blagosklonny, M.V. Flavopiridol induces p53 via initial inhibition of Mdm2 and p21 and, independently of p53, sensitizes apoptosis-reluctant cells to tumor necrosis factor. *Cancer Res.* **2004**, *64*, 3653–3660. [CrossRef] [PubMed]
27. Choi, A.M.; Ryter, S.W.; Levine, B. Autophagy in human health and disease. *N. Engl. J. Med.* **2013**, *368*, 651–662. [CrossRef] [PubMed]

28. Jin, S. Autophagy, mitochondrial quality control, and oncogenesis. *Autophagy* **2006**, *2*, 80–84. [CrossRef] [PubMed]
29. Villar, V.H.; Nguyen, T.L.; Delcroix, V.; Teres, S.; Bouchecareilh, M.; Salin, B.; Bodineau, C.; Vacher, P.; Priault, M.; Soubeyran, P.; et al. mTORC1 inhibition in cancer cells protects from glutaminolysis-mediated apoptosis during nutrient limitation. *Nat. Commun.* **2017**, *8*, 14124. [CrossRef] [PubMed]
30. Lamoureux, F.; Thomas, C.; Crafter, C.; Kumano, M.; Zhang, F.; Davies, B.R.; Gleave, M.E.; Zoubeidi, A. Blocked autophagy using lysosomotropic agents sensitizes resistant prostate tumor cells to the novel Akt inhibitor AZD5363. *Clin. Cancer Res.* **2013**, *19*, 833–844. [CrossRef] [PubMed]
31. Nguyen, H.G.; Yang, J.C.; Kung, H.J.; Shi, X.B.; Tilki, D.; Lara, P.N., Jr.; DeVere White, R.W.; Gao, A.C.; Evans, C.P. Targeting autophagy overcomes Enzalutamide resistance in castration-resistant prostate cancer cells and improves therapeutic response in a xenograft model. *Oncogene* **2014**, *33*, 4521–4530. [CrossRef] [PubMed]
32. Blessing, A.M.; Rajapakshe, K.; Reddy Bollu, L.; Shi, Y.; White, M.A.; Pham, A.H.; Lin, C.; Jonsson, P.; Cortes, C.J.; Cheung, E.; et al. Transcriptional regulation of core autophagy and lysosomal genes by the androgen receptor promotes prostate cancer progression. *Autophagy* **2017**, *13*, 506–521. [CrossRef]
33. Fiskus, W.; Sharma, S.; Shah, B.; Portier, B.P.; Devaraj, S.G.; Liu, K.; Iyer, S.P.; Bearss, D.; Bhalla, K.N. Highly effective combination of LSD1 (KDM1A) antagonist and pan-histone deacetylase inhibitor against human AML cells. *Leukemia* **2014**, *28*, 2155–2164. [CrossRef] [PubMed]
34. Gupta, S.; Weston, A.; Bearrs, J.; Thode, T.; Neiss, A.; Soldi, R.; Sharma, S. Reversible lysine-specific demethylase 1 antagonist HCI-2509 inhibits growth and decreases c-MYC in castration- and docetaxel-resistant prostate cancer cells. *Prostate Cancer Prostatic Dis.* **2016**, *19*, 349–357. [CrossRef] [PubMed]
35. Hahm, E.R.; Singh, S.V. Honokiol causes G0-G1 phase cell cycle arrest in human prostate cancer cells in association with suppression of retinoblastoma protein level/phosphorylation and inhibition of E2F1 transcriptional activity. *Mol. Cancer Ther.* **2007**, *6*, 2686–2695. [CrossRef] [PubMed]
36. Roy, S.; Kaur, M.; Agarwal, C.; Tecklenburg, M.; Sclafani, R.A.; Agarwal, R. p21 and p27 induction by silibinin is essential for its cell cycle arrest effect in prostate carcinoma cells. *Mol. Cancer Ther.* **2007**, *6*, 2696–2707. [CrossRef]
37. Wang, M.; Liu, X.; Guo, J.; Weng, X.; Jiang, G.; Wang, Z.; He, L. Inhibition of LSD1 by Pargyline inhibited process of EMT and delayed progression of prostate cancer in vivo. *Biochem. Biophys. Res. Commun.* **2015**, *467*, 310–315. [CrossRef]
38. Perillo, B.; Di Santi, A.; Cernera, G.; Ombra, M.N.; Castoria, G.; Migliaccio, A. Nuclear receptor-induced transcription is driven by spatially and timely resticted waves of ROS. The role of Akt, IKKα, and DNA damage repair enzymes. *Nucleus* **2014**, *5*, 482–491. [CrossRef]
39. Perillo, B.; Di Santi, A.; Cernera, G.; Ombra, M.N.; Castoria, G.; Migliaccio, A. Phosphorylation of H3 serine 10 by IKKα governs cyclical production of ROS in estrogen-induced transcription and ensures DNA wholeness. *Cell Death Differ.* **2014**, *9*, 1503. [CrossRef] [PubMed]
40. Chen, Z.Y.; Chen, H.; Qiu, T.; Weng, X.D.; Guo, J.; Wang, L.; Liu, X.H. Effects of cisplatin on the LSD1-mediated invasion and metastasis of prostate cancer cells. *Mol. Med. Rep.* **2016**, *14*, 2511–2517. [CrossRef] [PubMed]
41. Dehm, S.M.; Schmidt, L.J.; Heemers, H.V.; Vessella, R.L.; Tindall, D.J. Splicing of a novel androgen receptor exon generates a constitutively active androgen receptor that mediates prostate cancer therapy resistance. *Cancer Res.* **2008**, *68*, 5469–5477. [CrossRef] [PubMed]
42. Hu, R.; Dunn, T.A.; Wei, S.; Isharwal, S.; Veltri, R.W.; Humphreys, E.; Han, M.; Partin, A.W.; Vessella, R.L.; Isaacs, W.B.; et al. Ligand-independent androgen receptor variants derived from splicing of cryptic exons signify hormone-refractory prostate cancer. *Cancer Res.* **2009**, *69*, 16–22. [CrossRef] [PubMed]
43. Hornberg, E.; Ylitalo, E.B.; Crnalic, S.; Antti, H.; Stattin, P.; Widmark, A.; Bergh, A.; Wikstrom, P. Expression of androgen receptor splice variants in prostate cancer bone metastases is associated with castration-resistance and short survival. *PLoS ONE* **2011**, *6*, e19059. [CrossRef] [PubMed]
44. Qu, Y.; Dai, B.; Ye, D.; Kong, Y.; Chang, K.; Jia, Z.; Yang, X.; Zhang, H.; Zhu, Y.; Shi, G. Constitutively active AR-V7 plays an essential role in the development and progression of castration-resistant prostate cancer. *Sci. Rep.* **2015**, *5*, 7654. [CrossRef] [PubMed]

45. Antonarakis, E.S.; Lu, C.; Wang, H.; Luber, B.; Nakazawa, M.; Roeser, J.C.; Chen, Y.; Mohammad, T.A.; Fedor, H.L.; Lotan, T.L.; et al. AR-V7 and resistance to enzalutamide and abiraterone in prostate cancer. *N. Engl. J. Med.* **2014**, *371*, 1028–1038. [CrossRef] [PubMed]
46. Scher, H.I.; Lu, D.; Schreiber, N.A.; Louw, J.; Graf, R.P.; Vargas, H.A.; Johnson, A.; Jendrisak, A.; Bambury, R.; Danila, D.; et al. Association of AR-V7 on Circulating Tumor Cells as a Treatment-Specific Biomarker With Outcomes and Survival in Castration-Resistant Prostate Cancer. *JAMA Oncol.* **2016**, *2*, 1441–1449. [CrossRef] [PubMed]
47. Regufe da Mota, S.; Bailey, S.; Strivens, R.A.; Hayden, A.L.; Douglas, L.R.; Duriez, P.J.; Borrello, M.T.; Benelkebir, H.; Ganesan, A.; Packham, G.; et al. LSD1 inhibition attenuates androgen receptor V7 splice variant activation in castration resistant prostate cancer models. *Cancer Cell Int.* **2018**, *18*, 71. [CrossRef]
48. Naiki, T.; Asamoto, M.; Toyoda-Hokaiwado, N.; Naiki-Ito, A.; Tozawa, K.; Kohri, K.; Takahashi, S.; Shirai, T. Organ specific Gst-pi expression of the metastatic androgen independent prostate cancer cells in nude mice. *Prostate* **2012**, *72*, 533–541. [CrossRef]
49. Naiki, T.; Naiki-Ito, A.; Asamoto, M.; Kawai, N.; Tozawa, K.; Etani, T.; Sato, S.; Suzuki, S.; Shirai, T.; Kohri, K.; et al. GPX2 overexpression is involved in cell proliferation and prognosis of castration-resistant prostate cancer. *Carcinogenesis* **2014**, *35*, 1962–1967. [CrossRef] [PubMed]
50. Berman, E.; Duigou-Osterndorf, R.; Krown, S.E.; Fanucchi, M.P.; Chou, J.; Hirsch, M.S.; Clarkson, B.D.; Chou, T.C. Synergistic cytotoxic effect of azidothymidine and recombinant interferon alpha on normal human bone marrow progenitor cells. *Blood* **1989**, *74*, 1281–1286. [PubMed]
51. Schmukler, E.; Wolfson, E.; Haklai, R.; Elad-Sfadia, G.; Kloog, Y.; Pinkas-Kramarski, R. Chloroquine synergizes with FTS to enhance cell growth inhibition and cell death. *Oncotarget* **2014**, *5*, 173–184. [CrossRef] [PubMed]

12

Obesity Inflammation and Prostate Cancer

Kazutoshi Fujita *, Takuji Hayashi, Makoto Matsushita, Motohide Uemura and Norio Nonomura

Department of Urology, Osaka University Graduate School of Medicine, Suita, Osaka 565-0871, Japan; takujihayashi0929@gmail.com (T.H.); matsushita@uro.med.osaka-u.ac.jp (M.M.); uemura@uro.med.osaka-u.ac.jp (M.U.); nono@uro.med.osaka-u.ac.jp (N.N.)

* Correspondence: kazu.fujita2@gmail.com

Abstract: The prevalence of obesity is increasing in the world, and obesity-induced disease, insulin-resistance, cardiovascular disease, and malignancies are becoming a problem. Epidemiological studies have shown that obesity is associated with advanced prostate cancer and that obese men with prostate cancer have a poorer prognosis. Obesity induces systemic inflammation via several mechanisms. High-fat diet-induced prostate cancer progresses via adipose-secretory cytokines or chemokines. Inflammatory cells play important roles in tumor progression. A high-fat diet or obesity changes the local profile of immune cells, such as myeloid-derived suppressor cells and macrophages, in prostate cancer. Tumor-associated neutrophils, B cells, and complements may promote prostate cancer in the background of obesity. Interventions to control systemic and/or local inflammation and changes in lifestyle may also be viable therapies for prostate cancer.

Keywords: obesity; inflammation; prostate cancer; immune cells; cytokine; high-fat diet

---

1. Introduction

Since 1980, the prevalence of obesity has doubled in the world. Obesity is caused by genetic factors, neuroendocrine factors, psychological factors, and environmental factors [1]. In the United States, almost 40% of people suffer from obesity, and the present situation is a "pandemic" of obesity. In Asian countries such as Korea and Japan, the prevalence of obesity is still low, approximately less than 10%, but the prevalence of obesity has increased over the last decade [1,2]. The incidence rate of prostate cancer is also increasing and is now highest in Japan. Overfeeding with a high-fat and/or high-calorie diet and less physical activity result in an energy imbalance and adiposity. Obesity causes insulin resistance, type 2 diabetes, cardiovascular diseases, and several malignancies via systemic inflammation. The resulting medical costs due to obesity are increasing and becoming an important issue worldwide.

Prostate cancer has had high morbidity among elderly men. Many patients with prostate cancer are in the early stage and have good prognosis after several treatments including prostatectomy, radiation therapy, hormonal therapy, and even active surveillance. However, some progressive prostate cancer patients in the late stage with poorly-differentiated cancer cells, or local invasion, or metastatic lesion are more resistant to several treatments including hormonal therapy or chemotherapy, and have poor prognosis. (Figure 1) It is important to elucidate the mechanism of the factors inducing prostate cancer progression.

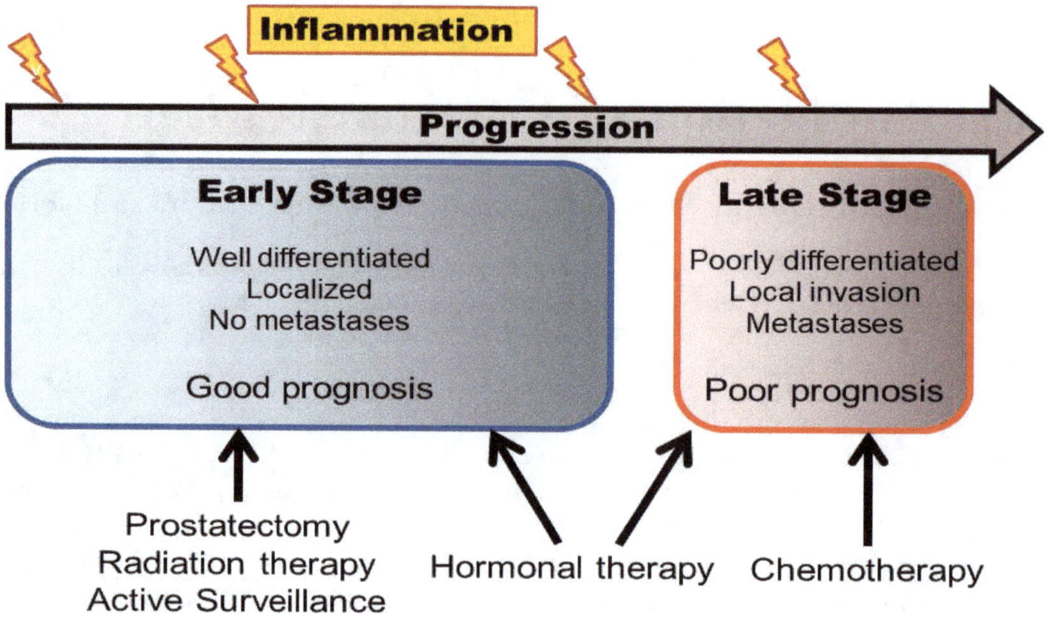

Figure 1. The scheme of different stages and progression of prostate cancer.

Chronic inflammation is the major etiology behind the development of several cancers, such as hepatocellular carcinoma, squamous cell carcinoma in the urinary bladder, colorectal cancer, and gastric cancer. Inflammatory cells migrating to the local area generate reactive oxygen species and reactive nitrogen species that induce mutations of DNA in normal epithelia [3]. Acute or chronic inflammation is a common histological finding in both benign and malignant tissues in prostatectomy specimens [4,5]. The causes of inflammation in the prostate vary among bacteria causing prostatitis and sexually-transmitted disease, hormonal changes of estrogen [6], physical trauma caused by corpora amylacea [7,8], urine reflux to the prostate gland, and environmental factors such as dietary habits [9,10]. Dietary habits cause inflammation of the prostate and can result in carcinogenesis in the early stage [11,12]. Dietary-induced inflammation could last for the entire life, and chronic inflammation can also stimulate the progression of prostate cancer in the late stage. However, the association of immune cells in tumor microenvironments with prostate cancer is still unclear.

In this review, the link between obesity and prostate cancer is discussed based on the recent findings related to inflammation.

2. Obesity and Prostate Cancer

Several studies reported that obesity was associated with the increased risk of several cancers, such as colon, breast, endometrial, kidney, gastric, esophagus, pancreas, liver, and gall bladder [13,14]. Several studies have shown the association of obesity with the risk of prostate cancer. A prospective study of 3673 men in the United States showed that greater body mass index (BMI) was an independent predictor of prostate cancer (relative risk = 1.7 for BMI > 27.8 kg/m^2 compared with <23.6 kg/m^2; $p = 0.1$). The percent change in BMI from baseline to age 50 was also positively associated with risk ($p = 0.01$) [15]. Another prospective study in the United States showed that BMI was weakly and positively associated with prostate cancer, and the association of obesity with the risk of clinically-significant prostate cancer strengthened after the exclusion of well-differentiated, localized tumors [16]. However, a prospective study of 36,959 Swedish men showed that the incidence of localized prostate cancer was inversely associated with BMI in middle-to-late adulthood (the rate ratio for 35 kg/m^2 when compared with 22 kg/m^2 was 0.69 (95% confidence interval (CI) 0.52–0.92)), but not in early adulthood. BMI in middle-to-later adulthood was associated with a non-statistically significant increase in the risk of fatal prostate cancer (rate ratio for every five-unit increase: 1.12 (0.88–1.43)) and BMI in early adulthood with a decreased risk of fatal prostate cancer (rate ratio for every five-unit increase: 0.72 (0.51–1.01)) [17]. A prospective study of 141,896 men in the European Prospective Investigation into Cancer and Nutrition (EPIC) cohort showed that high BMI at a young age was inversely associated with the overall risk of prostate cancer (relative risk = 0.89, 95% CI 0.80–0.98, BMI $\geq$ 26 vs. 20–21.9, p = 0.01) and with fatal and advanced disease [18]. Obesity at a young age causes the delayed onset of puberty and may result in the lower lifetime exposure of insulin-like growth factor 1 (IGF-I), which may affect the development of prostate cancer later in life [18,19]. A meta-analysis of 12 prospective studies of localized prostate cancer (1,033,009 men, 19,130 cases) and 13 of advanced prostate cancer (1,080,790 men, 7067 cases) showed an inverse linear relationship with BMI for localized prostate cancer ($p \leq 0.001$, relative risk: 0.94 for every 5-kg/m^2 increase) and a positive linear relationship with BMI for advanced prostate cancer ($p = 0.001$, relative risk: 1.09 for every 5-kg/m^2 increase) [20]. Obesity thus could affect the incidence of the risk of prostate cancer in the early stage in the opposite direction according to the type of prostate cancer. The underlying mechanisms of this inverse association of obesity with localized prostate cancer could be the low testosterone levels in obese men. Obese men have a lower concentration of free testosterone due to a decrease of lutenizing hormone (LH) pulse amplitude and serum LH levels [21]. Plasma total testosterone and free testosterone were positively associated with increased risk of low-grade prostate cancer [22]. However, the association of testosterone, free testosterone, and the free-to-total testosterone ratio with prostate cancer is still controversial [23]. Furthermore, the impact of obesity-induced systemic inflammation on the inverse relationship of localized prostate cancer to BMI is still unknown.

Obesity may also affect the prognosis of prostate cancer in the late stage. An analysis of 4123 men treated by radical prostatectomy showed that higher BMI was associated with biochemical recurrence after radical prostatectomy (hazard ratio (HR) 1.02, 95% CI 1.00–1.02, p = 0.008) [24]. A retrospective analysis of 4268 radical prostatectomy patients within the Shared Equal Access Regional Cancer Hospital (SEARCH) database showed that being overweight and obesity were associated with prostate cancer-specific mortality (HR 1.88, p = 0.061 and HR 2.05, p = 0.039, respectively) [25]. A prospective study of 404,576 men showed a positive linear trend in the prostate cancer death rate with higher BMI ($p < 0.001$) [14]. These epidemiological studies showed obvious evidence of the association of obesity with advance prostate cancer.

3. Obesity and Inflammation

Many studies have shown that obesity causes systemic inflammation through the action of various mechanisms. Adipocytes secrete tumor necrosis factor (TNF)-α in obese mice that causes systemic inflammation [26]. A high-fat diet (HFD) changes the intestinal microbiota and increases the translocation of live Gram-negative bacteria through the intestinal mucosa into the bloodstream and mesenteric adipose tissue, which results in continuous bacteremia [27]. Fatty acids activate toll like receptor 4 (TLR4) signaling in adipocytes and macrophages. Female mice lacking TLR4 show increased obesity, but are partially protected against HFD-induced insulin resistance, possibly due to reduced inflammatory gene expression in the liver and fat [28]. Obesity induces activation of the innate immune system. Adipose depots contain multiple immune cells. Macrophages in adipose tissues are increased in the obese, skewing to the M1-polalized macrophages. These macrophages show a pro-inflammatory phenotype and secrete inflammatory cytokines such as TNF-α [29].

It is still unclear how such systemic inflammation affects local inflammation of the prostate (Figure 2). Several chemokines and cytokines secreted from prostate cancer cells may recruit immune cells to the prostate. Which organ are these immune cells activated in? Some immune cells could be "taught" in the intestinal wall [30], but there has been no evidence of the homing of these intestinal immune cells to a distant organ. In bone marrow or regional lymph nodes, the immune cells might be activated by factors related to obesity and subsequently recruited to the prostate. Otherwise, the local immune cells recruited by prostate cancer cells might be activated by the obesity-related factors. The elucidation of these factors related to obesity could lead to the development of new treatments or the prevention of prostate cancer in the early stage.

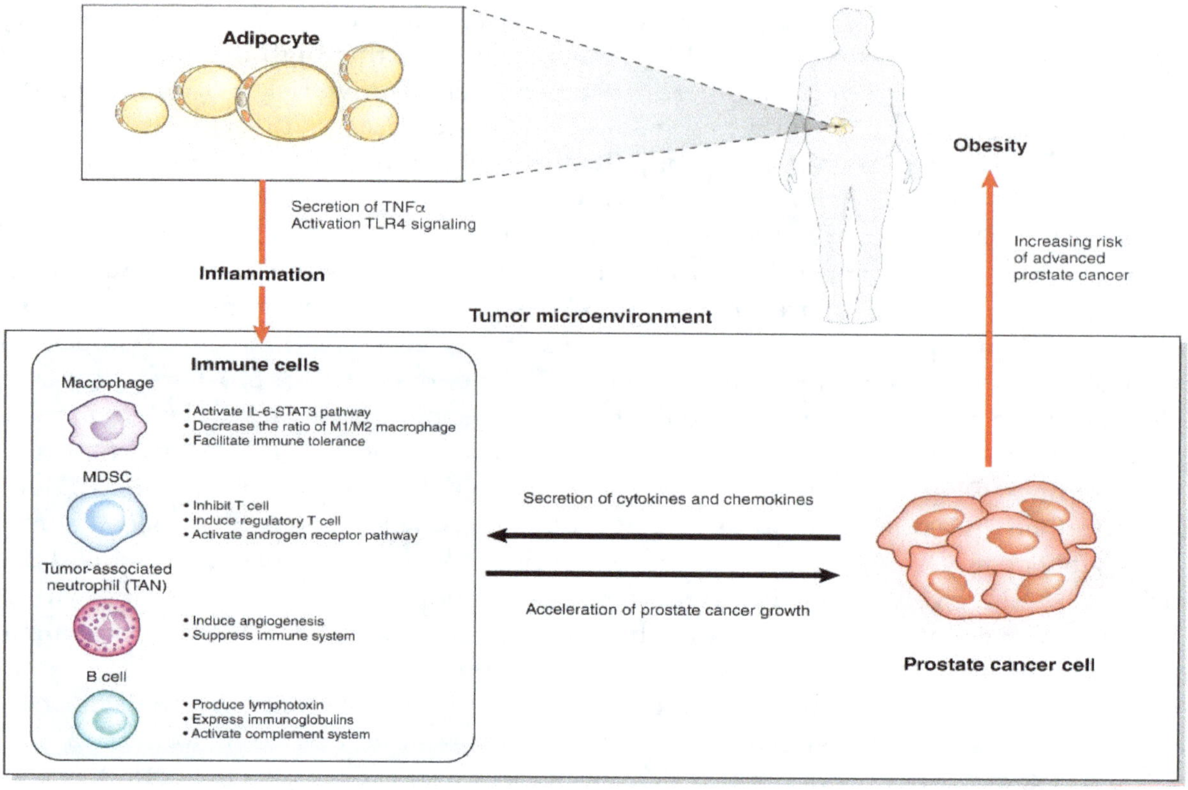

Figure 2. Interaction of immune cells with adipocytes and prostate cancer cells.

4. Obesity Promotes Prostate Cancer Growth

Although the link between obesity and prostate cancer has not been definitively determined, several studies focusing on the cytokines and/or chemokines have been reported. In a mouse xenograft model of the prostate cancer cell line LNCaP, serum monocyte chemoattractant protein-1 (MCP-1) was significantly increased, and tumor growth was promoted in HFD-fed mice [31]. Palmitic acid is one of the saturated free fatty acids abundantly included in HFDs. The addition of palmitic acid induced the expression of macrophage inhibitory cytokine 1 (MIC1) in vitro, and serum levels of MIC1 were increased in the HFD-fed mice xenograft model. Obese patients with prostate cancer were also found to have higher serum levels of MIC1 than those in healthy controls [32]. HFDs also modulate miRNA expression in prostate cancer cells. Prostate cancer cells cultured in the serum of HFD-fed mice showed a marked increase in cell proliferation and the attenuation of miR-130a. miR-130a modulated MET expression in prostate cancer cell lines, and furthermore, cytoplasmic MET in prostate cancer tissues was overexpressed in patients with higher BMI [33]. An HFD also induced increases in leptin, C-C motif ligand (CCL)3, CCL4, CCL5, and C-X-C motif ligand (CXCL)10 in the sera of transgenic adenocarcinoma of mouse prostate (TRAMP) mice. The conditioned medium of sera from HFD-fed TRAMP mice promoted the proliferation, migration, and invasion of DU-145 cells [34]. Obese patients with prostate cancer showed increased expression of epithelial CXCL1, which induces the recruitment of adipose stromal cells from white adipose tissue to the tumor and promotes the tumor's growth [35]. These reports showed that cytokines and chemokines could play important roles in the obesity-associated progression of prostate cancer in the early and late stage. Because TRAMP mice lacking expression of androgen receptor are thought to be models for a very advanced stage with neuroendocrine cancer cells and independent from androgen receptor, the findings using TRAMP mice might be compatible with prostate cancer patients in only the late stage. Moreover, the detailed mechanisms including the tumor microenvironments are still unknown.

5. Inflammation in Prostate Cancer

In the tumor microenvironments, the interactions among cancer cells, immune cells, endothelial cells, and fibroblasts can play important roles. Inflammatory cells consist of innate immune cells and acquired immune cells. Acquired immune cells include B cells and T cells, which act based on antigen recognition. While innate immune cells are the main players in inflammation, innate immune cells and acquired immune cells also orchestrate the inflammation. Innate immune cells including neutrophils, myeloid cells, mast cells, and macrophages are different from acquired immune cells by receptor-mediated activation and their rapid response to invading pathogens and foreign bodies [36]. Macrophages and neutrophils are the most abundant immune cells in the tumor microenvironment [37].

To reveal the relationship between HFD-induced inflammation and tumor progression in the prostate, we used two genetically-engineered prostate cancer mouse models, prostate-specific *Pten* knockout mice (Pb-Cre$^+$; *Pten*(fl/fl)) and *Pten* and *Tp53*-double knockout mice (Pb-Cre$^+$; *Pten*(fl/fl); *Tp53*(fl/fl)) on the C57BL/6 genetic background. The prostate weights and the ratio of Ki67-positive cells to tumor cells, which indicates the proliferative capacity of the tumor, of the mice in the HFD-fed double knockout mouse model were significantly higher than those of the control diet (CD)-fed model mice ($p = 0.011$, $p = 0.005$, respectively) (Figure 3A,B). Total RNA was isolated from prostatic tissues of both the CD-fed mice and HFD-fed double knockout mice, and transcriptome analysis of the two groups was performed using mRNA microarray technology. Gene ontology analysis revealed that many processes related to inflammation and the immune response were ranked in the top 22 processes expressed in the prostate of the HFD-fed double knockout mice (Figure 3C). This finding strongly suggests that local inflammation of the prostate is one of the most important factors for the progression of prostate cancer in obese or HFD-fed mice in the early and late stages. The profiles of the local immune cells in prostate cancer were analyzed in the *Pten* knockout mouse model fed with a CD or HFD. Although the number of B cells, T cells, macrophages, and mast cells and the ratio of CD8/CD4 T cells were not changed by the HFD, the number of myeloid-derived suppressor cells (MDSCs) and the M2/M1 macrophage ratio were significantly increased in the HFD-fed mice compared with the CD-fed mice. The promotion of tumor growth by the HFD was completely cancelled by the administration of celecoxib, a cyclooxygenase 2 (COX-2) inhibitor, which suggests that inflammation plays a central role in tumor progression caused by an HFD. IL-6 expression in prostate tissues was increased in HFD-fed mice, as were the amounts of phosphorylated signal transducer and activator of transcription 3 (STAT3) in prostate cancer cells. Inhibition of the IL-6 pathway resulted in the suppression of tumor growth by an HFD [38]. The HFD and subsequent obesity caused the increased secretion of IL-6 from local macrophages in the prostate tumor via unclear mechanisms. IL-6 might increase the number of local MDSCs and promote the proliferation of prostate cancer cells via signal transducer and activator of transcription 3 (STAT3) pathways. Because transcriptome analysis in double knockout mice resulted in different changes of gene expressions from *Pten* knockout mice after administration of HFD, *Tp53* may have many functions regarding inflammation. In addition, it might result in different findings if model mice on the other genetic background were to be examined.

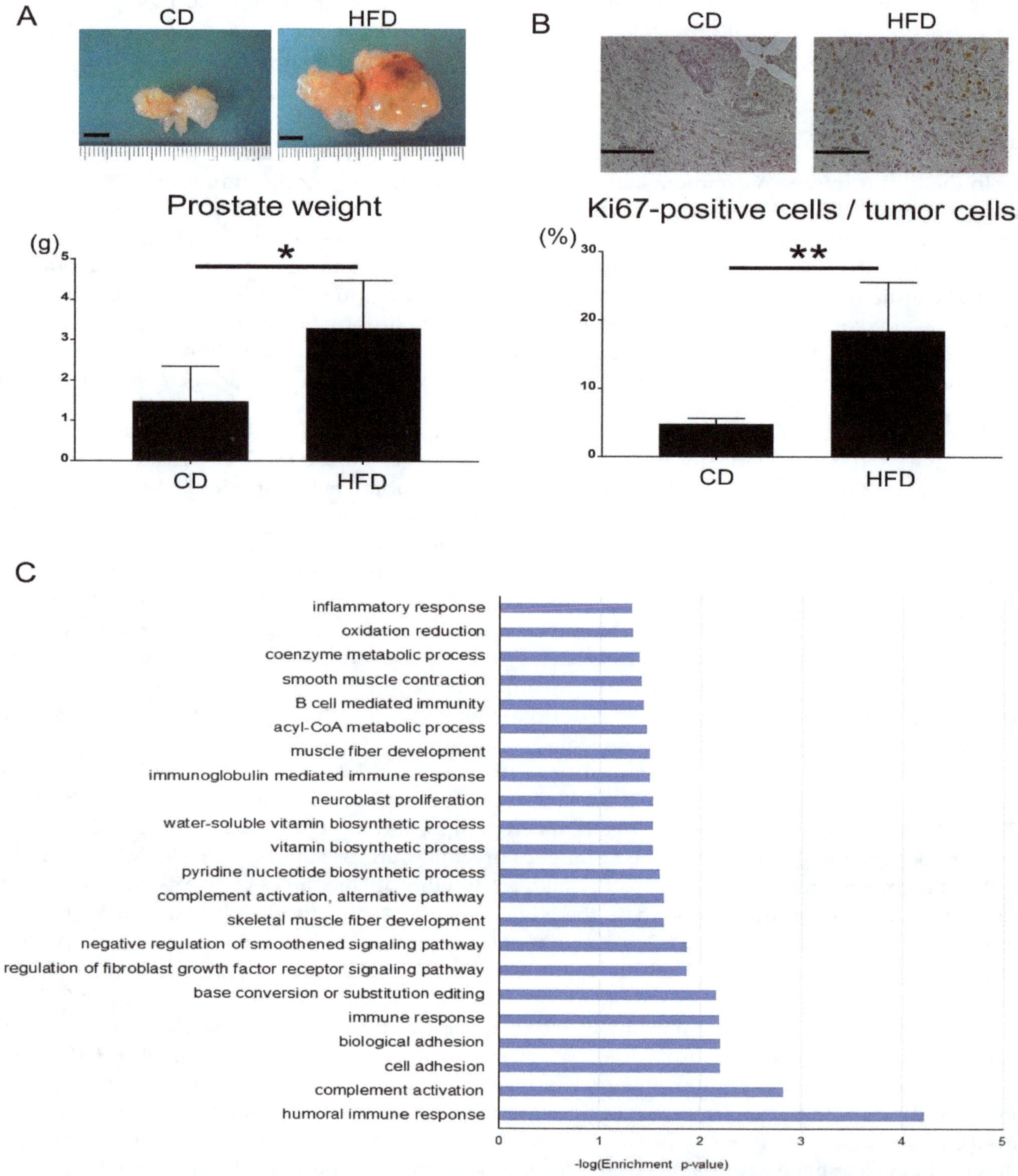

Figure 3. (**A**) Representative gross findings of the prostatic tissues (the black bar indicates 5 mm) (top) and prostate weights ($n = 6$ and 4, respectively) (bottom) of the model mice at 22 weeks of age. (**B**) Representative images of Ki67 staining for the prostatic tissues (top) and the ratio of Ki67-positive cells to tumor cells ($n = 4$ and 3, respectively) (bottom) of the model mice at 22 weeks of age. (**C**) Gene ontology analysis using mRNA microarray technology of the prostatic tissues of the model mice at 22 weeks of age (HFD-fed vs. CD-fed, $n = 3$, respectively; fold change >2.0, $p < 0.05$, biological process). CD, control diet; HFD, high-fat diet. * $p < 0.05$, ** $p < 0.01$.

6. Macrophages

Macrophages, one of the most abundant types of immune cells in tumor microenvironments, change the phenotype to promote tumor growth and metastasis. Macrophages are divided into classic macrophages (M1) and alternative macrophages (M2). M1 macrophages act in microbicidal and tumoricidal activity, and M2 macrophages act in tissue remodeling, immune tolerance, and tumor progression [39]. M1 macrophages are characterized by the secretion of interleukin-1β (IL-1β), IL-6, IL-12, and TNF-α, whereas M2 macrophages are characterized by the secretion of IL-4, IL-10, and TGF-β. At early stages of tumor development, macrophages undergo classic activation and exhibit an M1 phenotype [36,40]. Cytokines secreted from M1 macrophages play roles in tumor initiation and early promotion [36]. Exposure of macrophages to IL-4, colony-stimulating factor-1 (CSF1), granulocyte-macrophage colony-stimulating factor (GM-CSF), and TGFβ secreted by cancer cells polarize macrophages to the M2 phenotype, which acts to induce immunosuppressive microenvironments. Inflammatory cytokines secreted from adipocytes, such as TNF-α, IL-6, IL1β, and CCL2, recruit macrophages to the adipose tissues. Diet-induced obesity leads to a shift of the macrophage phenotype from M2 to M1 in mice [41,42]. In contrast, in mammary adipose tissue of breast cancer in obese women, macrophages showed a decrease in the expression of IL-10 and CD11c, which are characteristic of an M1 polarization phenotype. However, they also showed an increase in the expression of CD206, which is a surface marker of the M2 polarization phenotype, suggesting a mixed polarization phenotype in tumor microenvironments [39,43]. Macrophages are known to promote cancer growth and metastasis in prostate cancer, but the association of macrophages with obesity in prostate cancer is still unclear. Different from breast cancer, adipocytes are located in the area surrounding the prostate and are not found within the prostate tissues. Prostate cancer and stromal cells secrete CCL2, which strongly recruits macrophages [44,45]. CCL2 levels were increased in the sera of HFD-fed mice with an LNCaP xenograft. It is also reported that the number of tumor-infiltrating macrophages is not associated with BMI [46]. The role of macrophages in prostate cancer with a background of obesity will require further study.

7. Myeloid-Derived Suppressor Cells (MDSCs)

MDSCs have a strong immunosuppressive function that enables the regulation of immune response and suppresses overt inflammatory responses [47]. MDSCs represent a non-lymphoid immune suppressor cell population of myeloid origin that is enriched in cancer [48]. MDSCs are a heterogeneous population and express a mixture of surface markers typical for myeloid cells, but they lack the markers of lymphocytes, natural killer cells, macrophages, and dendritic cells [47]. MDSCs were originally found in mice, and their counterparts in humans are not well defined. MDSCs in mice are characterized by the surface marker Gr-1+CD11b+. MDSCs are divided into two major groups: cells with a morphology and surface markers characteristic of monocytes (monocytic (M)-MDSCs) and cells with surface markers characteristic of polymorphonuclear (PMN)-MDSCs). In mice, M-MDSCs are characterized by the surface markers of $CD11b^{+}Ly6C^{high}$ $Ly6G^{-}$, and PMN-MDSCs are characterized by $CD11b^{+}Ly6C^{low}$ $Ly6G^{+}$. In humans, the equivalent cells to PMN-MDSCs are defined as $CD11b^{+}CD14^{-}CD15^{+}$ or $CD11b^{+}CD14^{-}CD66b^{+}$ and M-MDSCs as $CD11b^{+}CD14^{+}HLA\text{-}DR^{-/low}$ CD15 [49]. MDSCs are characterized by the suppression of T cell response by ARG1, iNOS, and reactive oxygen species. MDSCs inhibit T cells via arginase-1, iNOS, and ROS and induce regulatory T cells by IL-10 and TGF-β. MDSCs also modulate the cytokine production of macrophages and promote tumor angiogenesis and eventually metastasis [47]. In a prostate cancer mouse model (TRAMP mouse), IL-23 secreted from MDSCs can activate the androgen receptor pathway and promote cell survival and proliferation under an androgen-deprived condition. Blockade of IL-23 can oppose MDSC-mediated resistance to castration in prostate cancer [50]. CXCL5 secreted from prostate cancer cells attracts MDSCs expressing CXCR2 in a mouse model of prostate cancer. Elimination of MDSCs or the blocking of CXCL5-CXCR2 signaling elicits an antitumor response for prostate cancer [51]. In humans, CD14+HLA-$DR^{-/low}$ M-MDSCs and Treg were significantly increased in peripheral blood from patients with prostate cancer compared with healthy donors. High levels of M-MDSCs in the blood were associated with a shorter median overall survival [52]. In patients with prostate cancer, MDSCs accumulate in the blood as prostate cancer progresses and inhibit the proliferation of autologous CD8+ T cells and the production of interferon-γ (IFN-γ) and granzyme-B [53]. MDSCs could be a new target in the prevention and treatment of prostate cancer and/or castration-resistant prostate cancer.

8. Neutrophils

Neutrophils primarily work as an antibacterial immune response, but tumor-associated neutrophils (TANs) also play important roles in tumor microenvironments. Similar to M1 and M2 macrophages, terms for antitumoral N1 neutrophils and protumoral N2 neutrophils were proposed [54]. The chemokines CXCL1, CXCL2, CXCL5, CXCL6, and CXCL8 secreted from tumor cells attract neutrophils in the blood to the tumor microenvironment via CXCR1 and CXCR2 on the surface of neutrophils [55]. TANs share a similar surface marker with PMN-MDSCs. Murine neutrophils are defined as CD11b+/GR1+/Ly6G+cells, whereas PMN-MDSCs are defined as $CD11b^{+}/GR1^{high}/Ly6G^{+}$ cells. PMN-MDSCs were named based on the function of immunosuppression. However, neutrophils can work in immunosuppression, but also have the opposite function of anti-tumor activity. N1 TANs function in tumor cell cytotoxicity, CD8+ T cell recruitment, and antibody-dependent cell-mediated cytotoxicity. In contrast, N2 TANs play roles in angiogenesis, immunosuppression, and tumor growth via several cytokines or proteins released from TANs [55]. In a mouse model, obesity caused the increase of neutrophils in the lung and promoted the metastasis of breast cancer cells to the lung in a GM-CSF- and IL-5-dependent manner [56]. In HFD-fed mice, cholesterol metabolites promoted the metastasis of breast cancer via neutrophils and $\gamma\delta$-T cells [57]. In another mouse model, obesity promoted the progression of pancreatic cancer and resistance to chemotherapy via TANs recruited by adipocyte-secreted IL1β [58]. Murine neutrophils are different from human neutrophils. Thus, it is still unclear whether the TANs play roles in prostate cancer progression in the late stage. The administration of cabozantinib resulted in the clearance of prostate cancer in mice by recruiting neutrophils to the tumor [59].

In humans, no markers equivalent to the mouse Gr1 marker exist, and human neutrophils are defined as $CD14^{-}/CD15^{+}/CD66b^{+}/CD16^{+}$. The neutrophil-lymphocyte ratio in peripheral blood is associated with a high Gleason score and the poor prognosis of men with prostate cancer [60,61]. The ratio is also a prognostic factor of abiraterone and docetaxel treatment in men with castration-resistant prostate cancer [62,63]. Low serum neutrophil count is a predictor of positive prostate biopsy results [64]. The presence of neutrophils in the epithelial lining of the prostate gland indicate prostatic inflammation and is a predictive factor of benign biopsy [65]. The protumor roles of neutrophils in human prostate cancer have not been confirmed yet, and further studies are warranted.

9. B Cells and Complements

In the mRNA microarray analysis of prostate tumors in CD- and HFD-fed double knockout mice, the expressions of immune-related genes including splice variants of immunoglobulins, complements (*Hc*, *C4b*), *Ccl8*, and *Cd52* were significantly higher in HFD-fed double knockout model mice compared with CD-fed double knockout model mice (Table 1). Gene ontology analysis revealed that humoral immune responses were key factors of HFD-induced tumor progression (Figure 3C). These results suggested that B cell-mediated and immunoglobulin-mediated immune responses could be key factors of HFD-induced prostate cancer growth. B cells play important roles in diet-induced obesity, chronic inflammation, and humoral immunity, the latter two of which are influenced by some kinds of fatty acids and lipid mediators [66]. B cells are also related with tumor progression in various types of cancer, including prostate cancer [67]. Tumor-infiltrating B cells produce lymphotoxin, a cytokine belonging to the TNF family, that leads to activation of IκB kinase α and STAT3, which promote the survival and proliferation of androgen-deprived prostate cancer cells that result in the development of a castration-resistant state in experiments using the TRAMP mice model [68]. It was reported that higher B cell infiltration was present within the intra-tumoral prostate cancer regions compared to the extra-tumoral benign prostate tissue regions in prostatectomy sections [69]. Immunoglobulins are expressed by B cells and a variety of tumor tissues and cancer cell lines [70,71]. Immunoglobulins are suggested to play important roles in promoting cancer progression. Immunoglobulin G silencing induced apoptosis and suppressed proliferation, migration, and invasion in LNCaP prostate cancer cells [72].

Table 1. The list of the gene symbols that were highly expressed in the prostatic tissues in HFD-fed mice compared to CD-fed mice (fold-change >2.0, $p < 0.05$, mRNA microarray).

Gene Symbol	Fold-Change (HFD-fed vs. CD-fed)	*p*-Value (HFD-fed vs. CD-fed)
LOC238440	17.739	0.0163
Ighv6-6	12.429	0.0170
Ighv3-8	10.111	0.0048
Ighv14-3	9.678	0.0395
Igkv10-96	8.240	0.0248
Mug1	5.773	0.0068
Snord13	5.003	0.0002
Igkv10-94	4.652	0.0260
Adck1	4.644	0.0261
Itm2a	4.559	0.0223
Igh-VJ558	4.509	0.0371
Gm830	4.480	0.0304
Igkv10-95	4.070	0.0303
Igj	4.029	0.0152
Igh-V3660	3.978	0.0244
Ighj4	3.787	0.0016
Igh-VJ558	3.765	0.0035
Igkv4-55	3.733	0.0171
Igkv4-59	3.716	0.0132
Igkv16-104	3.596	0.0194
Igkv4-91	3.542	0.0456
Ccl8	3.450	0.0002
Ighv1-76	3.350	0.0055
Slc17a4	3.333	0.0124
LOC637260	3.242	0.0237
Hc	3.179	0.0337
Tm4sf4	3.040	0.0424
Ighv1-42	3.018	0.0333
Igkv5-45	3.005	0.0095
Ighv14-4	3.003	0.0255
Ighv1-80	2.997	0.0095
Igkv4-72	2.952	0.0091
Ms4a12	2.946	0.0289
A1cf	2.911	0.0180
Ighv1-77	2.876	0.0240
Adamts5	2.800	0.0341
Gm13307	2.793	0.0064
Clstn2	2.784	0.0280
Igkj5	2.766	0.0091
Ighv5-17	2.765	0.0324
Pdlim3	2.763	0.0099
Ighj3	2.728	0.0045
Myh11	2.709	0.0351
Ighm	2.708	0.0376
Tcrg-V4	2.676	0.0445
Svep1	2.673	0.0410

Table 1. *Cont.*

Gene Symbol	Fold-Change (HFD-fed vs. CD-fed)	*p*-Value (HFD-fed vs. CD-fed)
Ighj1	2.651	0.0290
Iglv1	2.632	0.0041
Pcp4	2.626	0.0476
Cpxm2	2.617	0.0333
Maob	2.616	0.0196
Igkv4-70	2.596	0.0396
Pgm5	2.594	0.0453
Cyp2c68	2.583	0.0168
Igkv4-53	2.582	0.0451
Ighv1-62-2	2.526	0.0011
Ppef1	2.516	0.0118
Acnat1	2.511	0.0460
Gm13304	2.500	0.0403
Igkv12-89	2.463	0.0495
Igh-VX24	2.457	0.0235
Snord14e	2.437	0.0266
Gm13304	2.413	0.0414
Thbs4	2.352	0.0425
Mylk	2.352	0.0185
Cd52	2.345	0.0066
Abca8a	2.313	0.0066
Kcnab2	2.291	0.0122
Inmt	2.290	0.0449
Igh-V3660	2.272	0.0399
Igsf23	2.260	0.0138
Cd200	2.250	0.0255
Dkk2	2.236	0.0418
Acta1	2.225	0.0401
Hhip	2.212	0.0093
Ecm2	2.208	0.0109
Lgi2	2.180	0.0343
Igkv4-62	2.167	0.0373
Prelp	2.151	0.0243
Igkj1	2.143	0.0113
Nlrp6	2.112	0.0417
Gm5485	2.062	0.0280
Serpini1	2.016	0.0088
LOC102642448	2.014	0.0095
Kmo	2.009	0.0386
C4b	2.009	0.0254
Igkv4-53	2.001	0.0105

A complement system is also related to cancer progression [73,74]. Complement activation in the tumor microenvironment enhances tumor growth and increases metastasis. The hemolytic complement encoded by the *Hc* gene in mice, the expressions of which were increased in the prostatic tissues of HFD-fed mice in our results, corresponds to C5 in human. C5, one of the complements, was suggested to promote tumor progression controlling the tumor microenvironment [75,76]. A humoral immune response including B cells, immunoglobulins, and complements could be key factors of prostate cancer progression induced by inflammation in the late stage. The detailed mechanism of the phenomenon remains unclear, and further investigations are necessary to explore the causative mechanism.

10. Conclusions

Inflammation and immune responses play important roles in the progression of prostate cancer. Other inflammatory cells and immune cells could be also involved in the prostate cancer progression. T cells are also accumulated in prostate cancer of a diet-induced obese Hi-Myc mice [77]. The cytotoxic function of NK cells to prostate cancer cells is inhibited by humoral factors from adipocytes [78]. These local inflammatory cells are orchestrated by several signalings from immune cells, adipocytes, or prostate cancers. Prostate cancer cells stimulated by adipokines or saturated fatty acid could change the local immune profile in the backgrounds of obesity [79]. The interplay between prostate cancer and immune cells is a "chicken and egg" situation. Another possible mechanism to affect prostate cancer in obesity could be an intestinal microbiome. High-fat diet changes the intestinal microbiome and enhances colorectal cancer and liver cancer [80,81]. The microbiome could modulate the host immune system, and these changes in the immune system might have an effect on distant prostate cancer. Murine immune systems are different from human, and all the findings in mice model could not be extrapolated to human prostate cancer. However, common mechanisms would exist also in human prostate cancer. Further analysis in mice model would give new insights into the mechanisms of the progression of prostate cancer enhanced by obesity and inflammation. Interventions to address systemic and/or local inflammation and a change in lifestyle may be therapeutic for prostate cancer.

Author Contributions: Conceptualization, K.F. and N.N.; methodology, K.F.; formal analysis, T.H.; investigation, M.M.; resources, K.F.; data curation, T.H.; writing—original draft preparation, K.F.; writing—review and editing, T.H. and M.M.; visualization, K.F. and T.H.; supervision, N.N.; project administration, K.F.; funding acquisition, K.F., T.H., M.U. and N.N.

References

1. Arroyo-Johnson, C.; Mincey, K.D. Obesity Epidemiology Worldwide. *Gastroenterol. Clin. North Am.* **2016**, *45*, 571–579. [CrossRef]
2. Wang, Y.C.; McPherson, K.; Marsh, T.; Gortmaker, S.L.; Brown, M. Health and economic burden of the projected obesity trends in the USA and the UK. *Lancet* **2011**, *378*, 815–825. [CrossRef]
3. Weitzman, S.A.; Gordon, L.I. Inflammation and cancer: Role of phagocyte-generated oxidants in carcinogenesis. *Blood* **1990**, *76*, 655–663. [PubMed]
4. Cohen, R.J.; Shannon, B.A.; McNeal, J.E.; Shannon, T.; Garrett, K.L. Propionibacterium acnes associated with inflammation in radical prostatectomy specimens: A possible link to cancer evolution? *J. Urol.* **2005**, *173*, 1969–1974. [CrossRef] [PubMed]
5. Fujita, K.; Ewing, C.M.; Sokoll, L.J.; Elliott, D.J.; Cunningham, M.; De Marzo, A.M.; Isaacs, W.B.; Pavlovich, C.P. Cytokine profiling of prostatic fluid from cancerous prostate glands identifies cytokines associated with extent of tumor and inflammation. *Prostate* **2008**, *68*, 872–882. [CrossRef] [PubMed]
6. Ellem, S.J.; Wang, H.; Poutanen, M.; Risbridger, G.P. Increased endogenous estrogen synthesis leads to the sequential induction of prostatic inflammation (prostatitis) and prostatic pre-malignancy. *Am. J. Pathol.* **2009**, *175*, 1187–1199. [CrossRef] [PubMed]
7. Sfanos, K.S.; Wilson, B.A.; De Marzo, A.M.; Isaacs, W.B. Acute inflammatory proteins constitute the organic matrix of prostatic corpora amylacea and calculi in men with prostate cancer. *Proc. Natl. Acad. Sci. USA* **2009**, *106*, 3443–3448. [CrossRef]
8. DuPre, N.C.; Flavin, R.; Sfanos, K.S.; Unger, R.H.; To, S.; Gazeeva, E.; Fiorentino, M.; De Marzo, A.M.; Rider, J.R.; Mucci, L.A.; et al. Corpora amylacea in prostatectomy tissue and associations with molecular, histological, and lifestyle factors. *Prostate* **2018**, *78*, 1172–1180. [CrossRef]
9. Nakai, Y.; Nonomura, N. Inflammation and prostate carcinogenesis. *Int. J. Urol.* **2013**, *20*, 150–160. [CrossRef]
10. De Marzo, A.M.; Platz, E.A.; Sutcliffe, S.; Xu, J.; Grönberg, H.; Drake, C.G.; Nakai, Y.; Isaacs, W.B.; Nelson, W.G. Inflammation in prostate carcinogenesis. *Nat. Rev. Cancer* **2007**, *7*, 256–269. [CrossRef]
11. Norrish, A.E.; Ferguson, L.R.; Knize, M.G.; Felton, J.S.; Sharpe, S.J.; Jackson, R.T. Heterocyclic amine content of cooked meat and risk of prostate cancer. *J. Natl. Cancer Inst.* **1999**, *91*, 2038–2044. [CrossRef] [PubMed]

12. Nakai, Y.; Nelson, W.G.; De Marzo, A.M. The dietary charred meat carcinogen 2-Amino-1-Methyl-6-Phenylimidazo (4,5- *b*) Pyridine acts as both a tumor initiator and promoter in the rat ventral prostate. *Cancer Res.* **2007**, *67*, 1378–1384. [CrossRef] [PubMed]
13. Calle, E.E.; Kaaks, R. Overweight, obesity and cancer: Epidemiological evidence and proposed mechanisms. *Nat. Rev. Cancer* **2004**, *4*, 579–591. [CrossRef] [PubMed]
14. Calle, E.E.; Rodriguez, C.; Walker-Thurmond, K.; Thun, M.J. Overweight, obesity, and mortality from cancer in a prospectively studied cohort of U.S. adults. *N. Engl. J. Med.* **2003**, *348*, 1625–1638. [CrossRef] [PubMed]
15. Cerhan, J.R.; Torner, J.C.; Lynch, C.F.; Rubenstein, L.M.; Lemke, J.H.; Cohen, M.B.; Lubaroff, D.M.; Wallace, R.B. Association of smoking, body mass, and physical activity with risk of prostate cancer in the Iowa 65+ Rural Health Study (United States). *Cancer Causes Control.* **1997**, *8*, 229–238. [CrossRef] [PubMed]
16. Putnam, S.D.; Cerhan, J.R.; Parker, A.S.; Bianchi, G.D.; Wallace, R.B.; Cantor, K.P.; Lynch, C.F. Lifestyle and anthropometric risk factors for prostate cancer in a cohort of Iowa men. *Ann. Epidemiol.* **2000**, *10*, 361–369. [CrossRef]
17. Discacciati, A.; Orsini, N.; Andersson, S.-O.; Andrén, O.; Johansson, J.-E.; Wolk, A. Body mass index in early and middle-late adulthood and risk of localised, advanced and fatal prostate cancer: A population-based prospective study. *Br. J. Cancer* **2011**, *105*, 1061–1068. [CrossRef] [PubMed]
18. Möller, E.; Wilson, K.M.; Batista, J.L.; Mucci, L.A.; Bälter, K.; Giovannucci, E. Body size across the life course and prostate cancer in the Health Professionals Follow-up Study. *Int. J. Cancer* **2016**, *138*, 853–865. [CrossRef]
19. Juul, A.; Bang, P.; Hertel, N.T.; Main, K.; Dalgaard, P.; Jørgensen, K.; Müller, J.; Hall, K.; Skakkebaek, N.E. Serum insulin-like growth factor-I in 1030 healthy children, adolescents, and adults: Relation to age, sex, stage of puberty, testicular size, and body mass index. *J. Clin. Endocrinol. Metab.* **1994**, *78*, 744–752.
20. Discacciati, A.; Orsini, N.; Wolk, A. Body mass index and incidence of localized and advanced prostate cancer-a dose-response meta-analysis of prospective studies. *Ann. Oncol.* **2012**, *23*, 1665–1671. [CrossRef]
21. Lima, N.; Cavaliere, H.; Knobel, M.; Halpern, A.; Medeiros-Neto, G. Decreased androgen levels in massively obese men may be associated with impaired function of the gonadostat. *Int. J. Obes. Relat. Metab. Disord.* **2000**, *24*, 1433–1437. [CrossRef] [PubMed]
22. Platz, E.A.; Leitzmann, M.F.; Rifai, N.; Kantoff, P.W.; Chen, Y.-C.; Stampfer, M.J.; Willett, W.C.; Giovannucci, E. Sex steroid hormones and the androgen receptor gene CAG repeat and subsequent risk of prostate cancer in the prostate-specific antigen era. *Cancer Epidemiol. Biomarkers Prev.* **2005**, *14*, 1262–1269. [CrossRef] [PubMed]
23. Ujike, T.; Uemura, M.; Kawashima, A.; Nagahara, A.; Fujita, K.; Miyagawa, Y.; Nonomura, N. A novel model to predict positive prostate biopsy based on serum androgen level. *Endocr. Relat. Cancer* **2018**, *25*, 59–67. [CrossRef] [PubMed]
24. Freedland, S.J.; Branche, B.L.; Howard, L.E.; Hamilton, R.J.; Aronson, W.J.; Terris, M.K.; Cooperberg, M.R.; Amling, C.L.; Kane, C.J. Obesity, Risk of Biochemical Recurrence, and PSADT after Radical Prostatectomy: Results from the SEARCH Database. *BJU Int.* **2018**. [CrossRef] [PubMed]
25. Vidal, A.C.; Howard, L.E.; Sun, S.X.; Cooperberg, M.R.; Kane, C.J.; Aronson, W.J.; Terris, M.K.; Amling, C.L.; Freedland, S.J. Obesity and prostate cancer-specific mortality after radical prostatectomy: Results from the Shared Equal Access Regional Cancer Hospital (SEARCH) database. *Prostate Cancer Prostatic Dis.* **2017**, *20*, 72–78. [CrossRef]
26. Hotamisligil, G.S.; Shargill, N.S.; Spiegelman, B.M. Adipose expression of tumor necrosis factor-alpha: Direct role in obesity-linked insulin resistance. *Science* **1993**, *259*, 87–91. [CrossRef] [PubMed]
27. Amar, J.; Chabo, C.; Waget, A.; Klopp, P.; Vachoux, C.; Bermúdez-Humarán, L.G.; Smirnova, N.; Bergé, M.; Sulpice, T.; Lahtinen, S.; et al. Intestinal mucosal adherence and translocation of commensal bacteria at the early onset of type 2 diabetes: Molecular mechanisms and probiotic treatment. *EMBO Mol. Med.* **2011**, *3*, 559–572. [CrossRef] [PubMed]
28. Shi, H.; Kokoeva, M.V.; Inouye, K.; Tzameli, I.; Yin, H.; Flier, J.S. TLR4 links innate immunity and fatty acid–induced insulin resistance. *J. Clin. Investig.* **2006**, *116*, 3015–3025. [CrossRef] [PubMed]
29. Saltiel, A.R.; Olefsky, J.M. Inflammatory mechanisms linking obesity and metabolic disease. *J. Clin Investig.* **2017**, *127*, 1–4. [CrossRef]
30. Gensollen, T.; Iyer, S.S.; Kasper, D.L.; Blumberg, R.S. How colonization by microbiota in early life shapes the immune system. *Science* **2016**, *352*, 539–544. [CrossRef]

31. Huang, M.; Narita, S.; Numakura, K.; Tsuruta, H.; Saito, M.; Inoue, T.; Horikawa, Y.; Tsuchiya, N.; Habuchi, T. A high-fat diet enhances proliferation of prostate cancer cells and activates MCP-1/CCR2 signaling. *Prostate* **2012**, *72*, 1779–1788. [CrossRef] [PubMed]
32. Huang, M.; Narita, S.; Inoue, T.; Tsuchiya, N.; Satoh, S.; Nanjo, H.; Sasaki, T.; Habuchi, T. Diet-induced macrophage inhibitory cytokine 1 promotes prostate cancer progression. *Endocr. Relat. Cancer* **2014**, *21*, 39–50. [CrossRef] [PubMed]
33. Nara, T.; Narita, S.; Mingguo, H.; Yoshioka, T.; Koizumi, A.; Numakura, K.; Tsuruta, H.; Maeno, A.; Saito, M.; Inoue, T.; et al. Altered miRNA expression in high-fat diet-induced prostate cancer progression. *Carcinogenesis* **2016**, *37*, 1129–1137. [CrossRef]
34. Hu, M.; Xu, H.; Zhu, W.; Bai, P.; Hu, J.; Yang, T.; Jiang, H.; Ding, Q. High-fat diet-induced adipokine and cytokine alterations promote the progression of prostate cancer in vivo and in vitro. *Oncol. Lett.* **2017**, *15*, 1607–1615. [CrossRef]
35. Zhang, T.; Tseng, C.; Zhang, Y.; Sirin, O.; Corn, P.G.; Li-Ning-Tapia, E.M.; Troncoso, P.; Davis, J.; Pettaway, C.; Ward, J.; et al. CXCL1 mediates obesity-associated adipose stromal cell trafficking and function in the tumour microenvironment. *Nat. Commun.* **2016**, *7*, 11674. [CrossRef]
36. Shalapour, S.; Karin, M. Immunity, inflammation, and cancer: An eternal fight between good and evil. *J. Clin. Investig.* **2015**, *125*, 3347–3355. [CrossRef]
37. Galdiero, M.R.; Bonavita, E.; Barajon, I.; Garlanda, C.; Mantovani, A.; Jaillon, S. Tumor associated macrophages and neutrophils in cancer. *Immunobiology* **2013**, *218*, 1402–1410. [CrossRef]
38. Hayashi, T.; Fujita, K.; Nojima, S.; Hayashi, Y.; Nakano, K.; Ishizuya, Y.; Wang, C.; Yamamoto, Y.; Kinouchi, T.; Matsuzaki, K.; et al. High-fat diet-induced inflammation accelerates prostate cancer growth via IL6 signaling. *Clin. Cancer Res.* **2018**, *24*, 1–11. [CrossRef] [PubMed]
39. Corrêa, L.H.; Corrêa, R.; Farinasso, C.M.; de Sant'Ana Dourado, L.P.; Magalhães, K.G. Adipocytes and macrophages interplay in the orchestration of tumor microenvironment: New implications in cancer progression. *Front. Immunol.* **2017**, *8*, 1–12. [CrossRef]
40. Coussens, L.M.; Zitvogel, L.; Palucka, A.K. Neutralizing tumor-promoting chronic inflammation: A magic bullet? *Science* **2013**, *339*, 286–291. [CrossRef] [PubMed]
41. Lumeng, C.N.; Bodzin, J.L.; Saltiel, A.R. Obesity induces a phenotypic switch in adipose tissue macrophage polarization. *J. Clin. Investig.* **2007**, *117*, 175–184. [CrossRef] [PubMed]
42. Mclaughlin, T.; Shen, L.; Engleman, E.; Mclaughlin, T.; Ackerman, S.E.; Shen, L.; Engleman, E. Role of innate and adaptive immunity in obesity-associated metabolic disease. *J. Clin. Investig.* **2017**, *127*, 5–13. [CrossRef] [PubMed]
43. Arendt, L.M.; McCready, J.; Keller, P.J.; Baker, D.D.; Naber, S.P.; Seewaldt, V.; Kuperwasser, C. Obesity promotes breast cancer by CCL2-mediated macrophage recruitment and angiogenesis. *Cancer Res.* **2013**, *73*, 6080–6093. [CrossRef] [PubMed]
44. Loberg, R.D.; Day, L.L.; Harwood, J.; Ying, C.; St. John, L.N.; Giles, R.; Neeley, C.K.; Pienta, K.J. CCL2 is a potent regulator of prostate cancer cell migration and proliferation. *Neoplasia* **2006**, *8*, 578–586. [CrossRef] [PubMed]
45. Fujita, K.; Ewing, C.M.; Getzenberg, R.H.; Parsons, J.K.; Isaacs, W.B.; Pavlovich, C.P. Monocyte chemotactic protein-1 (MCP-1/CCL2) is associated with prostatic growth dysregulation and benign prostatic hyperplasia. *Prostate* **2010**, *70*, 473–481. [CrossRef] [PubMed]
46. Zeigler-Johnson, C.; Morales, K.H.; Lal, P.; Feldman, M. The relationship between obesity, prostate tumor infiltrating lymphocytes and macrophages, and biochemical failure. *PLoS ONE* **2016**, *11*. [CrossRef] [PubMed]
47. Millrud, C.R.; Bergenfelz, C.; Leandersson, K. On the origin of myeloid derived suppressor cells. *Oncotarget* **2017**, *8*, 3649–3665. [CrossRef]
48. Gabrilovich, D.I.; Bronte, V.; Chen, S.-H.; Colombo, M.P.; Ochoa, A.; Ostrand-Rosenberg, S.; Schreiber, H. The terminology issue for myeloid-derived suppressor cells. *Cancer Res.* **2007**, *67*, 425. [CrossRef]
49. Bronte, V.; Brandau, S.; Chen, S.H.; Colombo, M.P.; Frey, A.B.; Greten, T.F.; Mandruzzato, S.; Murray, P.J.; Ochoa, A.; Ostrand-Rosenberg, S.; et al. Recommendations for myeloid-derived suppressor cell nomenclature and characterization standards. *Nat. Commun.* **2016**, *7*, 1–10. [CrossRef]

50. Calcinotto, A.; Spataro, C.; Zagato, E.; Di Mitri, D.; Gil, V.; Crespo, M.; De Bernardis, G.; Losa, M.; Mirenda, M.; Pasquini, E.; et al. IL-23 secreted by myeloid cells drives castration-resistant prostate cancer. *Nature* **2018**, *559*, 363–369. [CrossRef]
51. Wang, G.; Lu, X.; Dey, P.; Deng, P.; Wu, C.C.; Jiang, S.; Fang, Z.; Zhao, K.; Konaparthi, R.; Hua, S.; et al. Targeting YAP-dependent MDSC infiltration impairs tumor progression. *Cancer Discov.* **2016**, *6*, 80–95. [CrossRef] [PubMed]
52. Idorn, M.; Køllgaard, T.; Kongsted, P.; Sengeløv, L.; thor Straten, P. Correlation between frequencies of blood monocytic myeloid-derived suppressor cells, regulatory T cells and negative prognostic markers in patients with castration-resistant metastatic prostate cancer. *Cancer Immunol. Immunother.* **2014**, *63*, 1177–1187. [CrossRef] [PubMed]
53. Hossain, D.M.S.; Pal, S.K.; Moreira, D.; Duttagupta, P.; Zhang, Q.; Won, H.; Jones, J.; D'Apuzzo, M.; Forman, S.; Kortylewski, M. TLR9-Targeted STAT3 silencing abrogates immunosuppressive activity of myeloid-derived suppressor cells from prostate cancer patients. *Clin. Cancer Res.* **2015**, *21*, 3771–3782. [CrossRef] [PubMed]
54. Eruslanov, E.B.; Singhal, S.; Albelda, S.M. Mouse versus human neutrophils in cancer: A major knowledge gap. *Trends Cancer Res.* **2017**, *3*, 149–160. [CrossRef] [PubMed]
55. Shaul, M.E.; Fridlender, Z.G. Cancer related circulating and tumor-associated neutrophils-subtypes, sources and function. *FEBS J.* **2018**, *285*, 4316–4342. [CrossRef] [PubMed]
56. Quail, D.F.; Olson, O.C.; Bhardwaj, P.; Walsh, L.A.; Akkari, L.; Quick, M.L.; Chen, I.-C.; Wendel, N.; Ben-Chetrit, N.; Walker, J.; et al. Obesity alters the lung myeloid cell landscape to enhance breast cancer metastasis through IL5 and GM-CSF. *Nat. Cell. Biol.* **2017**, *19*, 974–987. [CrossRef] [PubMed]
57. Baek, A.E.; Yu, Y.-R.A.; He, S.; Wardell, S.E.; Chang, C.-Y.; Kwon, S.; Pillai, R.V.; McDowell, H.B.; Thompson, J.W.; Dubois, L.G.; et al. The cholesterol metabolite 27 hydroxycholesterol facilitates breast cancer metastasis through its actions on immune cells. *Nat. Commun.* **2017**, *8*, 864. [CrossRef]
58. Incio, J.; Liu, H.; Suboj, P.; Chin, S.M.; Chen, I.X.; Pinter, M.; Ng, M.R.; Nia, H.T.; Grahovac, J.; Kao, S.; et al. Obesity-induced inflammation and desmoplasia promote pancreatic cancer progression and resistance to chemotherapy. *Cancer Discov.* **2016**, *6*, 852–869. [CrossRef]
59. Patnaik, A.; Swanson, K.D.; Csizmadia, E.; Solanki, A.; Landon-Brace, N.; Gehring, M.P.; Helenius, K.; Olson, B.M.; Pyzer, A.R.; Wang, L.C.; et al. cabozantinib eradicates advanced murine prostate cancer by activating antitumor innate immunity. *Cancer Discov.* **2017**, *7*, 750–765. [CrossRef]
60. Özsoy, M.; Moschini, M.; Fajkovic, H.; Soria, F.; Seitz, C.; Klatte, T.; Gust, K.; Briganti, A.; Karakiewicz, P.I.; Roupret, M.; et al. Elevated preoperative neutrophil–lymphocyte ratio predicts upgrading at radical prostatectomy. *Prostate Cancer Prostatic Dis.* **2018**, *21*, 100–105. [CrossRef]
61. Jang, W.S.; Cho, K.S.; Kim, M.S.; Yoon, C.Y.; Kang, D.H.; Kang, Y.J.; Jeong, W.S.; Ham, W.S.; Choi, Y.D. The prognostic significance of postoperative neutrophil-to-lymphocyte ratio after radical prostatectomy for localized prostate cancer. *Oncotarget* **2017**, *8*, 11778–11787. [CrossRef] [PubMed]
62. Fan, L.; Wang, R.; Chi, C.; Cai, W.; Zhang, Y.; Qian, H.; Shao, X.; Wang, Y.; Xu, F.; Pan, J.; et al. Systemic immune-inflammation index predicts the combined clinical outcome after sequential therapy with abiraterone and docetaxel for metastatic castration-resistant prostate cancer patients. *Prostate* **2018**, *78*, 250–256. [CrossRef] [PubMed]
63. Boegemann, M.; Schlack, K.; Thomes, S.; Steinestel, J.; Rahbar, K.; Semjonow, A.; Schrader, A.; Aringer, M.; Krabbe, L.-M. The role of the neutrophil to lymphocyte ratio for survival outcomes in patients with metastatic castration-resistant prostate cancer treated with abiraterone. *Int. J. Mol. Sci.* **2017**, *18*, 380. [CrossRef] [PubMed]
64. Fujita, K.; Imamura, R.; Tanigawa, G.; Nakagawa, M.; Hayashi, T.; Kishimoto, N.; Hosomi, M.; Yamaguchi, S. Low serum neutrophil count predicts a positive prostate biopsy. *Prostate Cancer Prostatic Dis.* **2012**, *15*, 386–390. [CrossRef] [PubMed]
65. Fujita, K.; Hosomi, M.; Tanigawa, G.; Okumi, M.; Fushimi, H.; Yamaguchi, S. Prostatic inflammation detected in initial biopsy specimens and urinary pyuria are predictors of negative repeat prostate biopsy. *J. Urol.* **2011**, *185*, 1722–1727. [CrossRef] [PubMed]
66. Shaikh, S.R.; Haas, K.M.; Beck, M.A.; Teague, H. The effects of diet-induced obesity on B cell function. *Clin. Exp. Immunol.* **2015**, *179*, 90–99. [CrossRef] [PubMed]

67. Flynn, N.J.; Somasundaram, R.; Arnold, K.M.; Sims-Mourtada, J. The multifaceted roles of b cells in solid tumors: Emerging treatment opportunities. *Target. Oncol.* **2017**, *12*, 139–152. [CrossRef]
68. Ammirante, M.; Luo, J.-L.; Grivennikov, S.; Nedospasov, S.; Karin, M. B-cell-derived lymphotoxin promotes castration-resistant prostate cancer. *Nature* **2010**, *464*, 302–305. [CrossRef]
69. Woo, J.R.; Liss, M.A.; Muldong, M.T.; Palazzi, K.; Strasner, A.; Ammirante, M.; Varki, N.; Shabaik, A.; Howell, S.; Kane, C.J.; et al. Tumor infiltrating B-cells are increased in prostate cancer tissue. *J. Transl. Med.* **2014**, *12*, 30. [CrossRef]
70. Chen, Z.; Gu, J. Immunoglobulin G expression in carcinomas and cancer cell lines. *FASEB J.* **2007**, *21*, 2931–2938. [CrossRef]
71. Qiu, X.; Zhu, X.; Zhang, L.; Mao, Y.; Zhang, J.; Hao, P.; Li, G.; Lv, P.; Li, Z.; Sun, X.; et al. Human epithelial cancers secrete immunoglobulin g with unidentified specificity to promote growth and survival of tumor cells. *Cancer Res.* **2003**, *63*, 6488–6495. [PubMed]
72. Xu, Y.; Chen, B.; Zheng, S.; Wen, Y.; Xu, A.; Xu, K.; Li, B.; Liu, C. IgG silencing induces apoptosis and suppresses proliferation, migration and invasion in LNCaP prostate cancer cells. *Cell. Mol. Biol. Lett.* **2016**, *21*, 27. [CrossRef] [PubMed]
73. Afshar-Kharghan, V. The role of the complement system in cancer. *J. Clin. Investig.* **2017**, *127*, 780–789. [CrossRef] [PubMed]
74. Macor, P.; Capolla, S.; Tedesco, F. Complement as a biological tool to control tumor growth. *Front. Immunol.* **2018**, *9*, 2203. [CrossRef] [PubMed]
75. Corrales, L.; Ajona, D.; Rafail, S.; Lasarte, J.J.; Riezu-Boj, J.I.; Lambris, J.D.; Rouzaut, A.; Pajares, M.J.; Montuenga, L.M.; Pio, R. Anaphylatoxin C5a creates a favorable microenvironment for lung cancer progression. *J. Immunol.* **2012**, *189*, 4674–4683. [CrossRef] [PubMed]
76. Chen, J.; Yang, W.-J.; Sun, H.-J.; Yang, X.; Wu, Y.-Z. C5b-9 staining correlates with clinical and tumor stage in gastric adenocarcinoma. *Appl. Immunohistochem. Mol. Morphol.* **2016**, *24*, 470–475. [CrossRef] [PubMed]
77. Blando, J.; Moore, T.; Hursting, S.; Jiang, G.; Saha, A.; Beltran, L.; Shen, J.; Repass, J.; Strom, S.; DiGiovanni, J. Dietary energy balance modulates prostate cancer progression in Hi-Myc mice. *Cancer Prev. Res.* **2011**, *4*, 2002–2014. [CrossRef] [PubMed]
78. Xu, L.; Shen, M.; Chen, X.; Zhu, R.; Yang, D.-R.; Tsai, Y.; Keng, P.C.; Chen, Y.; Lee, S.O. Adipocytes affect castration-resistant prostate cancer cells to develop the resistance to cytotoxic action of NK cells with alterations of PD-L1/NKG2D ligand levels in tumor cells. *Prostate* **2018**, *78*, 353–364. [CrossRef]
79. Landim, B.C.; de Jesus, M.M.; Bosque, B.P.; Zanon, R.G.; da Silva, C.V.; Góes, R.M.; Ribeiro, D.L. Stimulating effect of palmitate and insulin on cell migration and proliferation in PNT1A and PC3 prostate cells: Counteracting role of metformin. *Prostate* **2018**, *78*, 731–742. [CrossRef]
80. Singh, V.; Yeoh, B.S.; Chassaing, B.; Xiao, X.; Saha, P.; Aguilera Olvera, R.; Lapek, J.D.; Zhang, L.; Wang, W.-B.; Hao, S.; et al. Dysregulated Microbial Fermentation of Soluble Fiber Induces Cholestatic Liver Cancer. *Cell* **2018**, *175*, 679–694. [CrossRef]
81. Dai, Z.; Zhang, J.; Wu, Q.; Chen, J.; Liu, J.; Wang, L.; Chen, C.; Xu, J.; Zhang, H.; Shi, C.; et al. The role of microbiota in the development of colorectal cancer. *Int. J. Cancer* **2019**. [CrossRef] [PubMed]

13

Consecutive Prostate Cancer Specimens Revealed Increased Aldo–Keto Reductase Family 1 Member C3 Expression with Progression to Castration-Resistant Prostate Cancer

Yu Miyazaki [1], Yuki Teramoto [2], Shinsuke Shibuya [2], Takayuki Goto [1], Kosuke Okasho [1], Kei Mizuno [1], Masayuki Uegaki [1], Takeshi Yoshikawa [1], Shusuke Akamatsu [1], Takashi Kobayashi [1], Osamu Ogawa [1] and Takahiro Inoue [1,*]

1 Department of Urology, Kyoto University Graduate School of Medicine, Kyoto 606-8507, Japan; urozaki@kuhp.kyoto-u.ac.jp (Y.M.); goto@kuhp.kyoto-u.ac.jp (T.G.); k_okasho@kuhp.kyoto-u.ac.jp (K.O.); km1207@kuhp.kyoto-u.ac.jp (K.M.); uegaki57@kuhp.kyoto-u.ac.jp (M.U.); urotake9@kuhp.kyoto-u.ac.jp (T.Y.); akamats@kuhp.kyoto-u.ac.jp (S.A.); selecao@kuhp.kyoto-u.ac.jp (T.K.); ogawao@kuhp.kyoto-u.ac.jp (O.O.)

2 Department of Diagnostic Pathology, Kyoto University Hospital, Kyoto 606-8507, Japan; tera1980@kuhp.kyoto-u.ac.jp (Y.T.); sshibuya@kuhp.kyoto-u.ac.jp (S.S.)

* Correspondence: takahi@kuhp.kyoto-u.ac.jp

Abstract: Aldo-keto reductase family 1 member C3 (AKR1C3) is an enzyme in the steroidogenesis pathway, especially in formation of testosterone and dihydrotestosterone, and is believed to have a key role in promoting prostate cancer (PCa) progression, particularly in castration-resistant prostate cancer (CRPC). This study aims to compare the expression level of AKR1C3 between benign prostatic epithelium and cancer cells, and among hormone-naïve prostate cancer (HNPC) and CRPC from the same patients, to understand the role of AKR1C3 in PCa progression. Correlation of AKR1C3 immunohistochemical expression between benign and cancerous epithelia in 134 patient specimens was analyzed. Additionally, correlation between AKR1C3 expression and prostate-specific antigen (PSA) progression-free survival (PFS) after radical prostatectomy was analyzed. Furthermore, we evaluated the consecutive prostate samples derived from 11 patients both in the hormone-naïve and castration-resistant states. AKR1C3 immunostaining of cancer epithelium was significantly stronger than that of the benign epithelia in patients with localized HNPC ($p < 0.0001$). High AKR1C3 expression was an independent factor of poor PSA PFS ($p = 0.032$). Moreover, AKR1C3 immunostaining was significantly stronger in CRPC tissues than in HNPC tissues in the same patients ($p = 0.0234$). Our findings demonstrate that AKR1C3 is crucial in PCa progression.

Keywords: AKR1C3; hormone-naïve prostate cancer; castration-resistant prostate cancer; immunohistochemistry; tissue microarray

---

1. Introduction

Prostate cancer (PCa) is one of the most commonly diagnosed malignancies and the second leading cause of cancer deaths in the United States [1]. In Japan, the mortality rate of PCa is the sixth among those of all male malignancies, although the estimated incidence rates have slightly declined, possibly due to reduced prostate-specific antigen (PSA) screening, the same as in the United States [2]. In the early 1940s, Huggins and Hodges demonstrated growth and survival of PCa to depend on androgens [3]. Therefore, androgen deprivation therapy (ADT) has been a standard clinical procedure for the control of PCa growth, with patients mostly responsive to ADT at the beginning of the therapy, which is also called hormone-naïve prostate cancer (HNPC); however, most of those patients relapse thereafter, developing castration-resistant prostate cancer (CRPC). ADT is required to treat advanced PCa and biochemical recurrent cases after curative radical treatment; however, despite the reduction of serum testosterone (T) to castration levels and an observed tumor response in 80%–90% of the patients, residual concentrations of intratumoral 5α-dihydrotestosterone (DHT) remain at 10%–40% of the pre-ADT levels in castration-resistant and hormone-naïve states [4,5]. The amount of residual androgens is substantial for triggering androgen receptor (AR) signaling, AR target gene expression, and cancer cell proliferation [6]. The de-novo pathway, which commences with cholesterol requiring multiple androgen synthetic enzymes, may be a result of the intratumoral androgen function; however, whether the complete repertoire of synthetic enzymes is required to generate androgens from cholesterol remains to be fully elucidated [7,8]. Circulating adrenal androgens, which are abundant in the form of dehydroepiandrosterone (DHEA) and with sulfated modification, dehydroepiandrosterone sulfate (DHEA-S), are other significant points of origin of the androgens. The acquired capacity of converting the adrenal androgens to more potent forms is a characteristic of CRPC. Therefore, abiraterone acetate, which is a potent CYP17A1 (17-hydroxylase/17, 20-lyase) inhibitor, is effective against CRPC, and has been recently approved for treating metastatic HNPC. Type 5 17α-hydroxysteroid dehydrogenase, in another name, aldo-keto reductase family 1 member C3 (AKR1C3) is a crucial enzyme in the steroidogenesis pathway. It catalyzes Δ^4-androsetene-3, 17-dione to T, DHT to 5α-androstane-3α, 17β-diol (3α-diol), and 3α-diol to androsterone; thus, it plays an important role in the formation of T and DHT [9]. Additionally, AKR1C3 can also reduce the weak estrogen, estrone, to the potent estrogen, 17β-estradiol, which might induce local estrogen production, contributing to PCa occurrence [10,11]. Estrogen and estrogen receptor (ER) (ER alpha: ERα and ER beta: ERβ) axes play an important role in both prostate carcinogenesis and progression to CRPC [12,13]. Although PCa co-expresses classical ERs, ERα and ERβ, and also non-genomic receptor, GRP30, complex interactions between ERs and AR, and those among various ligands in PCa cells need further investigation [12,13]. Moreover, AKR1C3 is known as prostaglandin (PG) F synthase that catalyzes the conversion of PGD_2 to 11-βPGF2α and PGF2α prostanoids, hence contributing to proliferation and radio-resistance in PCa cells [14,15]. All these issues imply that AKR1C3 could have a potential role in PCa biology. Several studies have demonstrated that AKR1C3 expression levels are elevated in PCa cells than in benign cells; moreover, it is highly expressed in the CRPC cell lines and human CRPC tissues rather than in the hormone-naïve ones [9,16–22]. Nevertheless, most reports have focused on the PCa tissues derived from different patients and compared the expression levels in normal/benign prostate hyperplasia tissues, localized cancer, and metastatic CRPC. This study aimed to compare the expression level of AKR1C3 between normal prostatic epithelium and cancer cells in the same patients. Moreover, we evaluated AKR1C3 expression and PSA progression-free survival after radical prostatectomy. We also investigated the expression level in hormone-naïve cancer and advanced CRPC in the same patients, to better understand the role of AKR1C3 in PCa progression.

2. Materials and Methods

2.1. Human Prostate Tissue Samples

All PCa patients included in this study were Japanese patients. Tissue-microarrays (TMAs) consisted of 175 radical prostatectomy (RP) specimens of patients with hormone-naïve PCa, who received RP between December 2004 and October 2012 at Kyoto University Hospital [23]. The TMA was developed with one core from each case. This study included 134 cases, which had both cancer and non-cancer tissues in each TMA core. We defined PSA failure as two consecutive measurements of PSA levels of ≥0.2 ng/mL, and the date of PSA failure as the time of the first measurement of PSA level of ≥0.2 ng/mL. When PSA levels after surgery did not decline below

0.2 ng/mL, we defined the date of PSA failure by the time of surgery. Consecutive prostate samples derived from 11 patients, both at hormone-naïve and castration-resistant states, were evaluated. HNPC specimens consisted of samples from needle biopsy or from transurethral resection of the prostate (TUR-P). CRPC specimens were collected from the TUR-P samples against urinary retention or gross hematuria, penectomy for pain control, and spinal laminectomy against spinal cord compression due to bone metastases.

2.2. Immunohistochemistry

Immunohistochemistry was carried out using anti-AKR1C3 antibody (at a dilution of 1:200; Abcam (ab49680, Abcam plc, Cambridge, UK)). As a positive control of anti-AKR1C3 antibody, we used surgical specimens of breast cancer (estrogen receptor (+) and progesterone receptor (+)) [24]. Immunohistostainings were performed using Ventana Discovery Ultra system (Roche diagnostics) as an automatic immunohistostaining apparatus. All specimens were evaluated by two urological pathologists (S.S. and T.Y.). AKR1C3 immunostaining of benign epithelium was relatively homogenous, whereas that of cancer epithelium was heterogeneous. Thus, the strongest immunostaining intensity of AKR1C3 was compared between benign epithelium and cancer cells at each spot. In order to compare AKR1C3 immunostaining in consecutive specimens of cancer cells in each individual, and to evaluate the correlation of AKR1C3 immunostaining of cancer cells with PSA progression-free survival (PFS) after RP, the pathologists evaluated each of the staining proportion and intensity, and the sum of these evaluation scores was considered as the total score (TS). "Proportion score (PS)" was evaluated according to the expression rate of stained tumor cells as: <1% (score 0), 1%–10% (score 1), 11%–33% (score 2), 34%–66% (score 3), and >67% (score 4). "Intensity score (IS)" was evaluated as none (score 0), weak (score 1), intermediate (score 2), and strong (score 3) in most immunostained cells. The Gleason score (GS) of hematoxylin and eosin staining was also evaluated by the urological pathologists.

2.3. Statistical Analysis

Results were analyzed with JMP13 software (SAS Institute Inc., Cary, NC, USA); *p*-values were calculated with the Kruskal–Wallis test, Pearson's chi-squared test, and Wilcoxon signed-rank test. PSA PFS was estimated by Kaplan–Meier analysis and groups compared with the log-rank test. Cox proportional hazard analysis was used to examine the factors associated with PSA PFS. A p value less than 0.05 was considered to be statistically significant.

3. Results

3.1. AKR1C3 Immunostaining of Cancer Epithelium Is Significantly Stronger than That of Benign Epithelia in Patients with Localized Hormone-Naïve Prostate Cancer

Clinical and pathological features are demonstrated and the results of statistical analysis of correlation between demographic features and AKR1C3 expression are presented as *p*-values in Table 1. Representative immunostaining of AKR1C3 is presented in Figure 1 (Figures S1–S4, Supplementary Materials). The distribution of AKR1C3 immunostaining scores are presented in Figure 2. AKR1C3 immunostaining was significantly stronger in cancer epithelia than in benign ones within the same spots ($p < 0.0001$). No correlation was observed between GS and AKR1C3 immunostaining in each spot (Table 1). These results suggested that AKR1C3 might play a role in PCa occurrence.

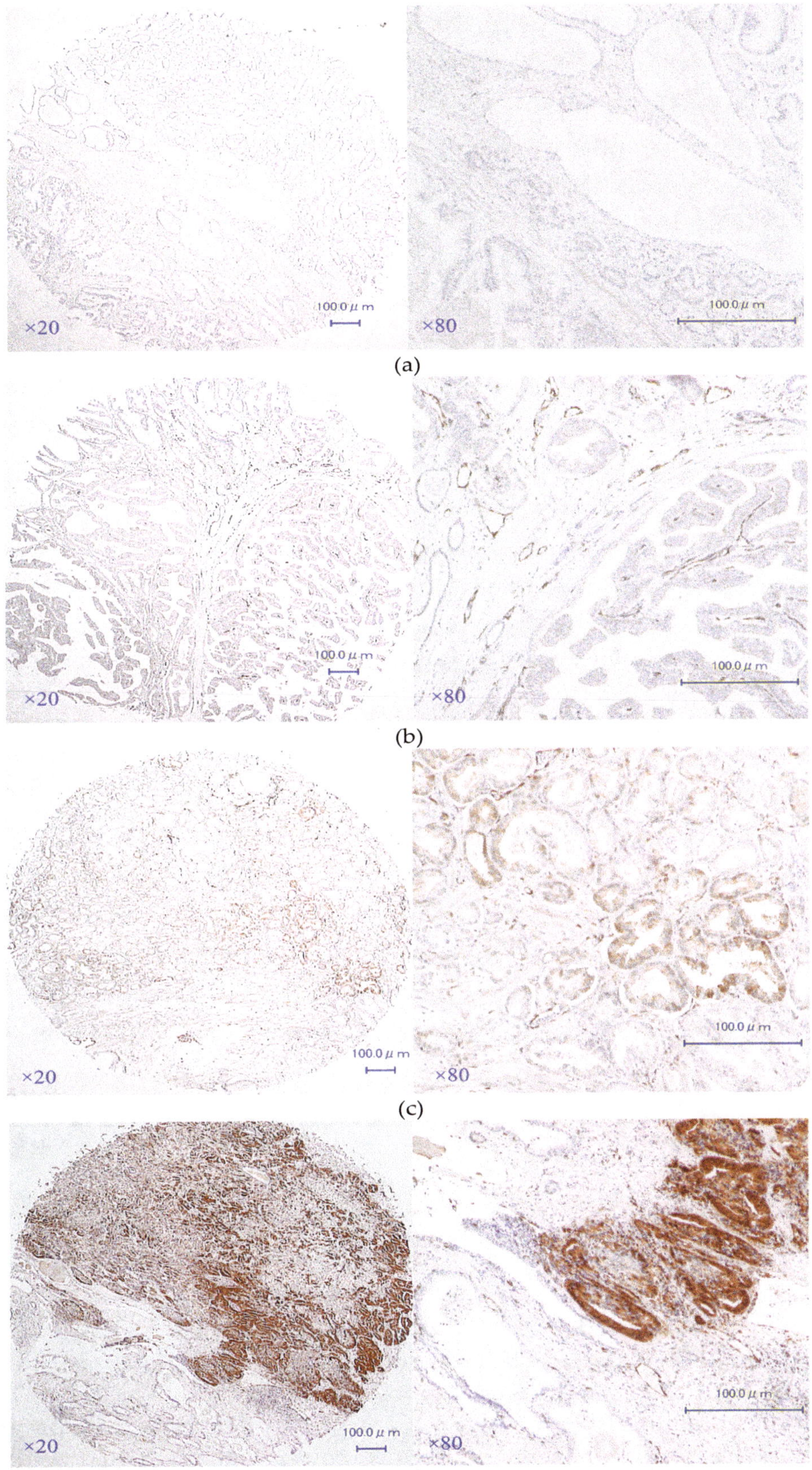

Figure 1. Representative immunostainings of aldo-keto reductase family 1 member C3 (AKR1C3): (**a**) score 0 (none staining), (**b**) score 1 (weak staining), (**c**) score 2 (intermediate staining), and (**d**) score 3 (strong staining).

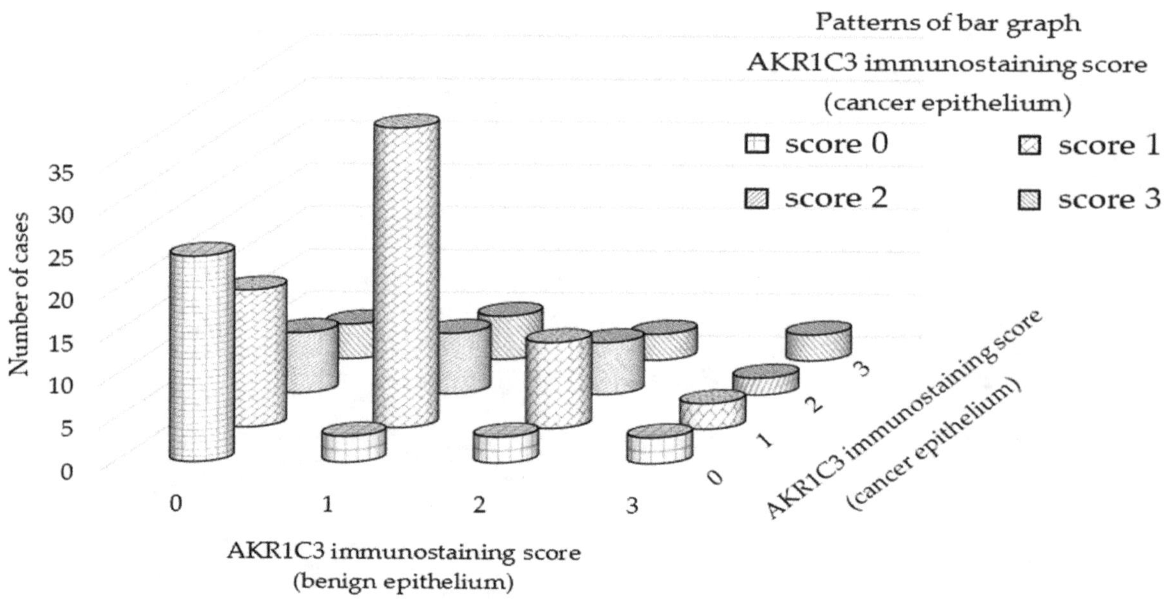

Figure 2. Difference of AKR1C3 immunostaining score between benign and cancer epithelia in the same individuals. AKR1C3 immunostaining was significantly stronger in the cancer epithelia than in the benign ones at the same spots ($p < 0.0001$, Pearson's chi-squared test).

Table 1. Clinicopathological features of tissue-microarray (TMA) specimens with both benign epithelium and cancer cells in the same spot.

$n = 134$		n	p Value
Age (mean ± SD)		65.6 ± 6.31	0.6087 †
PSA, ng/mL (median)		7.25 (IQR 5.40–9.88)	0.9429 †
Pathological T stage	T2a	7	N.A.
	T2b	1	
	T2c	77	
	T3a	37	
	T3b	12	
Grade group (pathological)	1	47	0.4119 ††
	2	38	
	3	37	
	4	9	
	5	3	

p-values indicate correlation of expression intensity with AKR1C3 total score; † Kruskal–Wallis test, †† Pearson's chi-squared test.

3.2. AKR1C3 Immunostaining of Cancer Cells Is Statistically Associated with PSA Progression-Free Survival after Radical Prostatectomy

The distribution of TS of AKR1C3 immunostaining in cancer cells from RP specimens is listed in Table 2. RP cases were dichotomized according to the median TS of the AKR1C3 immunostainings as: AKR1C3 TS ≤ 2 and AKR1C3 TS ≥ 3. AKR1C3 immunostainings and PSA PFS after RP were statistically correlated, and cases with a high AKR1C3 immunostaining TS had lower PSA PFS than those with a low AKR1C3 immunostaining TS ($p = 0.042$) (Figure 3). In order to evaluate prognostic factors for PSA PFS after RP, cox proportional hazards regression analysis was conducted with PSA at diagnosis, Gleason grade group, and AKR1C3 expression. AKR1C3 expression was an independent risk factor of PSA failure among our cohorts ($p = 0.032$, hazard ratio = 2.19) (Table 3). These results showed that AKR1C3 expression of cancer cells may be a prognostic marker of patients who received RP.

Table 2. AKR1C3) total score distribution of TMA specimens.

AKR1C3 Immunostaining	Score	n = 134
AKR1C3 total score	0	45
	2	37
	3	11
	4	19
	5	14
	6	4
	7	4
AKR1C3 total score (median)	2 (IQR 0–4)	

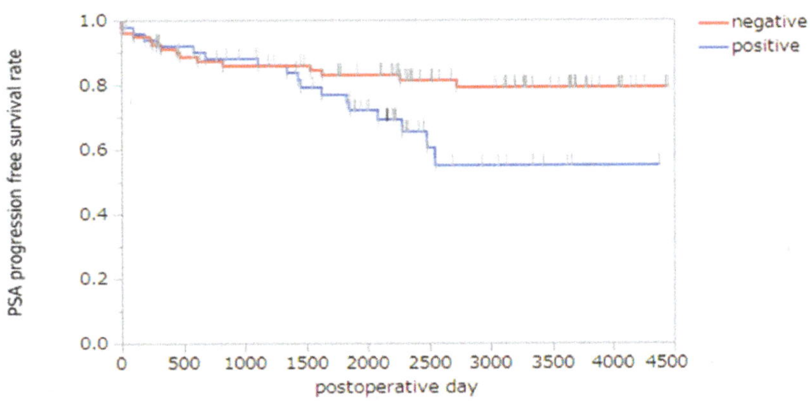

Number at risk										
Post-operative day	0	500	1000	1500	2000	2500	3000	3500	4000	4500
AKR1C3 negative (TS ≤ 2)	82	69	65	63	57	42	37	24	10	0
AKR1C3 positive (TS ≥ 3)	52	48	44	36	27	12	9	5	2	0

Figure 3. Kaplan–Meier survival curves revealed that the AKR1C3 positive group (TS ≥ 3) had a significantly lower PSA PFS rate than the negative group (TS ≤ 2) ($p = 0.042$, log-rank test).

Table 3. Cox proportional hazard regression analysis of prostate-specific antigen progression free survival (PSA PFS) and clinical and pathological variables.

Variables	PSA PFS Rate		
	HR	95% CI	p Value †††
PSA level before RP	1.12	1.05–1.18	0.0003
Grade group	1.66	1.16–2.36	0.0053
AKR1C3 (TS)	2.19	1.07–4.55	0.032

HR: hazard ratio, CI: confidence interval, and ††† Wald test.

3.3. AKR1C3 Immunostaining Is Significantly Stronger in CRPC Tissues rather than in Hormone-Naïve Ones in the Same Cases

We obtained HNPC and CRPC tissues from the same patients in 11 cases; clinical and pathological characteristics are presented in Table 4. CRPC tissues revealed significantly stronger AKR1C3 immunostaining than hormone-naïve tissues in the same cases ($p = 0.0234$, Wilcoxon signed-rank test; Table 4). Interestingly, the longitudinal specimens at hormone-naïve, hormone-sensitive, and castration-resistant states were evaluated in one patient. Immunostainings of AKR1C3 are presented in Figure 4; AKR1C3 was gradually up-regulated with disease progression. These results implied that up-regulation of AKR1C3 might be required for the progression to CRPC in some cases, and it could be a therapeutic target for this complicated disease.

Table 4. Clinicopathological features and results of AKR1C3 immunostaining total score in 11 cases, including both hormone-naïve and castration-resistant specimens.

Case	Age at Diagnosis	Clinical Stage at Diagnosis	Gleason Score at Diagnosis	Excised CRPC Organ	Age at Excision	PSA at Excision	Days from Diagnosis to Castration	Treatment until Excision of CRPC Specimens	AKR1C3 Immunostaining Total Score	
									at HNPC	at CRPC
Case 1	57	cT3bN1M0	3 + 3	Prostate	62	25	786	CAB + DOC	6	7
Case 2	73	cT3bN0M0	3 + 4	Prostate	84	5	2860	CAB	4	7
Case 3	60	cT3aN0M1c	4 + 4	Prostate	65	392	603	CAB + DOC	4	4
Case 4	79	cT3aN0M0	4 + 4	Penis	82	16.1	1051	CAB	2	4
Case 5	68	cT3aN0M1c	4 + 4	Thoracic vertebra	75	39.21	702	CAB + DOC + Abi	0	5
Case 6	70	cT3aN0M0	4 + 3	Prostate	78	408.2	2112	CAB	6	6
Case 7	76	cT4N1M1b	4 + 5	Prostate	85	14.07	2320	CAB + Enz	4	7
Case 8	78	cT4N1M1b	5 + 4	Thoracic vertebra	79	7.58	321	CAB	7	7
Case 9	69	cT4N0M0	4 + 5	Prostate	71	46.74	567	CAB	2	7
Case 10	80	cT3bN1M0	4 + 5	Prostate	82	6.21	544	CAB	1	1
Case 11	69	cT4N0M1b	4 + 5	Prostate	70	127	318	CAB + Enz	6	5

CAB: combined androgen blockade, DOC: docetaxel, Abi: abiraterone, and Enz: enzalutamide.

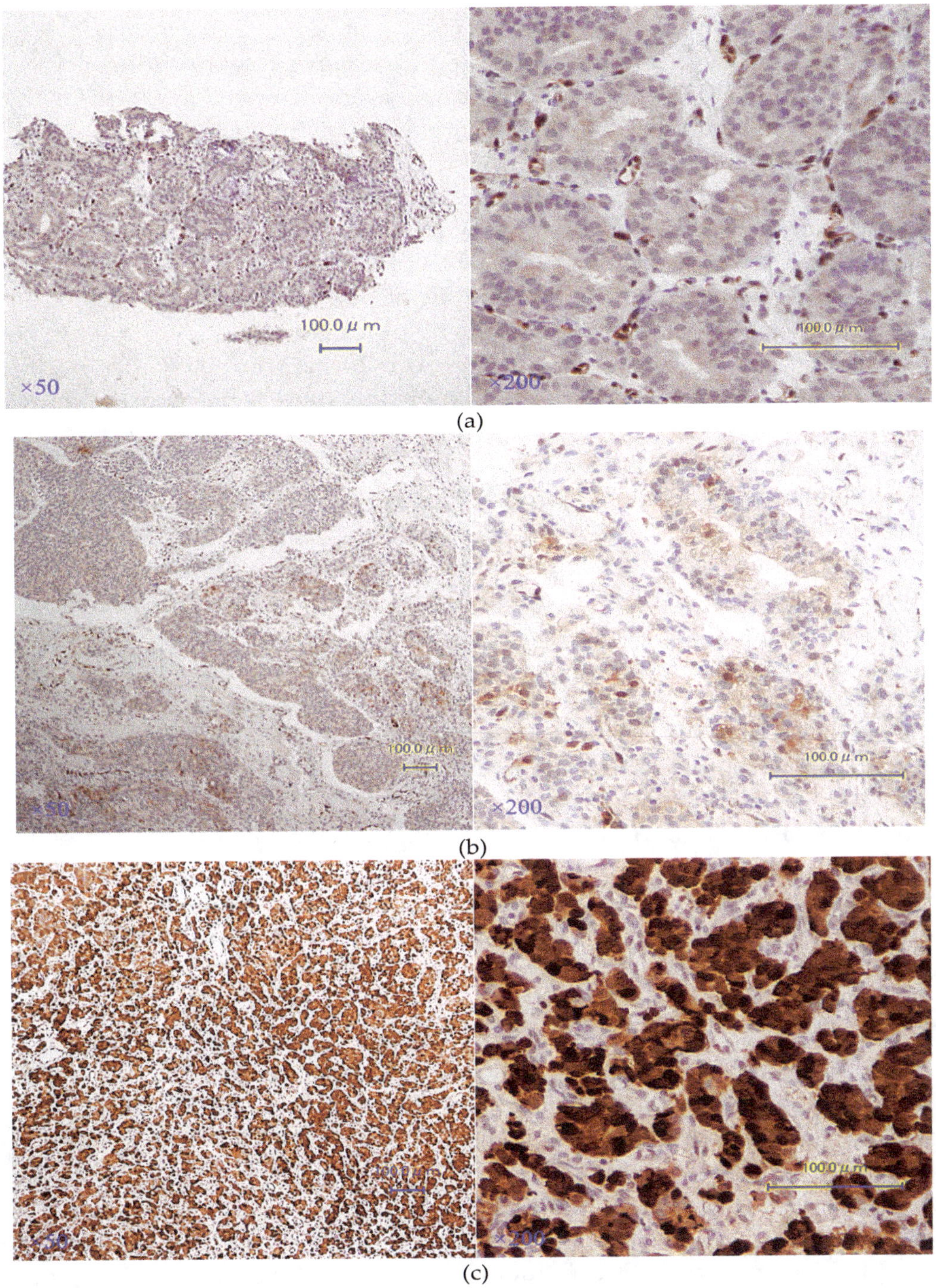

Figure 4. (**a**–**c**) AKR1C3 immunostaining of HNPC, Hormone-sensitive prostate cancer (HSPC), and CRPC specimens in the same case (case 9). CAB (leuprolide acetate + bicalutamide) was initiated for case 9 after diagnosis (Figure 3a): (**a**) HNPC (biopsy) specimens at diagnosis, PSA: 29.9 ng/mL (normal reference range 0-4.0 ng/mL), AKR1C3: Proportion score (PS), 1; Intensity score (IS), 1; Total score (TS), 2; (**b**) HSPC specimens on day 45 after commencing CAB, PSA: 2.89 ng/mL, AKR1C3: PS, 2; IS, 2; TS, 4; (**c**) CRPC specimens, PSA: 46.74 ng/mL, AKR1C3: PS, 4; IS, 3; TS 7. After CAB initiation, transurethral lithotomy (TUL) and TUR-P were performed due to repeated urinary retention resulting from bladder stone (Figure 4b). After 1.5 years of bicalutamide, 2 months of flutamide, and 3 months of ethinylestradiol, together with continuous luteinizing hormone-releasing hormone (LHRH) agonist administration, TUR-P was performed due to urinary retention caused by enlargement of the local tumor (Figure 4c). The immunostaining results suggested increased expression of AKR1C3 in PCa tissues with disease progression.

4. Discussion

Based on our immunohistochemical analysis of human prostate tissues, we confirmed that AKR1C3 might be crucial in PCa occurrence and progression. In particular, this is the first study to report that AKR1C3 immunostaining increases along the treatment course, that is, AKR1C3 expression elevates from the hormone-naïve status to the CRPC stage in the same patient. Moreover, our study is unique in showing that the PSA PFS rate of patients with high AKR1C3 expression in cancer cells derived from RP specimens was lower than that with low AKR1C3 expression. The majority of our CRPC cases received LHRH analog collectively with bicalutamide as a primary hormonal therapy. Nevertheless, we could not evaluate the correlation of pharmacological treatment with AKR1C3 immunostaining, since CRPC specimens were mostly obtained just during the time of transition to CRPC, before administration of docetaxel or androgen receptor-axis-targeted agents (ARATs), including enzalutamide and abiraterone. To the best of our knowledge, there is no report of AKR1C3 expression after ARAT treatment; however, AKR1C3 activation, both in vitro and in vivo, using PCa cell lines, has been shown as a factor of resistance against ARATs [25,26].

Lin et al. was the first to report a high-titer isoform-specific monoclonal antibody for AKR1C3 and demonstrated AKR1C3 expression in stromal cells, though only faintly in epithelial cells in normal prostate; however, in PCa cells, elevated expression was observed by immunostaining [23]. The same group also reported AKR1C3 to be positive in immunostaining, in 9 out of 11 PCa cases, and showed variation from strong to negative immunostaining within the same tumors, as in our study [26]. They also found no correlation in staining patterns between AR and AKR1C3 expression, consistent with our study (data not shown) [27]. Tian et al. examined the primary PCa biopsy specimens and showed that AKR1C3 expression by immunostaining gradually increases with an elevated GS in PCa epithelium [16]. In our cases, there was no correlation between GS and AKR1C3 immunostaining, which is incompatible with Tian's results. This might be because our cases underwent RP and most of them had GSs of less than 8, while in Tian's report, half of the PCa cases had GSs of 8 or higher. Stanbrough et al. revealed that AKR1C3 expression, as per immunohistochemistry, showed negative-to-heterogeneously weak staining in most primary PCa, but intermediate-to-strong AKR1C3 staining in CRPC specimens, which is in agreement with our results [22]. AKR1C3 expression analysis of cancer cells derived from RP showed immunostaining to be correlated with PSA PFS after RP. The result is reproducible even if we adopt IS of AKR1C3 as a representative of its expression (data not shown). In our knowledge, this is the first report of correlation of AKR1C3 expression with PSA PFS after RP. Additionally, AKR1C3 expression was an independent factor of poor PSA PFS when we analyzed the prognostic factors by multivariate analysis including the initial PSA level and Gleason grade group, which were previously considered to be significant predictors of recurrence-free survival after RP [28]. In future, we should analyze AKR1C3 expression and survival after RP in a much larger cohort to understand its role in clinical practice.

AKR1C3 is a multifunctional steroid-metabolizing enzyme that catalyzes androgen, estrogen, progesterone, and PG metabolism [29]. It reduces DHT to form 3α-diol, which is a neurosteroid that acts as a positive allosteric modulator of the gamma-aminobutyric acid type A receptor ($GABA_AR$) [30]. The 3α-diol stimulates AR-negative PCa cells through $GABA_AR$. Further, it up-regulates the epidermal growth factor (EGF) family members in AR-negative PCa cells and stimulates EGF receptor and Src. These results together suggest that AKR1C3 modulates intraprostatic neurosteroid that, in turn, activates AR-negative PCa progression. AKR1C3 is known to regulate its expression by ERG via direct binding to *AKR1C3* gene [31]. Furthermore, ERG and AKR1C3 expression in human metastatic PCa tissues was revealed to positively correlate with each other by immunohistochemistry. AKR1C3 is known to regulate the stability of ubiquitin ligase Siah2, and thus enhance the Siah2-dependent regulation of AR activity via non-catalytic function [21]. Wang et al. reported that AKR1C3 could drive epithelial–mesenchymal transition (EMT) by activating the ERK signaling pathways and up-regulating transcription factors such as ZEB1, TWIST1, and SLUG, thereby facilitating PCa metastasis [19]. Therefore, AKR1C3 might be crucial in PCa progression.

There are several limitations in our study. The samples analyzed were relatively small, in particular, in consecutive PCa with CRPC progression. We were unaware of the role of AKR1C3 overexpression, in particular, in progression to CRPC, since we did not analyze the tissue concentrations of T and other androgens. Moreover, we did not know how AKR1C3 overexpression correlated in response to new ARATs, namely enzalutamide and abiraterone. Only three cases received ARATs before CRPC tissue extraction. Case 5 received abiraterone 3 months before laminectomy, case 7 was administered enzalutamide 8 months before TUR-P, and case 11 acquired enzalutamide 10 days before TUR-P.

In conclusion, expression of multifunctional AKR1C3, which is known to be the most up-regulated steroidogenic enzyme in patients with CRPC, might increase with the occurrence and longitudinal progression of the tumor in certain cases [17,32]. Targeting AKR1C3 might overcome the complicated disease, CRPC, and also control resistance against ARATs. We plan to further reveal the functions of AKR1C3 in PCa progression and discover potent and specific drugs to inhibit AKR1C3 function.

Author Contributions: Y.M. and T.I. conceived the study. Y.M. and T.I. drafted the manuscript. Y.M. performed statistical analysis. Y.T. and S.S. performed pathological specimen evaluation and scoring. All the authors participated in patient enrolment and data collection, helped to draft the manuscript, and read and approved the final manuscript.

Acknowledgments: We thank all members of Department of Urology and Diagnostic Pathology Kyoto University Graduate School of Medicine, for their devoted support and constructive advice.

Abbreviations

AKR1C3	Aldo-Keto Reductase Family 1 Member C3
PCa	prostate cancer
CRPC	castration-resistant prostate cancer
HNPC	hormone-naïve prostate cancer
ADT	androgen deprivation therapy
T	testosterone
DHT	5α-dihydrotestosterone
AR	androgen receptor
DHEA	dehydroepiandrosterone
ER	estrogen receptor
PG	prostaglandin
TMA	tissue-microarray
RP	radical prostatectomy
TUR-P	transurethral resection of the prostate
PFS	progression-free survival
TS	total score
IS	intensity score
PS	proportion score
GS	Gleason score
CAB	combined androgen blockade
DOC	docetaxel
Abi	abiraterone
Enz	enzalutamide
LHRH	luteinizing hormone-releasing hormone
ARATs	androgen receptor-axis-targeted agents
$GABA_AR$	gamma-aminobutyric acid type A receptor

References

1. Siegel, R.L.; Miller, K.D.; Jemal, A. Cancer statistics, 2019. *CA Cancer J. Clin.* **2019**, *69*, 7–34. [CrossRef]
2. Cancer Statistics in Japan-2017. Available online: https://ganjoho.jp/en/professional/statistics/brochure/2017_en.html (accessed on 1 April 2019).
3. Huggins, C.; Hodges, C.V. Studies on Prostatic Cancer. I. The Effect of Castration, of Estrogen and of Androgen Injection on Serum Phosphatases in Metastatic Carcinoma of the Prostate. *Cancer Res.* **1941**, *1*, 293–297.
4. Page, S.T.; Lin, D.W.; Mostaghel, E.A.; Hess, D.L.; True, L.D.; Amory, J.K.; Nelson, P.S.; Matsumoto, A.M.; Bremner, W.J. Persistent intraprostatic androgen concentrations after medical castration in healthy men. *J. Clin. Endocrinol. MeTable* **2006**, *91*, 3850–3856. [CrossRef]
5. Nishiyama, T.; Hashimoto, Y.; Takahashi, K. The influence of androgen deprivation therapy on dihydrotestosterone levels in the prostatic tissue of patients with prostate cancer. *Clin. Cancer Res.* **2004**, *10*, 7121–7126. [CrossRef]
6. Dai, C.; Heemers, H.; Sharifi, N. Androgen Signaling in Prostate Cancer. *Cold Spring Harb. Perspect. Med.* **2017**, *7*. [CrossRef] [PubMed]
7. Locke, J.A.; Guns, E.S.; Lubik, A.A.; Adomat, H.H.; Hendy, S.C.; Wood, C.A.; Ettinger, S.L.; Gleave, M.E.; Nelson, C.C. Androgen levels increase by intratumoral de novo steroidogenesis during progression of castration-resistant prostate cancer. *Cancer Res.* **2008**, *68*, 6407–6415. [CrossRef]
8. Hofland, J.; van Weerden, W.M.; Dits, N.F.; Steenbergen, J.; van Leenders, G.J.; Jenster, G.; Schroder, F.H.; de Jong, F.H. Evidence of limited contributions for intratumoral steroidogenesis in prostate cancer. *Cancer Res.* **2010**, *70*, 1256–1264. [CrossRef]
9. Penning, T.M. Mechanisms of drug resistance that target the androgen axis in castration resistant prostate cancer (CRPC). *J. Steroid Biochem. Mol. Biol.* **2015**, *153*, 105–113. [CrossRef]
10. Byrns, M.C.; Jin, Y.; Penning, T.M. Inhibitors of type 5 17beta-hydroxysteroid dehydrogenase (AKR1C3): Overview and structural insights. *J. Steroid Biochem. Mol. Biol.* **2011**, *125*, 95–104. [CrossRef]
11. Ricke, W.A.; McPherson, S.J.; Bianco, J.J.; Cunha, G.R.; Wang, Y.; Risbridger, G.P. Prostatic hormonal carcinogenesis is mediated by in situ estrogen production and estrogen receptor alpha signaling. *FASEB J.* **2008**, *22*, 1512–1520. [CrossRef]
12. Di Zazzo, E.; Galasso, G.; Giovannelli, P.; Di Donato, M.; Di Santi, A.; Cernera, G.; Rossi, V.; Abbondanza, C.; Moncharmont, B.; Sinisi, A.A.; et al. Prostate cancer stem cells: The role of androgen and estrogen receptors. *Oncotarget* **2016**, *7*, 193–208. [CrossRef] [PubMed]
13. Di Zazzo, E.; Galasso, G.; Giovannelli, P.; Di Donato, M.; Castoria, G. Estrogens and Their Receptors in Prostate Cancer: Therapeutic Implications. *Front. Oncol.* **2018**, *8*, 2. [CrossRef]
14. Komoto, J.; Yamada, T.; Watanabe, K.; Takusagawa, F. Crystal structure of human prostaglandin F synthase (AKR1C3). *Biochemistry* **2004**, *43*, 2188–2198. [CrossRef]
15. Sun, S.Q.; Gu, X.; Gao, X.S.; Li, Y.; Yu, H.; Xiong, W.; Yu, H.; Wang, W.; Li, Y.; Teng, Y.; et al. Overexpression of AKR1C3 significantly enhances human prostate cancer cells resistance to radiation. *Oncotarget* **2016**, *7*, 48050–48058. [CrossRef]
16. Tian, Y.; Zhao, L.; Zhang, H.; Liu, X.; Zhao, L.; Zhao, X.; Li, Y.; Li, J. AKR1C3 overexpression may serve as a promising biomarker for prostate cancer progression. *Diagn. Pathol.* **2014**, *9*, 42. [CrossRef]
17. Hamid, A.R.; Pfeiffer, M.J.; Verhaegh, G.W.; Schaafsma, E.; Brandt, A.; Sweep, F.C.; Sedelaar, J.P.; Schalken, J.A. Aldo-keto reductase family 1 member C3 (AKR1C3) is a biomarker and therapeutic target for castration-resistant prostate cancer. *Mol. Med.* **2013**, *18*, 1449–1455. [CrossRef]
18. Jernberg, E.; Thysell, E.; Bovinder Ylitalo, E.; Rudolfsson, S.; Crnalic, S.; Widmark, A.; Bergh, A.; Wikstrom, P. Characterization of prostate cancer bone metastases according to expression levels of steroidogenic enzymes and androgen receptor splice variants. *PLoS ONE* **2013**, *8*, e77407. [CrossRef]
19. Wang, B.; Gu, Y.; Hui, K.; Huang, J.; Xu, S.; Wu, S.; Li, L.; Fan, J.; Wang, X.; Hsieh, J.T.; et al. AKR1C3, a crucial androgenic enzyme in prostate cancer, promotes epithelial-mesenchymal transition and metastasis through activating ERK signaling. *Urol. Oncol.* **2018**, *36*, 472.e11–472.e20. [CrossRef]
20. Barnard, M.; Quanson, J.L.; Mostaghel, E.; Pretorius, E.; Snoep, J.L.; Storbeck, K.H. 11-Oxygenated androgen precursors are the preferred substrates for aldo-keto reductase 1C3 (AKR1C3): Implications for castration resistant prostate cancer. *J. Steroid Biochem. Mol. Biol.* **2018**, *183*, 192–201. [CrossRef]

21. Fan, L.; Peng, G.; Hussain, A.; Fazli, L.; Guns, E.; Gleave, M.; Qi, J. The Steroidogenic Enzyme AKR1C3 Regulates Stability of the Ubiquitin Ligase Siah2 in Prostate Cancer Cells. *J. Biol. Chem.* **2015**, *290*, 20865–20879. [CrossRef]
22. Stanbrough, M.; Bubley, G.J.; Ross, K.; Golub, T.R.; Rubin, M.A.; Penning, T.M.; Febbo, P.G.; Balk, S.P. Increased expression of genes converting adrenal androgens to testosterone in androgen-independent prostate cancer. *Cancer Res.* **2006**, *66*, 2815–2825. [CrossRef]
23. Uegaki, M.; Kita, Y.; Shirakawa, R.; Teramoto, Y.; Kamiyama, Y.; Saito, R.; Yoshikawa, T.; Sakamoto, H.; Goto, T.; Akamatsu, S.; et al. Downregulation of RalGTPase-activating protein promotes invasion of prostatic epithelial cells and progression from intraepithelial neoplasia to cancer during prostate carcinogenesis. *Carcinogenesis* **2019**, in press.
24. Lin, H.K.; Steckelbroeck, S.; Fung, K.M.; Jones, A.N.; Penning, T.M. Characterization of a monoclonal antibody for human aldo-keto reductase AKR1C3 (type 2 3alpha-hydroxysteroid dehydrogenase/type 5 17beta-hydroxysteroid dehydrogenase); immunohistochemical detection in breast and prostate. *Steroids* **2004**, *69*, 795–801. [CrossRef]
25. Liu, C.; Lou, W.; Zhu, Y.; Yang, J.C.; Nadiminty, N.; Gaikwad, N.W.; Evans, C.P.; Gao, A.C. Intracrine Androgens and AKR1C3 Activation Confer Resistance to Enzalutamide in Prostate Cancer. *Cancer Res.* **2015**, *75*, 1413–1422. [CrossRef]
26. Liu, C.; Armstrong, C.M.; Lou, W.; Lombard, A.; Evans, C.P.; Gao, A.C. Inhibition of AKR1C3 Activation Overcomes Resistance to Abiraterone in Advanced Prostate Cancer. *Mol. Cancer Ther.* **2017**, *16*, 35–44. [CrossRef]
27. Fung, K.M.; Samara, E.N.; Wong, C.; Metwalli, A.; Krlin, R.; Bane, B.; Liu, C.Z.; Yang, J.T.; Pitha, J.V.; Culkin, D.J.; et al. Increased expression of type 2 3alpha-hydroxysteroid dehydrogenase/type 5 17beta-hydroxysteroid dehydrogenase (AKR1C3) and its relationship with androgen receptor in prostate carcinoma. *Endocr. Relat. Cancer* **2006**, *13*, 169–180. [CrossRef]
28. Nelson, C.P.; Dunn, R.L.; Wei, J.T.; Rubin, M.A.; Montie, J.E.; Sanda, M.G. Contemporary preoperative parameters predict cancer-free survival after radical prostatectomy: A tool to facilitate treatment decisions. *Urol. Oncol.* **2003**, *21*, 213–218. [CrossRef]
29. Penning, T.M. Aldo-Keto Reductase Regulation by the Nrf2 System: Implications for Stress Response, Chemotherapy Drug Resistance, and Carcinogenesis. *Chem. Res. Toxicol.* **2017**, *30*, 162–176. [CrossRef]
30. Xia, D.; Lai, D.V.; Wu, W.; Webb, Z.D.; Yang, Q.; Zhao, L.; Yu, Z.; Thorpe, J.E.; Disch, B.C.; Ihnat, M.A.; et al. Transition from androgenic to neurosteroidal action of 5alpha-androstane-3alpha, 17beta-diol through the type A gamma-aminobutyric acid receptor in prostate cancer progression. *J. Steroid Biochem. Mol. Biol.* **2018**, *178*, 89–98. [CrossRef]
31. Powell, K.; Semaan, L.; Conley-LaComb, M.K.; Asangani, I.; Wu, Y.M.; Ginsburg, K.B.; Williams, J.; Squire, J.A.; Maddipati, K.R.; Cher, M.L.; et al. ERG/AKR1C3/AR Constitutes a Feed-Forward Loop for AR Signaling in Prostate Cancer Cells. *Clin. Cancer Res.* **2015**, *21*, 2569–2579. [CrossRef]
32. Verma, K.; Gupta, N.; Zang, T.; Wangtrakluldee, P.; Srivastava, S.K.; Penning, T.M.; Trippier, P.C. AKR1C3 Inhibitor KV-37 Exhibits Antineoplastic Effects and Potentiates Enzalutamide in Combination Therapy in Prostate Adenocarcinoma Cells. *Mol. Cancer Ther.* **2018**, *17*, 1833–1845. [CrossRef]

14

KIFC1 Inhibitor CW069 Induces Apoptosis and Reverses Resistance to Docetaxel in Prostate Cancer

Yohei Sekino [1,2], Naohide Oue [1,*], Yuki Koike [1], Yoshinori Shigematsu [1,2], Naoya Sakamoto [1], Kazuhiro Sentani [1], Jun Teishima [2], Masaki Shiota [3], Akio Matsubara [2] and Wataru Yasui [1]

1 Department of Molecular Pathology, Graduate School of Biomedical and Health Sciences, Hiroshima University, Hiroshima 734-8551, Japan; akikosekino@gmail.com (Y.S.); b145456@hiroshima-u.ac.jp (Y.K.); yoshis413@yahoo.co.jp (Y.S.); nasakamoto@hiroshima-u.ac.jp (N.S.); kzsentani@hiroshima-u.ac.jp (K.S.); wyasui@hiroshima-u.ac.jp (W.Y.)

2 Department of Urology, Graduate School of Biomedical and Health Sciences, Hiroshima University, Hiroshima 734-8551, Japan; teishima@hiroshima-u.ac.jp (J.T.); matsua@hiroshima-u.ac.jp (A.M.)

3 Department of Urology, Graduate School of Medical Sciences, Kyushu University, Fukuoka 812-8582, Japan; shiota@uro.med.kyushu-u.ac.jp

* Correspondence: naoue@hiroshima-u.ac.jp

Abstract: Kinesin family member C1 (KIFC1) is a minus end-directed motor protein that plays an essential role in centrosome clustering. Previously, we reported that KIFC1 is involved in cancer progression in prostate cancer (PCa). We designed this study to assess the involvement of KIFC1 in docetaxel (DTX) resistance in PCa and examined the effect of KIFC1 on DTX resistance. We also analyzed the possible role of a KIFC1 inhibitor (CW069) in PCa. We used DTX-resistant PCa cell lines in DU145 and C4-2 cells to analyze the effect of KIFC1 on DTX resistance in PCa. Western blotting showed that KIFC1 expression was higher in the DTX-resistant cell lines than in the parental cell lines. Downregulation of KIFC1 re-sensitized the DTX-resistant cell lines to DTX treatment. CW069 treatment suppressed cell viability in both parental and DTX-resistant cell lines. DTX alone had little effect on cell viability in the DTX-resistant cells. However, the combination of DTX and CW069 significantly reduced cell viability in the DTX-resistant cells, indicating that CW069 re-sensitized the DTX-resistant cell lines to DTX treatment. These results suggest that a combination of CW069 and DTX could be a potential strategy to overcome DTX resistance.

Keywords: KIFC1; prostate cancer; docetaxel resistance; apoptosis; CW069

---

1. Introduction

Prostate cancer (PCa) is the most prevalent cancer among men and the second leading cause of cancer-related death in developed countries [1]. Androgen deprivation therapy is initially effective for advanced PCa. However, most of these patients eventually progress to castration-resistant PCa (CRPC), which is a life-threatening disease [2,3]. Docetaxel (DTX) is the standard chemotherapy for CRPC [4]. However, nearly all patients who are treated with DTX become refractory. Therefore, clarifying new molecular mechanisms underlying DTX resistance is necessary to overcome DTX resistance in CRPC.

Increased centrosome number, called centrosome amplification (CA), is a hallmark of human cancer [5]. Recent reports have shown that CA correlates with aneuploidy and malignant behavior in some human cancers including uterine cervical cancer, breast cancer, and PCa [6,7]. Although aneuploidy might cause multipolar spindles and lead to apoptosis, cancer cells overcome these lethal effects of CA by using centrosome clustering. Centrosome clustering, defined as the reshaping of transient multipolar spindles into pseudo-bipolar structures, is a well-studied mechanism that allows cancer cells to avoid apoptosis [8,9]. Kinesin family member C1 (KIFC1) is a minus end-directed motor protein that plays an essential role in centrosome clustering [10–12]. Several reports show that KIFC1 is upregulated and is involved in cancer progression in some cancers [13–15]. In addition, the overexpression of KIFC1 suppresses DTX-mediated apoptosis in breast cancer cells [16]. Previously, we showed that KIFC1 was associated with a poor prognosis after radical prostatectomy or after DTX treatment in PCa [17]. Additionally, knockdown of KIFC1 improved DTX sensitivity in LNCaP cells and DU145 cells. However, the role of KIFC1 in DTX resistance in PCa is not well known. In this study, we used DTX-resistant PCa cells from C4-2 cells and DU145 cells to analyze the involvement of KIFC1 in DTX resistance. We examined the expression and functional role of KIFC1 and analyzed the effect of KIFC1 knockdown on DTX resistance in DTX-resistant PCa cell lines. We also investigated the effect of the KIFC1 inhibitor CW069 in PCa cell lines.

2. Materials and Methods

2.1. Cell Lines

Two PCa cell lines (DU145 and C4-2) and DTX-resistant DU145 cells (DU145-DR) and DTX-resistant C4-2 cells (C4-2-DR) were kindly provided by Dr. Masaki Shiota (Kyushu University, Fukuoka, Japan). The DU145 cell lines were maintained in MEM (Nissui Pharmaceutical Co. Ltd., Tokyo, Japan), and the C4-2 cells were maintained in RPMI 1640 (Nissui Pharmaceutical Co. Ltd.) containing 10% fetal bovine serum (BioWhittaker, Walkersville, MD, USA), 2 mM L-glutamine, 50 U/mL penicillin, and 50 g/mL streptomycin in a humidified atmosphere of 5% CO_2 at 37 °C. DTX cell lines were cultured under DTX at a dose of 2 ng/mL for DU145-DR and 5 ng/mL for C4-2-DR.

2.2. DTX and CW069 Treatment

DTX was obtained from Sanofi-Aventis and handled according to the manufacturer's recommendations [18]. CW069 was obtained from Funakoshi (Tokyo, Japan). Twenty-four hours after transfection of siRNAs for KIFC1 or negative control, these cells were exposed to DTX for 48 hr. Cell viability was measured by an MTT assay. An MTT assay was performed 48 h after DTX or CW069 treatment. Drug sensitivity curves and IC50 values were calculated using GraphPad Prism 4.0 software (GraphPad Software) [17].

2.3. Western Blotting Analysis

For Western blotting analysis, cells were lysed as described previously [19]. Primary antibody, KIFC1 (H00003833-M01, Abnova, Taipei, Taiwan), Bcl-2 (sc-7382, Santa Cruz Biotechnology, Santa Cruz, CA, USA), Bax (sc-7480, Santa Cruz Biotechnology, Santa Cruz, CA, USA), cleaved PARP (c-PARP) (#5625, Cell Signaling Technology, Inc., Danvers, MA, USA), cleaved caspase-3 (c-caspase-3) (#9661, Cell Signaling Technology, Inc., Danvers, MA, USA) were used. β-Actin (Sigma-Aldrich, St. Louis, MO, USA) was used as a loading control.

2.4. qRT-PCR Analysis

Total RNA was isolated from frozen cancer cell lines using Isogen (Nippon Gene, Tokyo, Japan), and 1 μg of total RNA was converted to cDNA with a first-strand cDNA synthesis kit (Amersham Biosciences Corp., Piscataway, NJ, USA). The qPCR was performed with a SYBR Select Master Mix (Applied Biosystems, Austin, TX, USA) as described previously [20]. ACTB-specific PCR products, which were amplified from the same RNA samples, served as internal controls. KIFC1 primer sequence: forward primer GACGCCCTGCTTCATCTG; reverse primer CCAGGTCCACAAGACTGAGG.

2.5. RNA Interference

Silencer® Select (Ambion, Austin, TX, USA) against KIFC1 was used for RNA interference. Two independent oligonucleotides and negative control small interfering RNA (siRNA) (Invitrogen, Carlsbad, CA, USA) were used. Transfection was performed using Lipofectamine RNAiMAX (Invitrogen) according to the manufacturer's instructions. Cells were used 48 h after transfection in each of the experiments and assays [17].

2.6. Cell Death ELISA

Cells were seeded in 12-well plates (1×10^5 cells) and treated as indicated. Mono- and oligonucleosomes in the cytoplasmic fraction were measured by a cell death detection ELISA kit (Roche, Basel, Switzerland) according to the manufacturer's instructions. Absorbance was determined at 405 nm.

2.7. Statistical Analysis

Statistical differences were evaluated using a two-tailed Student *t*-test or Mann-Whitney U-test. A p-value of <0.05 was considered statistically significant. Statistical analyses were conducted primarily using GraphPad Prism software (GraphPad Software Inc., La Jolla, CA, USA). The combination index (CI) was calculated by the Chou–Talalay method. A combination index (CI) < 1 indicates synergism, CI = 1 an additive effect, and CI > 1 an antagonistic effect [21].

3. Results

3.1. Characterization of DTX-Resistant PCa Cell Lines

We used C4-2-DR and DU145-DR cells to analyze the involvement of KIFC1 in DTX resistance. MTT assays were performed to measure cell viability under various concentrations of DTX in the C4-2-DR and DU145-DR cells. The IC50 values of the C4-2-DR and DU145-DR cells were significantly higher than those of the parental DU145 and C4-2 cells, which was consistent with previous results (Figure 1A) [22,23]. We compared the expression of c-PARP, which was used as a marker of apoptosis in the parental and DTX-resistant cell lines. Western blotting showed that the expression of c-PARP and c-caspase-3 was induced by DTX treatment in the parental DU145 and C4-2 cells. On the contrary, the expression of c-PARP was not changed by DTX treatment in DU145-DR and C4-2-DR cells (Figure 1B). These results suggest that the DU145-DR and C4-2-DR cells were resistant to DTX treatment.

3.2. KIFC1 is Overexpressed in DTX-Resistant Cell Lines

To verify whether KIFC1 is involved in DTX resistance, we investigated the expression of KIFC1 in DU145-DR and C4-2-DR cells. Western blotting and qRT-PCR showed that KIFC1 was overexpressed in DU145-DR and C4-2-DR cells compared with the parental DU145 and C4-2 cells at both mRNA and protein levels (Figure 2A,B).

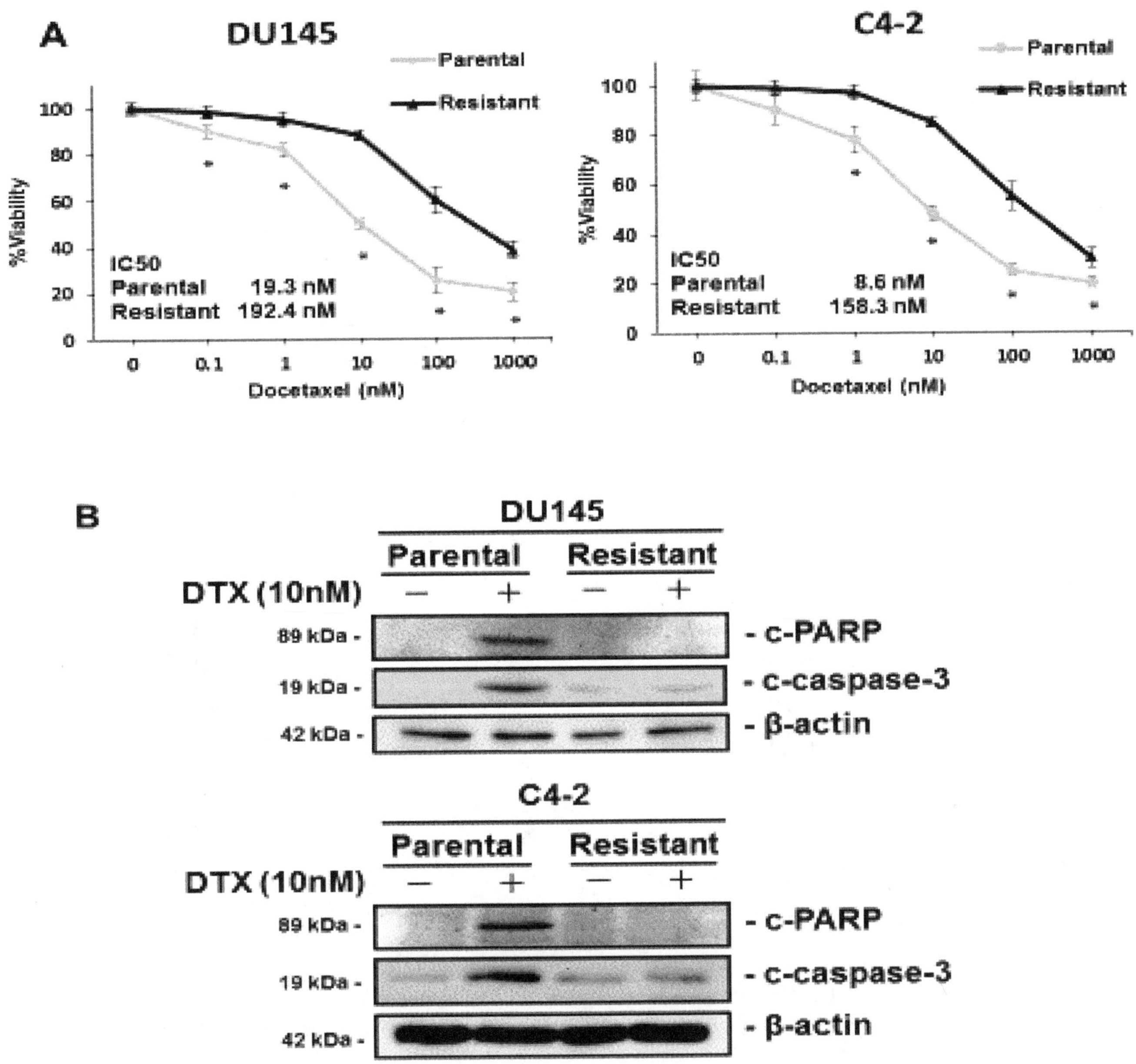

Figure 1. Characterization of docetaxel (DTX)-resistant prostate cancer cell lines. (**A**) The dose-dependent effects of DTX on the viability of parental and DTX-resistant cell lines in DU145 and C4-2 cells. The results are expressed as the mean and S.D. of triplicate measurements. * $p < 0.01$. (**B**) Western blotting of c-PARP and c-caspase-3 in parental and DTX-resistant cell lines in DU145 and C4-2 cells in the presence of DTX (10 nM) or vehicle (ethanol). β-actin was used as a loading control. c-PARP: cleaved PARP; c-caspase-3: cleaved caspase-3.

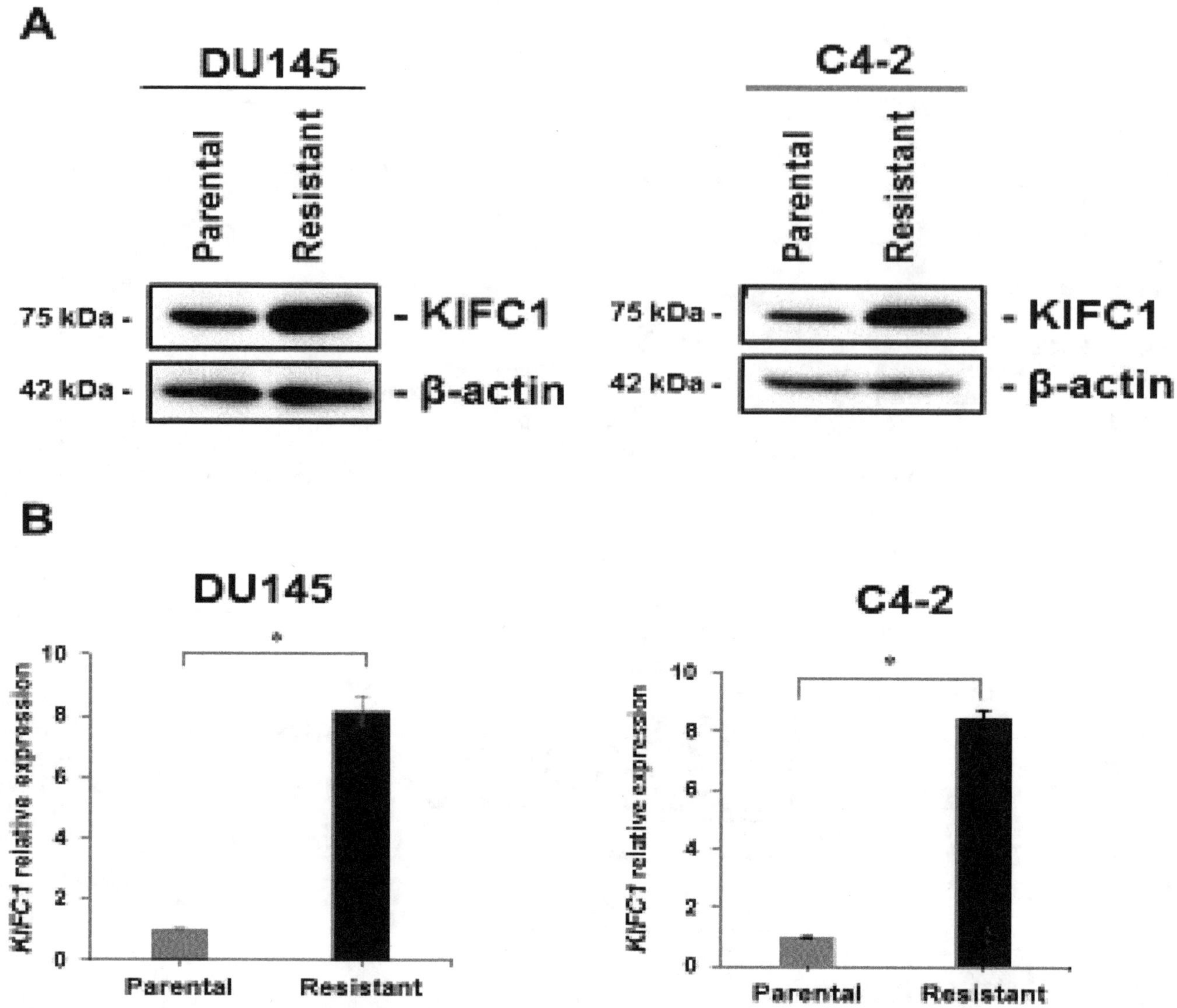

Figure 2. KIFC1 is overexpressed in docetaxel (DTX)-resistant cell lines and in a castration-resistant prostate cancer (CRPC) patient. (**A**) Western blotting of KIFC1 in parental and DTX-resistant cell lines. β-actin was used as a loading control. (**B**) qRT-PCR of KIFC1 in parental and DTX-resistant cell lines. The results are expressed as the mean and S.D. of triplicate measurements. * $p < 0.01$.

3.3. *Inhibition of KIFC1 Induces Apoptosis Pathway and Reverses DTX Resistance In Vitro*

Several studies have shown that KIFC1 is associated with an apoptosis pathway [24,25]. We used RNA interference targeting KIFC1 in DU145-DR and C4-2-DR cells and confirmed the efficiency of KIFC1 knockdown by Western blotting (Figure 3A). Western blotting showed that inhibition of KIFC1 enhanced the expression of Bax2, c-PARP, and c-caspase-3 and reduced the expression of Bcl-2 in DU145-DR and C4-2-DR cells (Figure 3A). Given that KIFC1 was overexpressed in the DTX-resistant cell lines and is involved in the apoptosis pathway, we next analyzed whether the knockdown of KIFC1 improves DTX sensitivity in DU145-DR and C4-2-DR cells. We measured cell viability in DU145-DR and C4-2-DR cells with knockdown of KIFC1 under various concentrations of DTX. We found that downregulation of KIFC1 re-sensitized DU145-DR and C4-2-DR cells to DTX treatment (Figure 3B).

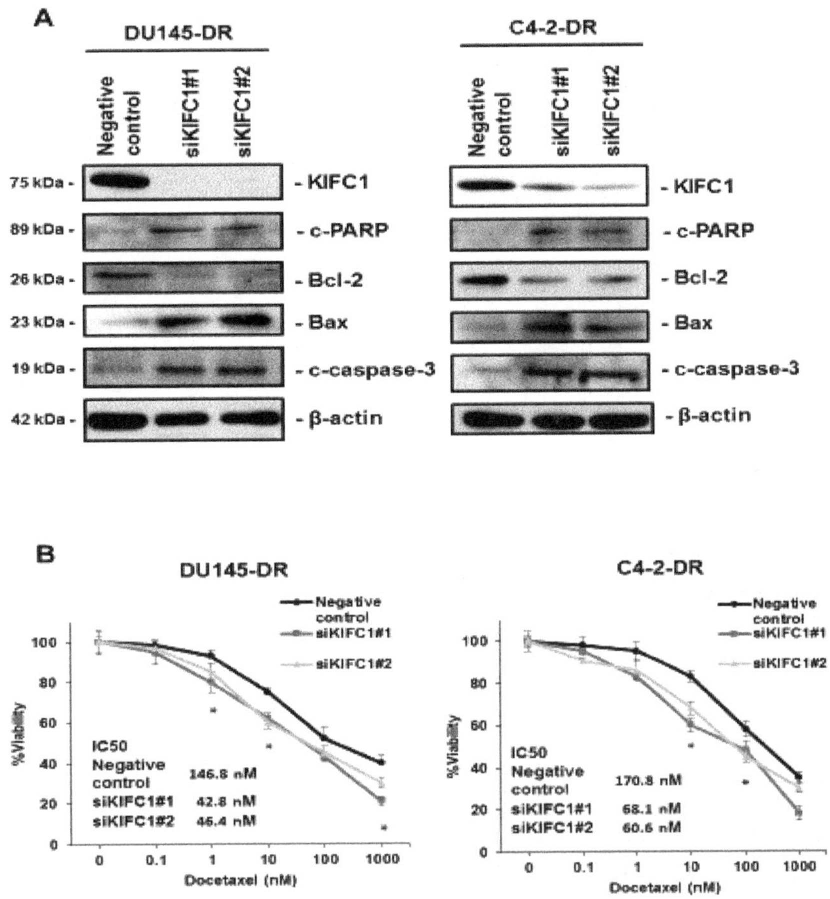

Figure 3. Inhibition of KIFC1 induces an apoptosis pathway and reverses docetaxel (DTX) resistance in vitro. (**A**) Western blotting of KIFC1, c-PARP, Bcl-2, Bax, and c-caspase-3 in DU145-DR and C4-2-DR cells transfected with a negative control or two different siRNAs for KIFC1. β-actin was used as a loading control. c-PARP: cleaved PARP; c-caspase-3: cleaved caspase-3 (**B**) The dose-dependent effects of DTX on the viability of DU145-DR and C4-2-DR cells transfected with negative control or two different siRNAs for KIFC1. The results are expressed as the mean and S.D. of triplicate measurements. * $p < 0.01$.

3.4. Effect of KIFC1 Inhibitor CW069 on Cell Viability

A recent study reported that CW069 is a novel and allosteric inhibitor of KIFC1 [26]. To clarify the effect of CW069 on cell viability in PCa, we measured cell viability under various concentrations of CW069 in both parental and DTX-resistant cell lines. CW069 treatment suppressed cell viability in both the parental and DTX-resistant cell lines (Figure 4A). The IC50 values of the DTX-resistant cell lines treated with CW069 were significantly lower than those of the parental cell lines, suggesting that the effect of CW069 on cell viability may depend on the expression of KIFC1. Next, to test whether CW069 could selectively suppress cell viability in cancer cells, we investigated the effect of CW069 in RWPE-1 cells, which is a normal prostate epithelial cell line [27]. Western blotting demonstrated that the expression of KIFC1 was not detected in RWPE-1 cells (Figure 4B). As we expected, CW069 treatment had little effect on cell viability in RWPE-1 cells compared with the DU145 and C4-2 cells (Figure 4C). Furthermore, we performed a cell death ELISA assay to analyze the ability of CW069 to induce apoptotic cell death in RWPE-1, DU145, and C4-2 cells. CW069 treatment had little effect on apoptotic cell death in the RWPE-1 cells but had a significant effect on the DU145 and C4-2 cells (Figure 4D).

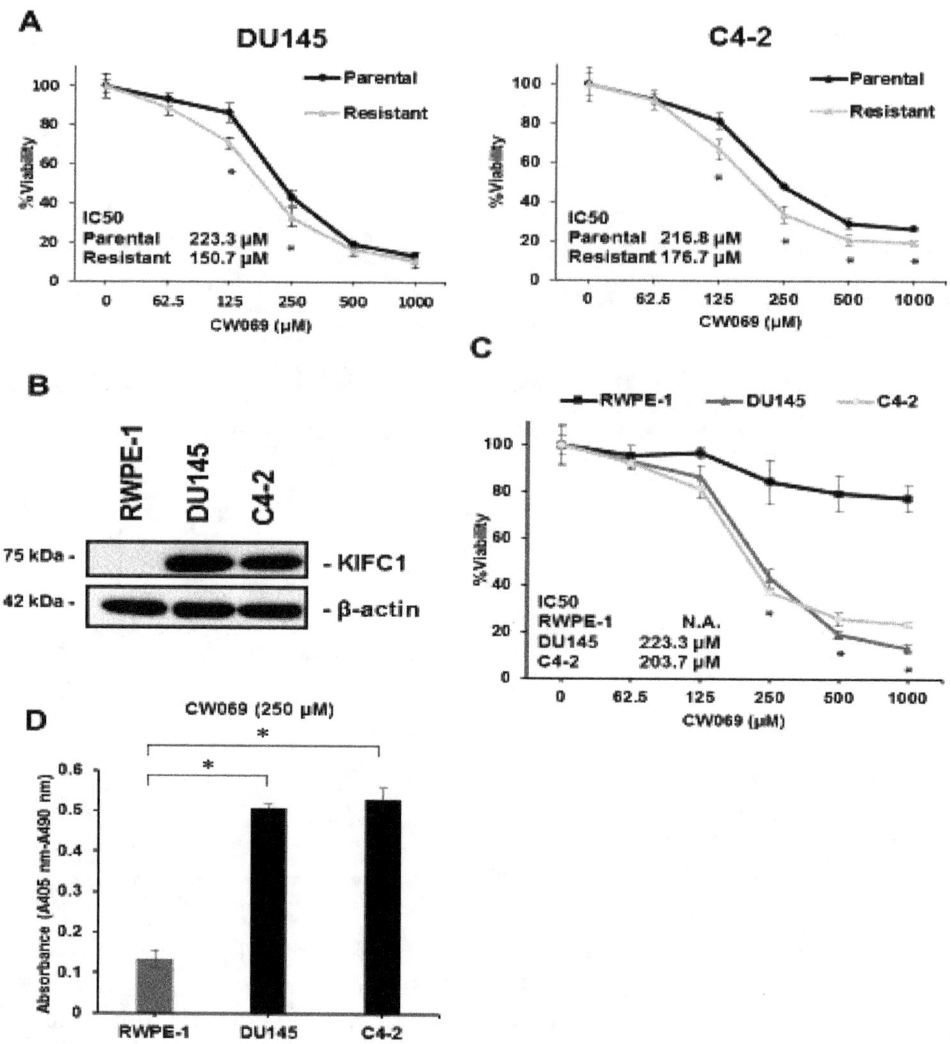

Figure 4. The effect of the KIFC1 inhibitor CW069 on cell viability (**A**) The dose-dependent effects of CW069 on cell viability in parental and docetaxel-resistant cell lines in DU145 and C4-2 cells. The results are expressed as the mean and S.D. of triplicate measurements. * $p < 0.01$. (**B**) Western blotting of KIFC1 in RWPE-1, DU145, and C4-2 cells. (**C**) The dose-dependent effects of CW069 on cell viability in RWPE-1, DU145, and C4-2 cells. The results are expressed as the mean and S.D. of triplicate measurements. * $p < 0.01$. (**D**) The cell death ELISA in RWPE-1, DU145, and C4-2 cells treated with CW069 (250 µM). The results are expressed as the mean and S.D. of triplicate measurements. * $p < 0.01$.

3.5. CW069 Re-Sensitizes DTX-Resistant Cell Lines to DTX Treatment

As shown in Figure 3B, knockdown of KIFC1 reversed DTX resistance. Therefore, we investigated the effect of combination therapy with DTX and CW069. We measured cell viability under DTX alone or in combination with CW069 in parental and DTX-resistant cell lines. DTX alone had little effect on cell viability in the DTX-resistant cell lines. However, the combination of DTX and CW069 significantly reduced cell viability in the DTX-resistant cell lines (Figure 5A). The cell death ELISA assay showed that the combination of DTX and CW069 led to significant induction of apoptosis compared to DTX alone in the DTX-resistant cell lines (Figure 5B). In addition, we analyzed the dose response for the combination of DTX and CW069 in the DTX-resistant cell lines and calculated the combination index to assess whether the combination of DTX and CW069 is synergistic or additive. A synergistic effect was observed in the DTX-resistant cell lines (Table 1).

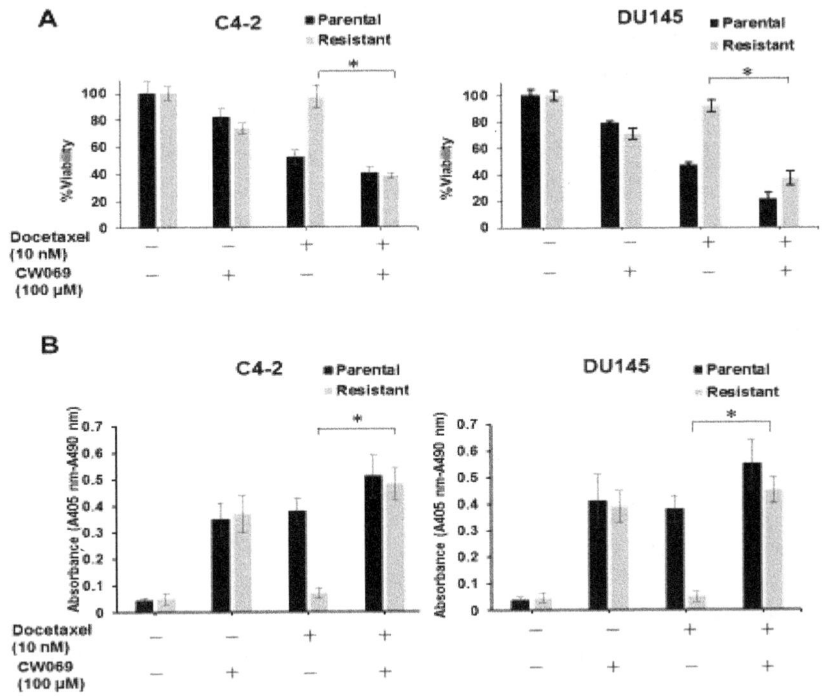

Figure 5. CW069 re-sensitizes DTX-resistant cell lines to docetaxel (DTX) treatment. (**A**) The effect of the combination of DTX and CW069 on cell viability in parental and DTX-resistant cell lines in DU145 and C4-2 cells. The results are expressed as the mean and S.D. of triplicate measurements. * $p < 0.01$. (**B**) The effect of the combination of DTX and CW069 on apoptosis in parental and DTX-resistant cell lines in DU145 and C4-2 cells. The results are expressed as the mean and S.D. of triplicate measurements. * $p < 0.01$.

Table 1. The combination index (CI) values for the combination of docetaxel (DTX) and CW069 in DTX-resistant cell lines.

DU145-DR		Docetaxel (nM)		
		5	10	20
CW069 (μM)	50	0.55	0.45	0.45
	100	0.73	0.48	0.54
	200	0.72	0.71	0.47
C4-2-DR		**Docetaxel (nM)**		
		5	10	20
CW069 (μM)	50	0.66	0.51	0.62
	100	0.77	0.46	0.53
	200	0.66	0.67	0.42

CI: combination index; DU145-DR: docetaxel-resistant DU145; C4-2-DR: docetaxel-resistant C4-2; CI = 1: an additive effect; CI < 1: a synergistic effect.

4. Discussion

DTX has been the first-line therapy for metastatic CRPC patients since 2004. Recent clinical studies have reported that early DTX treatment combined with androgen deprivation therapy results in improved overall survival in comparison to androgen deprivation therapy alone in patients with metastatic hormone-sensitive PCa [28,29]. This finding suggests that the beneficial effect of DTX may not be restricted to CRPC and that DTX treatment is becoming increasingly more important in PCa [30]. Although DTX treatment improves overall survival, disease relapse eventually occurs due to the development of DTX resistance [31]. Several factors have been shown to be involved in DTX resistance [30,32]. Loss of p53 leads to DTX resistance, and p53 status is an essential determinant of

DTX sensitivity [33]. Recent evidence has shown that alteration of β-tubulin isotypes is correlated with DTX resistance [34]. In addition, the expression of multidrug-resistant proteins such as ABCB1 is upregulated in a DTX-resistant PCa cell line. However, these above molecules have not been utilized clinically. Therefore, there is an urgent need to clarify the mechanisms of DTX resistance. A recent study reported that the expression of KIFC1 is upregulated in DTX-resistant breast cancer cell lines compared with that of DTX-sensitive cell lines. What is more, overexpression of KIFC1 increased the pools of free tubulin and promoted DTX resistance in breast cancer [16]. This evidence suggests that KIFC1 may antagonize the effect of DTX at least through the dissociation of tubulin from microtubules. In the present study, the expression of KIFC1 was upregulated in DTX-resistant PCa cell lines. Knockdown of KIFC1 re-sensitized the DTX-resistant cells to DTX treatment in DU145 and C4-2 cells. To date, some preclinical studies have addressed the finding that anti-apoptotic proteins regain sensitivity to DTX [30]. ABT-263, which is a Bcl-2 inhibitor, restored DTX sensitivity in DTX-resistant cells in PCa [35]. Furthermore, glucocorticoid receptor antagonism also re-sensitizes DTX resistance through a reduction of BcL-xL expression [36]. In the present study, knockdown of KIFC1 suppressed the expression of Bcl-2, cleaved PARP and cleaved caspase-3, and enhanced the expression of Bax. This result indicates a potential mechanistic explanation for the restoration of DTX sensitivity in PCa.

A recent study reported that CW069 was identified as a highly selective small-molecule KIFC1 inhibitor using a chemogenomics-based approach [26]. CW069 increases multipolar spindle formation and inhibits cell viability in cancer cells in breast cancer. However, to date, there have been few reports on CW069 [26,37]. In the present study, CW069 treatment selectively damaged parental and DTX-resistant PCa cells but had little effect on cell viability in RWPE-1 cells. The result that CW069 re-sensitized DTX-resistant cell lines to DTX treatment has potential clinical implications. In addition, the synergistic effect was found in the combination of DTX and CW069. In current cancer treatments, different types of chemotherapeutic agents are combined to improve efficacy and to minimize toxicity. Our previous study showed that the expression of KIFC1 was higher in PCa tissues than in various normal tissue samples [17]. Collectively, these results suggest that a combination of DTX and CW069 may be a promising therapy for CRPC patients that causes fewer adverse effects.

There are some limitations in this study. First, so far, three KIFC1 inhibitors (CW069, AZ82, and SR31527) have been reported [26,38,39]. Although these three drugs have been shown to lead to multipolar mitosis and decrease cell viability in human cancer, their effects were somewhat different because each drug binds to a different allosteric site on KIFC1 [38]. Furthermore, using a unique assay, a recent study showed that these drugs might not be specific to KIFC1 [40] Therefore, further study using these three drugs in PCa will be necessary in the future to verify our current findings. Second, recent studies have shown that cross-resistance of DTX cells were resistant to both DTX and cabazitaxel [41,42]. However, in this study, we focused on the role of KIFC1 and KIFC1 inhibitor (CW069) on only DTX resistance in PCa. In the near future, we will investigate the role of KIFC1 on cross-taxan resistance in PCa.

5. Conclusions

In conclusion, we used DTX-resistant PCa cell lines to analyze the role of KIFC1 in DTX resistance. We found that the expression of KIFC1 was significantly upregulated in cells with DTX resistance. Inhibition of KIFC1 induced an apoptosis pathway and re-sensitized the cellular response to DTX. Additionally, CW069 re-sensitized DTX-resistant cells to DTX treatment. The data presented here emphasize the great potential of combination therapy with DTX and CW069 in the treatment of PCa.

Author Contributions: Y.S., N.S., N.O., K.S., and W.Y. designed the study. Y.S., J.T., M.S., and A.M. provided patients' clinical information. Y.S. and Y.K. performed the experiments and acquired data. Y.S., N.S., N.O., and W.Y. interpreted the results. Y.S., N.O., and M.S. drafted the manuscript. N.O., J.T., A.M., K.S., and W.Y. edited it. All authors approved the final content for journal submission and publication.

Acknowledgments: We thank M. Gleave (Vancouver Prostate Centre, Vancouver, BC, Canada) for originally providing the C4-2 cells. This work was carried out with the kind cooperation of the Research Center for Molecular Medicine of the Faculty of Medicine of Hiroshima University. We also thank the Analysis Center of Life Science of Hiroshima University for the use of their facilities.

References

1. Van Neste, L.; Herman, J.G.; Otto, G.; Bigley, J.W.; Epstein, J.I.; Van Criekinge, W. The epigenetic promise for prostate cancer diagnosis. *Prostate* **2012**, *72*, 1248–1261. [CrossRef] [PubMed]
2. Antognelli, C.; Cecchetti, R.; Riuzzi, F.; Peirce, M.J.; Talesa, V.N. Glyoxalase 1 sustains the metastatic phenotype of prostate cancer cells via EMT control. *J. Cell Mol. Med.* **2018**, *22*, 2865–2883. [CrossRef] [PubMed]
3. Molina, A.; Belldegrun, A. Novel therapeutic strategies for castration resistant prostate cancer: Inhibition of persistent androgen production and androgen receptor mediated signaling. *J. Urol.* **2011**, *185*, 787–794. [CrossRef] [PubMed]
4. Petrylak, D.P.; Tangen, C.M.; Hussain, M.H.; Lara, P.N., Jr.; Jones, J.A.; Taplin, M.E.; Burch, P.A.; Berry, D.; Moinpour, C.; Kohli, M.; et al. Docetaxel and estramustine compared with mitoxantrone and prednisone for advanced refractory prostate cancer. *N. Engl. J. Med.* **2004**, *351*, 1513–1520. [CrossRef] [PubMed]
5. Chan, J.Y. A clinical overview of centrosome amplification in human cancers. *Int. J. Biol. Sci.* **2011**, *7*, 1122–1144. [CrossRef] [PubMed]
6. Ganem, N.J.; Godinho, S.A.; Pellman, D. A mechanism linking extra centrosomes to chromosomal instability. *Nature* **2009**, *460*, 278–282. [CrossRef] [PubMed]
7. Kwon, M.; Godinho, S.A.; Chandhok, N.S.; Ganem, N.J.; Azioune, A.; Thery, M.; Pellman, D. Mechanisms to suppress multipolar divisions in cancer cells with extra centrosomes. *Genes Dev.* **2008**, *22*, 2189–2203. [CrossRef]
8. Quintyne, N.J.; Reing, J.E.; Hoffelder, D.R.; Gollin, S.M.; Saunders, W.S. Spindle multipolarity is prevented by centrosomal clustering. *Science* **2005**, *307*, 127–129. [CrossRef]
9. Godinho, S.A.; Kwon, M.; Pellman, D. Centrosomes and cancer: How cancer cells divide with too many centrosomes. *Cancer Metast. Rev.* **2009**, *28*, 85–98. [CrossRef]
10. Kwon, M.; Bagonis, M.; Danuser, G.; Pellman, D. Direct microtubule-binding by myosin-10 orients centrosomes toward retraction fibers and subcortical actin clouds. *Dev. Cell.* **2015**, *34*, 323–337. [CrossRef]
11. Rath, O.; Kozielski, F. Kinesins and cancer. *Nat. Rev. Cancer* **2012**, *12*, 527–539. [CrossRef] [PubMed]
12. Kleylein-Sohn, J.; Pöllinger, B.; Ohmer, M.; Hofmann, F.; Nigg, E.A.; Hemmings, B.A.; Wartmann, M.; et al. Acentrosomal spindle organization renders cancer cells dependent on the kinesin HSET. *J. Cell Sci.* **2012**, *125*, 5391–5402. [CrossRef]
13. Leber, B.; Maier, B.; Fuchs, F.; Chi, J.; Riffel, P.; Anderhub, S.; Wagner, L.; Ho, A.D.; Salisbury, J.L.; Boutros, M.; Krämer, A. Proteins required for centrosome clustering in cancer cells. *Sci. Transl. Med.* **2010**, *2*, 33ra8. [CrossRef] [PubMed]
14. Basto, R.; Brunk, K.; Vinadogrova, T.; Peel, N.; Franz, A.; Khodjakov, A.; Raff, J.W. Centrosome amplification can initiate tumorigenesis in flies. *Cell* **2008**, *133*, 1032–1042. [CrossRef] [PubMed]
15. Han, J.; Wang, F.; Lan, Y.; Wang, J.; Nie, C.; Liang, Y.; Song, R.; Zheng, T.; Pan, S.; Pei, T.; et al. KIFC1 regulated by miR-532-3p promotes epithelial-to-mesenchymal transition and metastasis of hepatocellular carcinoma via gankyrin/AKT signaling. *Oncogene* **2018**. [CrossRef] [PubMed]
16. De, S.; Cipriano, R.; Jackson, M.W.; Stark, G.R. Overexpression of kinesins mediates docetaxel resistance in breast cancer cells. *Cancer Res.* **2009**, *69*, 8035–8042. [CrossRef] [PubMed]
17. Sekino, Y.; Oue, N.; Shigematsu, Y.; Ishikawa, A.; Sakamoto, N.; Sentani, K.; Teishima, J.; Matsubara, A.; Yasui, W. KIFC1 induces resistance to docetaxel and is associated with survival of patients with prostate cancer. *Urol. Oncol.* **2017**, *35*, e31.e13–e31.e20. [CrossRef] [PubMed]
18. Sekino, Y.; Sakamoto, N.; Goto, K.; Honma, R.; Shigematsu, Y.; Sentani, K.; Oue, N.; Teishima, J.; Matsubara, A.; Yasui, W. Transcribed ultraconserved region Uc.63+promotes resistance to docetaxel through regulation of androgen receptor signaling in prostate cancer. *Oncotarget* **2017**, *8*, 94259–94270. [CrossRef]

19. Sekino, Y.; Sakamoto, N.; Goto, K.; Honma, R.; Shigematsu, Y.; Quoc, T.P.; Sentani, K.; Oue, N.; Teishima, J.; Kawakami, F.; et al. Uc.416 + A promotes epithelial-to-mesenchymal transition through miR-153 in renal cell carcinoma. *BMC Cancer* **2018**, *18*, 952. [CrossRef]
20. Sekino, Y.; Oue, N.; Mukai, S.; Shigematsu, Y.; Goto, K.; Sakamoto, N.; Sentani, K.; Hayashi, T.; Teishima, J.; Matsubara, A.; et al. Protocadherin B9 promotes resistance to bicalutamide and is associated with the survival of prostate cancer patients. *Prostate* **2019**, *79*, 234–242. [CrossRef]
21. Chou, T.C. Drug Combination Studies and Their Synergy Quantification Using the Chou-Talalay Method. *Cancer Res.* **2010**, *70*, 440–446. [CrossRef] [PubMed]
22. Shiota, M.; Kashiwagi, E.; Yokomizo, A.; Takeuchi, A.; Dejima, T.; Song, Y.; Tatsugami, K.; Inokuchi, J.; Uchiumi, T.; Naito, S. Interaction between docetaxel resistance and castration resistance in prostate cancer: Implications of Twist1, YB-1, and androgen receptor. *Prostate* **2013**, *73*, 1336–1344. [CrossRef] [PubMed]
23. Shiota, M.; Itsumi, M.; Yokomizo, A.; Takeuchi, A.; Imada, K.; Kashiwagi, E.; Inokuchi, J.; Tatsugami, K.; Uchiumi, T.; Naito, S. Targeting ribosomal S6 kinases/Y-box binding protein-1 signaling improves cellular sensitivity to taxane in prostate cancer. *Prostate* **2014**, *74*, 829–838. [CrossRef] [PubMed]
24. Fu, X.; Zhu, Y.; Zheng, B.; Zou, Y.; Wang, C.; Wu, P.; Wang, J.; Chen, H.; Du, P.; Liang, B.; et al. KIFC1, a novel potential prognostic factor and therapeutic target in hepatocellular carcinoma. *Int. J. Oncol.* **2018**, *52*, 1912–1922. [CrossRef] [PubMed]
25. Pannu, V.; Rida, P.C.; Ogden, A.; Turaga, R.C.; Donthamsetty, S.; Bowen, N.J.; Rudd, K.; Gupta, M.V.; Reid, M.D.; Cantuaria, G.; et al. HSET overexpression fuels tumor progression via centrosome clustering-independent mechanisms in breast cancer patients. *Oncotarget* **2015**, *6*, 6076–6091. [CrossRef] [PubMed]
26. Watts, C.A.; Richards, F.M.; Bender, A.; Bond, P.J.; Korb, O.; Kern, O.; Riddick, M.; Owen, P.; Myers, R.M.; Raff, J.; et al. Design, synthesis, and biological evaluation of an allosteric inhibitor of HSET that targets cancer cells with supernumerary centrosomes. *Chem. Biol.* **2013**, *20*, 1399–1410. [CrossRef] [PubMed]
27. Webber, M.M.; Bello, D.; Quader, S. Immortalized and tumorigenic adult human prostatic epithelial cell lines: Characteristics and applications. Part 3. Oncogenes, suppressor genes, and applications. *Prostate* **1997**, *30*, 136–142. [CrossRef]
28. Vale, C.L.; Burdett, S.; Rydzewska, L.H.M.; Albiges, L.; Clarke, N.W.; Fisher, D.; Fizazi, K.; Gravis, G.; James, N.D.; Mason, M.D.; et al. Addition of docetaxel or bisphosphonates to standard of care in men with localised or metastatic, hormone-sensitive prostate cancer: A systematic review and meta-analyses of aggregate data. *Lancet Oncol.* **2016**, *17*, 243–256. [CrossRef]
29. Sweeney, C.J.; Chen, Y.H.; Carducci, M.; Liu, G.; Jarrard, D.F.; Eisenberger, M.; Wong, Y.N.; Hahn, N.; Kohli, M.; Cooney, M.M.; et al. Chemohormonal therapy in metastatic hormone-sensitive prostate cancer. *N. Engl. J. Med.* **2015**, *373*, 737–746. [CrossRef]
30. Kroon, J.; Kooijman, S.; Cho, N.J.; Storm, G.; van der Pluijm, G. Improving taxane-based chemotherapy in castration-resistant prostate cancer. *Trends Pharmacol. Sci.* **2016**, *37*, 451–462. [CrossRef]
31. Huang, X.; Chau, C.H.; Figg, W.D. Challenges to improved therapeutics for metastatic castrate resistant prostate cancer: From recent successes and failures. *J. Hematol. Oncol.* **2012**, *5*, 35. [CrossRef] [PubMed]
32. Armstrong, C.M.; Gao, A.C. Drug resistance in castration resistant prostate cancer: Resistance mechanisms and emerging treatment strategies. *Am. J. Clin. Exp. Urol.* **2015**, *3*, 64–76. [PubMed]
33. Liu, C.; Zhu, Y.; Lou, W.; Nadiminty, N.; Chen, X.; Zhou, Q.; Shi, X.B.; deVere White, R.W.; Gao, A.C. Functional p53 determines docetaxel sensitivity in prostate cancer cells. *Prostate* **2013**, *73*, 418–427. [CrossRef] [PubMed]
34. Ploussard, G.; Terry, S.; Maillé, P.; Allory, Y.; Sirab, N.; Kheuang, L.; Soyeux, P.; Nicolaiew, N.; Coppolani, E.; Paule, B.; et al. Class III beta-tubulin expression predicts prostate tumor aggressiveness and patient response to docetaxel-based chemotherapy. *Cancer Res.* **2010**, *70*, 9253–9264. [CrossRef] [PubMed]
35. Tamaki, H.; Harashima, N.; Hiraki, M.; Arichi, N.; Nishimura, N.; Shiina, H.; Naora, K.; Harada, M. Bcl-2 family inhibition sensitizes human prostate cancer cells to docetaxel and promotes unexpected apoptosis under caspase-9 inhibition. *Oncotarget* **2014**, *5*, 11399–11412. [CrossRef] [PubMed]
36. Kroon, J.; Puhr, M.; Buijs, J.T.; van der Horst, G.; Hemmer, DM.; Marijt, K.A.; Hwang, M.S.; Masood, M.; Grimm, S.; Storm, G.; et al. Glucocorticoid receptor antagonism reverts docetaxel resistance in human prostate cancer. *Endocr. Relat. Cancer* **2016**, *23*, 35–45. [CrossRef]

37. Choe, M.H.; Kim, J.; Ahn, J.; Hwang, S.G.; Oh, J.S.; Kim, J.S. Centrosome clustering is a tumor-selective target for the improvement of radiotherapy in breast cancer cells. *Anticancer Res.* **2018**, *38*, 3393–3400. [CrossRef]
38. Zhang, W.; Zhai, L.; Wang, Y.; Boohaker, R.J.; Lu, W.; Gupta, V.V.; Padmalayam, I.; Bostwick, R.J.; White, E.L.; Ross, L.J.; et al. Discovery of a novel inhibitor of kinesin-like protein KIFC1. *Biochem. J.* **2016**, *473*, 1027–1035. [CrossRef]
39. Wu, J.; Mikule, K.; Wang, W.; Su, N.; Petteruti, P.; Gharahdaghi, F.; Code, E.; Zhu, X.; Jacques, K.; Lai, Z.; et al. Discovery and mechanistic study of a small molecule inhibitor for motor protein KIFC1. *ACS Chem. Biol.* **2013**, *8*, 2201–2208. [CrossRef]
40. Yukawa, M.; Yamauchi, T.; Kurisawa, N.; Ahmed, S.; Kimura, K.I.; Toda, T. Fission yeast cells overproducing HSET/KIFC1 provides a useful tool for identification and evaluation of human kinesin-14 inhibitors. *Fungal Genet. Biol.* **2018**, *116*, 33–41. [CrossRef]
41. Lombard, A.P.; Liu, C.; Armstrong, C.M.; Cucchiara, V.; Gu, X.; Lou, W.; Evans, C.P.; Gao, A.C. ABCB1 Mediates Cabazitaxel–Docetaxel Cross-Resistance in Advanced Prostate Cancer. *Mol. Cancer Ther.* **2017**, *16*, 2257–2266. [CrossRef] [PubMed]
42. Van Soest, R.J.; De Morrée, E.S.; Kweldam, C.F.; De Ridder, C.M.; Wiemer, E.A.; Mathijssen, R.H.; De Wit, R.; Van Weerden, W.M. Targeting the Androgen Receptor Confers In Vivo Cross-resistance Between Enzalutamide and Docetaxel, But Not Cabazitaxel, in Castration-resistant Prostate Cancer. *Eur. Urol.* **2015**, *67*, 981–985. [CrossRef] [PubMed]

15

Fibroblast Growth Factor Family in the Progression of Prostate Cancer

Jun Teishima *, Tetsutaro Hayashi, Hirotaka Nagamatsu, Koichi Shoji, Hiroyuki Shikuma, Ryoken Yamanaka, Yohei Sekino, Keisuke Goto, Shogo Inoue and Akio Matsubara

Department of Urology, Graduate School of Biomedical and Health Sciences, Hiroshima University, 1-2-3 Kasumi, Minami-ku, Hiroshima 734-8551, Japan; tetsutaro.hayashi@gmail.com (T.H.), k717171k@yahoo.co.jp (H.N.), urokshoji@yahoo.co.jp (K.S.), himuro.49.1026@gmail.com (H.S.), yamanaka_ryouken@hiro-hosp.jp (R.Y.), akikosekino@gmail.com (Y.S.), keigoto@hiroshima-u.ac.jp (K.G.), inosyogo@hiroshima-u.ac.jp (S.I.), matsua@hiroshima-u.ac.jp (A.M.)

* Correspondence: teishima@hiroshima-u.ac.jp

Abstract: Fibroblast growth factors (FGFs) and FGF receptors (FGFRs) play an important role in the maintenance of tissue homeostasis and the development and differentiation of prostate tissue through epithelial-stromal interactions. Aberrations of this signaling are linked to the development and progression of prostate cancer (PCa). The FGF family includes two subfamilies, paracrine FGFs and endocrine FGFs. Paracrine FGFs directly bind the extracellular domain of FGFRs and act as a growth factor through the activation of tyrosine kinase signaling. Endocrine FGFs have a low affinity of heparin/heparan sulfate and are easy to circulate in serum. Their biological function is exerted as both a growth factor binding FGFRs with co-receptors and as an endocrine molecule. Many studies have demonstrated the significance of these FGFs and FGFRs in the development and progression of PCa. Herein, we discuss the current knowledge regarding the role of FGFs and FGFRs—including paracrine FGFs, endocrine FGFs, and FGFRs—in the development and progression of PCa, focusing on the representative molecules in each subfamily.

Keywords: fibroblast growth factor; fibroblast growth factor receptor; prostate cancer

---

1. Introduction

Prostate cancer (PCa) is one of the most common hormone-dependent cancers. Androgen-deprived therapy (ADT) has been the standard option for PCa. It is initially effective in most PCa cases; however, PCa becomes refractory for ADT in spite of the castration level of serum testosterone, which is called "castration-resistant prostate cancer" (CRPC). There have been multiple studies on the efficacy of various agents that include androgen-receptor (AR) targeted agents and anticancer drugs.

Many studies have demonstrated the aberrant activation of fibroblast growth factor (FGF)/FGF receptor (FGFR) signaling in several cancers, including head and neck, lung, breast, endometrial, bladder, and prostate cancer [1,2]. Herein, we review the FGF family's involvement in the development and progression of prostate cancer. We mainly discuss the representative molecules in each subfamily: FGF9 as classic FGFs, FGF19 as endocrine FGFs, and FGFR2IIIb.

2. FGFs and FGFRs

The human *Fgf* gene family consists of 22 members, and they are classified into seven subfamilies based on phylogenetic analysis (Figure 1) [3]. FGFs are also classified into three subfamilies (intracrine, paracrine, and endocrine FGFs) based on their mechanism of action. Intracrine FGFs are not typical and does not bind FGFR. Their function is not mediated by a receptor. Therefore, we herein focused two subfamilies, paracrine FGFs and endocrine FGFs. Paracrine FGFs—which consist of 15 members—exert their biological function through binding to an extracellular domain of FGFRs with heparin/ heparan sulfate and activating tyrosine kinase signaling of FGFRs.

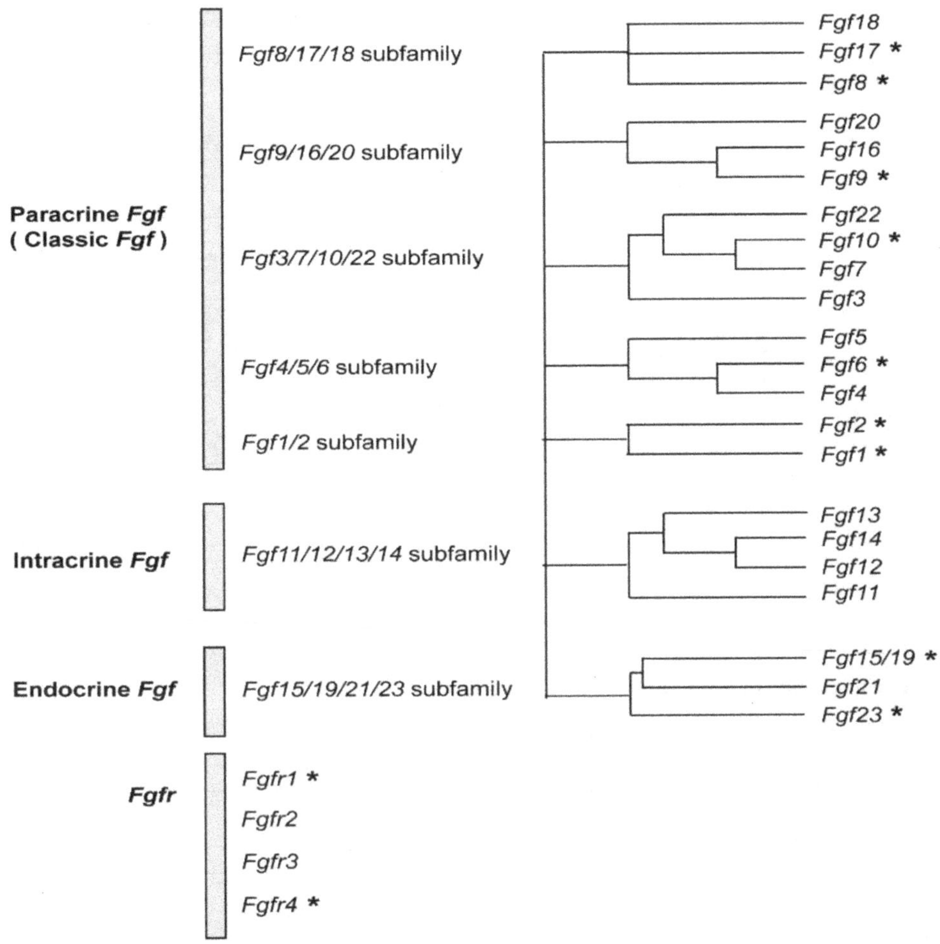

Figure 1. *Fgf* genes consisting of 7 subfamilies and *Fgfr* genes. Asterisks indicates fibroblast growth factor (FGF)/ fibroblast growth factor receptor (FGFR) whose expression are enhanced in prostate cancer cells and/or tissues.

Endocrine FGFs consist of FGF19, FGF21, and FGF23. These molecules have a low affinity to heparin/heparan sulfate in contrast to paracrine FGFs. They have to form complexes with co-receptors, α/β-Klotho, to bind to the extracellular domain of FGFR. Because the endocrine FGFs' affinity to FGFRs changes depending on the existence of α/β-Klotho, they have hormone-like activity beyond functioning as a growth factor (Figure 2). Several studies have demonstrated the physiological function of endocrine FGFs (Table 1) [4]. Among endocrine FGFs, FGF15 (the mouse orthologue of human FGF19) was the first molecule to be identified. The physiological activity of FGF19, including the regulation of glucose and bile acid metabolism, is exerted through the formation of a complex with FGFR4 and beta-klotho and follows the activation of a signaling cascade by recruiting adaptor proteins in cytoplasm [5–7]. FGF21 mainly acts as a metabolic regulator in the liver, adipose tissue, and the pancreas. The tissue-specific metabolic action of FGF21 depends on its specificity to the receptor. FGF21 binds with beta-klotho and FGFR1c [6,8,9]. FGF23 is a bone-derived endocrine hormone. Expression of FGF23 is induced by activation of the vitamin D receptor (VDR) with 1,25 dihydroxyvitamin D (1,25D), and FGF23 is a suppressor of 1,25D. So FGF23 and 1,25D are linked by mutual regulation. A signaling cascade of FGF23 activates through formation of a complex with alfa-klotho and FGFR1 for the kidney and FGFR3c for the parathyroid gland. FGF23 regulates phosphate and vitamin D metabolism in the kidney and inhibits parathyroid hormone secretion and vitamin D synthesis in the parathyroid gland [10,11].

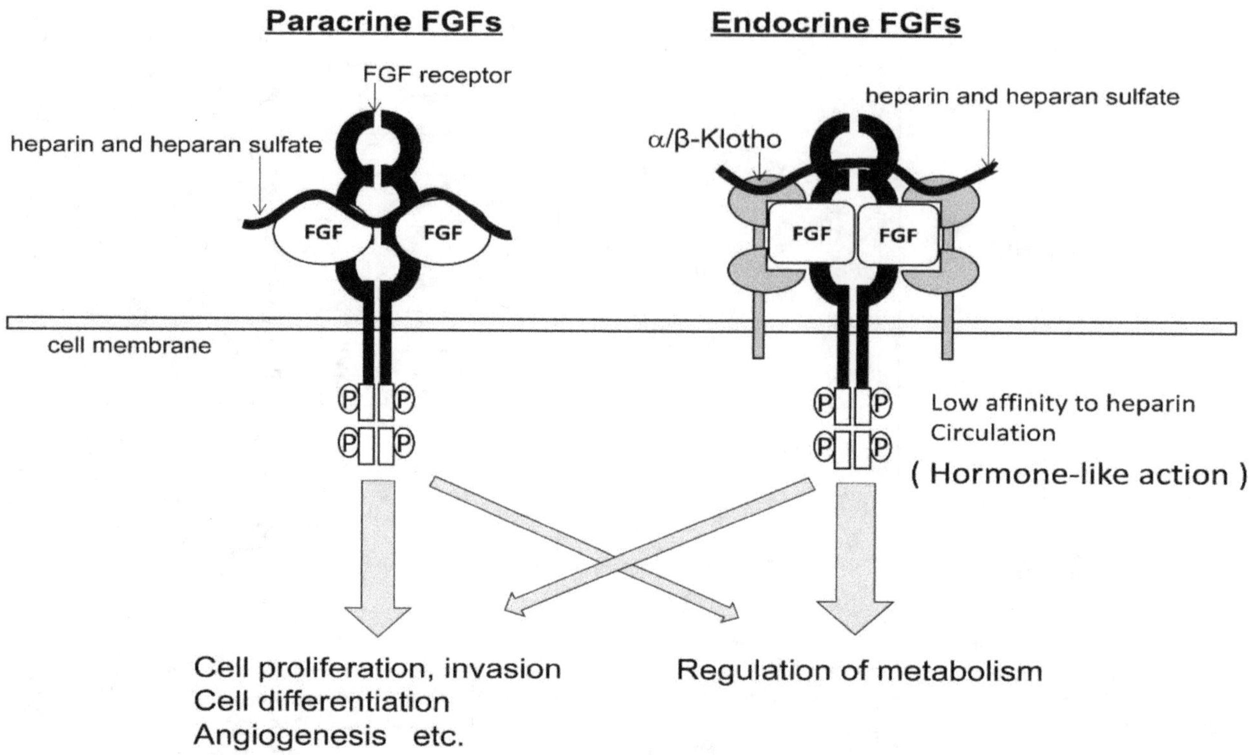

Figure 2. Difference in the mechanism in signal transduction between paracrine FGFs and endocrine FGFs.

Table 1. Pathophysiological activity of endocrine FGFs.

	Up-Regulation	Down-Regulation	Increase	Decrease
FGF19	Glycogen synthesis	Bile acid synthesis Gluconeogenesis	Extrahepatic cholestasis Chronic hemodialysis	IBD NAFLD Primary bile acid malabsorption Obesity
FGF21	Hepatic fatty acid oxidation Ketogenesis Glucogenesis Thermogenesis WAT browing Growth hormone resistance Weight loss Ovulation	Ovulation Growth hormone signaling	Type 2 diabetes Metabolic syndrome NAFLD Coronary heart disease	Anorexia Nervosa
FGF23	Calcium secretion Life span	Renal phosphate absorption Bone and renal calcium reabsorption Vitamin D synthesis PTH secretion	ADHR XLH rickets TIO Cardiac hypertrophy	Hemodialysis Familial tumoral calcinosis

ADHR, autosomal dominant hypophosphataemic rickets; FGF, fibroblast growth factor; IBD, irritable bowel disease; NAFLD, non-alcoholic fatty liver disease; PTH, parathyroid hormone; TIO, tumor-induced osteomalacia; WAT, white adipose tissue; and XLH, X-linked hypophosphataemic.

Four FGFRs, FGFR1-4, contain an extracellular ligand binding domain with three immunoglobulin (Ig)-like domains (I–III), a transmembrane domain, and a split intracellular tyrosine kinase domain. FGFR1-3 have two kinds of Ig-like III domains, IIIb and IIIc, which are generated by alternative splicing. The Ig-like domain is crucial for determining ligand-binding specificity, and as a result, seven FGFR with different ligand-binding specificities are derived from four *Fgfr* genes [12].

3. FGF-FGFR Signaling in Epithelial-Stromal Interaction in Prostate Tissue

Epithelial-stromal interaction plays an important role in maintaining the homeostasis in normal prostatic tissue [13,14]. Stromal tissues secrete paracrine factors that include FGF ligands, and they lead to stimulation of epithelial maintenance and growth. Huang et al. reported the significance of FGFR2 signaling for preserving stemness and preventing differentiation of prostate stem cells [15]. FGFR2IIIb, a splicing variant of the FGFR2, is a resident form of FGFRs expressed in normal prostate epithelial cells. FGFR2IIIb is specific to FGF7, and FGF7-FGFR2IIIb contributes epithelial-stromal interaction [16]. The loss of FGFR2 isoforms is shown in human PCa tissues, and the loss of FGFR2IIIb is associated with the characteristics of castration-resistant prostate cancer (CRPC) in particular [17].

The expression pattern of FGFR is different in each PCa cell line. For instance, FGFR1 expression is detectable and FGFR2IIIb expression is undetectable in PC3 cells that show androgen-independent growth and high potential of cell proliferation. On the other hand, FGFR2IIIb expression is detectable in LNCap cells that show expression of androgen receptor (AR), androgen-dependent growth, and relatively low potential of cell proliferation [18]. In addition to FGFs and FGFRs, FGF receptor substrate 2alpha (FRS2alpha), an FGFR interactive adaptor protein, involves multiple signaling pathways to the activated FGFR kinase. FRS2 alfa is required for prostate development and tumorigenesis [19], as well as in angiogenesis [20].

4. Effects of the Restoration of FGFR2IIIb in Prostate Cancer Cells

Many studies have demonstrated the association of aberrant FGFR signaling with the development and progression of PCa [21,22]. Binding ligands, FGFRs form functional dimerization and lead transphosphorylation and activation of downstream signaling pathways such as Ras, Src, PKCγ, MAPK, PI3K-AKT, and STAT [23–25]. The involvement of FGF signaling in various molecular mechanisms has been reported in PCa. Shao et al. reported that FGF-FGFR signaling plays an important role in transformation induced by the loss of a PTEN tumor suppressor when combined with the expression of the TMPRSS2/ERG fusion gene [26], and activation of FGF-FGFR signaling by FGF8b overexpression in PTEN deficiency is reported to be associated with prostate tumorigenesis [27]. FGF-FGFR signaling is also related to the induction of an inflammatory response in PCa tissues [28]. The involvement of aberrant FGFR1 signaling in the progression of PCa in particular was demonstrated in several studies. FGFR1 signaling promotes the reprogramming of energy metabolism from oxidative phosphorylation to aerobic glycolysis by regulating the expression of an LDH isoenzyme [29]. It also promotes an inflammatory response through activation of NF-κB signaling [30]. Furthermore, activation of FGFR1 signaling promotes epithelial to mesenchymal transition and androgen independency in PCa cells [31,32]. Loss of FGFR2IIIb and enhancement of ectopic expression of FGFR1 PCa progression have been reported as common events in the progression of PCa [21]. As FGFR2IIIb plays an important role in the maintenance and its disorder is found in PCa cell lines and tissues, several investigators have reported the effects of restoring FGFR2IIIb. In animal models and PCa cell lines, FGFR2IIIb's restoration also restored responsiveness to stroma and significantly reduced in vivo tumorigenesis. In castration-resistant human PCa cells, restoration of FGFR2IIIb showed the inhibition of cell proliferation, the induction of differentiation, and the enhancement of apoptosis in a ligand-independent manner [21,33,34]. In addition, in PCa cells overexpressing FGFR2IIIb, clonogenic cell death increased in concurrence with enhanced apoptosis and cell cycle arrest in the G2/M-phase and radiosensitivity by gamma-irradiation [35]. Another study reported the effect FGFR2IIIb's restoration had on the chemosensitivity in PCa cells. Restoration of FGFR2IIIb led to the enhanced chemosensitivity of several agents, especially docetaxel. The expression of N-cadherin, vimentin, survivin, and XIAP were induced by restoring FGFR2IIIb [36]. This data indicates that PCa cell lines are induced to a more differentiated phenotype when changing the pattern of gene expression that became sensitive to radiation and chemotherapy when FGFR2IIIb was restored.

5. Involvement of FGFs in the Development and Progression of Prostate Cancer

5.1. Paracrine FGFs

Upregulated expression of FGF1, FGF2, FGF8, FGF9, and FGF10 were shown in human PCa [37]. Murine studies demonstrated the epithelial and mesenchymal interaction using a FGF/FGFR complex in PCa [38]. Pecqueux et al. demonstrated the association between strong expression of FGF2 in tumor stroma and a high postoperative recurrence rate and that exogenous FGF2 can drive genomic instability to promote PCa progression through enhancement of DNA damage [39]. Cuevas et al. reported the linkage between altered micro-environmental signaling that includes FGF2 overexpression and mitotic instability [40]. In the study focused on a bone metastatic site, FGF2 was upregulated in osteoblast and promoted the proliferation of PCa cells under the loss of TGFβ signaling [41]. These reports indicated the significance of FGF/FGFR signaling through paracrine FGFs (especially FGF2) in a cancer micro-environment for the development and progression of PCa.

FGF9 is an abundant molecule in nervous tissue and soft tissue, while it has been reported to be a key molecule of epithelial-stromal interaction [42]. Several studies have demonstrated that FGF9 was associated with the proliferation of glia [43], regulation of the differentiation of astrocytes [44] and oligodendrocytes [45], and the regulation of joint formations [46,47]. FGF9's involvement in malignant neoplasms has been reported in glioma [48], ovarian cancer [49], and lung cancer [50]. In prostate epithelial cells, overexpression of FGF9 lead the augmentation of reactive stroma formation and promoted initiation and progression of PCa [51]. In our study, cell viability and invasion of LNCaP was significantly enhanced, and expression of MMP2, N-cadherin, and VEGF-A were induced in LNCaP incubated in medium with FGF9. In immuno-histochemical staining, the prevalence of both VEGF-A and N-cadherin-positive cells was significantly higher compared to FGF9-negative cases [52]. The biochemical relapse-free survival (bRFS) rate in cases with FGF9-positive cases was significantly lower than that in FGF9-negative cases [53]. "FGF9-positive" in this study was determined based on the findings of immunohistochemistry staining on just one representative section. And even in FGF9-positive cases, just a small population of FGF9-positive cells with very aggressive pathological features was present. In other words, FGF9 was only positive in especially high-grade cancer cells in cases with localized PCa. Furthermore, several studies have reported FGF9 in PCa at an advanced stage. Accumulation of FGF9 to the region of bone metastasis formed by AR-negative PCa cells indicated that FGF9 was a key factor in formation of bone metastasis [54]. In AR-negative CRPC cases, neuroendocrine differentiation (NE) is one of the most representative phenotypes. In a further dedifferentiated phenotype, "AR-negative and NE-negative", expression of FGF1, FGF8, and FGF9 increased, and MAPK pathway was activated [55]. The results of these studies indicated that FGF9 might be a key molecule for an advanced PCa that include CRPC rather than a localized one.

5.2. Endocrine FGFs

Endocrine FGFs have two different characteristics, a growth factor and a metabolic regulator. Investigating endocrine FGFs might clarify the molecular link between the progression of cancer and metabolism. Among endocrine FGFs, the association between FGF19 and malignant diseases such as liver cancer, colon cancer, and prostate cancer was demonstrated [56–59]. FGF19 induces the expression of markers of epithelial mesenchymal transition in hormone-sensitive prostate cancer cells [60,61]. Consistent with the results, accumulation of FGF19 in cytoplasm was shown in poorly differentiated prostate cancer cells in human prostate cancer tissues derived from radical prostatectomy, and the presence of FGF19-positive tissues correlated with positive immuno-histochemical staining with N-cadherin in prostate cancer tissues [60]. The bRFS rate in cases with FGF19-positive tissues was significantly lower than in cases with FGF19-negative tissues [60]. In prostate cancer cells, FGF19 stimulates cell proliferation and cell invasion through activation of MAP kinase and AKT pathways [62]. Expression of FGF19 increased in castration-resistant cell lines compared with castration-sensitive ones or immortalized normal prostate cells [62]. According to these reports, FGF19 might be associated with the risk of post-operative recurrence by enhancement of cell proliferation and epithelial-mesenchymal transition of PCa cells.

Since endocrine FGFs that include FGF19 act as circulating hormones related to several metabolic diseases, the impact of their serum concentration for metabolic diseases was investigated. These molecules are expected to be a potential serum biomarker for PCa. We measured serum FGF19 and beta-klotho level in cases with PCa. While there was no relationship between the serum klotho level and pathological findings, the results showed that patients with a high Gleason score had higher serum FGF19 levels than those with a low Gleason score [60]. This data indicates that FGF19 might be a potential serum biomarker in PCa. One limitation to using endocrine FGFs that include FGF19 as a serum biomarker is the change of serum concentration according to dietary conditions and blood sugar level because endocrine FGFs act as a metabolic regulator and their level is also regulated by a feedback mechanism. Further study will clarify the optimal timing and conditions for measuring endocrine FGFs to apply them as a serum biomarker for PCa. Besides FGF19, the increased expression of FGF23 in many PCa cell lines and PCa tissues was reported. FGF23 enhances proliferation, invasion through activation of AKT, and the MAPK pathway in PCa cell lines. These findings indicate that it can promote PCa progression in an autocrine, paracrine, and/or endocrine manner [63].

6. Conclusions

In this article, we described the impact of FGF-FGFR abnormalities on the development and/or the progression of PCa. Since we found the heterogeneity of PCa, it is important to clarify and understand various molecular mechanisms of PCa in order to determine the most appropriate therapeutic strategies.

Author Contributions: Conceptualization, J.T. and A.M.; methodology, J.T.; software, J.T.; formal analysis, K.G., R.Y., H.S., K.S. and H.N.; investigation, J.T., T.H., K.G., K.S. and H.N.; writing—original draft preparation, J.T.; writing—review and editing, J.T. and T.H.; visualization, S.I. and Y.S.; supervision, K.G., T.H., S.I., Y.S. and A.M.; funding acquisition, J.T. and A.M.

References

1. Corn, P.G.; Wang, F.; McKeehan, W.L.; Navone, N. Targeting fibroblast growth factor pathways in prostate cancer. *Clin. Cancer. Res.* **2013**, *19*, 5856–5866. [CrossRef] [PubMed]
2. Wesche, J.; Haglund, K.; Haugsten, E.M. Fibroblast growth factors and their receptors in cancer. *Biochem. J.* **2011**, *437*, 199–213. [CrossRef]
3. Itoh, N.; Ornitz, D.M. Fibroblast growth factors: From molecular evolution to roles in development, metabolism and disease. *J. Biochem.* **2011**, *149*, 121–130. [CrossRef] [PubMed]
4. Degirolamo, C.; Sabbà, C.; Moschetta, A. Therapeutic potential of the endocrine fibroblast growth factors FGF19, FGF21 and FGF23. *Nat. Rev. Drug Discov.* **2016**, *15*, 51–69. [CrossRef] [PubMed]
5. Lin, B.C.; Wang, M.; Blackmore, C.; Desnoyers, L.R. Liver-specific activities of FGF19 require Klotho beta. *J. Biol. Chem.* **2007**, *282*, 27277–27284. [CrossRef] [PubMed]
6. Kurosu, H.; Choi, M.; Ogawa, Y.; Dickson, A.S.; Goetz, R.; Eliseenkova, A.V.; Mohammadi, M.; Rosenblatt, K.P.; Kliewer, S.A.; Kuro-o, M. Tissue-specific expression of betaKlotho and fibroblast growth factor (FGF) receptor isoforms determines metabolic activity of FGF19 and FGF21. *J. Biol. Chem.* **2007**, *282*, 26687–26695. [CrossRef] [PubMed]
7. Xie, M.H.; Holcomb, I.; Deuel, B.; Dowd, P.; Huang, A.; Vagts, A.; Foster, J.; Liang, J.; Brush, J.; Gu, Q.; et al. FGF-19, a novel fibroblast growth factor with unique specificity for FGFR4. *Cytokine* **1999**, *11*, 729–735. [CrossRef]
8. Ogawa, Y.; Kurosu, H.; Yamamoto, M.; Nandi, A.; Rosenblatt, K.P.; Goetz, R.; Eliseenkova, A.V.; Mohammadi, M.; Kuro-o, M. β-Klotho is required for metabolic activity of fibroblast growth factor 21. *Proc. Natl. Acad. Sci. USA* **2007**, *104*, 7432–7437. [CrossRef]

9. Ding, X.; Boney-Montoya, J.; Owen, B.M.; Bookout, A.L.; Coate, K.C.; Mangelsdorf, D.J.; Kliewer, S.A. β-Klotho is required for fibroblast growth factor 21 effects on growth and metabolism. *Cell Metab.* **2012**, *16*, 387–393. [CrossRef]
10. Kolek, O.I.; Hines, E.R.; Jones, M.D.; LeSueur, L.K.; Lipko, M.A.; Kiela, P.R.; Collins, J.F.; Haussler, M.R.; Ghishan, F.K. 1alpha, 25-Dihydroxyvitamin D3 upregulates FGF23 gene expression in bone: The final link in a renal-gastrointestinal-skeletal axis that controls phosphate transport. *Am. J. Physiol. Gastrointest. Liver. Physiol.* **2005**, *289*, G1036–G1042. [CrossRef]
11. Yu, X.; Sabbagh, Y.; Davis, S.I.; Demay, M.B.; White, K.E. Genetic dissection of phosphate- and vitamin D-mediated regulation of circulating Fgf23 concentrations. *Bone* **2005**, *36*, 971–977. [CrossRef] [PubMed]
12. Zhang, X.; Ibrahimi, O.A.; Olsen, S.K.; Umemori, H.; Mohammadi, M.; Ornitz, D.M. Receptor specificity of the fibroblast growth factor family. The complete mammalian FGF family. *J. Biol. Chem.* **2006**, *281*, 15694–15700. [CrossRef] [PubMed]
13. Lin, Y.; Wang, F. FGF signaling in prostate development, tissue homoeostasis and tumorigenesis. *Biosci. Rep.* **2010**, *30*, 285–291. [CrossRef] [PubMed]
14. Cotton, L.M.; O'Bryan, M.K.; Hinton, B.T. Cellular signaling by fibroblast growth factors (FGFs) and their receptors (FGFRs) in male reproduction. *Endocr. Rev.* **2008**, *29*, 193–216. [CrossRef] [PubMed]
15. Huang, Y.; Hamana, T.; Liu, J.; Wang, C.; An, L.; You, P.; Chang, J.Y.; Xu, J.; Jin, C.; Zhang, Z.; et al. Type 2 Fibroblast Growth Factor Receptor Signaling Preserves Stemness and Prevents Differentiation of Prostate Stem Cells from the Basal Compartment. *J. Biol. Chem.* **2015**, *290*, 17753–17761. [CrossRef] [PubMed]
16. Lin, Y.; Liu, G.; Zhang, Y.; Hu, Y.P.; Yu, K.; Lin, C.; McKeehan, K.; Xuan, J.W.; Ornitz, D.M.; Shen, M.M.; et al. Fibroblast growth factor receptor 2 tyrosine kinase is required for prostatic morphogenesis and the acquisition of strict androgen dependency for adult tissue homeostasis. *Development* **2007**, *134*, 723–734. [CrossRef] [PubMed]
17. Matsubara, A.; Yasumoto, H.; Usui, T. Hormone Refractory Prostate Cancer and Fibroblast Growth Factor Receptor. *Breast Cancer* **1999**, *6*, 320–324. [CrossRef]
18. Carstens, R.P.; Eaton, J.V.; Krigman, H.R.; Walther, P.J.; Garcia-Blanco, M.A. Alternative splicing of fibroblast growth factor receptor 2 (FGF-R2) in human prostate cancer. *Oncogene* **1997**, *15*, 3059–3065. [CrossRef]
19. Zhang, Y.; Zhang, J.; Lin, Y.; Lan, Y.; Lin, C.; Xuan, J.W.; Shen, M.M.; McKeehan, W.L.; Greenberg, N.M.; Wang, F. Role of epithelial cell fibroblast growth factor receptor substrate 2alpha in prostate development, regeneration and tumorigenesis. *Development* **2008**, *135*, 775–784. [CrossRef]
20. Liu, J.; You, P.; Chen, G.; Fu, X.; Zeng, X.; Wang, C.; Huang, Y.; An, L.; Wan, X.; Navone, N.; et al. Hyperactivated FRS2α-mediated signaling in prostate cancer cells promotes tumor angiogenesis and predicts poor clinical outcome of patients. *Oncogene* **2016**, *35*, 1750–1759. [CrossRef]
21. Lu, W.; Luo, Y.; Kan, M.; McKeehan, W.L. Fibroblast growth factor-10. A second candidate stromal to epithelial cell andromedin in prostate. *J. Biol. Chem.* **1999**, *274*, 12827–12834. [CrossRef]
22. Feng, S.; Wang, F.; Matsubara, A.; Kan, M.; McKeehan, W.L. Fibroblast growth factor receptor 2 limits and receptor 1 accelerates tumorigenicity of prostate epithelial cells. *Cancer Res.* **1997**, *57*, 5369–5378. [PubMed]
23. Turner, N.; Grose, R. Fibroblast growth factor signaling: From development to cancer. *Nat. Rev. Cancer* **2010**, *10*, 116–129. [CrossRef] [PubMed]
24. Beenken, A.; Mohammadi, M. The FGF family: Biology, pathophysiology and therapy. *Nat. Rev. Drug Discov.* **2009**, *8*, 235–253. [CrossRef] [PubMed]
25. Li, Q.; Ingram, L.; Kim, S.; Beharry, Z.; Cooper, J.A.; Cai, H. Paracrine Fibroblast Growth Factor Initiates Oncogenic Synergy with Epithelial FGFR/Src Transformation in Prostate Tumor Progression. *Neoplasia* **2018**, *20*, 233–243. [CrossRef] [PubMed]
26. Shao, L.; Wang, J.; Karatas, O.F.; Feng, S.; Zhang, Y.; Creighton, C.J.; Ittmann, M. Fibroblast growth factor receptor signaling plays a key role in transformation induced by the TMPRSS2/ERG fusion gene and decreased PTEN. *Oncotarget* **2018**, *9*, 14456–14471. [CrossRef] [PubMed]
27. Zhong, C.; Saribekyan, G.; Liao, C.P.; Cohen, M.B.; Roy-Burman, P. Cooperation between FGF8b overexpression and PTEN deficiency in prostate tumorigenesis. *Cancer Res.* **2006**, *66*, 2188–2194. [CrossRef]
28. Ronca, R.; Tamma, R.; Coltrini, D.; Ruggieri, S.; Presta, M.; Ribatti, D. Fibroblast growth factor modulates mast cell recruitment in a murine model of prostate cancer. *Oncotarget* **2017**, *8*, 82583–82592. [CrossRef]
29. Liu, J.; Chen, G.; Liu, Z.; Liu, S.; Cai, Z.; You, P.; Ke, Y.; Lai, L.; Huang, Y.; Gao, H.; et al. Aberrant FGFR tyrosine kinase signaling enhances the Warburg effect by reprogramming LDH isoform expression and activity in prostate cancer. *Cancer Res.* **2018**, *78*, 4459–4470. [CrossRef]

30. Wang, C.; Ke, Y.; Liu, S.; Pan, S.; Liu, Z.; Zhang, H.; Fan, Z.; Zhou, C.; Liu, J.; Wang, F. Ectopic fibroblast growth factor receptor 1 promotes inflammation by promoting nuclear factor-κB signaling in prostate cancer cells. *J. Biol. Chem.* **2018**, *293*, 14839–14849. [CrossRef]
31. Acevedo, V.D.; Gangula, R.D.; Freeman, K.W.; Li, R.; Zhang, Y.; Wang, F.; Ayala, G.E.; Peterson, L.E.; Ittmann, M.; Spencer, D.M. Inducible FGFR-1 activation leads to irreversible prostate adenocarcinoma and an epithelial-to-mesenchymal transition. *Cancer Cell.* **2007**, *12*, 559–571. [CrossRef] [PubMed]
32. Kobayashi, M.; Huang, Y.; Jin, C.; Luo, Y.; Okamoto, T.; Wang, F.; McKeehan, W.L. FGFR1 abrogates inhibitory effect of androgen receptor concurrent with induction of androgen-receptor variants in androgen receptor-negative prostate tumor epithelial cells. *Prostate* **2011**, *71*, 1691–1700. [CrossRef] [PubMed]
33. Matsubara, A.; Kan, M.; Feng, S.; McKeehan, W.L. Inhibition of growth of malignant rat prostate tumor cells by restoration of fibroblast growth factor receptor 2. *Cancer Res* **1998**, *58*, 1509–1514. [PubMed]
34. Yasumoto, H.; Matsubara, A.; Mutaguchi, K.; Usui, T.; McKeehan, W.L. Restoration of fibroblast growth factor receptor2 suppresses growth and tumorigenicity of malignant human prostate carcinoma PC-3 cells. *Prostate* **2004**, *61*, 236–242. [CrossRef] [PubMed]
35. Matsubara, A.; Teishima, J.; Mirkhat, S.; Yasumoto, H.; Mochizuki, H.; Seki, M.; Mutaguchi, K.; Mckeehan, W.L.; Usui, T. Restoration of FGF receptor type 2 enhances radiosensitivity of hormone-refractory human prostate carcinoma PC-3 cells. *Anticancer Res.* **2008**, *28*, 2141–2146. [PubMed]
36. Shoji, K.; Teishima, J.; Hayashi, T.; Ohara, S.; Mckeehan, W.L.; Matsubara, A. Restoration of fibroblast growth factor receptor 2IIIb enhances the chemosensitivity of human prostate cancer cells. *Oncol. Rep.* **2014**, *32*, 65–70. [CrossRef] [PubMed]
37. Ronca, R.; Giacomini, A.; Di Salle, E.; Coltrini, D.; Pagano, K.; Ragona, L.; Matarazzo, S.; Rezzola, S.; Maiolo, D.; Torrella, R.; et al. Long-Pentraxin 3 derivative as a small-molecule FGF trap for cancer therapy. *Cancer Cell.* **2015**, *28*, 225–239. [CrossRef]
38. Memarzadeh, S.; Xin, L.; Mulholland, D.J.; Mansukhani, A.; Wu, H.; Teitell, M.A.; Witte, O.N. Enhanced paracrine FGF10 expression promotes formation of multifocal prostate adenocarcinoma and an increase in epithelial androgen receptor. *Cancer Cell.* **2007**, *12*, 572–585. [CrossRef]
39. Pecqueux, C.; Arslan, A.; Heller, M.; Falkenstein, M.; Kaczorowski, A.; Tolstov, Y.; Sültmann, H.; Grüllich, C.; Herpel, E.; Duensing, A. FGF-2 is a driving force for chromosomal instability and a stromal factor associated with adverse clinico-pathological features in prostate cancer. *Urol. Oncol.* **2018**, *36*, e15–e365. [CrossRef]
40. Cuevas, R.; Korzeniewski, N.; Tolstov, Y.; Hohenfellner, M.; Duensing, S. FGF-2 disrupts mitotic stability in prostate cancer through the intracellular trafficking protein CEP57. *Cancer Res.* **2013**, *73*, 1400–1410. [CrossRef]
41. Meng, X.; Vander Ark, A.; Daft, P.; Woodford, E.; Wang, J.; Madaj, Z.; Li, X. Loss of TGF-β signaling in osteoblasts increases basic-FGF and promotes prostate cancer bone metastasis. *Cancer Lett.* **2018**, *418*, 109–118. [CrossRef] [PubMed]
42. Jin, C.; Wang, F.; Wu, X.; Yu, C.; Luo, Y.; McKeehan, W.L. Directionally specific paracrine communication mediated by epithelial FGF9 to stromal FGFR3 in two-compartment premalignant prostate tumors. *Cancer Res.* **2004**, *64*, 4555–4562. [CrossRef] [PubMed]
43. Reuss, B.; Hertel, M.; Werner, S.; Unsicker, K. Fibroblast growth factors-5 and -9 distinctly regulate expression and function of the gap junction protein connexin43 in cultured astroglial cells from different brain regions. *Glia* **2000**, *30*, 231–241. [CrossRef]
44. Lum, M.; Turbic, A.; Mitrovic, B.; Turnley, A.M.J. Fibroblast growth factor-9 inhibits astrocyte differentiation of adult mouse neural progenitor cells. *Neurosci. Res.* **2009**, *87*, 2201–2210. [CrossRef] [PubMed]
45. Cohen, R.I.; Chandross, K.J. Fibroblast growth factor-9 modulates the expression of myelin related proteins and multiple fibroblast growth factor receptors in developing oligodendrocytes. *J. Neurosci. Res.* **2000**, *61*, 273–287. [CrossRef]
46. Harada, M.; Murakami, H.; Okawa, A.; Okimoto, N.; Hiraoka, S.; Nakahara, T.; Akasaka, R.; Shiraishi, Y.; Futatsugi, N.; Mizutani-Koseki, Y.; et al. FGF9 monomer/ dimer equilibrium regulates extracellular matrix affinity and tissue diffusion. *Nat. Genet.* **2009**, *41*, 289–298. [CrossRef] [PubMed]
47. Kalinina, J.; Byron, S.A.; Makarenkova, H.P.; Olsen, S.K.; Eliseenkova, A.V.; Larochelle, W.J.; Dhanabal, M.; Blais, S.; Ornitz, D.M.; Day, L.A.; et al. Homodimerization controls the fibroblast growth factor 9 subfamily's receptor binding and heparan sulfate-dependent diffusion in the extracellular matrix. *Mol. Cell. Biol.* **2009**, *29*, 4663–4678. [CrossRef]

48. Miyagi, N.; Kato, S.; Terasaki, M.; Aoki, T.; Sugita, Y.; Yamaguchi, M.; Shigemori, M.; Morimatsu, M. Fibroblast growth factor-9 (glia-activating factor) stimulates proliferation and production of glial fibrillary acidic protein in human gliomas either in the presence or in the absence of the endogenous growth factor expression. *Oncol. Rep.* **1999**, *6*, 87–92. [CrossRef]
49. Hendrix, N.D.; Wu, R.; Kuick, R.; Schwartz, D.R.; Fearon, E.R.; Cho, K.R. Fibroblast growth factor 9 has oncogenic activity and is a downstream target of Wnt signaling in ovarian endometrioid adenocarcinomas. *Cancer Res.* **2006**, *66*, 1354–1362. [CrossRef]
50. Wang, C.K.; Chang, H.; Chen, P.H.; Chang, J.T.; Kuo, Y.C.; Ko, J.L.; Lin, P. Aryl hydrocarbon receptor activation and overexpression upregulated fibroblast growth factor-9 in human lung adenocarcinomas. *Int. J. Cancer.* **2009**, *125*, 807–815. [CrossRef]
51. Huang, Y.; Jin, C.; Hamana, T.; Liu, J.; Wang, C.; An, L.; McKeehan, W.L.; Wang, F. Overexpression of FGF9 in prostate epithelial cells augments reactive stroma formation and promotes prostate cancer progression. *Int. J. Biol. Sci.* **2015**, *11*, 948–960. [CrossRef] [PubMed]
52. Teishima, J.; Yano, S.; Shoji, K.; Hayashi, T.; Goto, K.; Kitano, H.; Oka, K.; Nagamatsu, H.; Matsubara, A. Accumulation of FGF9 in prostate cancer correlates with epithelial-to-mesenchymal transition and induction of VEGF-A expression. *Anticancer Res.* **2014**, *34*, 695–700.
53. Teishima, J.; Shoji, K.; Hayashi, T.; Miyamoto, K.; Ohara, S.; Matsubara, A. Relationship between the localization of fibroblast growth factor 9 in prostate cancer cells and postoperative recurrence. *Prostate Cancer Prostatic Dis.* **2012**, *15*, 8–14. [CrossRef] [PubMed]
54. Li, ZG.; Mathew, P.; Yang, J.; Starbuck, M.W.; Zurita, A.J.; Liu, J.; Sikes, C.; Multani, A.S.; Efstathiou, E.; Lopez, A.; et al. Androgen receptor-negative human prostate cancer cells induce osteogenesis in mice through FGF9-mediated mechanisms. *J. Clin. Investig.* **2008**, *118*, 2697–2710. [CrossRef] [PubMed]
55. Bluemn, E.G.; Coleman, I.M.; Lucas, J.M.; Coleman, R.T.; Hernandez-Lopez, S.; Tharakan, R.; Bianchi-Frias, D.; Dumpit, R.F.; Kaipainen, A.; Corella, A.N.; et al. Androgen Receptor Pathway-Independent Prostate Cancer Is Sustained through FGF Signaling. *Cancer Cell.* **2017**, *32*, 474–489. [CrossRef] [PubMed]
56. Nicholes, K.; Guillet, S.; Tomlinson, E.; Hillan, K.; Wright, B.; Frantz, G.D.; Pham, T.A.; Dillard-Telm, L.; Tsai, S.P.; Stephan, J.P.; et al. A mouse model of hepatocellular carcinoma: Ectopic expression of fibroblast growth factor 19 in skeletal muscle of transgenic mice. *Am. J. Pathol.* **2002**, *160*, 2295–2307. [CrossRef]
57. Alvarez-Sola, G.; Uriarte, I.; Latasa, M.U.; Urtasun, R.; Bárcena-Varela, M.; Elizalde, M.; Jiménez, M.; Rodriguez-Ortigosa, C.M.; Corrales, F.J.; Fernández-Barrena, M.G.; et al. Fibroblast growth factor 15/19 in hepatocarcinogenesis. *Dig. Dis.* **2017**, *35*, 158–165. [CrossRef]
58. Desnoyers, L.R.; Pai, R.; Ferrando, R.E.; Hötzel, K.; Le, T.; Ross, J.; Carano, R.; D'Souza, A.; Qing, J.; Mohtashemi, I.; et al. Targeting FGF19 inhibits tumor growth in colon cancer xenograft and FGF19 transgenic hepatocellular carcinoma models. *Oncogene* **2008**, *27*, 85–97. [CrossRef]
59. Pai, R.; Dunlap, D.; Qing, J.; Mohtashemi, I.; Hotzel, K.; French, D.M. Inhibition of fibroblast growth factor 19 reduces tumor growth by modulating beta-catenin signaling. *Cancer Res.* **2008**, *68*, 5086–5095. [CrossRef]
60. Nagamatsu, H.; Teishima, J.; Goto, K.; Shikuma, H.; Kitano, H.; Shoji, K.; Inoue, S.; Matsubara, A. FGF19 promotes progression of prostate cancer. *Prostate* **2015**, *75*, 1092–1101. [CrossRef]
61. Liu, Z.; Zhang, H.; Ding, S.; Qi, S.; Liu, S.; Sun, D.; Dong, W.; Yin, L.; Li, M.; Zhao, X.; et al. βKlotho inhibits androgen/androgen receptor-associated epithelial-mesenchymal transition in prostate cancer through inactivation of ERK1/2 signaling. *Oncol. Rep.* **2018**, *40*, 217–225. [CrossRef] [PubMed]
62. Feng, S.; Dakhova, O.; Creighton, C.J.; Ittmann, M. Endocrine fibroblast growth factor FGF19 promotes prostate cancer progression. *Cancer Res.* **2013**, *73*, 2551–2562. [CrossRef] [PubMed]
63. Feng, S.; Wang, J.; Zhang, Y.; Creighton, C.J.; Ittmann, M. FGF23 promotes prostate cancer progression. *Oncotarget* **2015**, *6*, 17291–17301. [CrossRef] [PubMed]

16

Higher Serum Testosterone Levels Associated with Favorable Prognosis in Enzalutamide- and Abiraterone-Treated Castration-Resistant Prostate Cancer

Shinichi Sakamoto [1,*], Maihulan Maimaiti [1], Minhui Xu [1], Shuhei Kamada [2], Yasutaka Yamada [3], Hiroki Kitoh [4], Hiroaki Matsumoto [5], Nobuyoshi Takeuchi [1], Kosuke Higuchi [6], Haruhito A. Uchida [7], Akira Komiya [1], Maki Nagata [2], Hiroomi Nakatsu [3], Hideyasu Matsuyama [5], Koichiro Akakura [4] and Tomohiko Ichikawa [1]

1 Department of Urology, Chiba University Hospital, Chiba 260-8670, Japan; marghulanmaimaiti@gmail.com (M.M.); xuminhui198666@yahoo.co.jp (M.X.); nob.takeuchi1014@gmail.com (N.T.); akirakomiya@mac.com (A.K.); ichikawa@vmail.plala.or.jp (T.I.)

2 Department of Urology, Yokohama Rosai Hospital, Yokohama 222-0036, Japan; shu.ukmd.d@gmail.com (S.K.); makinagata1109@gmail.com (M.N.)

3 Department of Urology, Asahi Central Hospital, Aashi 289-2511, Japan; yasutaka1205@olive.plala.or.jp (Y.Y.); nakatsu@hospital.asahi.chiba.jp (H.N.)

4 Department of Urology, Japan Community Healthcare Organization Tokyo Shinjuku Medical Center, Shinjyuku 162-8543, Japan; hirokitoh@gmail.com (H.K.); akakurak@ae.auone-net.jp (K.A.)

5 Department of Urology, Graduate School of Medicine, Yamaguchi University, Ube 755-0046, Japan; hmatsumo@yamaguchi-u.ac.jp (H.M.); hidde@yamaguchi-u.ac.jp (H.M.)

6 Department of Urology, Funabashi Medical Center, Funabashi 273-8588, Japan; k_h1069k@yahoo.co.jp

7 Department of Chronic Kidney Disease and Cardiovascular Disease, Okayama University Graduate School of Medicine, Dentistry and Pharmaceutical Sciences, Okayama 700-0914, Japan; hauchida@okayama-u.ac.jp

* Correspondence: rbatbat1@gmail.com

Abstract: Testosterone plays a significant role in maintaining the tumor microenvironment. The role of the target serum testosterone (TST) level in enzalutamide- (Enza) and abiraterone (Abi)-treated castration-resistant prostate cancer (CRPC) patients was studied. In total, 107 patients treated with Enza and/or Abi at Chiba University Hospital and affiliated hospitals were studied. The relationships between progression-free survival (PFS), overall survival (OS), and clinical factors were studied by Cox proportional hazard and Kaplan–Meier models. In the Abi and Enza groups overall, TST $\geq$ 13 ng/dL (median) (Hazard Ratio (HR) 0.43, $p = 0.0032$) remained an independent prognostic factor for PFS. In the Enza group, TST $\geq$ 13 ng/dL (median) was found to be a significant prognostic factor (HR 0.28, $p = 0.0044$), while, in the Abi group, TST $\geq$ 12 ng/dL (median) was not significant (HR 0.40, $p = 0.0891$). TST showed significant correlation with PFS periods ($r = 0.\ 32$, $p = 0.0067$), whereas, for OS, TST $\geq$ 13 ng/dL (median) showed no significant difference in the Abi and Enza groups overall. According to Kaplan–Meier analysis, a longer PFS at first-line therapy showed a favorable prognosis in the Enza group ($p = 0.0429$), while no difference was observed in the Abi group ($p = 0.6051$). The TST level and PFS of first-line therapy may be considered when determining the treatment strategy for CRPC patients.

Keywords: abiraterone; enzalutamide; prostate cancer; androgen deprivation therapy; testosterone; castration resistant prostate cancer

---

1. Introduction

Prostate cancer is one of the most commonly diagnosed cancers in men [1]. Since the historical discovery of Dr. Huggins, androgen deprivation therapy (ADT) has been the mainstay of the therapy for locally advanced or metastatic prostate cancer [2]. According to the current guidelines, the target serum testosterone (TST) level during androgen deprivation therapy for prostate cancer was defined as <50 ng/dL [3]. However, we have recently reported the clinical significance of serum TST levels <20 ng/dL in prostate cancer patients who received combined androgen blockade (CAB) therapy in Japanese patients [4]. Despite an early response to ADT, the majority of patients with advanced disease progress and become refractory to ADT because of the emergence of castration-resistant prostate cancer (CRPC) cells. Although a number of mechanisms have been proposed, the androgen receptor (AR) plays a central role in the development of CRPC [5–7]. Evidence also indicate that estrogen receptor (ER) drives prostate growth [8,9]. Since the AR and ER axes play a major role in the development of CRPC, it is a classical treatment strategy to block either of the pathways. However, after a certain interval, the tumor relapses by acquiring treatment resistance. Thus, the establishment of the optimal treatment sequence in CRPC is the primary concern.

Some predictors were reported to be related to the response to enzalutamide (Enza) and abiraterone (Abi), such as the presence of AR splicing variants, early prostate-specific antigen (PSA) response, neutrophil-to-lymphocyte ratio (NLR), the presence of visceral metastases, and so on [10]. However, no definitive guideline to help determine which of the two drugs should be used has yet been established. Furthermore, a significant survival benefit was clinically identified in patients with high-volume castration-sensitive prostate cancer (CSPC) treated with ADT in combination with docetaxel in the CHAARTED and STAMPEDE trials [11–14], while the LATTITUDE and STAMPEDE trials indicated a significant survival advantage in patients with high-volume CSPC treated with ADT plus abiraterone [15,16]. The sequences of AR-targeted drugs and chemotherapeutic agents remain controversial. Therefore, it is of primary importance to establish useful prognostic factors to guide treatment strategies for individual CRPC patients.

Although a series of studies indicated the clinical significance of TST level and response to ADT, limited evidence exists related to TST and response to novel AR-targeted drugs. Classically, low TST related to favorable prognosis in patients received vintage ADT [3,4]. However, a recent study indicated a clinical advantage of high TST in patients who received Abi [17]. Furthermore, the prognostic significance of TST in patients who received Enza remains to be investigated.

Here, we studied the association between TST level and response to novel AR-targeted drugs. The present findings may thus help to determine the optimal treatment strategies for CRPC patients.

2. Materials and Methods

2.1. Patient Selection and Clinical Variables

A total of 107 patients treated with Enza and/or Abi for prostate cancer at Chiba University Hospital and affiliated hospitals between 2014 and 2017 were retrospectively analyzed.

The TST was defined as the total TST. The prognostic values of the were level and other clinical factors were evaluated in association with PSA levels and progression-free survival (PFS). Patients treated with radiation as first-line therapy or radical prostatectomy, having a history of radiation to the pelvis, systemic chemotherapy, and use of 5 alpha-reductase inhibitors was not included.

Age, body mass index (BMI), first-line PFS (PFS of patients treated with first-line ADT with LH-RH analogue/antagonist and bicalutamide), site of metastasis, Gleason score, PSA at baseline, TST, nadir TST, and time-to-nadir TST were included as clinical factors. The Architect Testosterone II® device (Abbot Diagnostics, Lake Forest, IL, USA) was used to determine TST levels.

2.2. Definition of PSA Progression

PSA failure was defined according to the definition of The Prostate Cancer Clinical Trials Working Group 2 (PCWG2): a rising PSA, >2 ng/mL higher than the nadir; the rise has to be at least 25% over the nadir and has to be confirmed by a second PSA determination at least three weeks later. Also, the patient is required to have castrated levels of testosterone (<50 ng/dL).

2.3. Definition of High-Volume Tumor

A high-volume was defined as the presence of visceral metastases or ≥4 bone lesions with ≥1 beyond the vertebral bodies and pelvis, on the basis of a previous report [11].

2.4. Institutional Approval

This study was approved by the Institutional Review Board of Chiba University Hospital (approval number 2279).

2.5. Statistical Analysis

Univariate and multivariate Cox proportional models and the Kaplan–Meier method were used for statistical analyses. Factors with $p < 0.05$ in univariate analysis were included in multivariate analysis when assessing Cox proportional models. Welch's *t*-test, Fisher test, and Wilcoxon's signed rank test were used to assess the associations of TST and other clinical variables. Statistical computations were carried out using the JMP 13.0.0 software program (SAS Institute, Cary, NC, USA). Significance was set at $p < 0.05$.

3. Results

The study population included 107 patients, of whom 89 were treated with Enza, and 46 were treated with Abi. Twenty-eight patients received sequential therapy with the novel AR-targeted drugs. The median follow-up time was 68.3 months from first-line ADT. The patients' characteristics are shown in Table 1. The patients' median age was 73.0 years. The median PSA levels were 30.1 ng/mL for Enza and 41.1 ng/dL for Abi. The rates of lymph node, lung, liver, and bone metastases were 31.78%, 8.41%, 5.61%, and 83.18%, respectively. The rates of previous steroid use and estramustine use in Enza and Abi were 40.19% and 46.73%, respectively. The median TST at the initiation of Enza was 13 ng/dL, and that of Abi was 12 ng/dL. Further information is shown in Table 1.

Table 1. Patient's backgrounds.

	Value	Range/%
Enza as initial therapy	82	
Abi as initial therapy	25	Total 107
Enza as second-line therapy	7	
Abi as second-line therapy	21	Total 28
Median age	73.0	54-88
Median BMI (kg/m^2)	23.4	16.09-34.06
Median TST at biopsy (ng/dL)	457.5	228-847
Median PSA at biopsy (ng/mL)	79.5	3.43-15332
Median PSA at Enza/Abi/total (ng/mL)	30.1/41.1/34.1	0.59–5942.62/3.52–13296/0.59–13296
Median TST at Enza/Abi/total (ng/dL)	13/12/13	2–92/3–31/2–92
Median previous treatment course number	3	1 to 5
Median EOD score 0/1/2/3/4	2	7/23/18/28/15
Median follow-up period (month)	68.3	11.81–241.60
Median Enza/Abi PFS period (month)	3.9/2.1	0–16.50/0–13.37
Median first-line PFS (month)	15.9	0.50–171.40

Table 1. *Cont.*

	Value	Range/%
Gleason Score sum (N)		
≤6	4	4.17%
7	15	15.63%
8	25	26.04%
≥9	52	54.17%
Bone mets	89	83.18%
Lymph mets	34	31.78%
Lung mets	9	8.41%
Liver mets	6	5.61%
No mets	12	11.21%
Patients who died	23	21.50%
Pre-/post-docetaxel	43/64	40.19/59.81%
Steroid use	43	40.19%
Estramustine use	50	46.73%
Enzalutamide dose	160 mg/80 mg	84/5
Abiraterone dose	1000 mg/750 mg	45/1

Enza: enzalutamide; Abi: abiraterone; BMI: body mass index; PFS: progression-free survival; Mets: metastasis; PSA: prostate-specific antigen; TST: target serum testosterone; EOD: extent of disease.

Table 2 shows the results of the univariate and multivariate analyses of the prognostic factors of PFS in Enza and Abi overall. By univariate analysis, first-line PFS ≥ 15.4 months (Hazard Ratio (HR) 0.60, $p = 0.0309$), previous docetaxel (HR 2.44, $p = <.0001$), high volume (HR 1.73, $p = 0.0198$), C-reactive protein (CRP) ≥ 0.15 mg/dL (HR 1.87, $p = 0.0173$), extent of disease (EOD) score (HR 2.79, $p = 0.0004$), PSA ≥ 34.1 ng/mL (HR 1.73, $p = 0.0143$), TST ≥ 13 ng/dL (HR 0.26, $p = <.0001$), steroid use (HR 2.24, $p = 0.0003$), and estramustine use (HR 1.69, $p = 0.0179$) were found to be significant prognostic factors. By multivariate analysis, only TST ≥ 13 ng/dL (HR 0.31, $p = 0.0365$) remained as an independent prognostic factor.

Table 2. Predictive factors of PFS in enzalutamide- and abiraterone- treated patients.

	Univariate Analysis				Multivariate Analysis		
	Cut off	HR	COI	*P*	HR	COI	*P*
Age	72	0.75	0.48–1.16	0.1932			
GS	9	0.93	0.59–1.48	0.7654			
First-line PFS (m)	15.4	0.60	0.37–0.95	0.0309	0.85	0.215–3.25	0.8126
Previous docetaxel	+/−	2.44	1.575–3.81	<0.0001	1.50	0.43–5.36	0.5222
Liver mets	+/−	1.69	0.70–3.46	0.2186			
Visceral mets	+/−	1.51	0.83–2.58	0.1682			
Lymph mets	0	1.67	1.07–2.63	0.0249	2.32	0.71–7.90	0.1607
High volume	+/−	1.73	1.09–2.83	0.0198	2.47	0.45–13.74	0.2927
EOD score	2	2.79	1.52–5.47	0.0004	2.79	0.65–14.57	0.1721
ALP (ng/dL)	254	1.20	0.78–1.86	0.4057			
ICTP (ng/mL)	6.6	1.40	0.71–2.78	0.3307			
Hb (g/dL)	11.9	0.65	0.42–1.01	0.0538			
LDH (mg/dL)	212	0.98	0.64–1.52	0.9446			
Alb (g/dL)	3.9	0.85	0.54–1.35	0.5001			
CRP (mg/dL)	0.15	1.87	1.11–3.30	0.0173	0.67	0.17–3.10	0.5871
NLR (ng/dL)	2.6	1.13	0.66–1.94	0.6593			
PSA (ng/mL)	34.1	1.73	1.12–2.71	0.0143	1.24	0.44–3.66	0.6899
TST (ng/dL)	13	0.26	0.13–0.51	<0.0001	0.31	0.10–0.93	0.0365 *
Steroid use	+/−	2.24	1.44–3.54	0.0003	1.10	0.40–2.95	0.8484
Estramustine use	+/−	1.69	1.09–2.65	0.0179	1.41	0.46–4.39	0.5442

Pre-treatment course: previous treatment course; GS: Gleason Score; ALP: alkaline phosphatase; ICTP: I collagen telopeptide; Hb: hemoglobin; LDH: Lactate dehydrogenase; Alb: albumin; CRP: C-reactive protein; NLR: Neutrophil/Lymphocyte ratio; PSA: prostate-specific antigen; * Statistical significance $p < 0.05$.

The prognostic value of TST was also confirmed in each Enza and Abi group.

As shown in Supplementary Table S1, TST $\geq$ 13 ng/dL (median) was a significant factor in univariate (HR 0.28, $p = 0.0044$) and multivariate analysis (HR0.08, $p = 0.0032$) in Enza-treated patients. However, as shown in Supplementary Table S2, TST $\geq$ 12 ng/dL (median) was not significant in univariate analysis (HR0.40, $p = 0.0891$) in Abi-treated patients.

Furthermore, the effect of te TST on the response to Enza and Abi was studied by the Kaplan–Meier model. As shown in Figure 1, higher TST levels ($\geq$13 ng/dL) were related to significantly better PFS in Enza and Abi groups overall ($p < 0.0001$) and in Enza-treated patients ($p = 0.0032$) (Figure 1a,b), while higher TST levels ($\geq$12 ng/dL) showed no prognostic difference in Abi-treated patients ($p = 0.0881$) (Figure 1c).

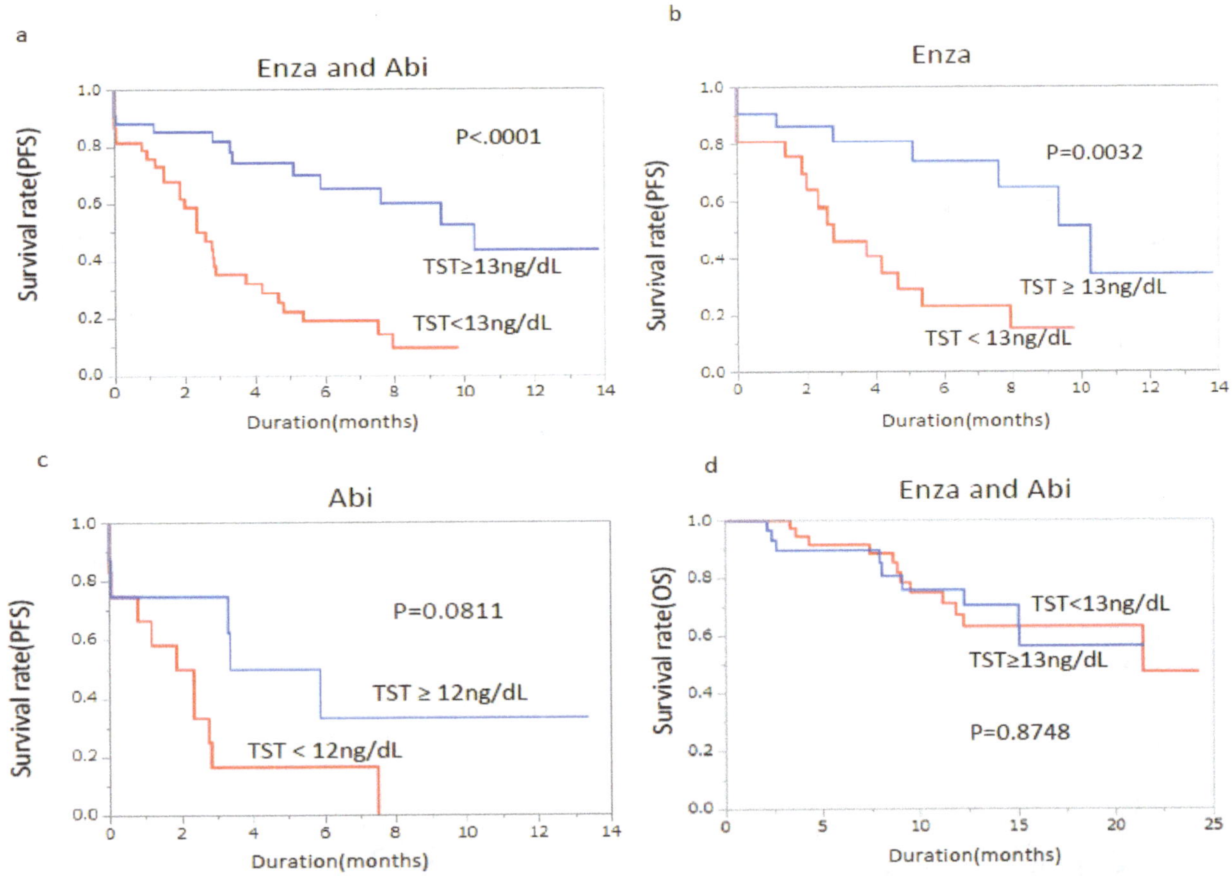

Figure 1. Progression-free survival according to TST levels. (**a**) Progression-free survival according to TST level <13 ng/dL or $\geq$13 ng/dL as the cut-off values in enzalutamide- and abiraterone-treated patients. (**b**) Progression-free survival according to TST level <13 ng/dL or $\geq$13 ng/dL as the cut-off values in enzalutamide-treated patients. (**c**) Progression-free survival according to TST level <12 ng/dL or $\geq$12 ng/dL as the cut-off values in abiraterone-treated patients. (**d**) Overall survival according to TST level <13 ng/dL or $\geq$13 ng/dL as the cut-off values in enzalutamide- and abiraterone-treated patients.

When the PFS periods were stratified by the TST level (8 to 20 ng/dL), higher TST levels were related to longer PFS in Enza and Abi groups, with a correlation coefficient of 0.32 ($p = 0.0067$) (Figure 2a,b).

To study the patients' characteristics, two groups (TST <13 ng/dL and TST $\geq$ 13 ng/dL) were compared with respect to various clinical factors. A level of TST $\geq$ 13 ng/dL was related to higher alkaline phosphatase (ALP) levels ($p = 0.0158$) (Table 3).

a

	PFS(month)			
TST(ng/dL)	<8	≥8	≥13	≥20
Enzalutamide	3.1	5.2	5.6	6.3
Abiraterone	3.8	3.5	5.4	5.4
Enza/Abi overall	3.3	4.7	5.7	6.3

b

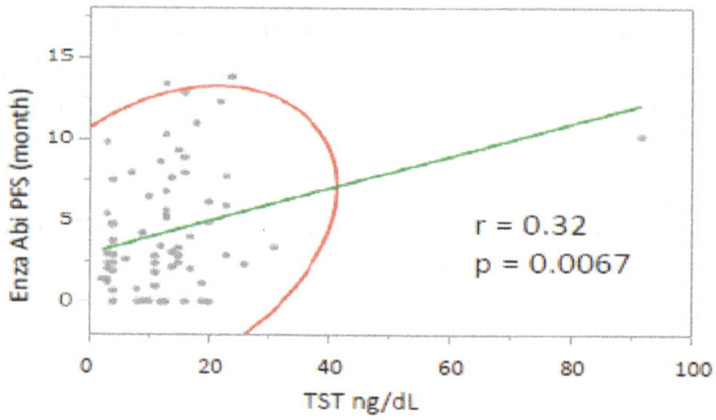

Figure 2. Progression-free survival based on the serum TST cut off level. (**a**) Progression-free survival according to TST level <8 ng/dL, ≥8 ng/dL, ≥13 ng/dL, or ≥20 ng/dL as the cut-off values in enzalutamide- and abiraterone-treated patients. (**b**) Correlation of serum TST level and progression-free survival in enzalutamide- and abiraterone-treated patients. The red circle indicates 95% of data used for the correlation analysis, to avoid errors.

Table 3. Comparison of clinical factors in patients with TST < 13 ng/dL and TST ≥ 13 ng/dL.

	TST < 13	TST ≥ 13	*P*-Value
	Median (Average)	**Median (Average)**	
Age	71.50(68.74)	70.00(70.88)	0.2298 †
1st-line PFS (month)	29.40(36.33)	14.63(32.755)	0.3576 ††
Pre-docetaxel	42.11%(16/38)	23.53%(8/34)	0.1336 †††
GS≥9	41.94%(13/31)	50%(16/30)	0.6159 †††
Lymph mets	54.29%(19/35)	38.24%(13/34)	0.2301 †††
Bone mets	92.11%(35/38)	88.24%(30/34)	0.7002 †††
Liver mets	5.56%(2/36)	14.71%(5/34)	0.2533 †††
Visceral mets	8.33%(3/36)	23.53%(8/34)	0.1062 †††
High volume	65.71%(23/35)	50.00%(17/34)	0.2270 †††
EOD score	2(2.24)	2(1.97)	0.3965 †
BSI	0.48(2.16)	0.59(2.33)	0.9342 ††
ALP (ng/dL)	209(647.87)	263.50(603.50)	0.0158 *,††
PSA (ng/mL)	33.34(710.73)	39.66(110.72)	0.8391 ††
CRP (mg/dL)	0.42(1.09)	0.50(1.53)	0.7866 ††
PSA at biopsy (ng/mL)	58.57(473.01)	72.95(563.79)	0.3903 ††
TST at biopsy (ng/dL)	4.42(4.22)	3.41(3.32)	0.1544 †
TST nadir at 1st line (ng/dL)	9(38.69)	12(13.23)	0.8974 ††

BSI: bone scan index; * Statistical significancer $p < 0.05$; † welch, †† wilcoxon, ††† fisher.

Table 4 shows the results of the univariate and multivariate analyses of the prognostic factors of overall survival (OS) in Enza and Abi groups overall. By multivariate analysis, previous docetaxel (HR 3.04, $p = 0.0038$), visceral mets (HR 7.88, $p = 0.0139$), ALP (HR 2.51, $p = 0.0090$) and lactate dehydrogenase (LDH) (HR 2.95, $p = 0.0033$) remained as independent prognostic factors. On the other hand, TST was not a significant factor neither in univariate analysis (HR 0.99, $p = 0.8750$) (Table 4), nor in Kaplan–Meier analysis ($p = 0.8748$) (Figure 1d). When patients who received initial AR-targeted therapy were selected, TST (HR 0.39, $p = 0.0086$) and CRP (HR 2.03, $p = 0.0487$) remained significant predictive factor for PFS in multivariate analysis (Supplementary Table S3), while, for OS, no factor remained significant in multivariate analysis (Supplementary Table S4).

Table 4. Predictive factors of overall survival (OS) in enzalutamide- and abiraterone-treated patients.

	Univariate Analysis				Multivariate Analysis		
	Cut off	HR	COI	*P*	HR	COI	*P*
Age	72	0.94	0.54–1.66	0.8371			
GS	9	0.81	0.45–1.46	0.4801			
First-line PFS (m)	15	0.70	0.36–1.23	0.2134			
Previous docetaxel	+/−	2.38	1.35–4.31	0.0025	3.04	1.41–7.175	0.0038 *
Liver mets	+/−	4.35	1.87–8.92	0.0014	0.35	0.07–1.905	0.2079
EOD score	2	2.87	1.305–7.56	0.0069	1.64	0.56–5.58	0.3778
Visceral mets	+/−	2.90	1.45–5.40	0.0038	7.88	1.62–29.52	0.0139 *
Lymph mets	+/−	2.13	1.21–3.80	0.0091	1.60	0.77–3.29	0.2041
High volume	+/−	2.65	1.38–5.63	0.0028	1.60	0.64–4.50	0.3266
ALP (ng/dL)	254	3.04	1.71–5.66	0.0001	2.51	1.25–5.265	0.0090 *
ICTP (ng/mL)	6.6	2.26	1.87–6.51	0.0941			
Hb (g/dL)	11.9	0.49	0.27–0.85	0.0109	0.87	0.39–1.86	0.7144
LDH (mg/dL)	212	1.92	1.10–3.43	0.0210	2.95	1.43–6.40	0.0033 *
Alb (g/dL)	3.9	0.61	0.33–1.10	0.0990			
CRP (mg/dL)	0.15	1.93	0.97–4.28	0.0607			
NLR (ng/dL)	2.6	1.93	0.93–4.18	0.0790			
PSA (ng/mL)	34.1	3.11	1.72–5.97	0.0001	1.76	0.88–3.69	0.1101
TST (ng/dL)	13	0.99	0.44–2.55	0.8750			
Steroid use	+/−	1.11	0.64–1.95	0.7011			
Estramustine use	+/−	0.95	0.55–1.66	0.8567			

* Statistical significance $p < 0.05$.

Next, the effects of the first-line therapy on the responses to Enza and Abi were evaluated. As shown in Figure 3a, first-line PFS (median ≥ 15.4 months) was related to a favorable response ($p = 0.0273$) for Enza and Abi overall (Figure 3a). However, the first-line PFS did not affect PFS in Abi-treated patients (median ≥ 12.6 months) ($p = 0.6105$), while they significantly affected PFS in Enza-treated patients (median ≥ 17.9 months) ($p = 0.0429$) (Figure 3b,c). Since the median first-line PFS in the Abi group was around 12 months, the patients were divided on the basis of the first-line PFS of 12.0 months for Enza. It was evident that a long first-line PFS (≥12.0 months) was related to significantly longer PFS in the Enza group ($p = 0.0046$) (Figure 3d).

The results of the sequential use of Enza and Abi were also examined. Although it was not significant, the subsequent usage of novel AR-targeted drugs reduced the PSA response rate in patients treated with both drug (Supplementary Table S5). Regarding PFS, a reduction in the response period was more evident in patients receiving Abi after Enza (first Abi 15.7 weeks vs. Enza before Abi 9.93 weeks) than in patients administered Enza after Abi (1st Enza 12 weeks vs Abi before Enza 11.1 weeks).

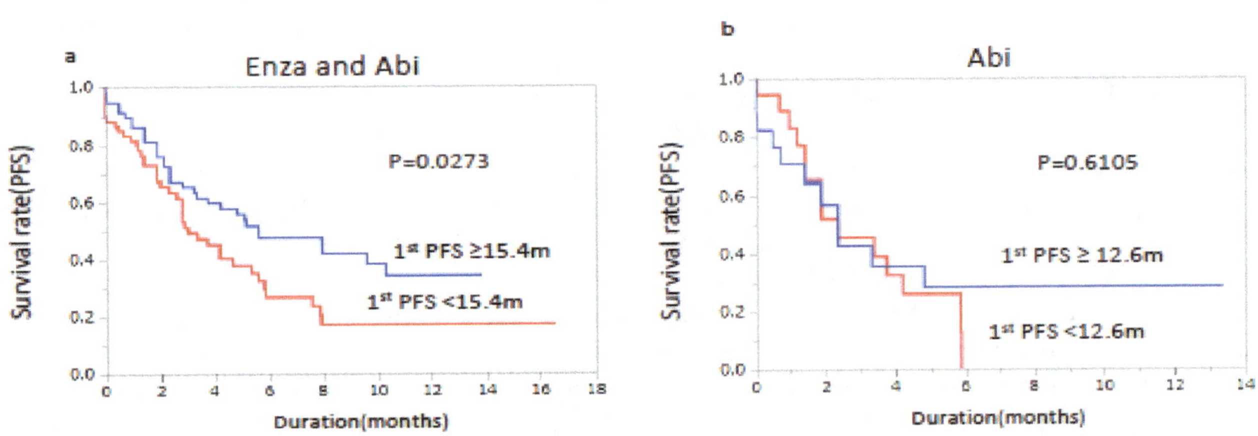

Figure 3. *Cont.*

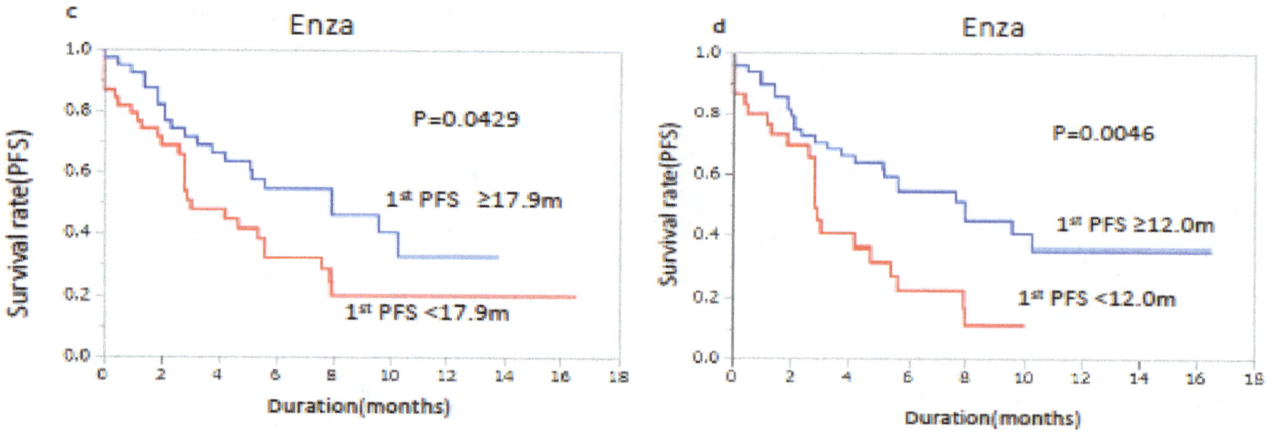

Figure 3. Progression-free survival according to the progression-free survival of first-line therapy. (**a**) Progression-free survival according to a median progression-free survival of first-line therapy <15.4 months or ≥15.4 months as the cut-off values in enzalutamide- and abiraterone-treated patients. (m, months). (**b**) Progression-free survival according to the progression-free survival of first-line therapy <12.0 months or ≥12.0 months as the cut-off values in enzalutamide-treated patients. (**c**) Progression-free survival according to a median progression-free survival of first-line therapy <17.9 months or ≥17.9 months as the cut-off values in enzalutamide-treated patients. (**d**) Progression-free survival according to a median progression-free survival of first-line therapy <12.6 months or ≥12.6 months as the cut-off values in abiraterone-treated patients.

4. Discussion

The current data indicate that a higher TST level at the stage of CRPC predicted a favorable response to Enza. When patients treated with Enza and Abi were combined, serum TST levels remained as independent prognostic factors for PFS. Moreover, the characteristics of the prognostic factors showed distinct differences between Enza and Abi. The response to Enza was more closely related to the response to first-line therapy, while the response to Abi was rather more independent from the response to first-line therapy. On the basis of the current data, TST level and the PFS of first-line therapy may be considered when choosing the treatment strategy for CRPC patients.

The reason why a higher TST level was associated with a favorable response to Enza and Abi may be related to the AR dependency of prostate cancer. When comparing clinical factors between higher and lower TST groups, higher TST was related to a relatively higher initial PSA (low TST 58.57 ng/mL vs high TST 81.65 ng/mL), although it was not significant. Since the expression of PSA is mediated by the transcriptional activity of nuclear AR, a higher initial PSA value may represent a higher basal AR activity inside the tumor microenvironment [18]. As both Enza and Abi work through AR, the higher AR dependency of the tumor may predict a higher response to AR-targeted drugs.

Another reason may be that a TST level of 13 ng/dL itself represents the remaining potential to target the AR-related pathway. The clinical significance of lowering the TST level below 50 ng/dL has been described in several reports [4,19]. Our group and others have previously reported that patients who achieved nadir TST < 20 ng/dL survived longer than those who did not [4,20–22]. Klotz et al. reported that patients with first-year nadir testosterone consistently >20 ng/dL had significantly higher risks of dying of prostate cancer [21]. These data indicated the clinical significance of lowering TST to <20 ng/dL in the tumor microenvironment [4]. However, these data were obtained when Enza and Abi were not on the market or were had a very limited use.

On the contrary, our data indicate that higher serum TST > 13 ng/dL represented the longer PFS for Enza and Abi groups, which is a novelty. The current cut-off value of serum TST of 13 ng/dL may represent the remaining AR dependency of the tumor that can only be blocked by treating with novel AR-targeted therapy. If tumor relapse occurs with a TST level >13 ng/dL, then this tumor may contain more AR-dependent cells, compared to tumors with a relapse occurring with TST < 13 ng/dL. Thus, even with a TST of 20 ng/dL, intensive blockade of the AR pathway through novel AR-targeted drugs would be effective, especially among patients with a higher TST level.

The clinical advantage of high TST was also reported by Ryan et al. [17]. Although the TST cut-off value was even lower (>8.6 ng/dL) compared to that in the present study, these authors found a higher TST to be related to a better response to Abi. On the other hand, in our study, a clinical advantage of high TST was not found for Abi-treated patients but was found for Enza-treated patients.

Interestingly enough, the higher TST group (TST $\geq$ 13 ng/dL) showed a relatively higher ALP, initial PSA, and PSA at the start of treatment with a similar rate of the high-volume tumor. The rate of visceral metastases was relatively higher in the TST $\geq$ 13 ng/dL group. Since the rate of visceral metastases was low (10–26%), it will be necessary to objectively assess the clinical significance of visceral metastases in a large number of patients. However, the present data may indicate that the response to novel AR-targeted drugs may not be related to a high or low tumor burden but may rather depend on the AR dependency of the tumor.

The current data indicate that the first-line PFS (12 months) predicted a favorable response to Enza (p = 0.0046), while the first-line PFS showed no association with the response to Abi (p = 0.6051). These data are in line with the findings of a previous report. Loriot et al. reported that the first-line ADT period predicted the response to novel AR-targeted drugs, mainly Enza [23]. Bellmunt et al. reported that the response to Abi was not related to the first-line ADT period [24]. The reason for this difference is not clear. However, one of the reasons may be that Enza shares a common mechanism with bicalutamide as an AR antagonist. Enza blocks AR with over a 30-fold higher affinity compared to bicalutamide [25]. The majority of the patients received CAB with bicalutamide and LH-RH agonist/antagonist as first-line ADT. Therefore, the patients who responded well to bicalutamide may also represent the patients who respond well to a potent AR antagonist, namely Enza. On the other hand, Abi works through Cyp17 inhibition, so the mechanism is distinct from that of bicalutamide and Enza. This mechanistic difference between Abi and Enza may represent the difference in response to the first-line therapy between Enza and Abi.

The present study provides several important findings. (1) Higher TST ($\geq$13 ng/dL) at the initiation of drug administration was associated with a favorable response to novel AR-targeted drugs, especially Enza; (2) The response to Enza was affected by the PFS of the first-line ADT, while the response to Abi was not affected by the PFS of the first-line ADT.

There are several limitations associated with this study. First, the sample size was rather small, which limits the reliability of the analysis, especially when assessing Abi and Enza independently. Second, because of the limited follow-up periods and limited outcomes regarding OS, the assessment of prognostic factors was mainly based on PFS. Third, because of the limited number of patients, the analysis was not divided in pre- or post-chemo settings. Although TST levels remain as predictors even among the factors that included the previous usage of chemotherapy, the response rate may also be affected by the previous usage of chemotherapy. Fourth, because of the prior introduction of Enza in the Japanese healthcare system, the majority of patients in the Enza group (92%) received the drug as a first-line novel AR-targeted drug, while half of the patients (54%) received Abi as a first-line AR-targeted drug. The different results observed for Enza and Abi, including association with TST levels, may possibly be affected by the background of the patients who received the drugs. We are currently performing a prospective study to re-assess the role of TST in the initial usage of Enza and Abi. The effects of the differences in the clinical features of the patients that induced the treatment choice in the current manuscript will be answered in the near future.

5. Conclusions

A higher TST level ($\geq$13 ng/dL) was associated with a favorable prognosis in Enza- and/or Abi-treated patients. A TST level of 13 ng/dL may predict a favorable response to novel AR-targeted drugs in CRPC patients.

Author Contributions: S.S. designed the study, collected the data, and completed the manuscript. M.X. and M.M. performed the statistical analysis. H.A.U. supervised the statistical analysis. S.K., Y.Y., H.K., H.M. (Hiroaki Matsumoto), N.T., K.H., M.N., H.N., H.M. (Hideyasu Matsuyama) collected the data. A.K., K.A., T.I. supervised the manuscript. All authors have read and approved the final version of the manuscript and agree with the order of presentation of the authors.

Acknowledgments: The authors are grateful to all members of Chiba University Hospital and the four related institutions for collecting and analyzing the patients' data. The authors particularly appreciate the help of Hisayo Karahi, a research assistant, for the documentation work.

Abbreviations

PSA	prostate-specific antigen
CRPC	castration-resistant prostate cancer
TST	testosterone
AR	androgen receptor
LH-RH	luteinizing hormone-releasing hormone
ALP	alkaline phosphatase
OS	overall survival
PFS	progression free survival
ADT	androgen deprivation therapy

References

1. Sakamoto, S.; McCann, R.O.; Dhir, R.; Kyprianou, N. Talin1 promotes tumor invasion and metastasis via focal adhesion signaling and anoikis resistance. *Cancer Res.* **2010**, *70*, 1885–1895. [CrossRef] [PubMed]
2. Huggins, C.; Hodges, C.V. Studies on prostatic cancer. I. The effect of castration, of estrogen and androgen injection on serum phosphatases in metastatic carcinoma of the prostate. *CA A Cancer J. Clin.* **1972**, *22*, 232–240. [CrossRef]
3. Gomella, L.G. Effective testosterone suppression for prostate cancer: Is there a best castration therapy? *Rev. Urol.* **2009**, *11*, 52–60.
4. Kamada, S.; Sakamoto, S.; Ando, K.; Muroi, A.; Fuse, M.; Kawamura, K.; Imamoto, T.; Suzuki, H.; Nagata, M.; Nihei, N.; et al. Nadir testosterone after long-term followup predicts prognosis in patients with prostate cancer treated with combined androgen blockade. *J. Urol.* **2015**, *194*, 1264–1270. [CrossRef]
5. Imamura, Y.; Sadar, M.D. Androgen receptor targeted therapies in castration-resistant prostate cancer: Bench to clinic. *Int. J. Urol.* **2016**, *23*, 654–665. [CrossRef]
6. Sakamoto, S. Editorial comment to androgen receptor targeted therapies in castration-resistant prostate cancer: Bench to clinic. *Int. J. Urol.* **2016**, *23*, 666. [CrossRef] [PubMed]
7. Sakamoto, S. Editorial comment to androgen receptor splice variant 7 in castration-resistant prostate cancer: Clinical considerations. *Int. J. Urol.* **2016**, *23*, 653. [CrossRef]
8. Di Zazzo, E.; Galasso, G.; Giovannelli, P.; Di Donato, M.; Castoria, G. Estrogens and their receptors in prostate cancer: Therapeutic implications. *Front. Oncol.* **2018**, *8*, 2. [CrossRef] [PubMed]
9. Di Zazzo, E.; Galasso, G.; Giovannelli, P.; Di Donato, M.; Di Santi, A.; Cernera, G.; Rossi, V.; Abbondanza, C.; Moncharmont, B.; Sinisi, A.A.; et al. Prostate cancer stem cells: The role of androgen and estrogen receptors. *Oncotarget* **2016**, *7*, 193–208. [CrossRef] [PubMed]
10. Chi, K.; Hotte, S.J.; Joshua, A.M.; North, S.; Wyatt, A.W.; Collins, L.L.; Saad, F. Treatment of mCRPC in the AR-axis-targeted therapy-resistant state. *Ann. Oncol.* **2015**, *26*, 2044–2056. [CrossRef]
11. Sweeney, C.J.; Chen, Y.H.; Carducci, M.; Liu, G.; Jarrard, D.F.; Eisenberger, M.; Wong, Y.N.; Hahn, N.; Kohli, M.; Cooney, M.M.; et al. Chemohormonal therapy in metastatic hormone-sensitive prostate cancer. *N. Engl. J. Med.* **2015**, *373*, 737–746. [CrossRef]
12. Sakamoto, S. Editorial comment to current status of primary pharmacotherapy and future perspectives toward upfront therapy for metastatic hormone-sensitive prostate cancer. *Int. J. Urol.* **2016**, *23*, 370. [CrossRef]
13. Shiota, M.; Eto, M. Current status of primary pharmacotherapy and future perspectives toward upfront therapy for metastatic hormone-sensitive prostate cancer. *Int. J. Urol.* **2016**, *23*, 360–369. [CrossRef]

14. James, N.D.; Sydes, M.R.; Clarke, N.W.; Mason, M.D.; Dearnaley, D.P.; Spears, M.R.; Ritchie, A.W.; Parker, C.C.; Russell, J.M.; Attard, G.; et al. Addition of docetaxel, zoledronic acid, or both to first-line long-term hormone therapy in prostate cancer (STAMPEDE): Survival results from an adaptive, multiarm, multistage, platform randomised controlled trial. *Lancet* **2016**, *387*, 1163–1177. [CrossRef]
15. Fizazi, K.; Tran, N.; Fein, L.; Matsubara, N.; Rodriguez-Antolin, A.; Alekseev, B.Y.; Ozguroglu, M.; Ye, D.; Feyerabend, S.; Protheroe, A.; et al. Abiraterone plus prednisone in metastatic, castration-sensitive prostate cancer. *N. Engl. J. Med.* **2017**, *377*, 352–360. [CrossRef]
16. James, N.D.; de Bono, J.S.; Spears, M.R.; Clarke, N.W.; Mason, M.D.; Dearnaley, D.P.; Ritchie, A.W.S.; Amos, C.L.; Gilson, C.; Jones, R.J.; et al. Abiraterone for prostate cancer not previously treated with hormone therapy. *N. Engl. J. Med.* **2017**, *377*, 338–351. [CrossRef]
17. Ryan, C.J.; Molina, A.; Li, J.; Kheoh, T.; Small, E.J.; Haqq, C.M.; Grant, R.P.; de Bono, J.S.; Scher, H.I. Serum androgens as prognostic biomarkers in castration-resistant prostate cancer: Results from an analysis of a randomized phase III trial. *J. Clin. Oncol.* **2013**, *31*, 2791–2798. [CrossRef]
18. Knudsen, K.E.; Penning, T.M. Partners in crime: Deregulation of AR activity and androgen synthesis in prostate cancer. *Trends Endocrinol. Metab.* **2010**, *21*, 315–324. [CrossRef]
19. Morote, J.; Orsola, A.; Planas, J.; Trilla, E.; Raventos, C.X.; Cecchini, L.; Catalan, R. Redefining clinically significant castration levels in patients with prostate cancer receiving continuous androgen deprivation therapy. *J. Urol.* **2007**, *178*, 1290–1295. [CrossRef]
20. Bertaglia, V.; Tucci, M.; Fiori, C.; Aroasio, E.; Poggio, M.; Buttigliero, C.; Grande, S.; Saini, A.; Porpiglia, F.; Berruti, A. Effects of serum testosterone levels after 6 months of androgen deprivation therapy on the outcome of patients with prostate cancer. *Clin. Genitourin. Cancer* **2013**, *11*, 325–330. [CrossRef]
21. Klotz, L.; O'Callaghan, C.; Ding, K.; Toren, P.; Dearnaley, D.; Higano, C.S.; Horwitz, E.; Malone, S.; Goldenberg, L.; Gospodarowicz, M.; et al. Nadir testosterone within first year of androgen-deprivation therapy (ADT) predicts for time to castration-resistant progression: A secondary analysis of the PR-7 trial of intermittent versus continuous ADT. *J. Clin. Oncol.* **2015**, *33*, 1151–1156. [CrossRef]
22. Yamamoto, S.; Sakamoto, S.; Minhui, X.; Tamura, T.; Otsuka, K.; Sato, K.; Maimaiti, M.; Kamada, S.; Takei, A.; Fuse, M.; et al. Testosterone reduction of >/= 480 ng/dL predicts favorable prognosis of japanese men with advanced prostate cancer treated with androgen-deprivation therapy. *Clin. Genitourin. Cancer* **2017**, *15*, e1107–e1115. [CrossRef]
23. Loriot, Y.; Eymard, J.C.; Patrikidou, A.; Ileana, E.; Massard, C.; Albiges, L.; Di Palma, M.; Escudier, B.; Fizazi, K. Prior long response to androgen deprivation predicts response to next-generation androgen receptor axis targeted drugs in castration resistant prostate cancer. *Eur. J. Cancer* **2015**, *51*, 1946–1952. [CrossRef]
24. Bellmunt, J.; Kheoh, T.; Yu, M.K.; Smith, M.R.; Small, E.J.; Mulders, P.F.; Fizazi, K.; Rathkopf, D.E.; Saad, F.; Scher, H.I.; et al. Prior endocrine therapy impact on abiraterone acetate clinical efficacy in metastatic castration-resistant prostate cancer: Post-hoc analysis of randomised phase 3 studies. *Eur. Urol.* **2016**, *69*, 924–932. [CrossRef]
25. Tran, C.; Ouk, S.; Clegg, N.J.; Chen, Y.; Watson, P.A.; Arora, V.; Wongvipat, J.; Smith-Jones, P.M.; Yoo, D.; Kwon, A.; et al. Development of a second-generation antiandrogen for treatment of advanced prostate cancer. *Science* **2009**, *324*, 787–790. [CrossRef]

17

Research Evidence on High-Fat Diet-Induced Prostate Cancer Development and Progression

Shintaro Narita *, Taketoshi Nara, Hiromi Sato, Atsushi Koizumi, Mingguo Huang, Takamitsu Inoue and Tomonori Habuchi

Department of Urology, Akita University School of Medicine, Akita 010-8543, Japan;
bkspt512@yahoo.co.jp (T.N.); hiromisato2002@yahoo.co.jp (H.S.); koizu3atsu4@yahoo.co.jp (A.K.);
huangmg0319@yahoo.co.jp (M.H.); takmitz@gmail.com (T.I.); thabuchi@gmail.com (T.H.)
* Correspondence: naritashintaro@gmail.com

Abstract: Although recent evidence has suggested that a high-fat diet (HFD) plays an important role in prostate carcinogenesis, the underlying mechanisms have largely remained unknown. This review thus summarizes previous preclinical studies that have used prostate cancer cells and animal models to assess the impact of dietary fat on prostate cancer development and progression. Large variations in the previous studies were found during the selection of preclinical models and types of dietary intervention. Subcutaneous human prostate cancer cell xenografts, such as LNCaP, LAPC-4, and PC-3 and genetic engineered mouse models, such as TRAMP and Pten knockout, were frequently used. The dietary interventions had not been standardized, and distinct variations in the phenotype were observed in different studies using distinct HFD components. The use of different dietary components in the research models is reported to influence the effect of diet-induced metabolic disorders. The proposed underlying mechanisms for HFD-induced prostate cancer were divided into (1) growth factor signaling, (2) lipid metabolism, (3) inflammation, (4) hormonal modulation, and others. A number of preclinical studies proposed that dietary fat and/or obesity enhanced prostate cancer development and progression. However, the relationship still remains controversial, and care should be taken when interpreting the results in a human context. Future studies using more sophisticated preclinical models are imperative in order to explore deeper understanding regarding the impact of dietary fat on the development and progression of prostate cancer.

Keywords: animal model; diet; fat; in vitro; in vivo; mouse; prostate cancer

---

1. Introduction

Prostate cancer is the most common type of cancer among men in 92 countries and the leading cause of cancer deaths among men in 48 countries [1]. In the United States as well, prostate cancer has been the most commonly diagnosed type of cancer among men, accounting for almost 1 in 5 new diagnoses [2]. While the incidence of latent prostate cancer has been similar between the United States and Japan, the incidence of clinically detected prostate cancer has been lower in Asia, including Japan [3,4]. Of note, the incidence of prostate cancer in Chinese and Japanese men has been reported to increase substantially after migration to the United States [5]. Furthermore, the morbidity and mortality due to prostate cancer in Asia increased remarkably in recent years [6]. Although the etiology of prostate cancer is multifocal, these epidemiological findings, including geographic and ethnic differences, suggest that lifestyle and/or environmental factors have a substantial influence on the development and progression of prostate cancer [7]. Epidemiological evidence suggested that among the acquired risk factors for prostate cancer development and progression, diet and obesity have a potential to cause prostate cancer initiation, promotion, and progression [8,9]. Several studies have implicated dietary fats as important factors of prostate cancer risk and its aggressive phenotype [9,10]. A number of clinical and preclinical studies have shown that total fat intake and specific fat composition play a potential role in prostate cancer, although their findings have remained inconclusive.

Considering these backgrounds, this study aimed to summarize previous preclinical studies regarding the relationship between dietary fat and prostate cancer development and progression, focusing on differences in preclinical models and dietary fat composition. Furthermore, potential mechanisms on dietary fat-induced prostate carcinogenesis were discussed by updating previous research evidence. To this end, previous preclinical studies investigating dietary fat and prostate cancer were identified using a PubMed search including only studies published in English. This review helps us to understand the current state of diet-induced prostate cancer research in order to guide future works exploring the association between dietary-fat and prostate cancer.

2. Various Preclinical Models

A number of animal models, including those involving prostate cancer cell xenografts and allografts, Transgenic Adenocarcinoma of the Mouse Prostate (TRAMP) mice, and other genetically engineered mice targeting oncogenes and tumor suppressor genes, were tested in order to assess the impact of dietary-fat intake on prostate cancer development and progression (Table 1). First, the models used in the previous studies were summarized as follows.

Table 1. Summary of preclinical models on dietary-fat induced prostate cancer development and progression.

Authors	Years	Animal Models	Tumors	Diet Summary	End Point	Summary of the Results
Wang [11]	1995	Nude mice	LNCaP	40.5%, 30.8%, 21.2%, 11.6%, or 2.31% fat	Tumor growth rates, tumor weights, ratios of final tumor weights to animal weights, PSA	Groups that continued to receive a 40.51% fat diet were substantially greater tumor growth rates, final tumor weights, and ratios of final tumor weights to animal weights than those whose diets were changed to 2.3 kcal%, 11.6 kcal%, or 21.2 kcal% fat.
Connoly [12]	1997	Nude mice	a) DU145 subcutaneous xenograft, b) DU145 into prostate	a) 18:2 ω-6-rich vs. 18:3 ω-3-rich vs. 20:5 and 22:6 ω-3-rich, b) ω-6-rich vs. a LF	Tumor growth	a) 18:2 ω-6-rich vs. 18:3 ω-3-rich mice were similar; a 30% reduction in tumor growth was observed in the 20:5 and 22:6 ω-3-rich groups. b) The mean tumor weight in the ω-6-rich group was twice that in the low-fat group.
Ngo [13]	2002		LNCaP cultured with human serum	Before and after residential diet and exercise	Cell growth, apoptosis, necrosis	Serum-stimulated LNCaP cell growth was reduced by 30% in post-11-day serum and by 44% in long-term serum relative to baseline. LNCaP cells incubated with post-diet and exercise serum showed higher apoptosis/necrosis, compared to baseline.
Barnard [14]	2003		LNCaP cultured with human serum	Volunteer serum (control, LF and exercise, exercise alone)	Cell growth	Both the LF/exercise and exercise alone groups had reduced LNCaP cell growth compared to control.
Ngo [15]	2003	CB17 SCID	a) LAPC-4 xenograft, b) LAPC-4 culture with 10% mouse serum	HFD (42%) vs. LFD (12%)	a) tumor growth, PSA, b) cell growth	LFD mice had significantly slower tumor growth rates and lower serum PSA levels compared to HFD mice. LAPC-4 cells cultured in vitro with media containing serum from LFD mice demonstrated slower growth than LAPC-4 cells cultured in media containing HFD mice serum.
Ngo [16]	2004	CB17 SCID	LAPC-4 xeograft	HFD (42%) vs LFD (12%)	Tumor growth, survival	Tumor latency and mouse survival were significantly longer in the LFD castration versus HFD castration group.
Venkateswaran [17]	2007	Swiss nu/nu	LNCaP xenograft	HC + HFD vs. LC + HFD	Tumor growth	Mice on the HC–HFD diet experienced increased tumor growth.
Berquin [18]	2007	Prostate-specific Pten deletion mouse		High ω-6 vs. ω-3 diet	Prostate weight, rate of invasive carcinoma	Prostate weight was significantly lower in mice fed high ω-3; half of the mice fed ω-3 develop invasive carcinoma, whereas 80% of the mice fed high ω-6 diet had invasive carcinoma.
Kobayashi [19]	2008	Prostate specific High-Myc transgenic mouse	a)LNCaP, b)MycCap with mice serum	HFD (42%) vs LFD (12%)	Rate of mPIN and cancer incidence	The number of mice that developed invasive adenocarcinoma at 7 months was 27 % less in the LFD group (12/28) compared to the HFD group (23/33, $p = 0.04$). Epithelial cells in PIN lesions in the LFD group had a significantly lower proliferative index compared to epithelial cells in the HFD group (21.7% vs. 28.9%, $p < 0.05$).
Freedland [20]	2008	SCID	LAPC-4 xenograft	NCKD (84% fat) vs. LFD (12% fat) vs. WD (40% fat)	Tumor growth, survival	NCKD mice tumor volumes were 33% smaller than WD mice (rank-sum, $p = 0.009$). No differences in tumor volume were observed between LFD and NCKD mice with the latter having the longest survival.
Narita [16]	2008	BALB/c-nu/nu	LNCap xenograft	HF (56.7%) vs. LF (10.2%)	Tumor volume, PSA	Tumor volume and serum PSA levels were significantly higher in the HFD group than in the LFD group.

Table 1. *Cont.*

Authors	Years	Animal Models	Tumors	Diet Summary	End Point	Summary of the Results
Mavropoulos [21]	2009	SCID	LNCaP xenograft	NCKD (83% fat) vs. LFD (12% fat) vs. WD (40% fat)	Tumor growth, survival	Tumor volumes in the WD group remained significantly larger than tumor volumes in the LFD and NCKD groups. Survival was significantly prolonged for the LF (hazard ratio, 0.49; 95% confidence interval, 0.29–0.79; $p = 0.004$) and NCKD groups (hazard ratio, 0.59; 95% confidence interval, 0.37–0.93; $p = 0.02$).
Tamura [22]	2009	Nude mice	LNCaP xenograft	HFD (14%) vs LFD (6%)	Tumor growth	LNCaP-Mock cells did not reveal any significantgrowth promotion by breeding with HFD. HFD breeding significantly promoted the growth of LNCaP-ELOVL7-1 cells in vivo ($p = 0.0081$).
Kalaany [23]	2009	Prostate-specific Pten deletion mouse		Ad libitum vs. CR	Percentage of proliferation and apoptosis	CR did not affect a PTEN-null mouse model of prostate cancer but significantly decreased tumor burden in a mouse model of lung cancer lacking constitutive PI3K signaling.
Bushemeyer [24]	2010	SCID	LAPC-4 xenograft	7 types of diet	Tumor growth, survival	No significant differences in tumor volume were observed among the various groups at any time point. Overall, the treatment group was not significantly related to survival.
Llaverias [25]	2010	TRAMP mouse		WD (21.2%) vs. chow (4.5%)	Prostate tumor incidence and progression	TRAMP mice fed a WD were shown to develop larger tumors compared to mice fed a chow diet. 67% (6 of 9 mice) of TRAMP mice fed a WD exhibited at least one metastatic focus, whereas 43% (3 of 7 mice) of mice fed a chow diet exhibited the same.
Lloyd [26]	2010	SCID	LAPC-4 xenograft	WD (40%) vs. chow (12%)	Tumor growth, survival	No difference in tumor growth or survival between chow and WD was observed.
Aronson [27]	2010		LNCaP cultured with human serum	PCa men with LF, high-fiber, soy protein-supplemented diet or WD for 4 weeks	Cell growth	LF, high-fiber, soy protein-supplement diet decreased LNCaP cancer cell growth.
Masko [28]	2010	SCID CB17	LAPC-4 xenograft	NCKD (84% fat), 10% carbohydrate diet (74% fat), or 20% carbohydrate diet (64% fat).	Tumor volume, PSA, survival	Tumors were significantly larger in the 10% carbohydrate group on days 52 and 59 ($p < 0.05$) and at no other point during the study. Diet did not affect survival ($p = 0.34$).
Akinsete [29]	2012	C3 (1) Tag transgenic mouse		High ω-6 vs. ω-3 diet	Tumor progression, apoptosis	Slower progression of tumorigenesis and enhanced apoptosis was observed in dorsalateral prostate of high ω-3 diet mice than in high ω-6 diet mice.
Mao [30]	2012	Homozygous prostate-specific RXRα knockout mouse		NWD (higher fat content, reduced calcium, vitamin D, and fiber) or AIN-76A		A significant joint effect of NWD and RXRα status in developing mPIN, but interaction was not significant owing to the small sample size.
Bonorden [31]	2012	a) TRAMP mouse, b) C57/BL6	b) TRAMP-C2 allograft	LFD (AIN-93M) vs. AIN-93M-HFD (33%)	a) tumor differentiation, percentage of metastasis, b) tumor weight and volume	No difference in the prostates of TRAMP mice. TRAMP-C2 cells grew faster when the mice were fed a HFD.

Table 1. *Cont.*

Authors	Years	Animal Models	Tumors	Diet Summary	End Point	Summary of the Results
Konijeti [32]	2012	SCID	22Rv1	HFD (43.3%) + saline, HFD + IGF-1R-Ab, LFD (12.4%) + saline, LFD + IGF-1R-Ab	Tumor volume	No significant differences in final tumor volumes or final tumor weights were observed between the treatment groups. At day 14 of the intervention, the mean tumor volume was significantly lower in the LFD + IGF-1R-Ab group than in the HF group.
Huang [33]	2012	BALB/c-nu/nu	LNCaP xenograft	HFD (59.9%) vs. HCD (9.5%) vs CD (41.2%)	Tumor volume	The tumor growth of LNCaP xenograft was significantly higher in the HFD group than in the HCD and CD groups.
Wang [34]	2012	a) nude mice, b) Prostate-specific Pten deletion mouse	a) pten-/- allograft	High ω-6 vs ω-3 diet	a) tumor volume and weight, b) body weight, invasion rate, Ki67	ω-3 PUFA resulted in slower growth of castration-resistant tumors compared to ω-6 PUFA.
Vandelsluis [35]	2013	Nu/nu athymic mice	LNCaP xenograft	HFD (23.8%) vs. SD (6.0%)	Tumor volume	The HF with exercise group showed significantly higher tumor growth rates compared to all other groups. The SD with exercise group had significantly lower tumor growth rates of compared to the HFD without exercise group.
Pommier [36]	2013	C57BL/6 Lxra and Lxrb double knockout mice		Normal or hypercholesterolemic diet	Presence of PIN, number of Ki-67 positive cells	High-cholesterol diet induced proliferation in LXR mutant mouse prostate.
Huang [37]	2014	BALB/c-nu/nu	LNCaP xenograft	HFD (59.9%) vs. LFD (9.5%)	Tumor volume	The tumor growth of LNCaP xenograft was significantly higher in the HFD group than the LFD groups.
Moiola [38]	2014	Swiss nu/nu	PC-3 xenograft	HFD (homemade) vs. CD	Tumor volume	No significant differences in tumor growth were observed in CD-fed mice; however, we found that only 60% of HFD-fed mice inoculated with CtBP1-depleted cells developed a tumor.
Chang [39]	2014	TRAMP mouse		HFD (45%) vs. CD (10%)	Histophathologica score	Histopathological scores in the dorsal and lateral lobes were higher in the 10-week HFD group than in the 10-week CD group.
Liu [40]	2015	Pten haploinsufficientmale mice		High calorie vs. regular diet	mPIN score	High-calorie diet caused neoplastic progression, angiogenesis, inflammation, and epithelial-mesenchymal transition
Cho [41]	2015	a) TRAMP, b) C57BL/6J	b) TRAMPC2 allograft	HFD (60%) vs. CD (10%)	Rate of poorly differentiated ca, tumor weight	In TRAMP mice, HFD feeding increased the incidence of poorly differentiated carcinoma. In the allograft model, HFD increased solid tumor growth and the expression of proteins related to proliferation/angiogenesis.
Xu [42]	2015	TRAMP		HFD (40%) vs. ND (16%)	Tumor formation rate, survival	The mortality of TRAMP mice from HFD group was significantly higher than that of normal diet group (23.81% and 7.14%, $p = 0.035$). The tumor incidence of HFD TRAMP mice at 20th week was significantly higher than normal diet group (78.57% and 35.71%, $p = 0.022$)
Xu [43]	2015	TRAMP		HF (40%) vs. ND (16%)	Tumor incidence, survival	TRAMP mice in HFD group had significantly higher mortality rates than those in the normal diet group ($p = 0.032$). The HFD group had a significantly higher tumor formation rate at age 20 weeks than the normal diet group ($p = 0.045$).

Table 1. *Cont.*

Authors	Years	Animal Models	Tumors	Diet Summary	End Point	Summary of the Results
Lo [44]	2016	SCID	PDX kidney capsule xenograft	HF (43%) vs LF (6%)	Pathology and biomarker expression	Prostate cancer tumorigenicity is not accelerated in the setting of diet-induced obesity or in the presence of human PPAT.
Liang [45]	2016	Immunocompetent FVB mice	MycCap alloraft	High ω-6 vs. ω-3 diet	Tumor volume	Tumor volumes were significantly smaller in the ω-3 than in the ω-6 group ($p = 0.048$).
Huang [46]	2016	BALB/c-nu/nu	LNCap xenograft	HFD (59.9%) vs LF (9.5%)	Intratumoral AKT and Extracellular Signal-regulated Kinase (ERK) activation, AMPK inactivation	HFD resulted in AKT and ERK activation and AMPK inactivation.
Labbe [47]	2016	Prostate specific Pten and Ptpn1 deletion mouse		HFD vs. chow	Microinvasive rate	PCa in Pten-/-Ptpn1-/- mice was characterized by increased cell proliferation and Akt activation, interpreted to reflect a heightened sensitivity to IGF-1 stimulation upon HFD feeding
Kwon [48]	2016	14K-creER PTEN (K14-CreER;Pten$^{fl/fl}$;mTmG (K14-Pten-mTmG) triple transgenic mice		HFD vs. RD	PIN 3/4 rate	HFD increased the number of PIN.
Zhang [49]	2016	C57BL6	RM1 mouse prostate cancer alloglaft	HFD (58%) vs. chow	Tumor growth	CXCL1 chemokine gradient was required for the obesity-dependent tumor ASC recruitment, vascularization and tumor growth promotion
Chang [50]	2017	C57BL6		HFD (45%) vs. chow	Cav-1 secretion from adipose tissue	Cav-1 secretion was evident in adipose tissues and were substantially promoted in HFD-fed mice.
Kim [51]	2017	SCID	PC-3 xenograft	10%, 45%, or 60% fat	Tumor size, tumor weight	The 45% and 60% fat diets significantly promoted the growth of xenografts comparison to the 10% fat diet
Nara [52]	2017	a) BALB/c-nu/nu	a) LNCap xenograft, b) PC-3 and DU145 cultured with mice serum	HFD (59.9%) vs. CD (9.5%)	a) Tumor volume, b) cell proliferation	The tumor growth of prostate cancer LNCaP xenograft was significantly higher in the HFD group than in the CD groups. Cells cultured with HFD mouse serum had higher proliferation.
Huang [53]	2017	BALB/c-nu/nu	Intraperitoneal injection PC-3M-luc-C6	HFD (59.9%) vs. LF (9.5%)	Luciferase activity (IVIS), number of metastasis	HFD and PrSC increased luciferase activity and number of metastasis.
Hayashi [54]	2018	Prostate-specific Pten deletion mouse		HFD (62.2%) vs. CD (12.5%)	Tumor growth	HFD accelerated tumor growth alogn with the inflammatory response.
Massillo [55]	2018	C57BL/6J	TRAMP C1 allograft	HFD (37%) vs. CD (5%)	Tumor volume	HFD significantly increased tumor growth and serum estradiol in mice.
Chen [56]	2018	Prostate specific Pten and Pml deletion mouse		HFD (60%) vs. chow (17%)	Rate of mice having metastases	A HFD-derived metastatic progression and increases lipid abundance in prostate tumors
Hu [57]	2018	TRAMP		HFD (40%) vs. CD (16%)	Proportion of poor tumor differentiation and tumor metastasis	A trend toward poorer PCa differentiation was observed in HFD-fed mice, while no statistical significance was detected.

Abbreviations: HFD: high-fat diet, LFD: low-fat diet, HC: high-calorie diet, LC: low-calorie diet, NKCD: high-fat/no-carbohydrate ketogenic diet, WD: Western-style diet, CR: calorie restriction, Ab: antibody, SD: standard diet, CD: control diet, PDX: patient-derived xenograft, NWD: new Western-style diet.

2.1. Human Cancer Cell Xenograft and Allograft Models

The most experienced models to assess the impact of dietary fats on prostate cancer growth were subcutaneous xenograft models [11,12,15,16,35,52,58].

Nude [16] and severe combined immunodeficient (SCID) mice [15] were frequently used as host mice for human prostate cancer cell xenografts. In 1995, Wand et al. first assessed the impact of five different fat percentages on human prostatic adenocarcinoma (LNCaP) xenograft growth using athymic nude mice [11]. Accordingly, mice who continued to receive a 40.5-kcal% fat diet had substantially greater tumor growth rates, final tumor weights, and final tumor weight to animal weight ratios compared to those whose diets were changed to 2.3 kcal%, 11.6 kcal%, or 21.2 kcal% fat, suggesting that those fed low-fat diets (LFDs) had decreased growth of established LNCaP tumors. An additional study demonstrated that an isocaloric LFD (12 kcal% fat) resulted in significantly slower tumor growth rates and lower serum prostate-specific antigen (PSA) levels compared to a high-fat diet (HFD) using LAPC-4 xenografts on SCID mice [15]. The same group also showed that reduced dietary fat intake delayed conversion from androgen-sensitive to androgen-insensitive prostate cancer and significantly prolonged survival of SCID mice bearing LAPC-4 xenografts [58]. Moreover, we had previously found that Balb/c-nu/nu mice receiving a HFD had significantly higher LNCaP xenograft tumor volumes and serum PSA levels than those receiving an LFD [52]. The impact of a HFD on xenograft tumor growth using other human prostate cancer cell lines, such as 22Rv-1 and PC-3, had also been investigated in previous literatures [32,51]. Although the significance of the effect varied, a number of studies proposed that a HFD accelerated tumor growth of human prostate cancer cell xenografts inoculated into immunodeficient mice. Conversely, several studies have found no relationship between a HFD and xenograft growth [22,26]. In a study comparing LAPC-4-xenografted SCID mice receiving an isocaloric Western diet (40% fat and 44% carbohydrate) and those receiving an LFD (12% fat and 72% carbohydrate), the authors found no difference in tumor growth or survival between both groups when saturated fat was used as the fat source [26]. Another study showed no difference in LNCaP tumor size between normal (6% fat) and high-fat (14% fat) diets [22]. Taken together, a number of studies involving subcutaneous human prostate cancer cell xenografts in immunodeficient mice suggested an association between HFD and xenograft growth, whereas several other studies showed no such relationship. The lack of standardization in terms of models and duration of specific diet feeding has remained problematic.

Given the variations in the genetic background of mouse strains, it is important to consider the importance of the immune system in tumor progression [59]. Several studies have investigated the impact of dietary fat on allografts using immunocompetent mice and mouse-derived prostate cancer cells [31,41,45,49,55]. Several groups have shown that a HFD significantly increased allograft tumor growth of TRAMP-derived prostate cancer cells, such as TRAMP-C1 and TRAMP-C2, in C57BL6 mice [31,41,55]. The study involving the largest number of allografts (low-fat; $n = 40$, high-fat; $n = 134$) revealed that mice receiving AIN-93M-high-fat diet had significantly heavier and significantly larger TRAMP-C2 allografts compared to those receiving AIN-93M, whereas no differences in prostate weight were observed among the groups [31]. This result suggests that TRAMP allografts derived from C57BL6 mice can be one of the promising allograft models when studying HFD-induced prostate cancer progression.

A unique study involving a peritoneal dissemination model established through intracorporeal injection of PC-3M-luc cells detected using the Xenogen IVIS™ system reported that a HFD increased tumor formation rates and total metastasis rates in the peritoneal organs [53].

In summary, given that most of the xenograft and allograft studies were performed using subcutaneous xenograft models, studies involving metastatic models and human patient-derived xenografts (PDXs) have been lacking. Although several studies using xenografts and allografts have shown that a HFD accelerated tumor growth, further validation is warranted.

2.2. TRAMP Mouse Models

Since its generation in 1996, the TRAMP mouse model has been one of the most widely used models in prostate cancer research [60]. This model represents a transgene comprising the minimal probasin promoter driving viral SV40 large-T and small-t antigens, which lead to prostate-specific inactivation of pRb and p53, specifically in the prostatic epithelium [61]. TRAMP mice develop prostatic intraepithelial neoplasia (PIN) by the time they are 6 weeks old; this progresses to high-grade PIN by the age of 12 weeks and poorly differentiated and invasive adenocarcinoma by the age of 24 weeks, with nearly 100% penetrance [61]. The impact of a HFD on the growth of TRAMP mouse tumors had been frequently evaluated [25,31,39,41–43,57,62]. Accordingly, Llaverias et al. showed that mice consuming a Western-type diet enriched in both fat and cholesterol had higher prostate tumor incidence and greater tumor burden compared to those fed a control chow diet [25]. After necropsy at 28 weeks, 33% of TRAMP mice fed a Western diet showed grossly evident spherical prostate tumors, whereas only 17% of TRAMP mice fed a chow diet exhibited the same [25]. In another study on TRAMP mice, Xu et al. revealed that the HFD group had significantly higher mortality than the normal diet group (23.81% and 7.14%, respectively, $p = 0.035$). Moreover, HFD-fed TRAMP mice had significantly higher tumor incidence at 20 weeks, as compared to the normal diet group (78.57% and 35.71%, $p = 0.022$, respectively) [43]. The same group also showed that HFD-fed mice suffered higher rates of extracapsular extension (20 weeks, 16.7% vs. 8.3%; 28 weeks, 66.7% vs. 50.0%, respectively) and distant metastasis (e.g., retroperitoneal lymph nodes or lung metastasis) (28 weeks, 41.7% vs. 25.0%, respectively) [62]. Bonorden et al. conducted a unique study involving the largest number of mice ($n = 25$ each) to assess the direct effect of diet and body weight on prostate tumors. TRAMP mice received low- and high-fat diets with the latter being divided into three groups: obesity-prone (the heaviest third), overweight (the middle third), and obesity-resistant (the lightest third). Accordingly, their results showed that body weight or diet had no effect of on either age at tumor detection, neuroendocrine status, or age at death [31]. Taken together, the impact of a HFD on tumor incidence and survival of TRAMP mice still remains controversial. The timing of diet change, selection of control diet, and diet ingredients may be important in establishing HFD-accelerated orthotopic prostate tumor models in TRAMP mice.

2.3. Other Genetically Engineered/Transgenic Mouse Models Targeting Oncogenes and Tumor Suppressor Genes

Several studies have investigated the effect of dietary fat on prostate cancer development and progression using genetically engineered mouse models (GEMMs) targeting oncogenes and tumor suppressor genes [19,29,30,36,40,47,48,54,56]. Designated Hi-myc uses a PB promoter coupled with a sequence of the ARR2 promoter, both of which lie upstream to the human c-Myc gene, in order to drive progression from mouse prostatic intraepithelial neoplasia (mPIN) to invasive adenocarcinoma [63]. Using this animal model, Kobayashi et al. showed that the HFD group (42 kcal% fat) had a greater number of invasive adenocarcinoma and a higher proliferative index in the PIN region compared to the LFD group (12 kcal% fat) [19]. Phosphatase And Tensin Homolog (Pten) alteration has been shown to be an early event in prostate cancer initiation and progression. Moreover, Pten-null mice that develop PIN have among the valuable animal models in prostate cancer research [64]. Kalaany et al. showed that 40% dietary restriction did not have any detectable effect on the extent or histological appearance of the PIN in Probasin-Cre; PTEN L/L prostate cancer models but significantly reduced tumor nodules in the lungs of K-RAS^{LA2}; P53 LSL/WT lung adenocarcinoma models [23], suggesting that the Phosphoinositide 3-kinase (PI3K)/ protein kinase B (AKT) pathway is critical for diet-induced cancer progression. Conversely, a high-calorie diet (45 kcal% fat) promoted prostate cancer progression in genetically susceptible Pten haploinsufficient mice with increasing inflammatory response in the presence of enhanced insulin response to chronically elevated insulin levels [40]. Hayashi et al. demonstrated that mice receiving a HFD for 17 weeks starting from an age of 5 weeks had significantly higher prostate weights of than those receiving control [54]. Moreover, HFD-fed model mice had a significantly higher Ki67-positive cell to tumor cell ratio than control mice, while no marked difference in glandular structures was observed between the control diet (CD)-fed and HFD-fed model mice [54]. An interesting study involving the basal cell-specific Pten-null model using K14-Pten-mTmG mice showed that HFD intake promoted the initiation and progression of PIN lesions [48]. Although dietary fat could potentially be associated with prostate cancer development of Pten-null mice, the impact may not be extensive. Additionally, the evaluation of prostate pathology in GEMMs needs to be standardized according to the Consensus Report from the Bar Harbor Meeting of the Mouse Models of Human Cancer Consortium Prostate Pathology Committee for accurate comparison among different studies [65].

With regard to other GEMMs, PTP1B (PTPN1), an androgen-regulated phosphatase, acts as a HFD-dependent tumor suppressor in prostate cancer driven by the absence of Pten, such as in the Pten-/-Ptpn1-/- mice model [47]. Deficiency in RXRα (a unique and important member of the nuclear receptor superfamily) in the prostates of mice receiving a new Western-style diet resulted in higher rates of mPIN and prostate cancer [30]. Pommier et al. showed that mice with double knockout of Liver X receptors (LXRa and LXRb), which belong to the nuclear receptor superfamily and are central mediators of cholesterol homeostasis, developed PIN under a diet high in cholesterol [36].

Reports regarding HFD-induced metastatic models using GEMMs have been rare. Chen et al. showed that among mice with Pten deletion and a double deletion of Pten and Promyelocytic Leukemia (PML), a suppressor of pp1α-dependent activation of MAPK signaling, those receiving a lard-based HFD displayed lymph node metastasis and lung metastasis, whereas those receiving a chow diet exhibited limited metastases [56].

Taken together, GEMM studies showed that a HFD enhanced tumor growth through the modulation of several genes, including those related to PTEN. Studies that assess the impact of a HFD using more aggressive, metastatic GEMMs while considering the effect of dual and/or multiple genes may be intriguing.

2.4. Others

Several studies have evaluated the proliferation of prostate cancer cell lines cultured with serum from mice and humans under different diet conditions [13,27]. Two mice studies proposed that a HFD serum enhanced cell proliferation of LAPC-4 and PC-3/DU145 cells in CB17 SCID and Balb-c/nu/nu mice, respectively [15,52]. With regard to in vitro studies using human sera, Barnard et al. assessed the growth of LNCaP cells cultured with healthy volunteer serum according to dietary fat and exercise condition [14]. Accordingly, they found that an LFD with exercise inhibited cell growth. Subsequently, after evaluating the growth of LNCaP cells cultured with sera from patients with prostate cancer receiving a low-fat, high-fiber, soy-protein supplement diet or Western diet for 4 weeks, Aronson et al. showed that the LFD induced changes in serum fatty acid levels with decreased LNCaP cancer cell growth [27]. In an interesting study by Lo et al., PDX models of prostate cancer cells implanted into the renal capsule of SCID mice were developed [44]. Histological analysis of the PDXs showed no differences in tumor pathology; PSA, androgen receptor, and homeobox protein Nkx-3.1 expression; or proliferation index between HFD- and LFD-fed mice. Furthermore, they also evaluated the impact of co-grafting human periprostatic adipose tissue (PPAT) with prostate cancer in PDX grafts. After harvesting the PDX tissues 10 weeks after grafting, histological analysis revealed no evidence of enhanced tumorigenesis with PPAT compared to prostate cancer grafts alone. It would be intriguing to assess the effects of a HFD on PDXs with a more aggressive prostate cancer phenotype obtained from metastatic disease and co-grafting this with PPAT from patients with severe obesity considering that the aforementioned model was established using tissues from patients with localized prostate cancer treated with surgery.

3. Differences in Diets

A number of studies have tried to assess the impact of a fat-enriched diet on prostate cancer development and progression. However, the dietary interventions had not been standardized, while distinct variations in phenotype had been observed among different studies using distinct HFD components. Differences in dietary components among research models had also been reported to affect the distinct effect of diet-induced metabolic disorders [66]. Therefore, in addition to the models used, the type of diet remains essential for studies to delineate diet-induced carcinogenesis.

3.1. *Direct Comparison between Two Different Diets Including the High-Fat Diet and Another Diet*

While conclusions have been frequently drawn from comparisons between a defined HFD and chow, specific details regarding the control diet are often lacking. Many studies have utilized a chow diet as the control treatment [25,26,49,50,56]. Regular chow is composed of agricultural byproducts, such as ground wheat, corn or oats, alfalfa, and soybean meals; a protein source, such as fish; and vegetable oil; it is supplemented with minerals and vitamins. Thus, chow can be considered a high-fiber diet containing complex carbohydrates with fats from various vegetable sources. Chow is inexpensive to manufacture and palatable for rodents. In contrast, defined HFDs consist of amino acid-supplemented casein, cornstarch, maltodextrin or sucrose, and soybean oil or lard and are supplemented with minerals and vitamins. Fiber is often provided by cellulose. Chow and defined diets may exert significant separate and independent unintended effects on the measured phenotypes in any research protocol [66]. In sum, multiple limitations may affect the results of the target groups.

A direct comparison between two different diets including the HFD, has been used extensively to understand the role of diet on prostate cancer development and progression [12,15,27,30,43,47,51,58]. Most of the studies showed that HFD-fed mice had greater body weight compared to controls, which leads one to consider whether diet has a direct or indirect (obesity-induced) effect on cancer development and progression. Although majority of the previous studies proposed an association between dietary fat and prostate cancer development/progression, several limitations need to be considered. First, a multitude of proportions per calories of fat have been observed with relative fat fractions ranging between 14% and 84% energy as fat. We need to consider the fact that the higher proportions of fat used in animal studies cannot be used in human diets. Second, we need to be careful about being misled by ignoring the impact of fat components, the control diet, and other elements in each diet. For instance, Lloyd et al. showed no difference in the growth and survival of LAPC-4 xenografts between SCID mice receiving a Western-style diet, including 40% kcal fat, and those fed an LFD (12% kcal) [26]. In this study, the fat consisted of 19% lard, 19% milk fat, and 1.9% corn oil. Conversely, another study demonstrated that HFD-fed SCID mice (42% kcal) had significantly faster LAPC-4 tumor xenograft growth and higher PSA levels compared to LFD-fed mice (12% kcal) [15]. In this study, the HFD was composed of corn oil. These lines of evidence suggest that different effects have been observed despite having similar percentages of fat components. Finally, publication bias should be taken into account for a comprehensive understanding, because negative data tend to remain unpublished.

3.2. *Comparison of the Impact of a High-Fat Diet using Multiple Diets*

A direct comparison between multiple diets using animal models is one method of identifying the diet having the most effect on tumor growth. In our previous study, LNCaP xenograft tumor growth in Balb/c-nu/nu mice were evaluated among three groups receiving a HFD (59.9 kcal% fat), Western-style diet (WD: 41.2 kcal% fat), and high carbohydrate diet (HCD: 9.5 kcal% fat) [33]. Accordingly, our results showed that the HFD group had significantly higher LNCaP xenograft tumor growth than the HCD and WD groups. In general, a ketogenic diet, which contains extremely high fat, is toxic to cancer [28]. Accordingly, the systematic review by Khodadadi et al. demonstrated that a ketogenic diet can potentially inhibit malignant cell growth and increase survival time [67]. Moreover, studies comparing the tumor growth and survival of LAPC-4 xenografts in SCID mice demonstrated that mice receiving a no-carbohydrate ketogenic diet (NCKD: 83% fat, 0% carbohydrate, 17% protein) had smaller tumors and higher survival than those receiving a low-fat/high-carbohydrate diet (LFD: 12% fat, 71% carbohydrate, 17% protein) or a high-fat/moderate carbohydrate diet (MCD: 40% fat, 43% carbohydrate, 17% protein) [20]. Another study also investigated the differences between three diets, namely a NCKD (84% fat–0% carbohydrate–16% protein kcal), 10% carbohydrate diet (74%

fat–10% carbohydrate–16% protein kcal), and 20% carbohydrate diet (64% fat–20% carbohydrate–16% protein kcal), with results showing significantly larger tumors in the 10% carbohydrate group but no difference in survival [28]. These lines of evidence suggested that extremely high fat percentages have a potential to exert an opposite effect on prostate cancer development and progression. Therefore, the proportion of total fat intake remains important.

One study using a Western-type diet (16% protein, 40% fat, 44% carbohydrate) evaluated the impact of seven diets: Group 1, ad libitum 7 days/week; Group 2, fasted 1 day/week and ad libitum 6 days/week; Group 3, fasted 1 day/week and fed 6 days/week via paired feeding to maintain isocaloric conditions similar to that in Group 1; Group 4, 14% calorie restriction (CR) 7 days/week; Group 5, fasted 2 days/week and ad libitum 5 days/week; Group 6, fasted 2 day/ week and fed 5 days/week via paired feeding to maintain isocaloric conditions similar to that in Group 1; and Group 7, 28% CR 7 days/week [24]. Accordingly, some of the groups did not exhibit trends toward tumor shrinkage and improved survival, although Groups 6 and 7 had lower lean body mass than Group 1 in a two-way comparison. The study implicated that intermittent calorie restriction via fasting with a Western-style diet had no impact on prostate cancer progression, despite the effect on body weight.

3.3. *Specific Components of Fat*

Each dietary fat has diverse physiological effects according to the different types and distributions of dietary fat components. Therefore, important relationships between specific types of dietary fat intake and prostate cancer development may be missed by merely evaluating the effect of total fat intake [68]. Fatty acids are classified based on whether or not the fatty acid carbon chain contains no double bond (saturated fatty acids (SFA)), one double bond (monounsaturated fatty acids (MUFA)), and more than one double bond (polyunsaturated fatty acids (PUFA)), as well as the configuration of the double bonds (*cis* or *trans*). In addition, PUFA are often classified based on the position of the first double bond from the fatty acid methyl terminus, creating omega-3 and -6 fatty acids. The primary sources for SFA, MUFA, and PUFA include animal fats such as lard and beef tallow, animal and certain vegetable fats such as olive oil, and vegetable oil such as corn and fish oils, respectively [66]. Corn oil and most vegetable oils contain omega-6 PUFA, whereas fish oils are high in omega-3 PUFA [69].

In general, a number of previous studies made use of a lard-based HFD, which is rich in SFA. Studies in human subjects have shown that SFA are more oncogenic than PUFA [70]. Moreover, several studies have shown that cancerous tissues exhibited elevated SFA and MUFA compared to adjacent normal tissues [71,72]. Mice receiving lard oil had been reported to have enhanced Toll-like receptor (TLR) activation and white adipose tissue inflammation, as well as reduced insulin sensitivity, compared to those receiving fish oil [69], suggesting that a diet rich in SFA accelerated metabolic inflammation. In general, MUFA, such as oleic acid and olive oil, are more likely to prevent or decrease the risk of carcinogenesis in other solid cancers, including breast and colon cancers [73]. Phenolic compounds, which prevent free radical-initiated peroxidation and regulate cancer-related oncogenes, have been considered to be associated with MUFA-induced chemoprevention [73]. Omega-3 and -6 PUFA are essential fatty acids that mammals can neither synthesize nor de novo interconvert, suggesting that they have to be obtained from the diet [18]. From an evolutionary standpoint, the human diet has had a 1:1 ratio of omega-6-to-omega-3 PUFA [74]. Over the past two centuries, however, this ratio has increased to nearly 10:1 due primarily to the increased use of vegetable oils in Western diets [8,45]. In general, the high consumption of omega-6 fatty acids leads to inflammation and cellular growth through the conversion of arachidonic acid (an omega-6 fatty acid) to hydroxyeicosatetraenoic and epoxyeicosatrienoic acids by cytochrome P450 oxygenases [75]. In contrast, omega-3 induces anti-inflammatory, pro-apoptotic, anti-proliferative, and anti-angiogenic pathways, providing antitumor effects against prostate cancer [76]. Fish oil, which contains omega-3 fatty acids, does not cause obesity because of peroxidization [77] and induces the activation of peroxisome proliferator-activated receptor alpha. These lines of evidence suggest that omega-3 and -6 PUFA have different effects on diet- and obesity-induced prostate cancer development and progression.

Three studies had reported on the difference in tumor growth between diets rich in omega-3 and -6 [18,29,45]. Accordingly, mice fed a high omega-3 diet had significantly lesser prostate weight gain than those fed a high omega-6 diet. Moreover, half of the mice fed a high omega-3 diet developed invasive carcinoma, whereas 80% of mice fed a high omega-6 diet had invasive carcinoma [18]. The second study revealed that fish oil slowed the progression of tumorigenesis in dorsolateral prostate C3 (1) tag transgenic mice [29]. The last study, which established MycCaP allografts in immunocompetent FVB mice, found that the ω-3 group had significantly smaller tumor volumes than the ω-6 group [45]. All three different models successfully confirmed that omega-3 inhibited tumor growth, which suggests the promising inhibitory effects of omega-3 fatty acid against prostate tumors.

Cholesterol, an organic compound, is a key component of membrane signaling microdomains. In humans, cholesterol can be either obtained from diet or synthesized de novo in the liver. Animal studies using the cholesterol uptake inhibitor ezetimibe for prostate cancer chemoprevention showed that lowering serum cholesterol level slows tumor growth and decreases angiogenesis and intratumoral androgens [78]. Pommier et al. demonstrated that a high-cholesterol diet induced proliferation in LXR mutant mouse prostate [36]. In a clinical setting, the meta-analysis performed by Bonovas et al. was the only study to find a significantly reduced incidence of advanced prostate cancer in subjects who were prescribed statins; however, no relationship between statin use and overall prostate cancer risk was demonstrated in other studies [79]. The observational study by Murtola et al. reported a dose-dependent, significant inverse association between overall prostate cancer incidence and statin use, with the strongest inverse association for early-stage prostate cancer [80]. However, clinical evidence on the protective effect of cholesterol-lowering drugs for prostate cancer chemoprevention is still weak and inconsistent; therefore, we are unable to draw a firm conclusion based on these results.

Finally, care should be taken when establishing how much of a role other nutrients contained in experimental diets have and the actual consumption of diets in each mouse given that the proportion of other ingredients changes when the percentage of fat components is modulated.

4. Potential Mechanisms

Previous studies have proposed several mechanisms in order to explain the possible association between dietary fat and prostate cancer development/progression. Accordingly, growth factor signaling, lipid accumulation, inflammation, and endocrine modulation had been hypothesized to be associated with HFD-induced prostate cancer development and/or progression (Figure 1). Certainly, a more thorough understanding of the possible association between dietary fat and prostate cancer risk requires further inquiry.

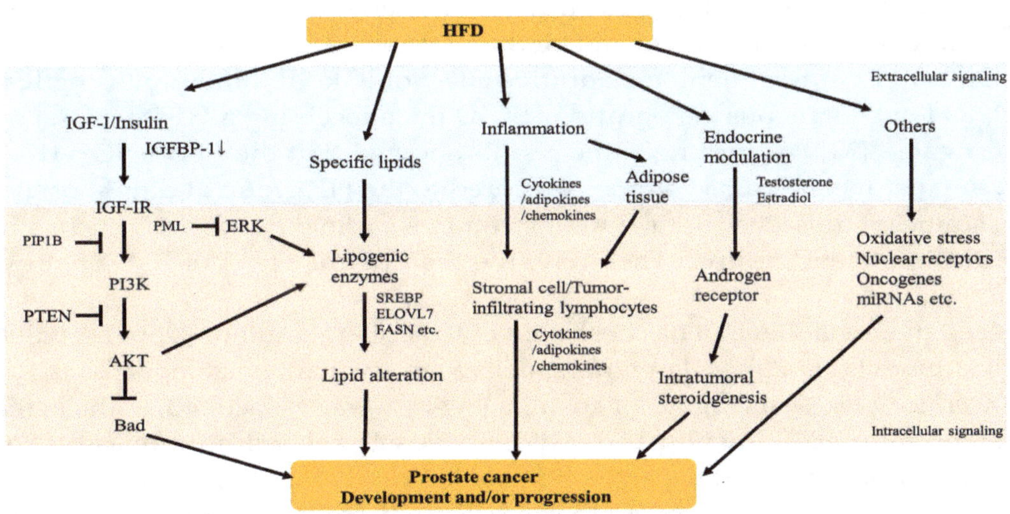

Figure 1. Scheme of potential mechanisms underlying high-fat diet induced prostate cancer development and/or progression.

4.1. Growth Factor Signaling

Obesity and hyperinsulinemia have been associated with increased amounts of circulating bioactive insulin-like growth factor-I (IGF-I), a growth factor determined to play a pathogenic role in many cancers [81]. Barnard et al. demonstrated that dietary fat reduction combined with a regular exercise intervention in men decreased serum IGF-I and increased serum IGFBP-1 levels, resulting in decreased growth of LNCaP human prostate cancer cells cultured in media containing volunteer serum [14]. The same group showed that LFD-fed mice had significantly slower tumor growth rates, lower levels of serum insulin, tumor IGF-I mRNA expression, and tumor IGFBP-2 immunostaining, and higher levels of serum IGFBP-1, which indicated that IGF-I signaling modulated fat-induced tumor growth in LAPC-4 xenografts [15]. We had previously demonstrated that IGF-I receptor (IGF-IR) mRNA levels were strikingly elevated in HFD-accelerated LNCaP xenografts and that the group having the lowest IGF-IR immunoreactivity tended to have the lowest body mass index in both human normal and prostate cancer epithelia [16]. Kobayashi et al. showed that an LFD reduced the development of prostate cancer in Hi-Myc mouse transgenic model with the suppression of the IGF-AKT pathway, which leads to higher serum IGFBP-1 levels, reduced serum mitogenicity, and lower AKT, GSK3beta, and S6K activities [19].

Several studies have demonstrated that hyperactivation of PI3K-AKT, which is one of the downstream targets for IGF-I signaling, desensitizes tumors to dietary modulations, including calorie restriction and a HFD [23,47]. The PI3K/AKT pathway is naturally inhibited by Pten, which is one of the most frequently lost or mutated tumor suppressor genes in prostate cancer [56]. Partial loss of *PTEN* is observed in 70% of localized prostate cancer, while complete loss thereof is associated with metastatic castration-resistant prostate cancer [56]. *PTEN* inactivation also induces aberrant activation of the PI3K/AKT pathway. As previously mentioned, conditional PTEN knockout produces indolent tumors in mouse prostates. One study that assessed the impact of diet restriction revealed that it does not affect a PTEN-null mouse model of prostate cancer but significantly decreases tumor burden in a mouse model of lung cancer lacking constitutive PI3K signaling, which suggests that PI3K signaling is strongly associated with diet-induced cancer progression [23]. Another study involving a GEMM mouse model showed that the loss of both PTEN and the protein tyrosine phosphatase Pypn1, a negative regulator of IGF-IR, enabled the development of a highly invasive prostate tumor, whereas PTEN deficiency alone resulted in tumors that were unresponsive to HFD [47]. Collectively, mechanisms involving PTEN and other related genes may have a higher impact on diet-induced prostate cancer development and progression.

Many other studies have proposed that IGF-I/PI3K/AKT signaling has an impact on diet-induced prostate cancer development and progression [21,28,43]. Therefore, IGF-I/PI3K/AKT signaling has been one of the promising pathways related to HFD-induced prostate cancer development and progression. To determine the impact of treatment, the additive effect of IGF-1R inhibition using IGF-IR blockade antibody on 22Rv1 subcutaneous xenografts in SCID mice receiving a HFD (43.3%) or LFD (12.4%) had been investigated [32]. Accordingly, the results showed that the LFD + IGF-1R-Ab group had a significantly smaller mean tumor volume compared to the HF group at day 14 of the intervention. However, no significant difference in final tumor volumes or final tumor weights had bene observed between the four treatment groups. Therefore, the therapeutic effect of IGF-I pathway inhibition remains unknown.

Diet-induced hyperinsulinemia has been shown to accelerate tumor growth in different prostate cancer xenograft models [17,20]. A large prospective survival analysis reported that higher serum C-peptide concentrations, a surrogate of insulin levels, were associated with increased prostate cancer-specific mortality [82]. Insulin and IGF-I are closely related hormones that act on specific tyrosine kinase receptors and elicit the activation of a cascade of intracellular proteins leading to the regulation of gene expression, protein synthesis, cell proliferation or death, and glucose and lipid metabolism. High insulin levels, as well as insulin receptor and IGF-I/IGF-IR axis activation, have been known to be associated with obesity induced cancer progression [83]. Regarding the impact of insulin levels on diet-induced prostate cancer growth, one study involving LAPC-4 xenografts in SCID mice receiving three different diets, NCKD (84% fat), 10% carbohydrate diet (74% fat), or 20% carbohydrate diet (64% fat), proposed that mice receiving a 10% carbohydrate diet had larger tumors than the other groups despite mice receiving a 20% carbohydrate diet having the lowest insulin levels [28]. As such, future studies need to elucidate the relationship between insulin levels and diet-induced prostate cancer carcinogenesis.

In addition, effects of different fat sources on the IGF/insulin axis have rarely been discussed and studied. It would be intriguing to know the varying impacts specific fats have on HFD-induced prostate cancer development and progression through growth factor signaling.

4.2. Lipid Accumulation

The changes in endogenously synthesized/exogenous lipid profiles and related enzymes have been linked to prostate cancer development and progression. Accordingly, Freedland et al. showed that mice with LAPC-4 xenografts receiving a NCKD diet had low hepatic fatty infiltration, which resulted in reduced tumor growth and longer survival [20]. Genome-wide gene expression analysis showed that the lipogenic gene ELOVL7, which possibly codes a long-chain fatty acid elongase, was overexpressed in clinical prostate cancer and regulated by SREBP1. Moreover, a HFD had been found to promote the growth of in vivo tumors of ELOLV7-expresssed prostate cancer [22]. In the aggressive and metastatic tumor progression observed in TRAMP mice receiving a Western-style diet, Llaveries et al. showed that the Western-style diet increased both the expression of the high density lipoprotein receptor SR-BI and angiogenesis [25]. Fatty acid synthase (FASN) is a cytosolic metabolic enzyme that catalyzes de novo fatty acid synthesis. Our previous study found that serum FASN levels were significantly lower and were inversely correlated with tumor volume in LNCaP xenograft mice receiving HFD [46].

A recent study has suggested that a Western-style HFD promotes metastatic prostate cancer through a prometastatic lipogenic program alteration [56]. In this study, the conditional inactivation of Pml (a suppressor of pp1α-dependent activation of MAPK signaling) in mouse prostates changed indolent PTEN-null prostate tumors into lethal metastatic tumors with MAPK reactivation, subsequent hyperactivation of an aberrant SREBP, and a lipidomic profile alteration. Pten-/-, Pml-/- mice receiving a chow diet displayed limited lymph node metastasis. However, most mice receiving a HFD developed lymph node metastasis, while half of them had lung metastasis. Moreover, Oil Red O staining showed that the tumors in mice receiving a HFD had higher lipid accumulation compared to those in mice receiving a chow diet. Sterol responsive element binding proteins (SREBPs) have been found to be a key regulator of lipogenic genes [56]. Studies have shown that HFD feeding stimulates SREBP expression, subsequent expression of genes encoding lipogenic enzymes, and lipid accumulation in nonadipose tissues [56,84]. Therefore, Pml and SREBP-dependent lipogenic alterations may be associated with HFD-enhanced prostate cancer progression.

Although de novo lipogenesis has emerged as an important player in prostate cancer, the impact of exogenous dietary fat on intraprostatic lipid profiles and activity of lipogenic enzymes remains largely unknown. Future studies are required to elucidate the mechanisms for endogenous lipid alterations and exogenous fat accumulation on dietary fat-induced prostate cancer development and progression.

4.3. Inflammation

Inflammations have been shown to promote the development and progression of prostate cancer [85]. A HFD with consequent obesity causes adipose tissue inflammation and cytokine secretion [86]. Accordingly, Liu et al. demonstrated that Pten +/- mice receiving a high-calorie diet exhibited neoplastic progression with stromal infiltration of inflammatory cells, such as macrophages, T cells, and inflammatory monocytes, into the prostates [40]. Additionally, the increased inflammatory response to a high-calorie diet was supported by the elevation in the expression of CD3, CD45, FoxP3, MCP-1, IL-6, and TNF alpha. Microarray analysis using TRAMP mice models showed that HFD feeding increased serum levels of MCP-1, MCP-5, TIMP-1, IL-16, CCL12, CXCL1, CXCL10, and CXCL13 [41].

Similar results were observed in the sera of TRAMP-C2 allograft models. Zhang et al. demonstrated that adipose stromal cell recruitment to tumors of RM1 mouse prostate cancer xenografts via CXCL1 and CXCL8 chemokines promoted prostate cancer progression [49]. Another study showed that MycCaP xenografted immunocompetent FVB mice receiving a diet rich in omega-3 exhibited tumor suppression, as well as lower gene expression of markers for M1 and M2 macrophages, associated cytokines (IL-6, TNF alpha, and IL-10), and the chemokine CCL-2. Hayashi et al. showed that a HFD increased the prostate weight and percentage of Ki67-positive MDSCs, as well as the M2/M1 macrophage ratio, in HFD-fed model mice with a higher serum IL-6 levels [54]. Furthermore, celecoxib suppressed tumor growth in HFD-fed but not CD-fed model mice, which suggested that HFD-induced tumor growth was associated with local inflammation. Taken together, tumor-infiltrating macrophages may perhaps be a key factor in HFD-induced prostate cancer progression [54]. One of our previous studies had demonstrated that the MCP-1/CCR2 pathway, a key regulator of macrophage infiltration, was highly associated with HFD-induced LNCaP xenograft tumor growth, supporting the results presented herein [33]. We also found that the expression of macrophage inhibitory cytokine 1 (MIC-1), a divergent member of the transforming growth factor beta, was stimulated by palmitic acid in vitro, while mice receiving a HFD containing high amounts of palmitic acid had LNCaP had significantly greater xenograft tumor growth, serum MIC1 levels, and fatty acid levels in xenograft tumors than those receiving an LFD in vivo [37]. Such lines of evidence suggesting the association between cytokines and tumor–macrophage interaction support the notion that tumor-associated macrophages play a role in HFD-induced prostate cancer development and progression. Another mouse xenograft experiment concluded that HFD enhanced prostate cancer metastasis and invasiveness through FABP4 and interleukin-8 upregulation [53]. FABP4, an abundant protein in adipocytes that is influenced by a HFD or obesity, may enhance prostate cancer progression and invasiveness by upregulating matrix metalloproteinases and cytokine production in the prostate cancer stromal microenvironment [53]. In one study demonstrating tumor growth decline among Pten KO mice receiving omega-3 fatty acid, the group with an omega-3-enriched diet exhibited a reduction in CD3+ lymphocyte levels and tumor microvessel density [18]. These lines of evidence suggest that a stromal microenvironment, including infiltration of immune cells, is associated with dietary-fat induced prostate cancer carcinogenesis. Moreover, other cytokines and chemokines, including TWEAK, (CCL)3, CCL4, and CCL5, had been found to be potentially associated with HFD-induced prostate cancer progression according to previous literatures [41,57,87].

The role of adipocyte function on HFD feeding was evaluated in a recent study [50]. Accordingly, Cav-1 secretion from fat tissue of HFD-fed mice was increased, while hypertrophied adipocytes were responsible for enhanced Cav-1 secretion in obese mice. Furthermore, secreted Cav-1 was taken up by the preadipocytes and LNCaP cells. The impact of hypertrophied adipocyte-induced Cav-1 secretion on prostate cancer progression and diet- and/or obesity-modulated adipose function could be associated with prostate inflammation and prostate carcinogenesis. Moreover, adipose tissues are known to enhance cancer progression via several underlying mechanisms, such as aromatization of adrenal androgens to estrogens in the adipocyte and deregulation of the expression and secretion of the adipokines [88,89]. Therefore, it is worthwhile to assess the role of quantitative and qualitative modulation of adipose tissues on prostate cancer progression using preclinical models.

In summary, the interaction among systematic and/or adipose secreted cytokines, inflammation, and immune cell infiltration into tumors may have a promising mechanistic role in HFD-induced prostate cancer development and progression.

4.4. *Endocrine Modulation*

Considerable epidemiological evidence has shown that fat-containing diets may increase the risk of certain hormone-dependent conditions in men via its effects on hormone metabolism [90]. Hormonal modulation has been one of the proposed mechanisms associated with diet- and/or obesity-induced prostate cancer carcinogenesis [8,9] given that sex hormones play a key role during normal and cancerous prostate growth and development. Two transgenic mouse studies showed that omega-3 fatty acids slowed prostate tumor growth through the modulation of sex steroid pathways. In the C3 (1) Tag transgenic mice study, the lowering testosterone, estradiol, and androgen receptor levels by the action of omega-3 fatty acids promoted apoptosis and suppressed prostate epithelial cell proliferation [29]. Another study demonstrated that omega-3 PUFA treatment slowed castration-resistant tumor growth and accelerated androgen receptor protein degradation [34]. A recent study showed that although serum cholesterol reduction did not significantly affect the rate of adenocarcinoma development in the PTEN-null transgenic mouse model of prostate cancer [78], it lowered intraprostatic androgens and slowed tumor growth. These results suggest that fat-containing diets, especially those that modulate of omega-3 fatty acid content, may potentially modulate intraprostatic hormonal status associated with cancerous tumor growth and progression. In the TRAMP-C1 allograft study, HFD increased tumor growth and serum estradiol levels [55]. The study also showed that intratumoral C-terminal-binding protein 1 (CtBP1) controls the transcription of aromatase (CYP19A), a key enzyme that converts androgens to estrogens, and was overexpressed with increased TRAMP-C1 allograft tumor growth in mice receiving a HFD. In another study, Moiola et al. found that mice with CtBP1-depleted PC-3 xenografts developed significantly smaller tumors than those inoculated with PC-3 control cells [38]. These results suggest that CtBP1 have the potential to be a key transcriptional factor associated with intratumoral hormonal modulation and HFD-induced prostate cancer growth.

4.5. Others

Using microarray analysis, our group had previously showed that several mRNAs and miRNAs become altered in HFD-induced LNCaP xenografts [16,52]. Therefore, complex mechanisms, including candidate pathways mentioned previously, may be considered to contribute to fat-diet induced prostate cancer development and progression. Nara et al. demonstrated that miR-130a was attenuated in HFD-induced prostate cancer progression with MET overexpression in vitro and in vivo and that cytoplasmic MET in prostate cancer tissues was overexpressed in patients with higher body mass index [52]. Kim et al. found that a HFD not only accelerated Src-induced prostate tumorigenesis, but also compromised the inhibitory effect of the anticancer drug dasatinib on Src kinase oncogenic potential in vivo [51]. Finally, the association between diet-induced prostate cancer progression and several pathways, including oxidative stress [39], epithelial–mesenchymal transition [40], and basal/luminal differentiation [48], have been proposed in previous studies.

5. Concluding Remarks

Over recent years, the molecular mechanisms behind HFD-induced prostate cancer development and progression have been studied using pre-clinical models. Although several lines of evidence have proposed its relationship with potential mechanisms, such as growth factor signaling, lipid accumulation, inflammation, and endocrine modulation, the current data still remains inconclusive. In addition, the studies presented herein have used various types of models and diet sources, suggesting the need for increases vigilance when communicating and interpreting information. Therefore, it is important to consider the predictability and limitation of each preclinical model when translating experimental results into clinical practice. Although information from pre-clinical models remain important for deeper understanding and exploration of novel treatment targets, further studies are needed to validate the impact of dietary fat and obesity on prostate cancer development and progression.

Author Contributions: Conceptualization, Data collection and Writing—Original Draft Preparation, S.N.; Reviewing: T.N., H.S., M.H. and A.K. Writing—Review and Editing, T.H.; Supervision, T.I. and T.H.

Acknowledgments: This study was supported in part by research grants from the MEXT/JSPS (Kakenhi, 16H02679, 16K10992, 19K09663) and the Japan Agency for Medical Research and Development (AMED) (grant 18gm0710002h0706).

References

1. Fitzmaurice, C.; Akinyemiju, T.F.; Al Lami, F.H.; Alam, T.; Alizadeh-Navaei, R.; Allen, C.; Alsharif, U.; Alvis-Guzman, N.; Amini, E.; Anderson, B.O.; et al. Global, Regional, and National Cancer Incidence, Mortality, Years of Life Lost, Years Lived With Disability, and Disability-Adjusted Life-Years for 29 Cancer Groups, 1990 to 2016: A Systematic Analysis for the Global Burden of Disease Study. *JAMA Oncol.* **2018**, *4*, 1553–1568. [CrossRef] [PubMed]
2. Siegel, R.L.; Miller, K.D.; Jemal, A. Cancer statistics, 2018. *CA Cancer J. Clin.* **2018**, *68*, 7–30. [CrossRef] [PubMed]
3. Watanabe, M.; Nakayama, T.; Shiraishi, T.; Stemmermann, G.N.; Yatani, R. Comparative studies of prostate cancer in Japan versus the United States. A review. *Urol. Oncol.* **2000**, *5*, 274–283. [CrossRef]
4. Breslow, N.; Chan, C.W.; Dhom, G.; Drury, R.A.; Franks, L.M.; Gellei, B.; Lee, Y.S.; Lundberg, S.; Sparke, B.; Sternby, N.H.; et al. Latent carcinoma of prostate at autopsy in seven areas. The International Agency for Research on Cancer, Lyons, France. *Int. J. Cancer* **1977**, *20*, 680–688. [CrossRef] [PubMed]
5. Shimizu, H.; Ross, R.K.; Bernstein, L.; Yatani, R.; Henderson, B.E.; Mack, T.M. Cancers of the prostate and breast among Japanese and white immigrants in Los Angeles County. *Br. J. Cancer* **1991**, *63*, 963–966. [CrossRef] [PubMed]
6. Kimura, T.; Egawa, S. Epidemiology of prostate cancer in Asian countries. *Int. J. Urol.* **2018**, *25*, 524–531. [CrossRef]
7. Ornish, D.; Weidner, G.; Fair, W.R.; Marlin, R.; Pettengill, E.B.; Raisin, C.J.; Dunn-Emke, S.; Crutchfield, L.; Jacobs, F.N.; Barnard, R.J.; et al. Intensive lifestyle changes may affect the progression of prostate cancer. *J. Urol.* **2005**, *174*, 1065–1069; discussion 1069–1070. [CrossRef] [PubMed]
8. Allott, E.H.; Masko, E.M.; Freedland, S.J. Obesity and prostate cancer: weighing the evidence. *Eur. Urol.* **2013**, *63*, 800–809. [CrossRef]
9. Venkateswaran, V.; Klotz, L.H. Diet and prostate cancer: mechanisms of action and implications for chemoprevention. *Nat. Rev. Urol.* **2010**, *7*, 442–453. [CrossRef]
10. Lin, P.H.; Aronson, W.; Freedland, S.J. An update of research evidence on nutrition and prostate cancer. *Urol. Oncol.* **2017**. [CrossRef]
11. Wang, Y.; Corr, J.G.; Thaler, H.T.; Tao, Y.; Fair, W.R.; Heston, W.D. Decreased growth of established human prostate LNCaP tumors in nude mice fed a low-fat diet. *J. Natl. Cancer Inst.* **1995**, *87*, 1456–1462. [CrossRef]
12. Connolly, J.M.; Coleman, M.; Rose, D.P. Effects of dietary fatty acids on DU145 human prostate cancer cell growth in athymic nude mice. *Nutri. Cancer* **1997**, *29*, 114–119. [CrossRef]
13. Ngo, T.H.; Barnard, R.J.; Tymchuk, C.N.; Cohen, P.; Aronson, W.J. Effect of diet and exercise on serum insulin, IGF-I, and IGFBP-1 levels and growth of LNCaP cells in vitro (United States). *Cancer Causes Control.* **2002**, *13*, 929–935. [CrossRef] [PubMed]
14. Barnard, R.J.; Ngo, T.H.; Leung, P.S.; Aronson, W.J.; Golding, L.A. A low-fat diet and/or strenuous exercise alters the IGF axis in vivo and reduces prostate tumor cell growth in vitro. *Prostate* **2003**, *56*, 201–206. [CrossRef]
15. Ngo, T.H.; Barnard, R.J.; Cohen, P.; Freedland, S.; Tran, C.; de Gregorio, F.; Elshimali, Y.I.; Heber, D.; Aronson, W.J. Effect of isocaloric low-fat diet on human LAPC-4 prostate cancer xenografts in severe combined immunodeficient mice and the insulin-like growth factor axis. *Clin. Cancer Res.* **2003**, *9*, 2734–2743.
16. Narita, S.; Tsuchiya, N.; Saito, M.; Inoue, T.; Kumazawa, T.; Yuasa, T.; Nakamura, A.; Habuchi, T. Candidate genes involved in enhanced growth of human prostate cancer under high fat feeding identified by microarray analysis. *Prostate* **2008**, *68*, 321–335. [CrossRef]
17. Venkateswaran, V.; Haddad, A.Q.; Fleshner, N.E.; Fan, R.; Sugar, L.M.; Nam, R.; Klotz, L.H.; Pollak, M. Association of diet-induced hyperinsulinemia with accelerated growth of prostate cancer (LNCaP) xenografts. *J. Natl. Cancer Inst.* **2007**, *99*, 1793–1800. [CrossRef] [PubMed]
18. Berquin, I.M.; Min, Y.; Wu, R.; Wu, J.; Perry, D.; Cline, J.M.; Thomas, M.J.; Thornburg, T.; Kulik, G.; Smith, A.; et al. Modulation of prostate cancer genetic risk by omega-3 and omega-6 fatty acids. *J. Clin. Invest.* **2007**, *117*, 1866–1875. [CrossRef]
19. Kobayashi, N.; Barnard, R.J.; Said, J.; Hong-Gonzalez, J.; Corman, D.M.; Ku, M.; Doan, N.B.; Gui, D.; Elashoff, D.; Cohen, P.; et al. Effect of low-fat diet on development of prostate cancer and Akt phosphorylation in the Hi-Myc transgenic mouse model. *Cancer Res.* **2008**, *68*, 3066–3073. [CrossRef]

20. Freedland, S.J.; Mavropoulos, J.; Wang, A.; Darshan, M.; Demark-Wahnefried, W.; Aronson, W.J.; Cohen, P.; Hwang, D.; Peterson, B.; Fields, T.; et al. Carbohydrate restriction, prostate cancer growth, and the insulin-like growth factor axis. *Prostate* **2008**, *68*, 11–19. [CrossRef]
21. Mavropoulos, J.C.; Buschemeyer, W.C., 3rd; Tewari, A.K.; Rokhfeld, D.; Pollak, M.; Zhao, Y.; Febbo, P.G.; Cohen, P.; Hwang, D.; Devi, G.; et al. The effects of varying dietary carbohydrate and fat content on survival in a murine LNCaP prostate cancer xenograft model. *Cancer Prev. Res.* **2009**, *2*, 557–565. [CrossRef]
22. Tamura, K.; Makino, A.; Hullin-Matsuda, F.; Kobayashi, T.; Furihata, M.; Chung, S.; Ashida, S.; Miki, T.; Fujioka, T.; Shuin, T.; et al. Novel lipogenic enzyme ELOVL7 is involved in prostate cancer growth through saturated long-chain fatty acid metabolism. *Cancer Res.* **2009**, *69*, 8133–8140. [CrossRef]
23. Kalaany, N.Y.; Sabatini, D.M. Tumours with PI3K activation are resistant to dietary restriction. *Nature* **2009**, *458*, 725–731. [CrossRef]
24. Buschemeyer, W.C., 3rd; Klink, J.C.; Mavropoulos, J.C.; Poulton, S.H.; Demark-Wahnefried, W.; Hursting, S.D.; Cohen, P.; Hwang, D.; Johnson, T.L.; Freedland, S.J. Effect of intermittent fasting with or without caloric restriction on prostate cancer growth and survival in SCID mice. *Prostate* **2010**, *70*, 1037–1043. [CrossRef]
25. Llaverias, G.; Danilo, C.; Wang, Y.; Witkiewicz, A.K.; Daumer, K.; Lisanti, M.P.; Frank, P.G. A Western-type diet accelerates tumor progression in an autochthonous mouse model of prostate cancer. *Am. J. Pathol.* **2010**, *177*, 3180–3191. [CrossRef]
26. Lloyd, J.C.; Antonelli, J.A.; Phillips, T.E.; Masko, E.M.; Thomas, J.A.; Poulton, S.H.; Pollak, M.; Freedland, S.J. Effect of isocaloric low fat diet on prostate cancer xenograft progression in a hormone deprivation model. *J. Urol.* **2010**, *183*, 1619–1624. [CrossRef]
27. Aronson, W.J.; Barnard, R.J.; Freedland, S.J.; Henning, S.; Elashoff, D.; Jardack, P.M.; Cohen, P.; Heber, D.; Kobayashi, N. Growth inhibitory effect of low fat diet on prostate cancer cells: results of a prospective, randomized dietary intervention trial in men with prostate cancer. *J. Urol.* **2010**, *183*, 345–350. [CrossRef]
28. Masko, E.M.; Thomas, J.A., 2nd; Antonelli, J.A.; Lloyd, J.C.; Phillips, T.E.; Poulton, S.H.; Dewhirst, M.W.; Pizzo, S.V.; Freedland, S.J. Low-carbohydrate diets and prostate cancer: How low is "low enough"? *Cancer Prev. Res.* **2010**, *3*, 1124–1131. [CrossRef]
29. Akinsete, J.A.; Ion, G.; Witte, T.R.; Hardman, W.E. Consumption of high omega-3 fatty acid diet suppressed prostate tumorigenesis in C3(1) Tag mice. *Carcinogenesis* **2012**, *33*, 140–148. [CrossRef]
30. Mao, G.E.; Harris, D.M.; Moro, A.; Heber, D.; Roy-Burman, P.; Zhang, Z.F.; Rao, J. A joint effect of new Western diet and retinoid X receptor alpha prostate-specific knockout with development of high-grade prostatic intraepithelial neoplasia in mice—A preliminary study. *Prostate* **2012**, *72*, 1052–1059. [CrossRef]
31. Bonorden, M.J.; Grossmann, M.E.; Ewing, S.A.; Rogozina, O.P.; Ray, A.; Nkhata, K.J.; Liao, D.J.; Grande, J.P.; Cleary, M.P. Growth and Progression of TRAMP Prostate Tumors in Relationship to Diet and Obesity. *Prostate Cancer* **2012**, *2012*, 543970. [CrossRef]
32. Konijeti, R.; Koyama, S.; Gray, A.; Barnard, R.J.; Said, J.W.; Castor, B.; Elashoff, D.; Wan, J.; Beltran, P.J.; Calzone, F.J.; et al. Effect of a low-fat diet combined with IGF-1 receptor blockade on 22Rv1 prostate cancer xenografts. *Mol. Cancer Ther.* **2012**, *11*, 1539–1546. [CrossRef]
33. Huang, M.; Narita, S.; Numakura, K.; Tsuruta, H.; Saito, M.; Inoue, T.; Horikawa, Y.; Tsuchiya, N.; Habuchi, T. A high-fat diet enhances proliferation of prostate cancer cells and activates MCP-1/CCR2 signaling. *Prostate* **2012**, *72*, 1779–1788. [CrossRef]
34. Wang, S.; Wu, J.; Suburu, J.; Gu, Z.; Cai, J.; Axanova, L.S.; Cramer, S.D.; Thomas, M.J.; Perry, D.L.; Edwards, I.J.; et al. Effect of dietary polyunsaturated fatty acids on castration-resistant Pten-null prostate cancer. *Carcinogenesis.* **2012**, *33*, 404–412. [CrossRef]
35. Vandersluis, A.D.; Venier, N.A.; Colquhoun, A.J.; Sugar, L.; Pollak, M.; Kiss, A.; Fleshner, N.E.; Klotz, L.H.; Venkateswaran, V. Exercise does not counteract the effects of a "westernized" diet on prostate cancer xenografts. *Prostate* **2013**, *73*, 1223–1232. [CrossRef]
36. Pommier, A.J.; Dufour, J.; Alves, G.; Viennois, E.; De Boussac, H.; Trousson, A.; Volle, D.H.; Caira, F.; Val, P.; Arnaud, P.; et al. Liver x receptors protect from development of prostatic intra-epithelial neoplasia in mice. *PLoS Genet.* **2013**, *9*, e1003483. [CrossRef]
37. Huang, M.; Narita, S.; Inoue, T.; Tsuchiya, N.; Satoh, S.; Nanjo, H.; Sasaki, T.; Habuchi, T. Diet-induced macrophage inhibitory cytokine 1 promotes prostate cancer progression. *Endocr. Relat. Cancer* **2014**, *21*, 39–50. [CrossRef]

38. Moiola, C.P.; De Luca, P.; Zalazar, F.; Cotignola, J.; Rodriguez-Segui, S.A.; Gardner, K.; Meiss, R.; Vallecorsa, P.; Pignataro, O.; Mazza, O.; et al. Prostate tumor growth is impaired by CtBP1 depletion in high-fat diet-fed mice. *Clin. Cancer Res.* **2014**, *20*, 4086–4095. [CrossRef]
39. Chang, S.N.; Han, J.; Abdelkader, T.S.; Kim, T.H.; Lee, J.M.; Song, J.; Kim, K.S.; Park, J.H. High animal fat intake enhances prostate cancer progression and reduces glutathione peroxidase 3 expression in early stages of TRAMP mice. *Prostate* **2014**, *74*, 1266–1277. [CrossRef]
40. Liu, J.; Ramakrishnan, S.K.; Khuder, S.S.; Kaw, M.K.; Muturi, H.T.; Lester, S.G.; Lee, S.J.; Fedorova, L.V.; Kim, A.J.; Mohamed, I.E.; et al. High-calorie diet exacerbates prostate neoplasia in mice with haploinsufficiency of Pten tumor suppressor gene. *Mol. Metab.* **2015**, *4*, 186–198. [CrossRef]
41. Cho, H.J.; Kwon, G.T.; Park, H.; Song, H.; Lee, K.W.; Kim, J.I.; Park, J.H. A high-fat diet containing lard accelerates prostate cancer progression and reduces survival rate in mice: Possible contribution of adipose tissue-derived cytokines. *Nutrients* **2015**, *7*, 2539–2561. [CrossRef]
42. Xu, H.; Hu, M.B.; Bai, P.D.; Zhu, W.H.; Liu, S.H.; Hou, J.Y.; Xiong, Z.Q.; Ding, Q.; Jiang, H.W. Proinflammatory cytokines in prostate cancer development and progression promoted by high-fat diet. *BioMed Res. Int.* **2015**, *2015*, 249741. [CrossRef]
43. Xu, H.; Jiang, H.W.; Ding, Q. Insulin-Like growth factor 1 related pathways and high-fat diet promotion of transgenic adenocarcinoma mouse prostate (TRAMP) cancer progression. *Actas Urol. Esp.* **2015**, *39*, 161–168. [CrossRef]
44. Lo, J.C.; Clark, A.K.; Ascui, N.; Frydenberg, M.; Risbridger, G.P.; Taylor, R.A.; Watt, M.J. Obesity does not promote tumorigenesis of localized patient-derived prostate cancer xenografts. *Oncotarget.* **2016**, *7*, 47650–47662. [CrossRef]
45. Liang, P.; Henning, S.M.; Schokrpur, S.; Wu, L.; Doan, N.; Said, J.; Grogan, T.; Elashoff, D.; Cohen, P.; Aronson, W.J. Effect of Dietary Omega-3 Fatty Acids on Tumor-Associated Macrophages and Prostate Cancer Progression. *Prostate* **2016**, *76*, 1293–1302. [CrossRef]
46. Huang, M.; Koizumi, A.; Narita, S.; Inoue, T.; Tsuchiya, N.; Nakanishi, H.; Numakura, K.; Tsuruta, H.; Saito, M.; Satoh, S.; et al. Diet-induced alteration of fatty acid synthase in prostate cancer progression. *Oncogenesis* **2016**, *5*, e195. [CrossRef]
47. Labbe, D.P.; Uetani, N.; Vinette, V.; Lessard, L.; Aubry, I.; Migon, E.; Sirois, J.; Haigh, J.J.; Begin, L.R.; Trotman, L.C.; et al. PTP1B Deficiency Enables the Ability of a High-Fat Diet to Drive the Invasive Character of PTEN-Deficient Prostate Cancers. *Cancer Res.* **2016**, *76*, 3130–3135. [CrossRef]
48. Kwon, O.J.; Zhang, B.; Zhang, L.; Xin, L. High fat diet promotes prostatic basal-to-luminal differentiation and accelerates initiation of prostate epithelial hyperplasia originated from basal cells. *Stem Cell Res.* **2016**, *16*, 682–691. [CrossRef]
49. Zhang, T.; Tseng, C.; Zhang, Y.; Sirin, O.; Corn, P.G.; Li-Ning-Tapia, E.M.; Troncoso, P.; Davis, J.; Pettaway, C.; Ward, J.; et al. CXCL1 mediates obesity-associated adipose stromal cell trafficking and function in the tumour microenvironment. *Nat. Commun.* **2016**, *7*, 11674. [CrossRef]
50. Chang, C.C.; Chen, C.Y.; Wen, H.C.; Huang, C.Y.; Hung, M.S.; Lu, H.C.; Chen, W.L.; Chang, C.H. Caveolin-1 Secreted from Adipose Tissues and Adipocytes Functions as an Adipogenesis Enhancer. *Obesity* **2017**, *25*, 1932–1940. [CrossRef]
51. Kim, S.; Yang, X.; Li, Q.; Wu, M.; Costyn, L.; Beharry, Z.; Bartlett, M.G.; Cai, H. Myristoylation of Src kinase mediates Src-induced and high-fat diet-accelerated prostate tumor progression in mice. *J. Biol. Chem.* **2017**, *292*, 18422–18433. [CrossRef]
52. Nara, T.; Narita, S.; Mingguo, H.; Yoshioka, T.; Koizumi, A.; Numakura, K.; Tsuruta, H.; Maeno, A.; Saito, M.; Inoue, T.; et al. Altered miRNA expression in high-fat diet-induced prostate cancer progression. *Carcinogenesis* **2016**, *37*, 1129–1137. [CrossRef]
53. Huang, M.; Narita, S.; Inoue, T.; Koizumi, A.; Saito, M.; Tsuruta, H.; Numakura, K.; Satoh, S.; Nanjo, H.; Sasaki, T.; et al. Fatty acid binding protein 4 enhances prostate cancer progression by upregulating matrix metalloproteinases and stromal cell cytokine production. *Oncotarget* **2017**, *8*, 111780–111794. [CrossRef]
54. Hayashi, T.; Fujita, K.; Nojima, S.; Hayashi, Y.; Nakano, K.; Ishizuya, Y.; Wang, C.; Yamamoto, Y.; Kinouchi, T.; Matsuzaki, K.; et al. High-Fat Diet-Induced Inflammation Accelerates Prostate Cancer Growth via IL6 Signaling. *Clin. Cancer Res.* **2018**, *24*, 4309–4318. [CrossRef]

55. Massillo, C.; Dalton, G.N.; Porretti, J.; Scalise, G.D.; Farre, P.L.; Piccioni, F.; Secchiari, F.; Pascuali, N.; Clyne, C.; Gardner, K.; et al. CTBP1/CYP19A1/estradiol axis together with adipose tissue impacts over prostate cancer growth associated to metabolic syndrome. *Int. J. Cancer* **2019**, *144*, 1115–1127. [CrossRef]
56. Chen, M.; Zhang, J.; Sampieri, K.; Clohessy, J.G.; Mendez, L.; Gonzalez-Billalabeitia, E.; Liu, X.S.; Lee, Y.R.; Fung, J.; Katon, J.M.; et al. An aberrant SREBP-dependent lipogenic program promotes metastatic prostate cancer. *Nat. Genet.* **2018**, *50*, 206–218. [CrossRef]
57. Hu, M.B.; Xu, H.; Zhu, W.H.; Bai, P.D.; Hu, J.M.; Yang, T.; Jiang, H.W.; Ding, Q. High-fat diet-induced adipokine and cytokine alterations promote the progression of prostate cancer in vivo and in vitro. *Oncol. Lett.* **2018**, *15*, 1607–1615. [CrossRef]
58. Ngo, T.H.; Barnard, R.J.; Anton, T.; Tran, C.; Elashoff, D.; Heber, D.; Freedland, S.J.; Aronson, W.J. Effect of isocaloric low-fat diet on prostate cancer xenograft progression to androgen independence. *Cancer Res.* **2004**, *64*, 1252–1254. [CrossRef]
59. Budhu, S.; Wolchok, J.; Merghoub, T. The importance of animal models in tumor immunity and immunotherapy. *Curr. Opin. Genet. Dev.* **2014**, *24*, 46–51. [CrossRef]
60. Gingrich, J.R.; Barrios, R.J.; Morton, R.A.; Boyce, B.F.; DeMayo, F.J.; Finegold, M.J.; Angelopoulou, R.; Rosen, J.M.; Greenberg, N.M. Metastatic prostate cancer in a transgenic mouse. *Cancer Res.* **1996**, *56*, 4096–4102.
61. Irshad, S.; Abate-Shen, C. Modeling prostate cancer in mice: something old, something new, something premalignant, something metastatic. *Cancer Metastasis Rev.* **2013**, *32*, 109–122. [CrossRef]
62. Hu, M.B.; Hu, J.M.; Jiang, L.R.; Yang, T.; Zhu, W.H.; Hu, Y.; Wu, X.B.; Jiang, H.W.; Ding, Q. Differential expressions of integrin-linked kinase, beta-parvin and cofilin 1 in high-fat diet induced prostate cancer progression in a transgenic mouse model. *Oncol. Lett.* **2018**, *16*, 4945–4952. [CrossRef]
63. Ellwood-Yen, K.; Graeber, T.G.; Wongvipat, J.; Iruela-Arispe, M.L.; Zhang, J.; Matusik, R.; Thomas, G.V.; Sawyers, C.L. Myc-driven murine prostate cancer shares molecular features with human prostate tumors. *Cancer Cell* **2003**, *4*, 223–238. [CrossRef]
64. Cunningham, D.; You, Z. In vitro and in vivo model systems used in prostate cancer research. *J. Biol. Methods* **2015**, *2*. [CrossRef]
65. Shappell, S.B.; Thomas, G.V.; Roberts, R.L.; Herbert, R.; Ittmann, M.M.; Rubin, M.A.; Humphrey, P.A.; Sundberg, J.P.; Rozengurt, N.; Barrios, R.; et al. Prostate pathology of genetically engineered mice: Definitions and classification. The consensus report from the Bar Harbor meeting of the Mouse Models of Human Cancer Consortium Prostate Pathology Committee. *Cancer Res.* **2004**, *64*, 2270–2305. [CrossRef]
66. Buettner, R.; Scholmerich, J.; Bollheimer, L.C. High-fat diets: modeling the metabolic disorders of human obesity in rodents. *Obesity* **2007**, *15*, 798–808. [CrossRef]
67. Khodadadi, S.; Sobhani, N.; Mirshekar, S.; Ghiasvand, R.; Pourmasoumi, M.; Miraghajani, M.; Dehsoukhteh, S.S. Tumor Cells Growth and Survival Time with the Ketogenic Diet in Animal Models: A Systematic Review. *Int. J. Pre. Med.* **2017**, *8*, 35. [CrossRef]
68. Di Sebastiano, K.M.; Mourtzakis, M. The role of dietary fat throughout the prostate cancer trajectory. *Nutrients* **2014**, *6*, 6095–6109. [CrossRef]
69. Heber, D.; Kritchevsky, D. Dietary Fats, Lipids, Hormones, and Tumorigenesis. Available online: https://www.springer.com/la/book/9780306453175 (accessed on 30 April 2019).
70. Hariri, N.; Thibault, L. High-fat diet-induced obesity in animal models. *Nutrition Res. Rev.* **2010**, *23*, 270–299. [CrossRef]
71. Hilvo, M.; Denkert, C.; Lehtinen, L.; Muller, B.; Brockmoller, S.; Seppanen-Laakso, T.; Budczies, J.; Bucher, E.; Yetukuri, L.; Castillo, S.; et al. Novel theranostic opportunities offered by characterization of altered membrane lipid metabolism in breast cancer progression. *Cancer Res.* **2011**, *71*, 3236–3245. [CrossRef]
72. Guo, S.; Wang, Y.; Zhou, D.; Li, Z. Significantly increased monounsaturated lipids relative to polyunsaturated lipids in six types of cancer microenvironment are observed by mass spectrometry imaging. *Sci. Rep.* **2014**, *4*, 5959. [CrossRef]
73. Othman, R. Dietary lipids and cancer. *Libyan J. Med.* **2007**, *2*, 180–184. [CrossRef]
74. Simopoulos, A.P. Evolutionary aspects of diet, the omega-6/omega-3 ratio and genetic variation: nutritional implications for chronic diseases. *Biomed. Pharmacother.* **2006**, *60*, 502–507. [CrossRef]
75. Berquin, I.M.; Edwards, I.J.; Kridel, S.J.; Chen, Y.Q. Polyunsaturated fatty acid metabolism in prostate cancer. *Cancer Metastasis Rev.* **2011**, *30*, 295–309. [CrossRef]

76. Spencer, L.; Mann, C.; Metcalfe, M.; Webb, M.; Pollard, C.; Spencer, D.; Berry, D.; Steward, W.; Dennison, A. The effect of omega-3 FAs on tumour angiogenesis and their therapeutic potential. *Eur. J. Cancer* **2009**, *45*, 2077–2086. [CrossRef]
77. Buettner, R.; Parhofer, K.G.; Woenckhaus, M.; Wrede, C.E.; Kunz-Schughart, L.A.; Scholmerich, J.; Bollheimer, L.C. Defining high-fat-diet rat models: Metabolic and molecular effects of different fat types. *J. Mol. Endocrinol.* **2006**, *36*, 485–501. [CrossRef]
78. Allott, E.H.; Masko, E.M.; Freedland, A.R.; Macias, E.; Pelton, K.; Solomon, K.R.; Mostaghel, E.A.; Thomas, G.V.; Pizzo, S.V.; Freeman, M.R.; et al. Serum cholesterol levels and tumor growth in a PTEN-null transgenic mouse model of prostate cancer. *Prostate Cancer Prostatic Dis.* **2018**, *21*, 196–203. [CrossRef]
79. Babcook, M.A.; Joshi, A.; Montellano, J.A.; Shankar, E.; Gupta, S. Statin Use in Prostate Cancer: An Update. *Nutr. Metab. Insights* **2016**, *9*, 43–50. [CrossRef]
80. Murtola, T.J.; Tammela, T.L.; Maattanen, L.; Huhtala, H.; Platz, E.A.; Ala-Opas, M.; Stenman, U.H.; Auvinen, A. Prostate cancer and PSA among statin users in the Finnish prostate cancer screening trial. *Int. J. Cancer* **2010**, *127*, 1650–1659. [CrossRef]
81. Roberts, D.L.; Dive, C.; Renehan, A.G. Biological mechanisms linking obesity and cancer risk: New perspectives. *Annu. Rev. Med.* **2010**, *61*, 301–316. [CrossRef]
82. Ma, J.; Li, H.; Giovannucci, E.; Mucci, L.; Qiu, W.; Nguyen, P.L.; Gaziano, J.M.; Pollak, M.; Stampfer, M.J. Prediagnostic body-mass index, plasma C-peptide concentration, and prostate cancer-specific mortality in men with prostate cancer: A long-term survival analysis. *Lancet. Oncol.* **2008**, *9*, 1039–1047. [CrossRef]
83. Poloz, Y.; Stambolic, V. Obesity and cancer, a case for insulin signaling. *Cell Death Dis.* **2015**, *6*, e2037. [CrossRef]
84. Lin, J.; Yang, R.; Tarr, P.T.; Wu, P.H.; Handschin, C.; Li, S.; Yang, W.; Pei, L.; Uldry, M.; Tontonoz, P.; et al. Hyperlipidemic effects of dietary saturated fats mediated through PGC-1beta coactivation of SREBP. *Cell* **2005**, *120*, 261–273. [CrossRef]
85. De Marzo, A.M.; Platz, E.A.; Sutcliffe, S.; Xu, J.; Gronberg, H.; Drake, C.G.; Nakai, Y.; Isaacs, W.B.; Nelson, W.G. Inflammation in prostate carcinogenesis. *Nat. Rev. Cancer* **2007**, *7*, 256–269. [CrossRef]
86. Lumeng, C.N.; Deyoung, S.M.; Bodzin, J.L.; Saltiel, A.R. Increased inflammatory properties of adipose tissue macrophages recruited during diet-induced obesity. *Diabetes* **2007**, *56*, 16–23. [CrossRef]
87. Huang, M.; Narita, S.; Tsuchiya, N.; Ma, Z.; Numakura, K.; Obara, T.; Tsuruta, H.; Saito, M.; Inoue, T.; Horikawa, Y.; et al. Overexpression of Fn14 promotes androgen-independent prostate cancer progression through MMP-9 and correlates with poor treatment outcome. *Carcinogenesis.* **2011**, *32*, 1589–1596. [CrossRef]
88. Di Zazzo, E.; Polito, R.; Bartollino, S.; Nigro, E.; Porcile, C.; Bianco, A.; Daniele, A.; Moncharmont, B. Adiponectin as Link Factor between Adipose Tissue and Cancer. *Int. J. Mol. Sci.* **2019**, *20*, 839. [CrossRef]
89. Di Zazzo, E.; Galasso, G.; Giovannelli, P.; Di Donato, M.; Castoria, G. Estrogens and Their Receptors in Prostate Cancer: Therapeutic Implications. *Front. Oncol.* **2018**, *8*, 2. [CrossRef]
90. Allen, N.E.; Key, T.J. The effects of diet on circulating sex hormone levels in men. *Nutr. Res. Rev.* **2000**, *13*, 159–184. [CrossRef]

Permissions

All chapters in this book were first published by MDPI; hereby published with permission under the Creative Commons Attribution License or equivalent. Every chapter published in this book has been scrutinized by our experts. Their significance has been extensively debated. The topics covered herein carry significant findings which will fuel the growth of the discipline. They may even be implemented as practical applications or may be referred to as a beginning point for another development.

The contributors of this book come from diverse backgrounds, making this book a truly international effort. This book will bring forth new frontiers with its revolutionizing research information and detailed analysis of the nascent developments around the world.

We would like to thank all the contributing authors for lending their expertise to make the book truly unique. They have played a crucial role in the development of this book. Without their invaluable contributions this book wouldn't have been possible. They have made vital efforts to compile up to date information on the varied aspects of this subject to make this book a valuable addition to the collection of many professionals and students.

This book was conceptualized with the vision of imparting up-to-date information and advanced data in this field. To ensure the same, a matchless editorial board was set up. Every individual on the board went through rigorous rounds of assessment to prove their worth. After which they invested a large part of their time researching and compiling the most relevant data for our readers.

The editorial board has been involved in producing this book since its inception. They have spent rigorous hours researching and exploring the diverse topics which have resulted in the successful publishing of this book. They have passed on their knowledge of decades through this book. To expedite this challenging task, the publisher supported the team at every step. A small team of assistant editors was also appointed to further simplify the editing procedure and attain best results for the readers.

Apart from the editorial board, the designing team has also invested a significant amount of their time in understanding the subject and creating the most relevant covers. They scrutinized every image to scout for the most suitable representation of the subject and create an appropriate cover for the book.

The publishing team has been an ardent support to the editorial, designing and production team. Their endless efforts to recruit the best for this project, has resulted in the accomplishment of this book. They are a veteran in the field of academics and their pool of knowledge is as vast as their experience in printing. Their expertise and guidance has proved useful at every step. Their uncompromising quality standards have made this book an exceptional effort. Their encouragement from time to time has been an inspiration for everyone.

The publisher and the editorial board hope that this book will prove to be a valuable piece of knowledge for researchers, students, practitioners and scholars across the globe.

List of Contributors

Malek Alassfar
Department of Urology, School of Medicine, Case Western Reserve University, Cleveland, OH 44106, USA

Shiv Verma and Eswar Shankar
Department of Urology, School of Medicine, Case Western Reserve University, Cleveland, OH 44106, USA
The Urology Institute, University Hospitals Cleveland Medical Center, Cleveland, OH 44106, USA

F. Naz Cemre Kalayci
The Hacettepe University Medical Center, Hacettepe University, Sihhiye, Ankara 06100, Turkey

Amrita Mukunda
Krieger School of Arts and Sciences, Johns Hopkins University, Baltimore, MD 21218, USA

Vaibhav Singh
Department of Inflammation and Immunity, Cleveland Clinic Foundation, Cleveland, OH 44195, USA

E. Ricky Chan
Department of Population and Quantitative Health Sciences, CaseWestern Reserve University, Cleveland, OH 44106, USA

Gregory T. MacLennan
Department of Pathology, Case Western Reserve University, Cleveland, OH 44106, USA

Sanjay Gupta
Department of Urology, School of Medicine, Case Western Reserve University, Cleveland, OH 44106, USA
The Urology Institute, University Hospitals Cleveland Medical Center, Cleveland, OH 44106, USA
Department of Nutrition, Case Western Reserve University, Cleveland, OH 44106, USA
Division of General Medical Sciences, Case Comprehensive Cancer Center, Cleveland, OH 44106, USA
Department of Urology, Louis Stokes Cleveland Veterans Affairs Medical Center, Cleveland, OH 44106, USA

Eunhee Yi
Department of Pharmacology and Biomedical Sciences, Seoul National University College of Medicine, Seoul 03080, Korea

Kum Hee Noh, Ae Jin Jeong, Haeri Lee and Song-Hee Lee
Department of Pharmacology and Biomedical Sciences, Seoul National University College of Medicine, Seoul 03080, Korea
Biomedical Science Project (BK21PLUS), Seoul National University College of Medicine, Seoul 03080, Korea

Pahn-Shick Chang
Department of Agricultural Biotechnology, Seoul National University, Seoul 08826, Korea
Center for Food and Bioconvergence, Seoul National University, Seoul 08826, Korea
Research Institute of Agriculture and Life Sciences, Seoul National University, Seoul 08826, Korea

Cheol Kwak
Department of Urology, Seoul National University Hospital, Seoul 03080, Korea
Department of Urology, Seoul National University Hospital, Seoul National University College of Medicine, Seoul 03080, Korea

Sang-Kyu Ye
Department of Pharmacology and Biomedical Sciences, Seoul National University College of Medicine, Seoul 03080, Korea
Biomedical Science Project (BK21PLUS), Seoul National University College of Medicine, Seoul 03080, Korea
Ischemic/Hypoxic Disease Institute, Seoul National University College of Medicine, Seoul 03080, Korea
Neuro-Immune Information Storage Network Research Center, Seoul National University College of Medicine, Seoul 03080, Korea

Tobias Busenbender, Moritz Kaune, Jessica Hauschild and Carsten Bokemeyer
Department of Oncology, Hematology and Bone Marrow Transplantation with Section Pneumology, HubertusWald-Tumorzentrum, University Medical Center Hamburg-Eppendorf, 20251 Hamburg, Germany

Gunhild von Amsberg
Department of Oncology, Hematology and Bone Marrow Transplantation with Section Pneumology, HubertusWald-Tumorzentrum, University Medical Center Hamburg-Eppendorf, 20251 Hamburg, Germany
Martini-Klinik, Prostate Cancer Center, University Hospital Hamburg-Eppendorf, 20251 Hamburg, Germany

Sergey A. Dyshlovoy
Department of Oncology, Hematology and Bone Marrow Transplantation with Section Pneumology, Hubertus Wald-Tumorzentrum, University Medical Center Hamburg-Eppendorf, 20251 Hamburg, Germany
Laboratory of Biologicaly Active Compounds, School of Natural Sciences, Far Eastern Federal University, Vladivostok 690091, Russia
Martini-Klinik, Prostate Cancer Center, University Hospital Hamburg-Eppendorf, 20251 Hamburg, Germany

Markus Graefen
Martini-Klinik, Prostate Cancer Center, University Hospital Hamburg-Eppendorf, 20251 Hamburg, Germany

Darya Tarbeeva, Sergey Fedoreyev, Marina Veselova, Anatoliy Kalinovskiy, Natalya Kim and Petr Gorovoy
G.B. Elyakov Pacific Institute of Bioorganic Chemistry, Far-East Branch, Russian Academy of Sciences, Vladivostok 690022, Russia

Valeria Grigorchuk
Federal Scientific Center of the East Asia Terrestrial Biodiversity (Institute of Biology and Soil Science), Far Eastern Branch, Russian Academy of Sciences, Vladivostok 690022, Russia

Sander Bekeschus
ZIK plasmatis, Leibniz Institute for Plasma Science and Technology (INP), 17489 Greifswald, Germany

Nadine Gelbrich
Department of Urology, University Medicine Greifswald, 17475 Greifswald, Germany

Verena Ressel
ZIK plasmatis, Leibniz Institute for Plasma Science and Technology (INP), 17489 Greifswald, Germany
Department of Urology, University Medicine Greifswald, 17475 Greifswald, Germany

Eric Freund
ZIK plasmatis, Leibniz Institute for Plasma Science and Technology (INP), 17489 Greifswald, Germany
Department of General, Visceral and Thoracic Surgery, University Medicine Greifswald, 17475 Greifswald, Germany

Alexander Mustea and Matthias B. Stope
Department of Gynecology and Gynecological Oncology, University Hospital Bonn, 53127 Bonn, Germany

Julia Ketteler, Alina Wittka, Verena Jendrossek and Diana Klein
Institute of Cell Biology (Cancer Research), University of Duisburg-Essen, University Hospital, Virchowstrasse 173, 45122 Essen, Germany

Andrej Panic
Institute of Cell Biology (Cancer Research), University of Duisburg-Essen, University Hospital, Virchowstrasse 173, 45122 Essen, Germany
Department of Urology and Urooncology, University of Duisburg-Essen, University Hospital, Essen, Hufelandstr. 55, 45122 Essen, Germany

Henning Reis
Institute of Pathology, University of Duisburg-Essen, University Hospital, Hufelandstr. 55, 45122 Essen, Germany

Patrick Maier and Carsten Herskind
Department of Radiation Oncology, University Medical Center Mannheim, Medical Faculty Mannheim, Heidelberg University, Theodor-Kutzer-Ufer 1-3, 68167 Mannheim, Germany

Ernesto Yagüe
Cancer Research Center, Division of Cancer, Imperial College London, Hammersmith Hospital Campus, London W12 0NN, UK

Kouji Izumi and Atsushi Mizokami
Department of Integrative Cancer Therapy and Urology, Kanazawa University Graduate School of Medical Science, 13-1 Takara-machi, Kanazawa, Ishikawa 920-8641, Japan

Takeshi Sasaki, Manabu Kato, Hideki Kanda, Kiminobu Arima and Yoshiki Sugimura
Department of Nephro-Urologic Surgery and Andrology, Mie University Graduate School of Medicine, Tsu, Mie 514-8507, Japan

Kenichiro Ishii
Department of Nephro-Urologic Surgery and Andrology, Mie University Graduate School of Medicine, Tsu, Mie 514-8507, Japan
Department of Oncologic Pathology, Mie University Graduate School of Medicine, Tsu, Mie 514-8507, Japan

Yoshifumi Hirokawa and Masatoshi Watanabe
Department of Oncologic Pathology, Mie University Graduate School of Medicine, Tsu, Mie 514-8507, Japan

Kazuhiro Iguchi
Laboratory of Community Pharmacy, Gifu Pharmaceutical University, Gifu, Gifu 501-1196, Japan

Izumi Matsuoka and Kohei Nishikawa
Department of Nephro-Urologic Surgery and Andrology, Mie University Graduate School of Medicine, Tsu, Mie 514-8507, Japan

Hiroshi Imai
Pathology Division, Mie University Hospital, Tsu, Mie 514-8507, Japan

Yasuomi Shimizu, Satoshi Tamada, Minoru Kato, Yuji Takeyama, Taro Iguchi and Tatsuya Nakatani
Department of Urology, Graduate School of Medicine, Osaka City University, Osaka 545-8585, Japan

Yukiyoshi Hirayama and Marianne D. Sadar
Genome Sciences Centre, BC Cancer, Vancouver, BC V5Z 1L3, Canada

Toshiki Etani, Keitaro Iida, Satoshi Nozaki, Noriyasu Kawai and Takahiro Yasui
Department of Nephro-Urology, Nagoya City University, Graduate School of Medical Sciences, Nagoya 467-8601, Japan

Taku Naiki
Department of Nephro-Urology, Nagoya City University, Graduate School of Medical Sciences, Nagoya 467-8601, Japan
Department of Experimental Pathology and Tumor Biology, Nagoya City University, Graduate School of Medical Sciences, Nagoya 467-8601, Japan

Aya Naiki-Ito, Hiroyuki Kato, Yuko Nagayasu, Shugo Suzuki and Satoru Takahashi
Department of Experimental Pathology and Tumor Biology, Nagoya City University, Graduate School of Medical Sciences, Nagoya 467-8601, Japan

Takayoshi Suzuki
Department of Chemistry, Kyoto Prefectural University of Medicine, Graduate School of Medical Science, Kyoto 606-0823, Japan
CREST, Japan Science and Technology Agency (JST), Kawaguchi 332-0015, Japan

Kazutoshi Fujita, Takuji Hayashi, Makoto Matsushita, Motohide Uemura and Norio Nonomura
Department of Urology, Osaka University Graduate School of Medicine, Suita, Osaka 565-0871, Japan

Yu Miyazaki, Takayuki Goto, Kosuke Okasho, Kei Mizuno, Masayuki Uegaki, Takeshi Yoshikawa, Shusuke Akamatsu, Takashi Kobayashi, Osamu Ogawa and Takahiro Inoue
Department of Urology, Kyoto University Graduate School of Medicine, Kyoto 606-8507, Japan

Yuki Teramoto and Shinsuke Shibuya
Department of Diagnostic Pathology, Kyoto University Hospital, Kyoto 606-8507, Japan

Naohide Oue, Yuki Koike, Naoya Sakamoto, Kazuhiro Sentani and Wataru Yasui
Department of Molecular Pathology, Graduate School of Biomedical and Health Sciences, Hiroshima University, Hiroshima 734-8551, Japan

Yohei Sekino and Yoshinori Shigematsu
Department of Molecular Pathology, Graduate School of Biomedical and Health Sciences, Hiroshima University, Hiroshima 734-8551, Japan
Department of Urology, Graduate School of Biomedical and Health Sciences, Hiroshima University, Hiroshima 734-8551, Japan

Jun Teishima and Akio Matsubara
Department of Urology, Graduate School of Biomedical and Health Sciences, Hiroshima University, Hiroshima 734-8551, Japan

Masaki Shiota
Department of Urology, Graduate School of Medical Sciences, Kyushu University, Fukuoka 812-8582, Japan

Tetsutaro Hayashi, Hirotaka Nagamatsu, Koichi Shoji, Hiroyuki Shikuma, Ryoken Yamanaka, Keisuke Goto and Shogo Inoue
Department of Urology, Graduate School of Biomedical and Health Sciences, Hiroshima University, 1-2-3 Kasumi, Minami-ku, Hiroshima 734-8551, Japan

Shinichi Sakamoto, Maihulan Maimaiti, Minhui Xu, Nobuyoshi Takeuchi, Akira Komiya and Tomohiko Ichikawa
Department of Urology, Chiba University Hospital, Chiba 260-8670, Japan

Shuhei Kamada and Maki Nagata
Department of Urology, Yokohama Rosai Hospital, Yokohama 222-0036, Japan

Yasutaka Yamada and Hiroomi Nakatsu
Department of Urology, Asahi Central Hospital, Aashi 289-2511, Japan

Hiroki Kitoh and Koichiro Akakura
Department of Urology, Japan Community Healthcare Organization Tokyo Shinjuku Medical Center, Shinjyuku 162-8543, Japan

Hiroaki Matsumoto and Hideyasu Matsuyama
Department of Urology, Graduate School of Medicine, Yamaguchi University, Ube 755-0046, Japan

Kosuke Higuchi
Department of Urology, Funabashi Medical Center, Funabashi 273-8588, Japan

Haruhito A. Uchida
Department of Chronic Kidney Disease and Cardiovascular Disease, Okayama University Graduate School of Medicine, Dentistry and Pharmaceutical Sciences, Okayama 700-0914, Japan

Shintaro Narita, Taketoshi Nara, Hiromi Sato, Atsushi Koizumi, Mingguo Huang, Takamitsu Inoue and Tomonori Habuchi
Department of Urology, Akita University School of Medicine, Akita 010-8543, Japan

Index

www.ingramcontent.com/pod-product-compliance
Lightning Source LLC
LaVergne TN
LVHW080946210926
848485LV00004B/785